Advanced Imaging in Biology and Medicine

Christoph W. Sensen • Benedikt Hallgrímsson

Editors

Advanced Imaging in Biology and Medicine

Technology, Software Environments, Applications

 Springer

Editors

Dr. rer. nat. Christoph W. Sensen
Sun Center of Excellence for Visual Genomics
Department of Biochemistry
and Molecular Biology
University of Calgary
3330 Hospital Drive NW
Calgary AB T2N 4N1
Canada
csensen@ucalgary.ca

Benedikt Hallgrímsson, Ph.D.
Department of Cell Biology and Anatomy
University of Calgary
3330 Hospital Drive NW
Calgary AB T2N 4N1
G503 Health Sciences Ctr.
Canada
bhallgri@ucalgary.ca

Marjan Eggermont is a senior instructor in The Schulich School of Engineering at the University of Calgary, teaching in the area of engineering design. She used to teach in the Fine Arts department in the areas of drawing, art fundamentals, and printmaking. Marjan has degrees in Military History (BA) and Fine Arts (BFA, MFA) and studied briefly at The Royal College of Art (UK). As an artist she has been in numerous art exhibitions nationally and internationally.

She was in 2004 one of the recipients of The Allan Blizzard Award, a national teaching award for collaborative projects that improve student learning. In 2005 she was part of the team that won the American Society for Mechanical Engineering Curriculum Innovation Award. She recently appeared in 'Printmaking at The Edge' as one of 45 international print artists. Marjan is represented by the Herringer Kiss gallery in Calgary and the Elissa Cristall gallery in Vancouver.

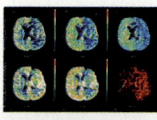

The image "The Origin of Thought" is an abstraction of the MRI (Acetazolamide stimulation/Resting condition) on the left. Artists are often trying to get to the essence of an image - the ultimate abstraction. In science the goal seems to be extreme realism, especially in digital imaging - for good reasons I might add. This abstraction can also be read as an extreme close-up. I am always interested in the similarities between macro and micro scale images. And if I try to imagine where thought is formed, perhaps it is - today - on the micro and macro scale.

ISBN 978-3-540-68992-8   e-ISBN 978-3-540-68993-5

Library of Congress Control Number: 2008934400

© 2009 Springer-Verlag Berlin Heidelberg

Printed on acid-free paper

springer.com

Preface

A picture says more than a thousand words. This is something that we all know to be true. Imaging has been important since the early days of medicine and biology, as seen in the anatomical studies of Leonardo Da Vinci or Andreas Vesalius. More than 100 years ago, the first noninvasive imaging technologies, such as Konrad Roentgen's X-ray technology, were applied to the medical field—and while still crude—revolutionized medical diagnosis. Today, every patient will be exposed to some kind of advanced imaging technology such as medical resonance imaging, computed tomography or four-dimensional ultrasound during their lifetime. Many diseases, such as brain tumors, are initially diagnosed solely by imaging, and most of the surgical planning relies on the patient imagery. 4D ultrasound is available to expecting parents who wish to create unique early memories of the new baby, and it may soon be used for the morphometric diagnosis of malformations that may one day be treatable—*in utero*!

Light and electron microscopy are unequal brethren, which have contributed to most of our knowledge about the existence and organization of cells, tissues and microorganisms. Every student of biology or medicine is introduced to the fascinating images of the microcosm. New advances have converted these imaging technologies, which were considered by many to be antiquated, into powerful tools for research in systems biology and related fields. The development of laser technology and advances in the development of computer systems have been instrumental in the improvement of imaging technologies, which will be utilized for many generations to gain new insight into complex biological and medical phenomena.

With the completion of the human genome, hopes were high that we would now be able to read the "blueprint" of life and understand how the human body works. Unfortunately, as is quite common in science, the complete genome has triggered more questions than it has helped to answer at this point. A simple approach to understanding how the body functions by reading the "blueprint" is not possible, as almost all of the bodily functions are spatiotemporal in nature. In addition, a protein modification which causes curled wings in *Drosophila melanogaster* will naturally have a completely different function in humans, and so a 1:1 transposition of knowledge from one organism to another is impossible. Genome researchers are now forced to conduct additional large-scale experiments, including gene expression, proteomic and metabolomic studies. Integrating the data from these complex experiments is an extremely difficult task, and displaying the results requires new approaches to imaging in order to allow a wide audience to make sense of the facts. Imaging technologies will be especially useful for the creation of spatiotemporal models, which can be used for the integration of "omics" data.

As we can see from the above three paragraphs, advanced imaging in medicine and biology is a very wide field. When we started to plan this book, we were unable to find a publication that provided a broad sampling of this rapidly expanding field. We hope that our readers will appreciate and benefit from the diversity of topics presented here.

We would like to thank Dr. Andrea Pillmann and Anne Clauss from Springer Verlag, Heidelberg, for their endless patience with the authors and editors. Without their constant support, this book would have been impossible to create.

Calgary,   *Benedikt Hallgrímsson*
December 2008   *Christoph W. Sensen*

Contents

Contributors

Matthias Amrein
Microscopy and Imaging Facility, Department of Cell Biology and Anatomy, Faculty of Medicine, University of Calgary, 3330 Hospital Drive N.W., Calgary, AB, Canada T2N 4N1, e-mail: mamrein@ucalgary.ca

Linda B Andersen
Radiology and Clinical Neurosciences, Hotchkiss Brain Institute, University of Calgary, Calgary, AB, Canada T2N 1N4
and
Seáman Family MR Research Centre, Foothills Medical Centre, Calgary Health Region, Calgary, AB, Canada T2N 2T9, e-mail: lbanders@ucalgary.ca

P.E. Andersen
Department of Photonics Engineering, Technical University of Denmark, Frederiksborgvej 399, Roskilde, DK-4000, Denmark, e-mail: peter.andersen@fotonik.dtu.dk

Christoph H. Borchers
Department of Biochemistry & Microbiology, University of Victoria—Genome British Columbia Protein Centre, University of Victoria, #3101-4464 Markham Street, Vancouver Island technology Park, Victoria, BC, Canada V8Z7X8, e-mail: christoph@proteincentre.com

Julia C. Boughner
Department of Cell Biology and Anatomy and the McCaig Bone and Joint Institute, Faculty of Medicine, University of Calgary, 3330 Hospital Drive NW, Calgary, AB, Canada T2N 4N1

Steven K. Boyd
Department of Mechanical and Manufacturing Engineering, Schulich School of Engineering, University of Calgary, 2500 University Drive, N.W., Calgary, Alberta, Canada T2N 1N4, e-mail: skboyd@ucalgary.ca

Nathan Cross
Department of Neurological Surgery, Case Western Reserve University,
10900 Euclid Avenue, Cleveland, OH 44106, USA

David Dean
Department of Neurological Surgery, Case Comprehensive Cancer Center, Case
Western Reserve University, 10900 Euclid Avenue, Cleveland, OH 44106, USA,
e-mail: daviddean@case.edu

Jos J. Eggermont
Department of Psychology, 2500 University Drive N.W, University of Calgary,
Calgary, AB, Canada, T2N 1N4, e-mail: eggermon@ucalgary.ca

Chris A. Flask
Case Comprehensive Cancer Center, Department of Radiology, Case Western
Reserve University, 10900 Euclid Avenue, Cleveland, OH 44106, USA

Richard Frayne
Radiology and Clinical Neurosciences, Hotchkiss Brain Institute, University of
Calgary, Calgary, AB, Canada T2N 1N4

and

Seaman Family MR Research Centre, Foothills Medical Centre, Calgary Health
Region, Calgary, AB, Canada T2N 2T9, e-mail: rfrayne@ucalgary.ca

Andrew H. Gee
Department of Engineering, University of Cambridge, Trumpington Street,
Cambridge CB2 1PZ, UK

Paul M.K. Gordon
Faculty of Medicine, University of Calgary, Sun Center of Excellence for Visual
Genomics, 3330 Hospital Drive NW, Calgary, AB, Canada, T2N 4N1

Benedikt Hallgrímsson
Department of Cell Biology and Anatomy and the McCaig Bone and Joint Institute,
Faculty of Medicine, University of Calgary, 3330 Hospital Drive NW, Calgary, AB,
Canada T2N 4N1, bhallgri@ucalgary.ca

Po-Wei Hsu
Department of Engineering, University of Cambridge, Trumpington Street,
Cambridge CB2 1PZ, UK

G.B.E. Jemec
Department of Photonics Engineering, Technical University of Denmark,
Frederiksborgvej 399, Roskilde, DK-4000, Denmark

Nicholas Jones
McCaig Bone and Joint Institute, Faculty of Medicine, University of Calgary, 3330
Hospital Drive NW, Calgary, AB, Canada T2N 4N1, e-mail: njones@ucalgary.ca

T.M. Jørgensen
Department of Photonics Engineering, Technical University of Denmark,
Frederiksborgvej 399, Roskilde, DK-4000, Denmark

Anton H.J. Koning
Erasmus MC University Medical Centre Rotterdam, The Netherlands

Cairine Logan
Department of Cell Biology and Anatomy and the Infant and Maternal Child Health
Institute, Faculty of Medicine, University of Calgary, 3330 Hospital Drive NW,
Calgary, AB, Canada T2N 4N1

John Robert Matyas
Faculty of Veterinary Medicine, 3330 Hospital Drive NW, Calgary, AB,
Canada T2N 4N1, e-mail: jmatyas@ucalgary.ca

M. Mogensen
Department of Photonics Engineering, Technical University of Denmark,
Frederiksborgvej 399, Roskilde, DK-4000, Denmark

Nancy L. Oleinick
Case Comprehensive Cancer Center, Department of Radiation Oncology, Case
Western Reserve University, 10900 Euclid Avenue, Cleveland, OH 44106, USA

Carol E. Parker
UNC-Duke Michael Hooker Proteomics Center, UNC—Chapel Hill 111 Glaxo
Building, CB #7028, Chapel Hill, NC 27599-7028, USA

Trish E. Parsons
Biological Anthropology Graduate Program and the McCaig Bone and Joint
Institute, University of Calgary, 3330 Hospital Drive NW, Calgary, AB,
Canada T2N 4N1

Jörg Peter
Department of Medical Physics in Radiology, German Cancer Research Center,
Im Neuenheimer Feld 280, Heidelberg 69120, Germany, e-mail: j.peter@dkfz.de

Richard W. Prager
Department of Engineering, University of Cambridge, Trumpington Street,
Cambridge CB2 1PZ, UK

B. Sander
Department of Photonics Engineering, Technical University of Denmark,
Frederiksborgvej 399, Roskilde, DK-4000, Denmark

Christoph W. Sensen
Department of Biochemistry and Molecular Biology, Faculty of Medicine,
University of Calgary, Sun Center of Excellence for Visual Genomics, 3330
Hospital Drive NW, Calgary, AB, Canada T2N 4N1, e-mail: csensen@ucalgary.ca

James Sharpe
EMBL-CRG Systems Biology Unit, Centre for Genomic Regulation, Dr. Aiguader
88 08003 Barcelona, Spain, e-mail: james.sharpre@crg.es

Derek Smith
Department of Biochemistry & Microbiology, University of Victoria—Genome
British Columbia Protein Centre, University of Victoria, #3101-4464 Markham
Street, Vancouver Island Technology Park, Victoria, BC, Canada V8Z7X8

Jung Soh
Department of Biochemistry and Molecular Biology, Faculty of Medicine,
University of Calgary, Sun Center of Excellence for Visual Genomics,
3330 Hospital Drive NW, Calgary, AB Canada, T2N 4N1

Detlev Suckau
Bruker Daltonik GmbH, Fahrenheitstr. 4, 28359 Bremen, Germany

J.B. Thomsen
Department of Photonics Engineering, Technical University of Denmark,
Frederiksborgvej 399, Roskilde, DK-4000, Denmark

L. Thrane
Department of Photonics Engineering, Technical University of Denmark,
Frederiksborgvej 399, Roskilde, DK-4000, Denmark

Matthew W. Tocheri
Human Origins Program, Department of Anthropology, National Museum of
Natural History, Smithsonian Institution, Washington DC 20013-7012, USA,
e-mail: tocherim@si.edu

Graham M. Treece
Department of Engineering, University of Cambridge, Trumpington Street,
Cambridge CB2 1PZ, UK

Ursula I. Tuor
Institute for Biodiagnostics (West), National Research Council of Canada,
Experimental Imaging Centre, Hotchkiss Brain Institute, University of Calgary,
3330 Hospital Dr, NW Calgary, AB, Canada T2N 4N1

Andrei L. Turinsky
Hospital for Sick Children, 555 University Avenue, Toronto, ON, Canada,
M5G 1X8, e-mail: turinsky@sickkids.ca

Davood Varghai
Department of Neurological Surgery, Case Western Reserve University,
10900 Euclid Avenue, Cleveland, OH 44106, USA

Part I
Imaging Technologies

Chapter 1
Micro-Computed Tomography

Steven K. Boyd

Abstract Micro-computed tomography (micro-CT) provides high-resolution three-dimensional (3D) geometry and density information for the analysis of biological materials, and is particularly well-suited for bone research. These detailed 3D data have provided significant insights into issues surrounding bone quality, and the recent advancement of this technology now provides the opportunity to perform these measurements in living subjects, including both experimental animal models and direct patient measurements. This chapter reviews the fundamental principles of micro-CT and caveats related to its use. It describes the current approaches for analyzing these rich 3D datasets, as well as leading-edge developments. These include analysis techniques such as structure classification, image registration, image-guided failure analysis, extrapolation of structure from time series, and high-throughput approaches for morphological analysis.

Introduction

Micro-computed tomography (micro-CT) is a noninvasive and nondestructive imaging method that provides high-resolution three-dimensional (3D) structure and density information. A principal application of micro-CT is for the measurement of the 3D architecture of bone (Fig. 1.1) since there is no need to perform sample preparation, but it can also be applied to other tissues such as cartilage, ligaments and vasculature, provided appropriate procedures are used (e.g., contrast-enhanced imaging). Since the introduction of micro-CT in the 1990s (Rüegsegger et al. 1996; Holdsworth et al. 1993; Feldkamp et al. 1989), more than 90% of research publications that have employed this technology have focused on bone

S.K. Boyd
Department of Mechanical and Manufacturing Engineering, Schulich School of Engineering, University of Calgary, 2500 University Drive, NW, Calgary, Alberta, Canada T2N 1N4, e-mail: skboyd@ucalgary.ca

C.W. Sensen and B. Hallgrímsson (eds.), *Advanced Imaging in Biology and Medicine.*
© Springer-Verlag Berlin Heidelberg 2009

S.K. Boyd

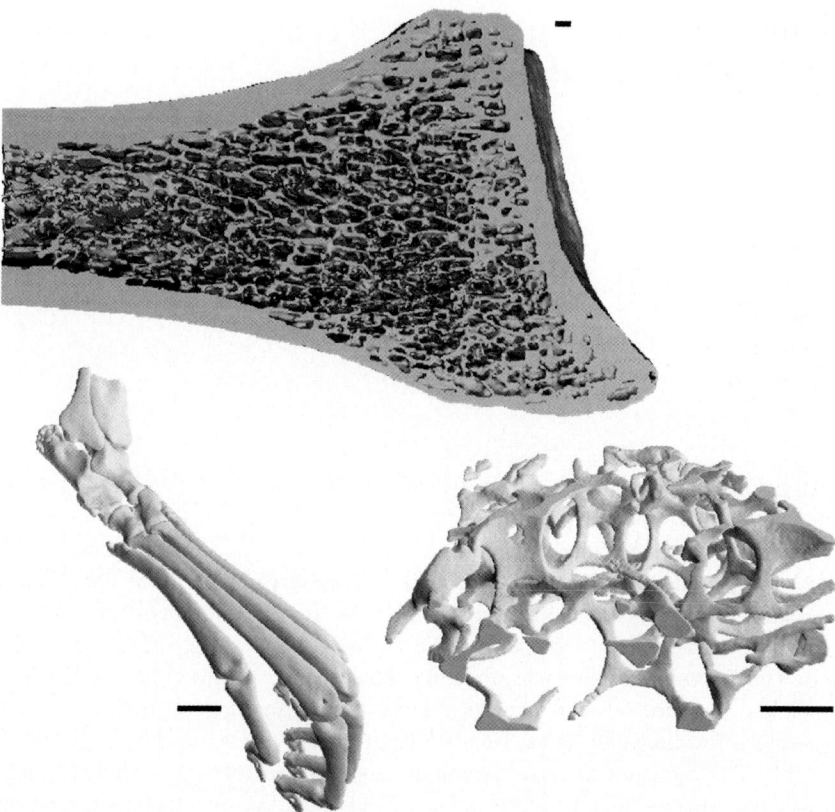

Fig. 1.1 Examples of micro-CT data, including a human distal radius, a mouse paw, and aluminum foam. The *black scale* bar adjacent to each image is 1 mm in length

research, and the reason for this is twofold. First, bone is a highly mineralized tissue consisting of approximately 75% inorganic material in the form of hydroxyapatite $(Ca_{10}(PO_4)_6(OH_2))$ and 25% matrix, which is primarily collagen. Because bone is a "hard" tissue, it is particularly well suited to providing high-contrast images from micro-CT. Second, bone plays an important functional role in the body, and the resolution of micro-CT (on the order of 1–50 μm) is ideally suited to measuring bone architectural features (i.e., trabeculae), which typically are in the size range of 30–120 μm. Bone is important in the skeleton because it provides mechanical support for the locomotion and protection of vital organs, and it also has physiological roles in mineral homeostasis (primarily calcium) and in the production of blood cells in red marrow.

There are many diseases that affect bone, and osteoporosis is one of the most prevalent. The clinical management of osteoporosis focuses largely on the prevention of fracture, and insight into fracture risk is often provided by medical imaging. The standard clinical approach has been to assess bone mineral density (BMD) using dual-energy X-ray absorptiometry (DXA), and BMD is currently the standard surrogate for estimating bone strength due to the reasonably strong (and intuitive) relation between "how much" mineralized bone there is and its strength. An important limitation of DXA (Bolotin 2007), however, is that it is measured in two dimensions (i.e., areal BMD), meaning that results can be biased by bone size and it cannot resolve information about bone architecture. Since the introduction of micro-CT, it has become increasingly clear that bone strength is not only dependent on "how much" mineralized bone there is, but also how well it is organized at the architectural level (Van Rietbergen et al. 1998). The orientation, thickness and connectedness of bone tissue are important in providing structural support. Although the measurement of architecture has been possible for decades by the analysis of histological sections, the procedure is destructive, time-consuming, and quantitative analysis is based on 2D stereological methods (Parfitt et al. 1983) which can provide biased results, particularly in the context of disease. The "breakthrough" of micro-CT is that it is nondestructive, fast and efficient, and quantitative analysis can be performed on 3D images representing the architecture, including assessments of morphological features as well as estimates of strength. Through the use of experimental animal models and human biopsies, insight into the contribution of bone architecture to bone strength and an understanding of the disease process and effects of intervention have been provided for a broad range of diseases such as osteoporosis, osteoarthritis, breast cancer and many more. Most of the knowledge gained to date has been based on in vitro measurements, but it is particularly exciting that in vivo micro-CT methods have been developed in the past five years, and these can be applied to both experimental animal models as well as direct measurements of patients.

The following sections will provide background on the medical physics for micro-CT imaging, and a discussion regarding maximizing image quality and caveats to avoid. This will be followed by a description of state-of-the-art approaches for assessing these data, example applications, and future directions.

Principles

Before any image analysis can begin, high-quality micro-CT data must be collected, and to provide the best data possible it is crucial to understand the principles behind micro-CT. The fundamental hardware components of a micro-CT system are the same as any CT scanner and include an X-ray source, the object stage, and the detector. Computed tomography techniques are based on collecting X-ray projection data in an ordered fashion from multiple projection angles, but the geometric arrangement of these components can vary. An excellent, detailed description of micro-CT principles can be found elsewhere (Bonse and Busch 1996). In typical in vitro micro-CT systems, the object to be scanned is placed on a rotating stage, while the source and detector are stationary (Fig. 1.2). In in vivo micro-CT systems, it is not feasible to rotate the object, so it is kept stationary while the source and detector are rotated. The field of CT in general has evolved through several

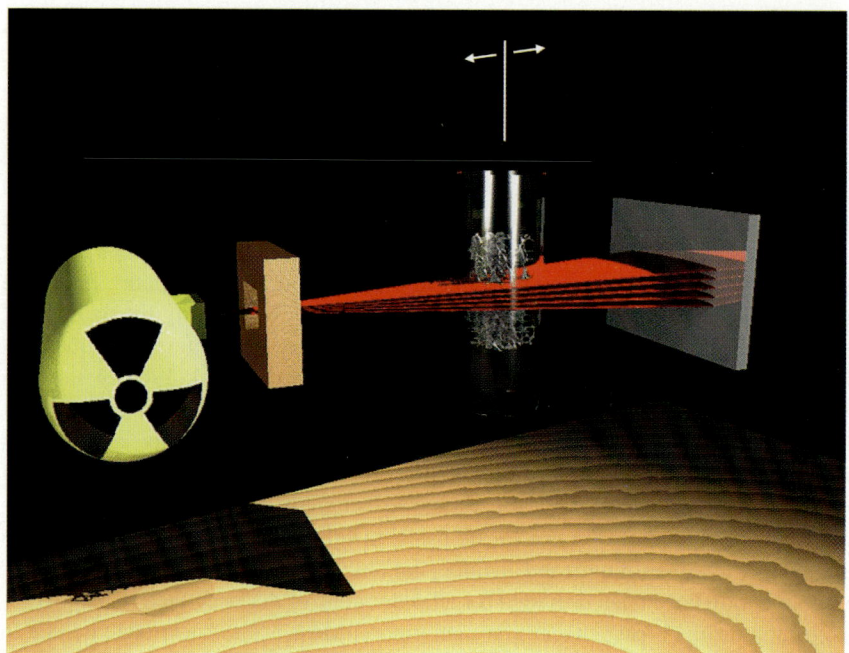

Fig. 1.2 The basic configuration of a cone beam micro-CT scanner. The object is placed between the X-ray source and the detector. In a standard in vitro system the object stage rotates, and in an in vivo system the object is stationary while the source-detector system rotates. The positioning of the object with respect to the X-ray source and detector changes the scan field of view and resolution. Multiple slices can be simultaneously collected using the cone-beam reconstruction. Image provided by Dr. M. Stauber; adapted with permission

generations of design, where the most recent is a helical scanning configuration used in clinical CT scanners—the object translates relative to the source and detectors to enable rapid scanning over long axial lengths. However, micro-CT uses earlier generation technology (third generation), where the point-source X-rays generated from an X-ray tube are emitted in a cone beam geometry to enable the collection of multiple slices of data in one rotation of the system, but without axial translation. The advantage of this configuration is that the mechanical stability of the system affords high-resolution acquisition, and allows flexibility of the geometric positioning of the object to be scanned relative to the X-ray source and detector. Micro-CT systems, where the object is placed on a rotating stage (typically in vitro micro-CT scanners), take advantage of this flexibility by allowing the translation of the stage relative to the source and detector before acquiring the sequential tomographic projections. Thus, moving the object closer to the X-ray source can result in increased image resolution, but there is a trade-off in limiting the field of view (FOV). Alternatively, moving the object closer to the detector provides a greater FOV, albeit with a lower image resolution.

Attenuation of X-Rays

The basis of computed tomography is the measurement of the attenuation of X-rays (photons) passing through an object. The attenuation depends on the energy of the X-rays and the composition and size of the object being scanned, and partial absorption is necessary in order to produce contrast. At high energy levels, the attenuation is dominated by Compton scattering, whereas at low energies the photoelectric effect is an important contributor and attenuation is highly dependent on the atomic number of the material (a desirable feature for characterizing object material characteristics). Although maximum contrast (i.e., mineralized bone vs. marrow) can be achieved when low energies are employed, the physical size of the specimen can result in too much attenuation overall, and insufficient X-rays at the detector for tomographic reconstruction—only small sized specimens can be scanned at low energies. Increasing the number of photons being emitted (i.e., longer times per projection, or increased X-ray current) can improve contrast, but it cannot compensate for the complete attenuation that occurs if the specimen is large. For large specimens, it is necessary to use higher energy X-rays to obtain a sufficient signal-to-noise ratio. In micro-CT, the X-ray source typically has a focal spot size of $5\,\mu m$, and the source energy and current can be adjusted and optimized when scanning different materials. The X-ray tube potentials can be adjusted typically in the range 20–100 kVp, and currents below $200\,\mu A$, depending on the manufacturer. The focal X-ray spot size, which effects image resolution, typically increases as X-ray power increases and depends on the energy and current selected. For example, on a typical commercial micro-CT system, the spot size is $5\,\mu m$ when the power is below 4 W, but increases to $8\,\mu m$ for powers of up to 8 W. Ideally, the smallest focal spot should be used, but because the production of X-rays (bremsstrahlung—radiation emission accompanying electron deceleration) in tubes is very inefficient ($\sim 1\%$ of the electrical energy supplied is converted to X-rays), too much power will cause the excessive heat at the focal spot and cause the target to melt.

Beam Hardening

It is important to note that X-ray energy from a desktop micro-CT is polychromatic —that is, there is a spectrum of X-ray energies emitted for any given potential setting. For example, a setting of 40 kVp may result in an X-ray spectrum that is peaked at 25 keV. The fact that there is a spectrum of energies has an important practical consequence referred to as *beam hardening*. The lower energy X-rays passing through an object are preferentially attenuated over higher energy X-rays, and the result is a shift in the energy spectrum toward higher energies after passing through the object (thus, the beam is "hardened"). Although beam hardening can be partially compensated for by post-processing, particularly in cases where there are only a few different materials in the object being scanned (i.e., bone and marrow), its correction is approximate and not trivial because it depends on material composition and path length through the sample. Although X-ray attenuation can be related to the density of the material being scanned, beam hardening results in errors in this conversion. Desktop micro-CT systems use tubes as the X-ray source, so polychromatic X-rays cannot be avoided. Approaches used to reduce the effects of beam hardening include using thin metallic plates (i.e., 0.5–1.0 mm aluminum or copper plates) to filter out low-energy X-rays from the beam prior to them penetrating into the measured object. Using the lowest possible energy to achieve partial attenuation combined with an appropriate filter provides the closest approximation to a single-peak spectrum, but this does not eliminate the problem.

The only way to completely eliminate beam hardening is to have monochromatic X-rays. Filtering can be performed to obtain a monochromatic beam by using Bragg reflection to pass only X-rays of a single energy; however, the intensity of the X-ray beam produced by an X-ray tube after filtering is reduced by several orders of magnitude, and as a result data acquisition is slow and impractical. However, with the advent of synchrotron radiation (SR), where X-rays are produced in highly specialized national laboratories by accelerating charged particles, the intensities of the X-rays are much higher, and monochromatization of the polychromatic beam can be achieved while still maintaining extremely high intensity. The filtering by Bragg reflection can be adjusted to output monochromatic X-rays of different energies, so the source can be tuned for specific materials and object sizes. The advantage of SR-based micro-CT is that the material properties can be characterized well, but the obvious disadvantage is the scarcity of these large and expensive facilities.

Detection and Geometry

The X-ray detector records the incident photons after they have passed through the object, and to do this the X-rays must be converted to visible light. Scintillation detectors perform this conversion to the visible light range based on the photo effects of materials called scintillators (e.g., $CaWO_4$, $CsI(Tl)$, etc.), and subsequently a CCD chip detects the analog signal. An analog-to-digital converter generates a digital signal for each picture element (pixel) which represents the total X-ray signal integrated over time at that pixel. A precise digital signal is required for micro-CT, and typically 8–16 bits are used, providing 256–65,536 signal levels. CCD pixel sizes typically found in modern micro-CT scanners are on the order of $25\,\mu m^2$, and the CCD normally provides 12-bit data.

Detector sizes may range from $1,024 \times 1,024$ square, but the use of asymmetrical CCD sizes where the largest dimension ranges from 2,048 to >4,000 pixels (and the other dimension is much smaller, i.e., $\sim$256) has certain advantages. Although the X-ray tube for micro-CT provides a point source, the beam can be collimated to a produce either a fan beam or a cone beam geometry. A square detector is appropriate for a cone beam micro-CT, and this geometry has the advantage of being efficient because it is able to collect multiple slices simultaneously with one scanner rotation. However, the trade-off is that geometric distortions can result from the cone beam geometry, particularly for the slices located furthest from the source-detector plane (Fig. 1.1). The fan beam geometry avoids this issue by collecting one slice per rotation, but has the disadvantage of being slow. Therefore, a compromise is to use a partial cone beam reconstruction where multiple slices are collected simultaneously, but not to the extent of requiring a square detector, thus minimizing geometric distortions. For this reason, detectors with asymmetric dimensions such as $3{,}072 \times 256$ may be utilized because they provide excellent resolution in the slice plane, but by limiting the number of simultaneously collected slices, the CCD can be quickly read and can provide slice data with minimal geometric distortions.

Reconstruction and Caveats

Cone beam reconstruction (Feldkamp et al. 1984) is used to convert the projection data collected by the CCD at multiple angle steps during a rotation (the sinogram data) into a stack of slices comprising the 3D image volume. Typically, projections are collected over $180°$ plus half the beam width at each end (i.e., a total rotation of $\sim220°$), and a high-resolution scan may contain 2,000 projections. There are other reconstruction methods that exist for including the use of an iterative or analytical method, but back projection is most commonly used due to its simplicity. This involves projecting each measured profile back over the reconstructed image area along rays oriented at the same angle as used for the projection. Prior to reconstruction, convolution (or filtering) is applied to the projection data to remove blurring artifacts and to enhance edges. A commonly used filter is the Shepp–Logan filter. Also, beam-hardening corrections (software) can be applied (Brooks and, Di Chiro 1976) and tailored to particular X-ray settings and assumed object compositions (i.e., bone with marrow).

The stacks of reconstructed slices provide the 3D dataset, and the scalar values represented in those data (termed CT numbers) are related to the density of the scanned object. Calibration is a necessary step where the CT numbers in the reconstructed image are related to the physical densities, and this is normally done using a phantom with known densities. Although the relation between density and CT number is highly linear and could be determined based on two data points, using a range of five known densities provides assurance. Liquid phantoms containing known concentrations of K_2HPO_4 can be used, but it is more common to use solid-state phantoms of calcium hydroxyapatite (HA) rods ranging from 0 to $800\,mg\,cm^{-3}$ due to their long-term stability (see http://www.qrm.de). Subsequent to calibration, the micro-CT data can be expressed in terms of an equivalent volumetric density of calcium hydroxyapatite ($mg\,cm^{-3}$ HA). Normal clinical CT uses Hounsfield units where air is designated $-1{,}000\,HU$, water $0\,HU$ and very dense materials $+1{,}000\,HU$; however, this system is not normally used in micro-CT. When interpreting the density data, it is important to keep in mind the influence of artifacts such as beam hardening, as already discussed, as well as problems associated with local tomography, partial volume effects and motion artifacts, as discussed in the following.

A common problem that can occur during measurement is that the object being scanned does not fit entirely within the field of view, which makes it necessary to perform *local tomography* (i.e., reconstruction of the region within the FOV only) to recover the image data. Although geometric features can accurately be determined with local tomography, density values are no longer accurate. The errors are largest when dense material is outside the FOV, and unfortunately the anatomic configuration of bone, where the cortex is at the periphery, leads to errors. Density data from bones that could not fit entirely within the FOV should be treated with caution.

Partial volume effects (PVE) are caused when more than one tissue type is represented by a single 3D voxel element. This typically occurs, for example, at the interface between bone and marrow when the voxel lies on the tissue boundary. The magnitude of PVE depends on the resolution of the micro-CT image relative to the sizes of the structures being measured. As the image resolution approaches the sizes of the structures (i.e., trabecular size), the PVE becomes more dominant. Although increasing image resolution may reduce PVE, it cannot be eliminated, as it is a natural consequence of a discrete representation of anatomical features by micro-CT.

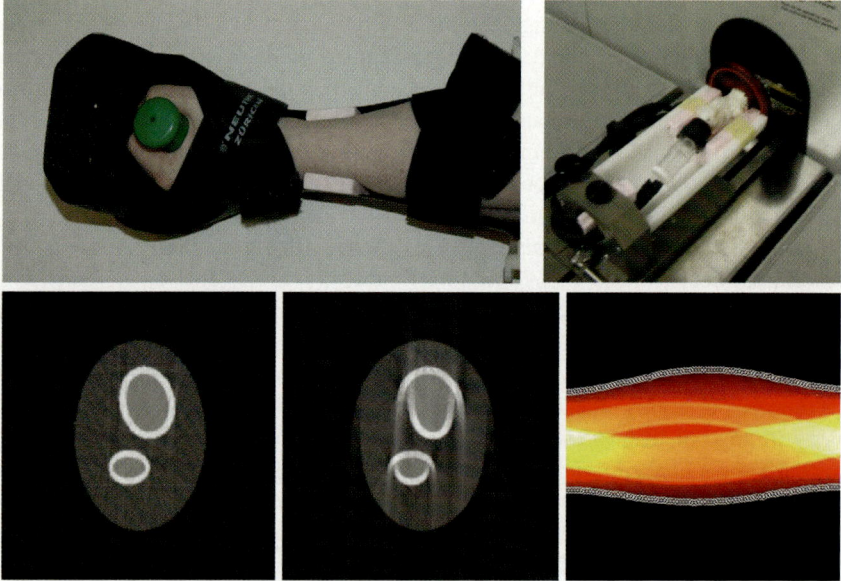

Fig. 1.3 Motion artifacts can be reduced when scanning in vivo through the use of a forearm cast (*top left*) and secured scanning bed (*top right*). Low-density foam (*pink*) provides additional support but is not visible in the resulting micro-CT scan. Errors due to motion artifacts can be simulated (*bottom row*) by manipulating the sinogram directly (adapted from Pauchard; with permission)

Motion artifacts during scanning will directly affect the quality of the image reconstruction and must be addressed for both in vitro and in vivo scanning. For in vitro specimens, a common approach is to tightly pack the specimen using a low-density material (i.e., foam). The problem is more challenging to address with in vivo measurements, and specialized beds for animal scanning or support casts for patient scanning have been developed (Fig. 1.3). The effects of movement artifacts can be subtle, and efforts are now underway to achieve their automatic detection and quantification (Pauchard and Boyd 2008).

Resolution

Resolution is fundamentally dependent on the number of photons produced by the X-ray source (Bonse and Busch 1996). For a given signal-to-noise ratio, doubling the resolution would require 16 times more photons. It is important to distinguish image resolution and voxel size. The voxel size of the reconstructed image applies during the filtered back-projection reconstruction procedure, and although there is flexibility in the actual setting it is typically ~ 1.5 times the spatial resolution as a rule of thumb. Spatial resolution represents the ability to detect physical features, and depends on many factors of the imaging system (source, detector, geometry, etc). It is defined by the measured modulation transfer function (MTF) and can be assessed on line pairs with varying frequencies ($lp\,mm^{-1}$). MTF is normalized to one when the spatial frequency is zero, and a common comparator of imaging systems is based on 10% MTF, which provides the smallest resolvable physical feature. It is measured on narrow sinusoidal line profiles, or approximated by measuring a thin object of known dimensions. The reporting of micro-CT resolution should ideally use the 10% MTF value, but this information is not always readily available (it may be provided by the manufacturer). The voxel size is most often reported, and it may not be representative of the true resolution. When it is reported, it should, at the very least, always be carefully distinguished from the true resolution by reporting it as the "nominal" isotropic resolution.

Sample Preparation

Minimal specimen preparation is required in order to perform micro-CT scanning. Often biological materials are stored in a phosphate buffer solution, alcohol or formaldehyde, and the specimens can be maintained in those solutions during scanning. One consideration may be that the liquid (i.e., alcohol) could cause damage to the specimen holder used for micro-CT. Obviously, no specimen preparation is required for in vivo scanning, although animal scanning requires use of an anesthetic (isoflurane).

Current Technologies

Micro-CT systems are currently used for a wide range of applications, and their designs are tuned in terms of image resolution, field of view, and radiation dose (important for in vivo measurements). It finds major application in the field of biomedicine, particularly for bone research (because no sample preparation is needed), but also for other biomaterials where preparation may be necessary (i.e., contrast-enhanced imaging) (Guldberg et al. 2003). Outside of biomedicine, other applications include measurements of small electronic parts, microdevices, porous materials (ranging from foams to rock cores), and even snow structure for avalanche research. The commercialization of this technology includes suppliers such as

- Scanco Medical (http://www.scanco.ch)
- Skyscan (http://www.skyscan.be)
- GE Healthcare (http://www.gehealthcare.com/usen/fun_img/pcimaging/)
- Xradia (http://www.xradia.com)

In the general category of in vitro measurement systems, resolutions range from as low as $<1\,\mu m$ (so-called nano-CT) to $100\,\mu m$. As the resolution improves, the field of view is reduced, and sometimes high resolutions can result in data sets that are so large that they are unmanageable and impractical. Therefore, most in vitro systems provide resolutions on the order of $5–10\,\mu m$, and fields of view ranging from 20 to 80 mm, and are designed for animal bone scans (i.e., mouse femur, mouse skull) and human biopsy data. Recently, in vivo systems have been introduced for both animal and human scanning. The resolution is limited in part by the radiation dose that can be safely administered, and for animal scanning it is possible to obtain $10–15\,\mu m$ resolutions in vivo without significant adverse effects after repeated scanning (Klinck et al. 2008; Brouwers et al. 2007). A typical delivered dose is $<1\,Gy$ for animal measurements in the mouse or rat. As for human measurements, although peripheral quantitative computed tomography (pQCT) systems that can obtain data at approximately $100\,\mu m$ (in-plane resolution; slice thickness of $>200\,\mu m$) have been available for over a decade, a new human micro-CT scanner providing high resolution data (HR-pQCT) has recently been developed. It can provide data with an isotropic resolution of $\sim100\,\mu m$ ($82\,\mu m$ voxel size) and a 130 mm FOV, and this can be obtained with multislice data at the distal radius and tibia while maintaining a safe, low-radiation dose ($<3\,\mu Sv$).

Finally, synchrotron radiation (SR) micro-CT is unique because rather than using an X-ray tube, it uses synchrotron facilities. Nonetheless, it offers the possibility of $<1\,\mu m$ resolution, and excellent accuracy for density measures due to the use of a monochromatic X-ray energy beam.

Based on developments in the past two decades, there are a wide variety of options for the collection of micro-CT data, and after the rich 3D data have been collected, the analysis of these data are then required.

Quantitative Assessment

The quantitative assessment typically entails describing the morphometric characteristics of the data, and these measurements can be generally categorized into either
metric or *nonmetric* measurements. Metric measurements include parameters that
have a specific unit of measure (e.g., trabecular thickness is measured in millimeters), and a nonmetric measurement normally measures a topological feature (e.g.,
connectivity of the structure). Although the focus here is on bone measurements,
these morphometric parameters can be applied to other porous materials measured
in the micro-CT scanner. Prior to performing morphometric analyses, it is common
practice to filter noise from the image data and segment it to extract the mineralized
phase.

Filtration is normally done using a Gaussian filter, but median filters can also
provide reasonable results. The general rule of thumb is to perform the least amount
of filtering possible. Many approaches have been used to extract the mineralized
phase of bone, but the most common is to simply threshold the data with a global
threshold value. This approach works well when micro-CT data is calibrated and
requires that the image resolution is good. For low-resolution data, methods based
on edge detection and local adaptation may provide better segmentation (Waarsing
et al. 2004). However, they often introduce variability into the segmentation process,
particularly in data that has not been density-calibrated, and therefore may confound
the outcome measurements from the study. For example, to test the effect of an
anti-osteoporosis treatment, the changes in bone tissue density may influence local
thresholding or edge-detection methods, leading to biased experimental findings.
In general, a good rule of thumb is to use the simplest segmentation method that
provides satisfactory results.

The identification of regions of analysis is often the rate-limiting step when
performing large analyses. This step is necessary, for example, to classify the trabecular and cortical compartments for subsequent analysis. Although simple geometric
shapes (i.e., a sphere, cube) can be used to extract ROIs, it is often advantageous
to include the entire compartment for analysis, and this process can be time-
consuming. Recently, however, there have been fully automated approaches for
identifying compartmental ROIs in both animal and human studies that have shown
promise (Buie et al. 2007; Kohler et al. 2007).

Stereological Methods

Before the advent of micro-CT, stereological methods were used to define morphometric parameters representing the trabecular structure (Parfitt et al. 1983), and the
concepts of these measurements form the basis for the morphological parameters
used today. Previously, using histological sections prepared from bone samples, primary measurements were made directly from the slices. These included the total
bone perimeter length in millimeters as P_b, the total bone area in square millimeters

Table 1.1 Definitions of 2D stereological measurements, their units, and the micro-CT stereological equivalent metrics

Stereological	Definition	Unit	Micro-CT
TBV	$(A_b/A_t)100$	[%]	BV/TV
S_v	$(P_b/A_t)1.199$	$[mm^2\,mm^{-3}]$	BS/BV
S/V	$(P_b/A_b)1.199$	$[mm^2\,mm^{-3}]$	BS/TV
MTPT	$2{,}000/1.199(A_b/P_b)$	[μm]	Tb.Th
MTPD	$1.199/2(P_b/A_t)$	[/mm]	Tb.N
MTPS	$(2{,}000/1.199)(A_t - A_b)/P_b$	[μm]	Tb.Sp

The value of 1.199 was determined experimentally for iliac trabecular bone and a plate model is assumed (Parfitt et al. 1983)

as A_b, and the total section area in square millimeters as A_t. Based on an assumption that the structure was primarily comprised of rod- or plate-like structures, indices such as trabecular bone volume (TBV) and mean trabecular plate thickness, density and separation (MTPT, MTPD, MTPS, respectively) were defined. The parameter descriptions more commonly used with micro-CT include bone volume ratio (BV/TV), bone surface to volume (BS/BV), bone surface ratio (BS/TV), trabecular thickness (Tb.Th), trabecular separation (Tb.Sp) and trabecular number (Tb.N). The definitions of the stereological measurements in terms of their primary 2D parameters, and assuming a plate-like bone architecture (derived from iliac crest data), are provided with units of measurement and current morphological names used in micro-CT (Table 1.1).

Although the development of new morphological measurement parameters is constantly evolving as new technologies are developed (i.e., SR micro-CT), and for specific applications (e.g., the characterization of micro-crack morphology) (Schneider et al. 2007), there are some basic parameters that are used consistently in the field of micro-CT to characterize porous structures, and bone in particular.

Metric Measurements

Bone volume ratio (BV/TV) [%] is a fundamental parameter that represents the amount of bone tissue compared to the total bone volume. It is calculated simply by summing the total number of voxels in the segmented image that represent bone divided by the total volume. Although it is similar to the bone mineral density (BMD) measured from QCT or DXA, it is not identical, because the calculation is based on the segmented data and density information has been removed. Variations in tissue density are not captured by BV/TV for any tissue that did not meet the threshold criteria of the segmentation process.

Bone surface ratio (BS/BV) $[mm^2\,mm^{-3}]$ measures the surface area of the bone relative to the amount of bone tissue. The surface area can be calculated using iso-surfacing methods (e.g., marching cubes) to extract the bone surface (Lorensen and Cline 1987). The triangles that represent the bone surface are added to obtain the total surface area. The BS/BV parameter indicates the proportion of bone surface that is available for remodeling.

Trabecular thickness (Tb.Th) [mm] and trabecular separation (Tb.Sp) [mm] parameters are both calculated by direct measurements of the 3D data. To distinguish these measurements from stereologically derived measures, they are often termed "direct" methods, and denoted by a "*" (e.g., Tb.Th*, Tb.Sp*). The direct measurement is akin to fitting spheres into the bone or marrow space of the 3D dataset (Fig. 1.4). The advantage of the so-called direct technique (Hildebrand and Rüegsegger 1997a) is that it not only provides the average thickness or separation, but also the variations of those measures throughout the structure, and is independent of the type of bone architecture (rod-like or plate-like). The Tb.Th

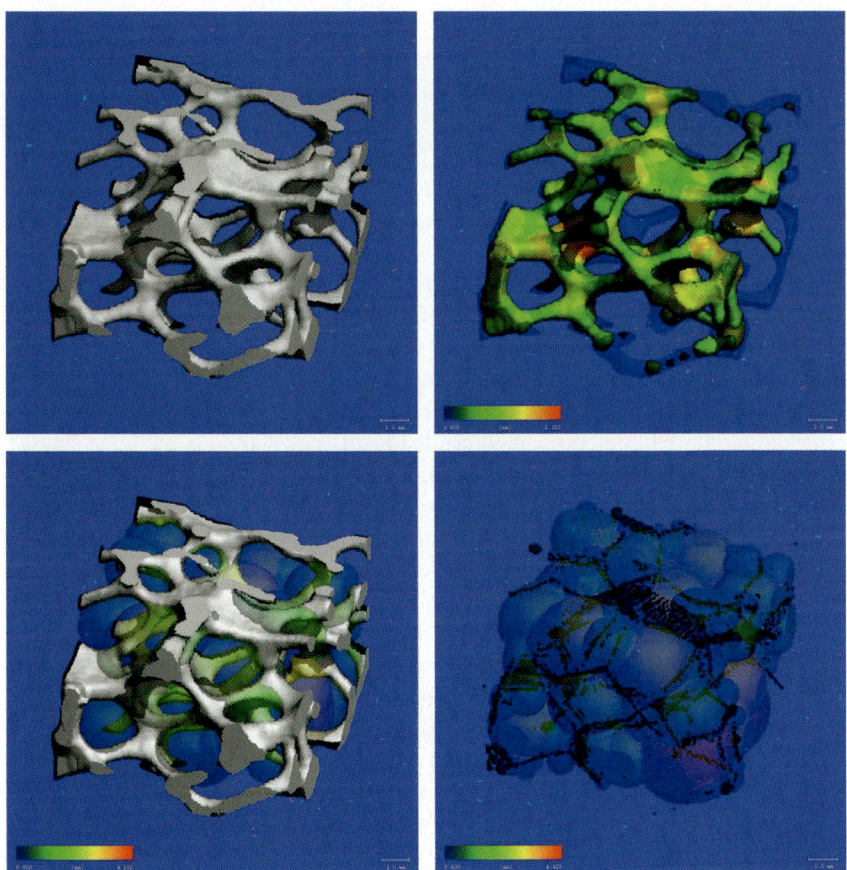

Fig. 1.4 A 3D structure is shown (*top right*) and used to illustrate the metric measurements for trabecular thickness (Tb.Th; *top right*), trabecular separation (Tb.Sp, *bottom left*) and trabecular number (Tb.N, *bottom right*). The color scale represents the size of the spheres used to calculate the parameters for these three parameters on the object (*bone*), background (*marrow*) and topology (*skeleton*), respectively

measurement can also be applied to the cortical compartment to provide a measurement of thickness (Ct.Th).

Trabecular number (Tb.N) [mm^{-1}] represents the number of trabeculae per unit length. It is calculated on the same basis as Tb.Th and Tb.Sp described above. The main difference is that the 3D structure is skeletonized (to extract the topology) (Fig. 1.4) and the distance between the skeletonized structures is inverted to provide the average number of trabeculae per unit length. This parameter provides an estimate of how many trabeculae exist in a bone structure, whereas Tb.Th and Tb.Sp provide the trabecular size and marrow spaces, respectively.

Nonmetric Measurements

Three common nonmetric morphological measurements include the structural model index, the measure of anisotropy, and the connectivity density. There are several other methods that are not covered here, and these include some well-established methods such as the trabecular pattern factor (Odgaard 1997), representing the number of concave vs. convex structures, and digital topological analysis (Wehrli et al. 2001), which characterizes the topological structure in terms of surfaces, profiles, curves and junctions.

The *structural model index* (SMI) characterizes the 3D bone structure in terms of the quantity of plates and rods in the structure (Hildebrand and Rüegsegger 1997b). The SMI ranges between 0 and 3, where 0 indicates an ideal plate structure and 3 an ideal rod structure—intermediate values indicate a mixture of rods and plates. Often this index is used to monitor the change in bone structure over time (i.e., osteoporosis progression). A basic component needed for its calculation is the bone surface, and since this measurement can be sensitive to resolution, the index is also somewhat dependent on image quality and resolution. A new approach to the concept of SMI has been to directly classify structures in terms of plates and rods (Stauber and Müller 2006), and this method may foresee more widespread use as an alternative to SMI due to it being less sensitive to image noise.

The *degree of anisotropy* (DA) reflects the preferred orientation of bone architecture, also often termed the bone fabric. The orientation of the structural components (i.e., trabeculae) is often related to the normal loading patterns for that bone, and when those normal loading patterns are disturbed (e.g., a joint injury), the degree of anisotropy may also be affected as the bone adapts to the new loading. The four most common methods of assessment include the mean intercept length (MIL) (Harrigan and Mann 1984), volume orientation (VO), star volume distribution (SVD), and star length distribution (SLD) (Odgaard 1997). Of these four, the MIL is the most established, and the standard against which all other techniques are compared. The procedure defines the three orthogonal orientations of the bone structure, and the result is normally presented as the ratio of the largest over the smallest magnitude of the orthogonal axes. A DA of one indicates that the structure is isotropic (no preferred orientation), but it is more common to find values that are greater than one, which represents a preferred orientation. For example, the cancellous bone in the lumbar spine is typically highly anisotropic (DA > 1). Defining the anisotropy of a structure is important, as it represents the organization of the bone, and this is closely related to the mechanical properties of the bone (Odgaard 1997).

Connectivity is defined as the degree to which a structure is multiply connected, and it is usually represented as *connectivity density* (Conn.D) [1 mm^{-3}]. Although connectivity can be defined in a number of ways, the most common approach in micro-CT is to use the topological approach (Odgaard and Gundersen 1993) because it is relatively stable and straightforward to determine from 3D segmented data. The number of connections between trabeculae is important in terms of bone strength because of the support provided. However, there is no direct correlation between connectivity density and strength because changes in connectivity can occur by two main mechanisms with very different implications for strength. For example, when bone loss occurs in plate-like bone it can lead to fenestrations (perforations) and result in an increase in connectivity; however, if bone loss occurs in rod-like bone, the connectivity will decrease. In both cases, the strength in the bone decreases, but the changes in connectivity are opposite. Nevertheless, connectivity is often used as an indicator of irreversible changes to bone architecture because, once trabeculae become disconnected, it seems that treatment to increase bone mass cannot restore those lost bridges. Although the inability to restore lost bridges has not been conclusively proven, new data with direct observations of individual trabeculae by in vivo micro-CT may clarify whether it can indeed occur.

Hybrid Approaches

The recent development of patient micro-CT (HR-pQCT) has provided unprecedented 3D images of human bone architecture (Fig. 1.1). However, even with a voxel size of only 82 μm, the structures are at the limit of accurate quantification by the standard direct morphometric procedures described above (MacNeil and Boyd 2007). For this reason, a conservative approach to the analysis of these data has been to use a hybrid analysis approach that combines stereological techniques and direct methods (Laib and Ruegsegger 1999). Metric quantities such as BV/TV, Tb.Th, Tb.Sp, and Ct.Th are not calculated directly from patient scans as would normally be done for micro-CT. The BV/TV is calculated by adding the CT density data (mg cm^{-3} HA) and dividing a value considered to define mineralized bone (1,200 mg cm^{-3} HA). A direct measure of BV/TV is not performed because partial volume effects leads to an overestimation of bone volume ratio. Similarly, for Tb.Th and Tb.Sp, these are derived based on the calculated BV/TV described here and the Tb.N by the direct method outlined earlier. The Tb.N is a stable calculation, even when the image resolution is low relative to the structures being measured; therefore, using the direct Tb.N and BV/TV, the thickness and separation parameters are determined standard stereological approaches (i.e., Tb.Th $= BV/TV/Tb.N^*$; Tb.Sp $= (1 - BV/TV)/Tb.N^*$). As was discussed earlier, the 2D stereological approaches suffer from the need to assume that the bone is plate-like or rod-like; however, in the hybrid approach used with HR-pQCT, this problem has been avoided by using a direct measurement of Tb.N from the 3D dataset. Mean cortical thickness (Ct.Th) is also not calculated directly, but instead estimated by dividing the mean cortical volume by the outer bone surface (periosteal surface).

Advanced Analysis Methods

There are several well-established analysis techniques that are used routinely to evaluate micro-CT data, and these are performed easily using widely available software packages (either supplied by the manufacturer or free). The establishment of new procedures always requires careful validation, and its success is largely dependent on the method being widely available and proven to provide new information that is distinct from other parameters. For example, the finite element method (which is discussed later in detail) must be shown to provide insight into bone strength above and beyond what can already be deduced by simple morphometric parameters such as bone volume ratio, or volumetric density.

One of the main drivers of new analysis methods is the development of new micro-CT hardware. The advent of SR micro-CT provides the ability to measure architectural features at resolutions never before possible, and the advent of in vivo micro-CT provides new opportunities to assess 3D data longitudinally. The following sections outline some recent developments in micro-CT analysis, and many of these new approaches have been driven by advances in technology.

Synchrotron Radiation Micro-CT

The ability to measure bone with submicrometer resolution using SR micro-CT (Peyrin et al. 1998) provides new insight into bone ultrastructure architecture and accordingly bone quality. The images generated with SR micro-CT provide outstanding spatial resolution and contrast, and it is possible to view cortical canal networks in mouse cortical bone, and even osteocyte lacunae (Schneider et al. 2007). Although several standard morphological indices have been designed to describe the trabecular and cortical structure of bone, the 3D quantification of canal networks and lacunae cannot be described by these parameters. Work focusing on cortical bone using standard micro-CT (Cooper et al. 2003) has been the basis for morphological indices describing new nomenclature for measurements based on SR micro-CT. For example, the number, volume and spacing of canals are expressed as N.Ca, Ca.V and Ca.Sp, respectively. Similarly, at the cellular level, parameters describing lacunae number, volume and volume density are expressed as N.Lc, Lc.V and Lc.V/Ct.TV, respectively. New nomenclature is also being developed for quantifying microcrack size in a similar fashion, as 3D data is only now becoming available for these highly detailed data (Voide et al. 2006). Although these morphological parameters introduced for SR micro-CT are new, they are based on the same analysis procedures described earlier for direct measurements of trabecular architecture. The versatility of the previously developed analysis procedures using direct method techniques (Hildebrand and Rüegsegger 1997a) has applications ranging from SR micro-CT-based measures to lower resolution magnetic resonance imaging (e.g., cartilage thickness from the Tb.Th* method).

Image Registration

An important development for in vivo analysis has been the application of 3D image registration for image alignment (Boyd et al. 2006b). Because repeated scans of the same animal or person can now be performed, the identification of a common region of interest between follow-up scans is important for two reasons. First, registration maximizes the reproducibility of the measurement technique by ensuring that only the volume of intersection between all sequential scans is assessed for morphological changes. This is important for sensitively detecting experimental effects over time (i.e., the apposition of new bone, or loss of bone structure). Secondly, through the use of registration with in vivo micro-CT data, it is possible to follow the developmental patterns of individual trabeculae and bones as a whole. Thus, for the first time, the disconnection of a trabecular structure can be physically observed, and the direct effects of treatments on the architecture can be assessed (Fig. 1.5). This represents an important advance from classical in vitro micro-CT studies where cross-sectional study designs are required, because only endpoint measurements are possible for in vitro micro-CT. In a cross-sectional study design, group-wise variations in bone architecture and the need to average quantitative results within

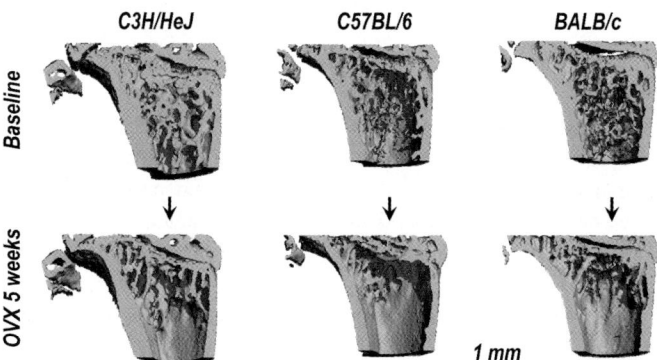

Fig. 1.5 The 3D bone architecture measured by in vivo micro-CT at the proximal tibia of three different inbred strains of mice at baseline, and the same three mice five weeks after ovariectomy (OVX). Image registration aligns the measurements from baseline to the measurements at five weeks, and detailed changes to the architecture can be observed

experimental groups can often hide subtle experimental effects. The advantage of in vivo micro-CT is that a repeated measures study design can be employed, and each animal or person acts as their own control. A repeated measures analysis not only provides better sensitivity to changes, but it also requires significantly fewer animals or patients to characterize the time course of a disease process (Boyd et al. 2006a).

In general, registration can be classified as either rigid registration or deformable (nonlinear) registration (Fitzpatrick et al. 2000). Typically, to align 3D micro-CT scans over time so that there is a common volume of interest (intersection of all images), a rigid registration approach is appropriate. It is based on an affine transformation that defines the linear rotation, shear, scale and translation of one image with respect to another (or at least the rotation and translation). The registration of two images involves defining one as the fixed (stationary) image and the other as the moving image. The determination of the affine transformation is done using an automated approach in order to find the best overlap between the fixed and moving 3D images. The measure of the "goodness of fit" is called the registration metric, and some examples of metrics include the measures of normalized correlation or mutual information. An optimizer determines the affine transformation that produces the best metric value, and the moving 3D image is subsequently transformed into the fixed image coordinate system via interpolation (usually at least linear interpolation). The use of 3D registration has also been used to improve the reproducibility of HR-pQCT measurements of the human radius by up to 40%, and this has important implications for improving the sensitivity of patient measurements and improving the minimum detectable change (MacNeil and Boyd 2008). When sequential images of bone deformation are available, it is possible to precisely identify local changes in tissue by finding the closest point from one surface to another and mapping that distance data (scalars expressed as colors) onto one of the images (Fig. 1.6).

Shape Variation

Rigid registration has also been the basis of a new technique for characterizing common morphological features within a cohort of animals and then comparing cohorts for significantly different physical shape differences. The method is a new high-throughput approach (Kristensen et al. 2008; Parsons et al. 2008) that avoids identifying physical landmarks, the approach traditionally used to quantify general shape. It is efficient and well suited as a screening tool for identifying important phenotypical features. Recently, it was applied to micro-CT measurements of the skulls of mice that develop cleft palate disease so that the developmental processes leading to the disease can be better understood (Fig. 1.7). The basis of the technique is to use 3D registration to superimpose the skulls of all mice from the same experimental group into a common reference frame, and then generate a "mean shape" image of the skull from that cohort. This can be done for many different strains of mice, and once the "mean shapes" for the mice have been determined, the superposition of those averages can be used to identify regions of shape differences. The

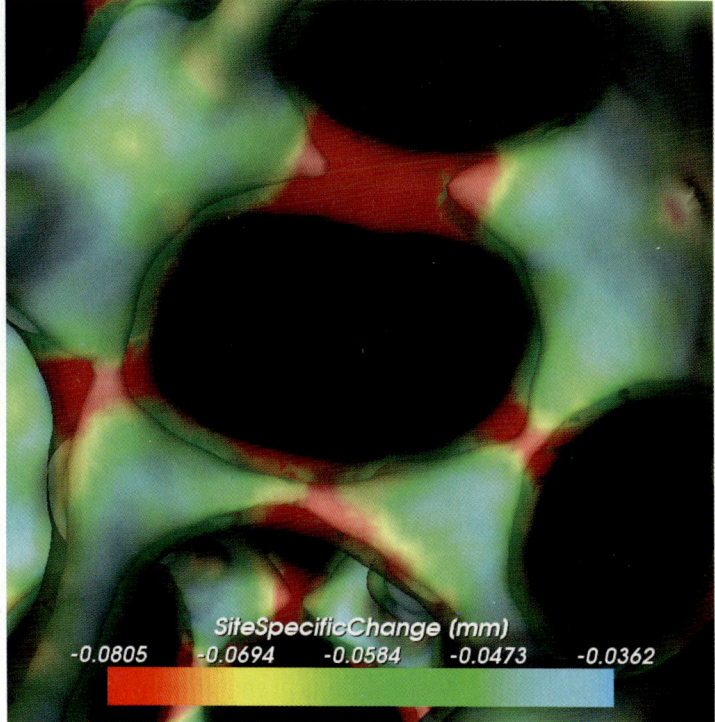

Fig. 1.6 Superposition by image registration of the trabecular architecture before and after osteoporotic-like changes is illustrated with simulated data as shown. The distance from one surface to another provides a measure of the local structural changes, and the colors represented the magnitude of the change

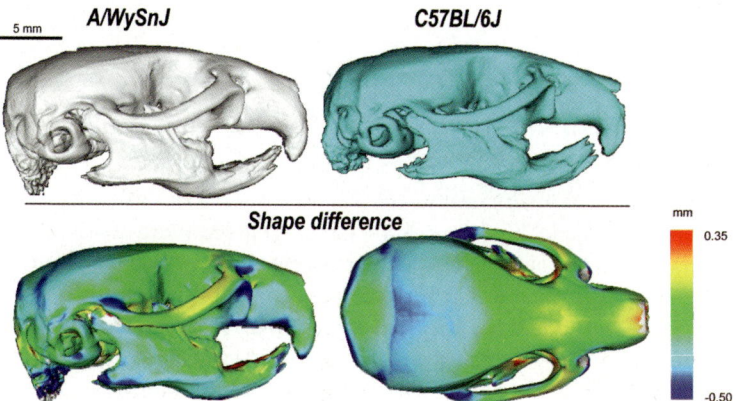

Fig. 1.7 Shape differences between two mouse strains (A/WySnJ and C57BL/6J) are illustrated using the high-throughput method, where a "mean shape" for each mouse strain is shown (*top row*), and the difference between mean shapes is represented on the A/WySnJ average shape, where colors represent the difference from the C57BL/6J skulls (*bottom*) shown from two viewpoints

differences are deemed statistically significant if the magnitude of the shape difference is larger than the variation in shape within each cohort. This new approach is still currently under development and offers a unique and efficient approach to accelerating research focused on functional genomics.

3D Structure Prediction

A deformable registration method is appropriate when a linear transformation between shapes cannot be defined (i.e., skeletal growth is typically nonlinear), or to define 3D maps that identify the relation between two different images. Similar to rigid registration, a fixed and moving image is defined, but rather than determining an affine transformation, a nonlinear transformation is determined. The most common approach employed to determine the nonlinear transformation is to use Demon's algorithm, which is an intensity-based registration approach. An output of the nonlinear deformation is the 3D deformation field, where a vector is produced for every image voxel. There are many interesting uses of the deformable registration method, including the use of the displacement map to calculate engineering strain (i.e., from two images of the same bone under different loads), and this could be used to validate the finite element method (discussed further in a separate chapter). Another novel approach is to use deformable registration to define the shape changes in trabecular architecture between successive in vivo micro-CT scans of an ovariectomized rat undergoing severe bone changes. The sequence of 3D deformation fields between each successive image pair can be extrapolated to generate an approximation of future 3D bone architecture (Boyd et al. 2004). While this method needs to be developed further, it offers the possibility of predicting bone architecture in the future based on the trend in the changes measured over time from in vivo micro-CT (Fig. 1.8).

Image-Guided Failure Analysis

A novel application of in vitro micro-CT is to monitor failures of bone structures as loads are applied. The technique is called image-guided failure assessment (IGFA) and was developed to better understand the mechanisms of bone failure (Nazarian and Müller 2004). A mechanical loading device compatible with micro-CT scanning encloses a cancellous bone specimen and applies load while a 3D image is acquired. Subsequently, the load is increased and the sample is re-scanned, and the procedure continues until bone failure occurs. The 3D images provide a method of monitoring deformations in the bone specimen and identifying the mechanism for failure. This approach has been applied on whole human vertebral specimens to monitor endplate deflection (Hulme et al. 2008), and recently in our lab with human radii under

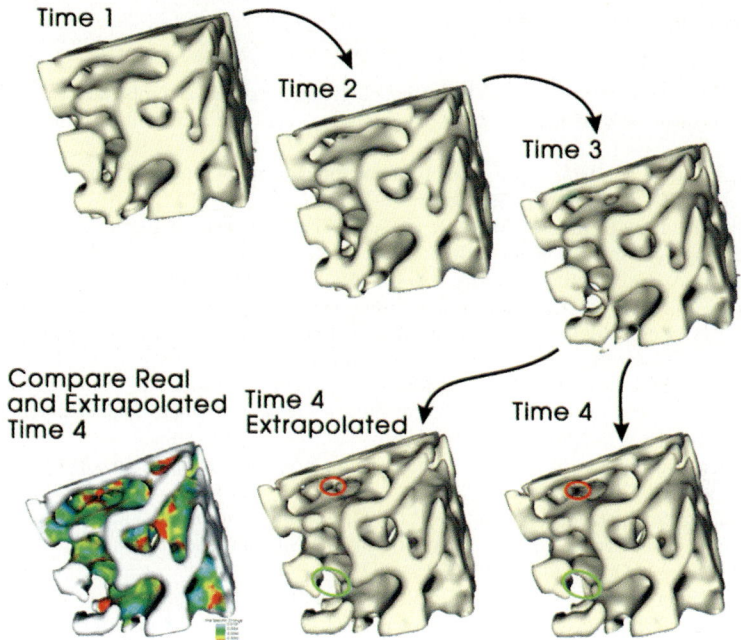

Fig. 1.8 Deformable registration of cancellous bone micro-CT images from a series of time points can be used to establish a trend in 3D architectural changes. In the example given here, the trend is established from three measurements from which the future 3D structure is extrapolated. The extrapolated and measured fourth time points can be compared visually and by using the site-specific method illustrating the magnitude of the prediction error

both axial and torsional loads to identify whether common modes of failure can be determined.

Future Directions

Micro-CT has proven to be an important biomedical research tool, particularly for bone research, since it provides insight into issues of bone quality in diseases and treatment. There are many other applications where micro-CT has an important impact, and these may include the measurement of other biological tissues (cartilage, ligaments, vasculature) using contrast imaging, as well as nonbiological materials. As the micro-CT methods push the boundaries for ultrahigh-resolution imaging from synchrotron radiation and patient applications for in vivo measurements, there will be the corresponding development of new approaches that can be used to analyze the data. While the collection of high-quality data critical to its success, so is the development of appropriate methods of extracting information. As new applications of micro-CT technology develop, so will appropriate analysis procedures.

References

Bolotin HH (2007) DXA in vivo BMD methodology: an erroneous and misleading research and clinical gauge of bone mineral status, bone fragility, and bone remodelling. Bone 41(1):138–154

Bonse U, Busch F (1996) X-ray computed microtomography (microCT) using synchrotron radiation (SR). Prog Biophys Mol Biol 65(1–2):133–169

Boyd SK, Mattmann C, Kuhn A, Müller R, Gasser JA (2004) A novel approach for monitoring and predicting bone microstructure in osteoporosis. 26th American Society of Bone and Mineral Research Annual Meeting, Seattle, USA, 1–5 Oct 2004. J Bone Miner Res 19(Suppl 1):S236–S237

Boyd SK, Davison P, Müller R, Gasser JA (2006a) Monitoring individual morphological changes over time in ovariectomized rats by in vivo micro-computed tomography. Bone 39(4):854–862

Boyd SK, Moser S, Kuhn M, Klinck RJ, Krauze PL, Müller R, Gasser JA (2006b) Evaluation of three-dimensional image registration methodologies for in vivo micro-computed tomography. Ann Biomed Eng 34(10):1587–1599

Brooks RA, Di Chiro G (1976) Beam hardening in X-ray reconstructive tomography. Phys Med Biol 21(3):390–398

Brouwers JE, van Rietbergen B, Huiskes R (2007) No effects of in vivo micro-CT radiation on structural parameters and bone marrow cells in proximal tibia of wistar rats detected after eight weekly scans. J Orthop Res 25(10):1325–1332

Buie HR, Campbell GM, Klinck RJ, MacNeil JA, Boyd SK (2007) Automatic segmentation based on a dual threshold technique for in vivo micro-CT bone analysis. Bone 41:505–515

Cooper DM, Turinsky AL, Sensen CW, Hallgrimsson B (2003) Quantitative 3D analysis of the canal network in cortical bone by micro-computed tomography. Anat Rec 274B(1):169–179

Feldkamp LA, Davis LC, Kress JW (1984) Practical cone-beam algorithm. J Opt Soc Am A1:612–619

Feldkamp LA, Goldstein SA, Parfitt AM, Jesion G, Kleerekoper M (1989) The direct examination of three-dimensional bone architecture in vitro by computed tomography. J Bone Miner Res 4(1):3–11

Fitzpatrick JM, Hill DLG, Maurer CR (2000) Chapter 8: image registration. In: Sonka M, Fitzpatrick JM (eds) Handbook of medical imaging, vol. 2: medical image processing and analysis. SPIE Press, Bellingham, WA, pp 447–513

Guldberg RE, Ballock RT, Boyan BD, Duvall CL, Lin AS, Nagaraja S, Oest M, Phillips J, Porter BD, Robertson G, Taylor WR (2003) Analyzing bone, blood vessels, and biomaterials with microcomputed tomography. IEEE Eng Med Biol Mag 22(5):77–83

Harrigan TP, Mann RW (1984) Characterization of microstructural anisotropy in orthotropic materials using a second rank tensor. J Mater Sci 19:761–767

Hildebrand T, Rüegsegger P (1997a) A new method for the model-independent assessment of thickness in three-dimensional images. J Microsc 185(1):67–75

Hildebrand T, Rüegsegger P (1997b) Quantification of bone microarchitecture with the structure model index. Comput Method Biomech Biomed Eng 1:15–23

Holdsworth DW, Drangova M, Fenster A (1993) A high-resolution XRII-based quantitative volume CT scanner. Med Phys 20(2 Pt 1):449–462

Hulme PA, Ferguson SJ, Boyd SK (2008) Determination of vertebral endplate deformation under load using micro-computed tomography. J Biomech 41(1):78–85

Klinck RJ, Campbell GM, Boyd SK (2008) Radiation effects on bone structure in mice and rats during in-vivo micro-CT scanning. Med Eng Phys 30(7):888–895

Kohler T, Stauber M, Donahue LR, Muller R (2007) Automated compartmental analysis for high-throughput skeletal phenotyping in femora of genetic mouse models. Bone 41(4):659–667

Kristensen E, Parsons TE, Gire J, Hallgrímsson B, Boyd SK (2008) A novel high-throughput morphological method for phenotypic analysis. IEEE Trans Biomed Eng (submitted for publication)

Laib A, Ruegsegger P (1999) Comparison of structure extraction methods for in vivo trabecular bone measurements. Comput Med Imaging Graph 23(2):69–74

Lorensen WE, Cline HE (1987) Marching cubes: a high resolution 3D surface construction algorithm. Comput Graphics 21(4):163–169

MacNeil JA, Boyd SK (2007) Accuracy of high-resolution peripheral quantitative computed tomography for measurement of bone quality. Med Eng Phys 29(10):1096–1105

MacNeil JA, Boyd SK (2008) Improved reproducibility of high-resolution peripheral quantitative computed tomography for measurement of bone quality. Med Eng Phys 30(6):792–799

Nazarian A, Müller R (2004) Time-lapsed microstructural imaging of bone failure behavior. J Biomech 37(1):55–65

Odgaard A (1997) Three-dimensional methods for quantification of cancellous bone architecture. Bone 20(4):315–328

Odgaard A, Gundersen HJ (1993) Quantification of connectivity in cancellous bone, with special emphasis on 3-D reconstructions. Bone 14(2):173–182

Parfitt AM, Mathews CH, Villanueva AR, Kleerekoper M, Frame B, Rao DS (1983) Relationships between surface, volume, and thickness of iliac trabecular bone in aging and in osteoporosis. Implications for the microanatomic and cellular mechanisms of bone loss. J Clin Invest 72(4):1396–1409

Parsons TE, Kristensen E, Hornung L, Diewert VM, Boyd SK, German RZ, Hallgrímsson B (2008) Phenotypic variability and craniofacial dysmorphology: increased shape variance in a mouse model for cleft lip. J Anat 212(2):135–143

Pauchard Y, Boyd SK (2008) Landmark based compensation of patient motion artifacts in computed tomography. SPIE Medical Imaging, San Diego, CA, 16–21 Feb. 2008

Peyrin F, Salome M, Cloetens P, Laval-Jeantet AM, Ritman E, Rüegsegger P (1998) Micro-CT examinations of trabecular bone samples at different resolutions: 14, 7 and 2 micron level. Technol Health Care 6(5–6):391–401

Rüegsegger P, Koller B, Müller R (1996) A microtomographic system for the nondestructive evaluation of bone architecture. Calcif Tissue Int 58(1):24–29

Schneider P, Stauber M, Voide R, Stampanoni M, Donahue LR, Müller R (2007) Ultrastructural properties in cortical bone vary greatly in two inbred strains of mice as assessed by synchrotron light based micro- and nano-CT. J Bone Miner Res 22(10):1557–1570

Stauber M, Müller R (2006) Volumetric spatial decomposition of trabecular bone into rods and plates—a new method for local bone morphometry. Bone 38(4):475–484

Van Rietbergen B, Odgaard A, Kabel J, Huiskes R (1998) Relationships between bone morphology and bone elastic properties can be accurately quantified using high-resolution computer reconstructions. J Orthop Res 16(1):23–28

Voide R, van Lenthe GH, Schneider P, Wyss P, Sennhauser U, Stampanoni M, Stauber M, Müller R (2006) Bone microcrack initiation and propagation—towards nano-tomographic imaging using synchrotron light. (5th World Congress of Biomechanics, Munich, Germany. July 29–August 4, 2006.) J Biomech 39(Suppl 1):S16

Waarsing JH, Day JS, Weinans H (2004) An improved segmentation method for in vivo microct imaging. J Bone Miner Res 19(10):1640–1650

Wehrli FW, Gomberg BR, Saha PK, Song HK, Hwang SN, Snyder PJ (2001) Digital topological analysis of in vivo magnetic resonance microimages of trabecular bone reveals structural implications of osteoporosis. J Bone Miner Res 16(8):1520–1531

Chapter 2
Advanced Experimental Magnetic Resonance Imaging

Ursula I. Tuor

Abstract Magnetic resonance imaging (MRI) is a noninvasive imaging technique with an extensive range of applications in biomedical diagnostic imaging. This chapter reviews selected aspects of advanced experimental MRI with a focus on brain imaging using animal MRI systems and their application to improving understanding of the pathophysiology of the brain. Anatomical MRI is advancing diagnostic applications through the increased use of quantitative morphology and MR relaxation times. Microscopic imaging is progressing with improvements in spatial resolution. Diffusion MRI imaging continues to enhance the information it provides on tissue and cellular or axonal structure using diffusion tensor imaging and diffusion tensor tractography. Magnetization transfer imaging is also providing supplementary information on pathophysiological changes in tissue, particularly white matter. Functional MRI in animals in conjunction with other invasive methods has improved our understanding of the fMRI response. Molecular MRI is a rapidly growing field that holds promise for the noninvasive imaging of molecular cellular processes using targeted or responsive contrast agents.

Introduction

Magnetic resonance imaging (MRI) has evolved rapidly from the acquisition of the first magnetic resonance images in the 1970s to an essential noninvasive method of producing body images of high quality for experimental or diagnostic use in the life sciences and medicine. MRI has a multitude of applications that continue to grow with the ongoing development of MRI hardware, software and contrast agents. Innovations in hardware involve the various components comprising an MRI

U.I. Tuor

Institute for Biodiagnostics (West), National Research Council of Canada, Experimental Imaging Centre, Hotchkiss Brain Institute, University of Calgary, 3330 Hospital Dr, NW, Calgary, Alberta, Canada T2N 4N1, e-mail: Ursula.Tuor@nrc-cnrc.gc.ca system, including new radiofrequency (rf) coils and types of magnets, whereas software innovation includes new sequence or image processing implementation. For example, scan times or signal-to-noise and fields of view have improved with the development of parallel imaging and multiple receive coils (de Zwart et al. 2006; Doty et al. 2007; Fujita 2007; Katscher and Bornert 2006; Ohliger and Sodickson 2006). The range of field strengths in use continues to expand (e.g., from 0.2 to 21 T), as do the configurations of the magnets, which vary from moving magnets that surround the patient during surgery to systems that combine MRI with other technologies such as magnetoencephalography or electroencephalography (Albayrak et al. 2004; Laufs et al. 2007; Ritter and Villringer 2006). MRI is a large and growing field; in order to limit its scope, this chapter will concentrate on providing an overview of advances in experimental MRI with a focus on brain imaging using animal MRI systems and their application to improving understanding of the pathophysiology of the brain. In many cases the reader will be referred to recent reviews of the topic.

MRI Principles

Briefly (and using simplified physics concepts), MRI is based on the fact that some atoms have nuclei that are unpaired and have a nonzero spin when placed in a strong uniform magnetic field. The most commonly imaged nuclei are protons, which are abundantly present within tissue *water*. Hydrogen protons have a simple spin of $1/2$, with the bulk collection of these nuclei aligned either parallel or antiparallel to the magnetic field, resulting in a net magnetization parallel to the field. Thus, within tissues in a strong magnetic field, there is a net magnetic dipole moment of nuclei precessing around the axial field. When the tissue is exposed to a brief pulse of electromagnetic energy produced by an appropriate rf pulse, the steady-stage equilibrium is perturbed so that some of the nuclei are temporarily flipped in phase to an unaligned state at higher energy. The net magnetization recovers as the spins return to align with the field with a time constant T_1 (spin–lattice relaxation). After the rf pulse, the excited spins also begin to dephase with time constants T_2 or T_2^*, a process called transverse or spin–spin relaxation. T_2 is acquired with a sequence where the dephasing is dependent on the microenvironment of the protons, whereas T_2^* is also affected by phase shifts caused by local field inhomogeneities and is shorter than T_2. The relaxation rates are dependent on tissue composition and thus result in MR signals that vary with tissue structure and content. The MR signals used to render the images are detected using rf coils which detect the rf signals that are retransmitted during the spin relaxations of the nuclei. Spatial information used to produce images is obtained using magnetic gradients that can apply rapid and precisely varying field strengths linearly in the x, y or z directions.

Anatomical MRI

Conventional MRI generally provides morphological information using contrast obtained from tissue differences in proton density, T_1 and T_2 relaxation. The most commonly used sequences are T_1- or T_2-weighted sequences that provide contrast in order to distinguish gray/white matter and CSF in the images according to weighting of the T_1 or T_2 relaxation of the tissue (Boulby and Rugg-Gunn 2003; Gowland and Stevenson 2003; Roberts and Mikulis 2007). T_1- and T_2-weighted images are the standard sequences that are used routinely for neuro-MRI, both clinically and experimentally, to detect anatomical or pathological tissue changes such as cerebral infarction or tumors. Although there have been innovations in image sequence design since the first standard spin-echo or gradient-echo methods, new sequences such as those that provide three-dimensional or faster echo-planar imaging sequences continue to employ some combination of T_1 or T_2/T_2^*-weighting for their contrast (Roberts and Mikulis 2007).

One major focus of work aimed at advancing traditional anatomical imaging has been to obtain objective or quantitative rather than qualitative assessments of anatomical changes. This has been achieved by the application of improved computing and image processing techniques that allow a quantitative *measure of morphological* changes in the brain. The contrast provided by regional brain differences in T_1- and T_2-weighted images, along with image analysis tools that segment structures, can allow an automated measurement of volumes of whole brain and brain structures of interest. Also, comparisons between imaging sessions, often using registration of the structures to an atlas, allow one to follow changes in brain morphometry over time (e.g., Makris et al. 2005; Pelletier et al. 2004; Whitwell and Jack 2005). Such quantitative assessment has resulted in an ability to measure the progression of morphometric changes such as those associated with development, aging or neurodegenerative diseases (e.g., Dubois et al. 2007; Makris et al. 2005; May and Matharu 2007; Pelletier et al. 2004; Whitwell and Jack 2005). Similar methods have been applied to animal studies in combination with the implementation of *MRI microscopy* (Benveniste and Blackband 2006; Benveniste et al. 2007; Jack et al. 2007; Johnson et al. 2007; Katscher and Bornert 2006; Nieman et al. 2007). MR microscopy has been driven by the need to identify the morphology associated with different transgenic mice in relation to their functional genetic and protein expression patterns. MR imaging of fixed specimens has progressed to the stage that a resolution of $10 \, \mu m^3$ has become feasible, thereby providing a nondestructive sectioning of samples at a resolution near that of conventional microscopy (Driehuys et al. 2008).

Quantitation of T_1 and T_2

Many pathologies or lesions of the brain result in local changes in T_1- or T_2-weighted images which provide diagnostic information, a prime example being stroke or cerebral infarction (e.g., Fig. 2.1). Gaining acceptance in experimental

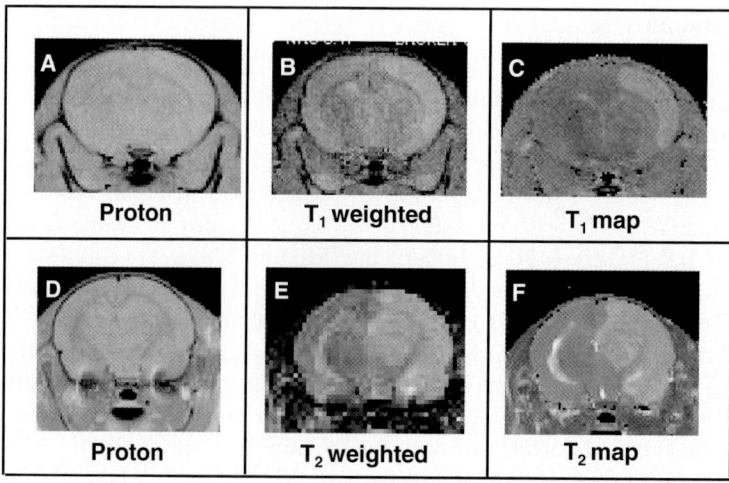

Fig. 2.1 (**a–f**) Differential contrast observed in proton, T_1 and T_2 magnetic resonance images of neonatal rat brains following cerebral hypoxia–ischemia. Intensity increases vary within the ischemic lesion depending on the sequence, i.e., proton maps (**a, d**), T_1-weighted (**b**), T_2-weighted (**e**), T_1 maps (**c**) or T_2 maps (**f**). Images were acquired using a 9.4 T Bruker MRI system, a SNAP-Shot Flash Inversion recovery sequence for T_1 imaging and a multiple-echo spin-echo sequence for T_2 imaging

studies and in some clinical applications such as multiple sclerosis or stroke is the use of MR sequences and image processing which provide quantitation of T_1 and T_2 (Boulby and Rugg-Gunn 2003; Dijkhuizen 2006; Gowland and Stevenson 2003; Neema et al. 2007; Weber et al. 2006a). Quantified T_1 or T_2 can provide a more consistent and potentially sensitive measure of pathological changes in tissue between subjects. Perhaps most extensively studied, particularly in animal models, are the quantitative T_1 and T_2 changes that occur following stroke or cerebral ischemia (e.g., Fig. 2.1). The challenge is to interpret the T_1 and T_2 changes according to the cellular and extracellular tissue changes. Fluid environments (e.g., CSF or tissue edema) are mobile and characterized by a slower loss of spin coherence and longer T_1 and T_2, whereas structured or constrained environments (e.g., myelinated white matter) undergo more spin–spin interactions and a more rapid loss of coherent spins, resulting in shorter T_2. Thus, T_1 should be sensitive to tissue water content, and indeed T_1-based sequences have been used to measure brain water content with MRI (Neeb et al. 2006). The increase in brain water or edema associated with stroke or cerebral hypoxia–ischemia is also reflected as an increase in T_1 (Barbier et al. 2005; Qiao et al. 2004). Vasogenic edema refers to an increase in brain water accompanied by dysfunction of the blood–brain barrier, and can be associated with increases in protein related to blood brain–barrier dysfunction. Edema with protein extravasation has been found to be associated most closely with increased T_2, supporting a greater sensitivity of spin coherence to proteins in the cellular microenvironment than increases in tissue water alone (Qiao et al. 2001). In general, standard quantitative MRI can be informative, but because T_1 and T_2 changes do not have unique tissue correlates, the assessment of tissue status or viability is enhanced by adding other imaging sequences, such as magnetization transfer or diffusion imaging, as discussed in detail below.

To complete the discussion of T_2 quantitation, one should note that additional potential information regarding tissue structure (including myelination of white matter) may be revealed using a multicomponent analysis of the T_2 relaxation (Laule et al. 2007; Wozniak and Lim 2006). When the T_2 relaxation in brain is followed for a large number of echo times with a spacing between echoes that is as short as possible, the resulting T_2 decay curve usually comprises two or three relaxation components considered to reveal different water environments. The shortest T_2 component within the analysis is interpreted to arise from water in the myelin sheath, whereas the longer components are attributed to axonal/intracellular and extracellular water. A very long T_2 component may be detected if the region of interest includes CSF. Compartmental T_2 changes have been detected with disorders such as multiple sclerosis and ischemia and hold promise regarding their ability to deliver insights into detection of demyelination and remyelination processes (Laule et al. 2007; Pirko and Johnson 2008) [T2 multiecho].

Diffusion Imaging

A wealth of information on cell or tissue structure, including axons in white matter, can also be inferred from MR sequences that probe the preferred directionality and/or diffusivity of water molecules within the tissue (Mori and Zhang 2006; Neil 2008; Wheeler-Kingshott et al. 2003). MR images (diffusion-weighted) are sensitive to the molecular motion of the water molecules, and such images are obtained by applying a pair of equal but opposing sequential field gradients. Signal intensities acquired are reduced according to the diffusivity of the water molecules, and if images are acquired for at least two different gradient strengths (e.g., zero and b1) then an apparent diffusion coefficient (ADC) can be calculated. Water motion in tissues is restricted by cell membranes, myelin sheaths or intracellular organelles, and the greater the restriction in motion the brighter the image in a diffusion-weighted image. Water diffusion in brain is altered by various processes, including development and aging or injury (Huppi and Dubois 2006; Wozniak and Lim 2006; Nucifora et al. 2007). For example, following acute stroke or cerebral ischemia, increases in DWI and decreases in ADC are observed (Fig. 2.2a, b). These are considered to reflect cell swelling and a reduction in extracellular space associated with ischemic reductions in ATP and ion pump failure (Sotak 2004; Talos et al. 2006; Weber et al. 2006a).

Diffusion tensor imaging (DTI) provides a measure of both the magnitude and direction of water diffusibility where this information is represented in a 3×3 matrix or tensor form for each voxel (Bammer et al. 2005; Mori and Zhang 2006; Nucifora et al. 2007; Talos et al. 2006). Such information is important in that water diffusion within tissues is generally unequal directionally. For example, in brain,

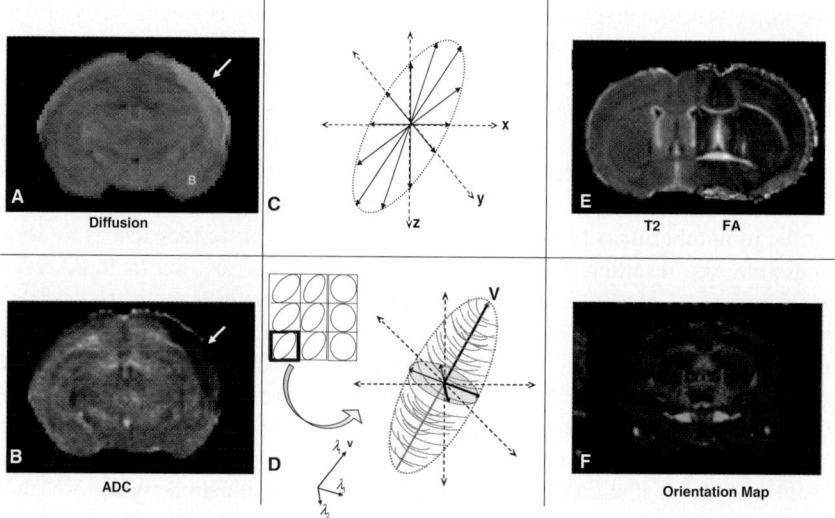

Fig. 2.2 (**a–f**) Applications of diffusion MRI using standard (**a, b**) and diffusion tensor imaging (**c–f**). Example of standard diffusion MR images from a neonatal rat brain soon after a transient cerebral hypoxia–ischemia depicts areas of ischemic injury as increases in intensity in a diffusion-weighted image (**a**) or a decrease in the apparent diffusion coefficient of water in the ADC map (**b**). Diffusion tensor MRI is performed using six or more diffusion sensitizing scans (**c**) that, for each voxel, provides a tensor with eigenvalues $(\lambda_1, \lambda_2, \lambda_3)$ that can be represented as an ellipsoid with direction V (**d**). Examples of a fractional anisotropy map (**e**) and of a color-coded orientation map (**f**) are from adult mice as reported by Mori and Zhang (reprinted from Mori and Zhang 2006, Figs. 8d and 10c, with permission from Elsevier)

diffusion of water within white matter axons can be highly aniosotropic or aligned in the direction of the axons. A full diffusion tensor (3×3 matrix) for each image voxel is obtained by acquiring at least six and often 12 or more noncollinear diffusion-sensitized images (Fig. 2.2c). This allows the determination of a magnitude and preferred direction of diffusivity for each voxel represented by an ellipsoid, and is generally used to study the integrity and directionality of white matter tracts (tractography). Limiting the number of diffusion directions to three orthogonal directions allows the determination of the trace or an average scalar measure of diffusion by determining the sum of the eigenvalues of the diagonalized matrix (Fig. 2.2d). The eigenvalues can also be used to calculate a measure of fractional anisotropy that consists of a value between 0 (dark on a map and indicative of complete isotropy) to 1 (bright on a map and fully aniosotropic) (Fig. 2.2e). The direction of the main eigenvector can be color-coded and provide a directionality map (Fig. 2.2f). As reviewed recently, the effects of patho/physiological processes such as injury or development on changes in relative anisotropy and axial diffusivity suggest that these diffusion measures hold promise as indicators of changes in myelination or axonal injury (Haku et al. 2006; MacDonald et al. 2007; Mori and Zhang 2006).

Diffusion tensor tractography utilizes the directionality of the main eigenvectors determined for each voxel and additional information concerning the relationships between voxels to generate images of connections between voxels that are considered to reflect white matter tracts. Diffusion tensor tractography is providing a promising new tool for exploring brain connections noninvasively in normal and diseased states (Mori and Zhang 2006; Neil 2008; Wheeler-Kingshott et al. 2003). Applications have been easier to implement first in humans considering the relatively large amounts of white matter compared to animals with lissencephalic brains. However, as the resolution achievable in experimental MR systems has increased, DTI imaging in animals has provided insights into the white matter tract changes associated with development and injury (Bockhorst et al. 2008; Mori et al. 2006). In contrast to clinical studies, animal results can be corroborated with histology. Indeed, interpretation of the tractography results is not exact and it is important to realize that the tracts visualized can be highly dependent on the algorithms and operator thresholds used. DTI analysis also has limitations when images are coarsely sampled, noisy or voxel-averaged, or when the tracts are curved or have multiple branches or crossings, and overcoming some of these limitations is an area of active research (Dauguet et al. 2007; Hua et al. 2008; Talos et al. 2006). One approach to address such issues is to combine fMRI with DTI tractography to aid in the appropriate choice of seed regions and to determine the functionality of tracts (Schonberg et al. 2006; Staempfli et al. 2008).

Magnetization Transfer Ratio (MTR) Imaging

Another less conventional MRI sequence is MT imaging that relies on tissue contrast from the interactions that occur between the protons in free fluid and those bound to macromolecular structures (Tofts et al. 2003). MT imaging probes the properties of the bound water by applying a radiofrequency pulse that is off-resonance from the free water protons yet still able to saturate the broader macromolecular proton pool. The constant exchange of magnetization between the free and bound water results in a decrease in signal intensity of off-resonance saturation images, and the ratio in intensity in MT images collected with and without the off-resonance saturation pulse provides an indication of the bound protons. The map of the magnetization transfer ratio (MTR) is produced for each voxel using $MTR = 100 \times (Mo - Ms)/Mo$ where Ms and Mo are the signal intensities obtained with and without MT saturation, respectively (Fig. 2.3a–c). Other approaches to quantitation have employed models describing compartments for the exchange and relaxation rates for the magnetization transfer process between bound and free pools of protons (Tofts et al. 2003). Such quantitative imaging has been complex to implement, requiring rather long scan times and potential specific absorption rate issues. However, there is substantial promise in its potential to provide pathologically specific relevant information regarding the spin exchange rate between the bound macromolecular proton pool and the directly measured free water protons along with their relative size (Tofts et al. 2003). Implementation of new techniques may facilitate this quantitation and accelerate clinical acceptance of such quantitative MT imaging (Gochberg and Gore 2007).

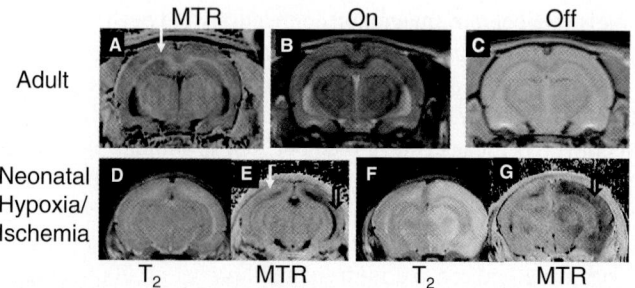

Fig. 2.3 (**a–g**) Representative magnetization transfer ratio (MTR) imaging in adult (**a–c**) and neonatal (**d–g**) rats. An MTR map (**a**) is determined using a ratio of the proton density-weighted image in the presence of an off-resonance saturation pulse to the bound proton pool (**b**) and in the absence of saturation (**c**). Developmental changes in MTR within white matter (*white arrows*) are apparent when comparing maps from an adult (**a**) and neonatal rat at postnatal day 8 (**d**). *Bright areas* in T_2-weighted images (**d, f**) and reductions in MTR (*black arrows*, E, G) are apparent 24 h following a transient unilateral cerebral hypoxic–ischemic insult in neonatal rats where the injury is either mild and rather selective for white matter (**d, e**) or more severe, resulting in hemispheric infarction (**f, g**)

To date, MTR imaging has been the most widely used MT method, and experimental studies have provided information regarding its biological significance and changes associated with diseases such as multiple sclerosis, Alzheimer's disease, brain trauma or tumors (Tofts et al. 2003). There is evidence that MTR can detect changes in axonal integrity and demyelination. Developmental increases of MTR in white matter are associated with myelination of axons (Engelbrecht et al. 1998; Wozniak and Lim 2006; Xydis et al. 2006) (Fig. 2.3a, d). Reductions in MTR in white matter occur in patients with multiple sclerosis within white matter lesions and regions of demyelination (Bagnato et al. 2006; Filippi and Agosta 2007; Horsfield 2005). Reductions in MTR also occur in animal models of experimental autoimmune encephalomyelitis and have been correlated with the inflammatory response and the appearance of specific macrophages (Blezer et al. 2007; Dijkhuizen 2006). MT imaging also detects tissue changes associated with stroke and reductions in MTR have been reported to occur following cerebral ischemia (Dijkhuizen 2006). We have found MT imaging to be particularly sensitive to ischemic changes observed in neonatal models of cerebral hypoxia–ischemia that produce either infarction or relatively selective white matter injury (Fig. 2.3e, f) (Tuor et al. 2008).

Thus, MTR imaging is sensitive to imaging changes in white matter related to changes in myelination or the inflammatory process. MTR also appears quite sensitive to the detection of cerebral edema, a not unexpected observation considering that increases in brain water would produce decreases in signal in the off-resonance saturation image. Thus, MT imaging can provide diagnostic information within the context of the disease process being investigated, and, although not definitive of the underlying cellular changes, it provides insights into tissue changes that are useful in combination with information obtained from other sequences.

Functional Magnetic Resonance Imaging

Another major application of MRI is functional MRI (fMRI) of the brain. This application uses MRI as a tool to study brain function and has transformed experimental and clinical neuroscience research. The field has seen an explosion of investigations using this technology to enhance our understanding of CNS processing under both normal and pathophysiological conditions (Jezzard and Buxton 2006; Matthews et al. 2006; Weiller et al. 2006). The human applications of fMRI range from classification and localization of diseases such as Alzheimer's, headache or epilepsy to understanding the mechanisms of reorganization of the brain following stroke or brain injury (Laatsch 2007; Sperling 2007; Ward 2006). There has been a great influence of fMRI on the cognitive neurosciences, including an improved understanding of CNS processing of pain and language pathways (Borsook and Becerra 2006, 2007; Crosson et al. 2007). The success of fMRI can be related to the wide availability of MRI scanners, the fact that MRI is noninvasive, its high temporal and spatial resolution, and the lack of any exposure to radioactivity such as occurs in positron emission tomography methods. However, fMRI technology is not trivial to implement, often requiring equipment and monitoring equipment to administer functional testing as well as specialists to provide expertise in a range of areas, including fMRI experimental design, image processing and interpretation.

The principle of fMRI is based on the fact that alterations in neuronal activity are closely coupled to changes in local blood flow within the activated part of the brain, resulting in small changes in MR signal intensity (Ferris et al. 2006; Nair 2005; Norris 2006). Although exact details of the neurophysiological origins of the fMRI signal contrast are still being clarified, it is apparent that there is a complex combination of changes in blood oxygenation in addition to changes in perfusion and blood volume within capillaries and veins, the magnitude and extent of which is dependant on factors such as the imaging sequence and field strength (Ferris et al. 2006; Nair 2005; Norris 2006). A major contributor to MR signal change is a reduction in deoxyhemoglobin, which during neuronal activation is related to the local increase in blood flow exceeding the corresponding increase in local oxygen extraction in magnitude. Since deoxyhemoglobin is paramagnetic, its decrease results in a signal increase in T_2^*-weighted images (and to a lesser degree T_2-weighted images), thereby providing a blood oxygenation level dependent (BOLD) MR signal contrast.

A major difference regarding fMRI in animals compared to humans is the need for anesthesia to be used in many experiments as a means of limiting movement of the subjects in the magnet. Some laboratories have had success in imaging awake animals by training them to be habituated to the imaging environment (e.g., Chin et al. 2008; Duong 2006; Ferris et al. 2006), but such studies are best suited to noninvasive or nonstressful stimuli, such as pharmacological fMRI studies investigating the effects of centrally acting drugs. A common anesthetic used for studies employing direct sensory or painful stimuli, such as electrical stimulation of the forepaw or formalin injection, is alpha-chloralose (e.g., Tuor et al. 2001, 2007). Under these conditions, a rather robust cortical response to forepaw stimulation is observed using either gradient-echo or spin-echo sequences sensitive to hemodynamic or BOLD changes associated with activation (e.g., Fig. 2.4a, b). A major disadvantage of alpha-chloralose is that it is not suitable for recovery studies. Other laboratories have successfully observed activation responses to forepaw stimulation in rats using inhalation anesthetics such as isoflurane at rather low concentrations (e.g. 0.74–1.3%) (Colonnese et al. 2008; Masamoto et al. 2007). However, success with such anesthetics may be dependent on additional factors. Using stimulation conditions considered optimal for forepaw stimulation under isoflurane anesthesia (Masamoto et al. 2007), and at minimum concentrations of isoflurane required for adequate sedation in our adult rats (i.e., 1.8%), we have been unable to detect an activation response. An alternate method of sedation, using the α_2 adrenoceptor agonist medetomidine, has also been employed successfully in recovery studies in rats (Weber et al. 2006b, 2008). It should be noted that this sedative is accompanied by increases in blood pressure (to greater than 180 mm Hg), bradycardia and reductions in ventilation rate. Thus, for certain fMRI studies in rodents, nonrecovery anesthetics such as alpha-chloralose will likely remain the anesthetic of choice.

Applications of fMRI in animals, although less extensive than humans, ranges from investigation of CNS pathways of cognitive or sensory processing such as pain (Borsook et al. 2006; Negus et al. 2006), to the response or plasticity/recovery of the brain or spinal cord following stroke or injury (Dijkhuizen 2006; Majcher et al. 2007; Weber et al. 2006a) (Fig. 2.4c). Experimental animals also allow the study of physiological changes such as blood gases or blood pressure demonstrating the sensitivity of the cerebral vasculature and fMRI to physiological status, which is of particular importance in diseased or injured brain (Tuor et al. 2007; Wang et al. 2006); see e.g., Fig. 2.4d, e. This provides a rationale for requiring physiological monitoring and/or control of blood gases and blood pressure in fMRI studies or the development of methods that can remove global effects unrelated to neuronal activation, for example by using the response in muscle tissue (Lowe et al. 2008). Pharmacological fMRI (phMRI) which uses fMRI to determine regional effects of drugs on the brain, is another growing application of this technology in preclinical drug testing (Borsook et al. 2006; Dijkhuizen 2006; Negus et al. 2006). Finally, an important area of research in experimental fMRI has been multimodal use of fMRI, i.e., its combination with other techniques such as optical or electrophysiological measures in order to better understand the physiological basis of fMRI and its interpretation with respect to altered neuronal activity (Colonnese et al. 2008; Huttunen et al. 2008; Logothetis et al. 2001; Qiao et al. 2006) or the effects of injury or disease (Dijkhuizen 2006; Weber et al. 2006a). Such studies have provided direct evidence for the association between increasing synaptic activity or local electrical field potential and the BOLD response.

Molecular MR Imaging

Another intensely active area of research within the evolving field of MRI is its extension to molecular MR imaging using novel contrast agents that are usually injected intravenously to image aspects of the metabolic, receptor and signal

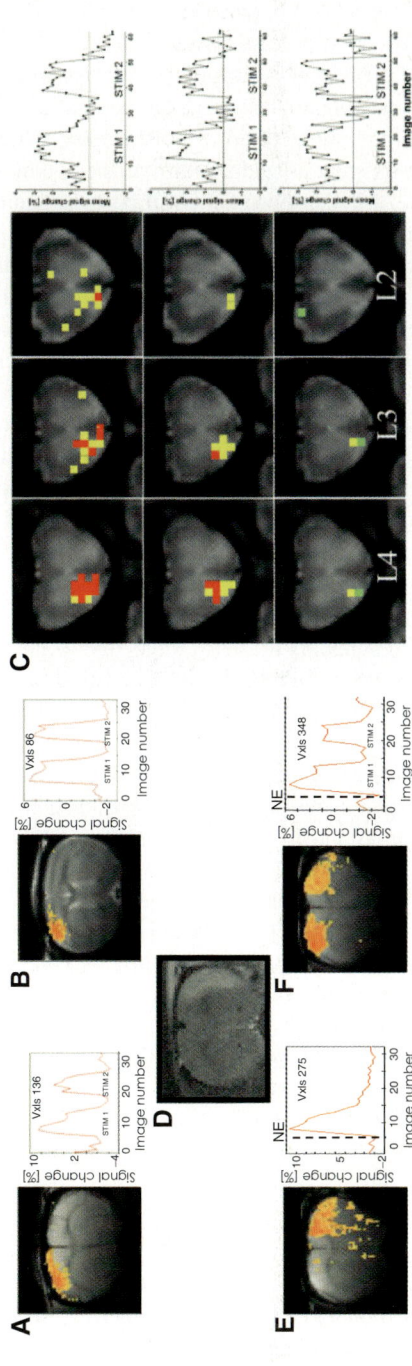

Fig. 2.4 (a–f) Functional MRI responses to electrical forepaw stimulation in rats anesthetized with alpha-chloralose. Voxels of activation (*orange to red*) correlate to the stimulation on(Stim)/off timecourse, with average changes in signal intensity shown in adjacent graphs. fMRI activation is detected in the sensory motor cortex using images acquired with either (**a**) a gradient-echo sequence (TR = 85 ms/TE = 10 ms) or (**b**) a fast spin-echo sequence (TR = 9.7 s/TE = 57 ms). Functional activation in the spinal cord (**c**) at different lumbar levels (*top row*) is inhibited by section of the L3 nerve (*middle row*) and prevented by section of both the L3 and L4 nerves entering the spinal cord (*bottom row*) (adapted from Majcher et al. 2007 with permission from Elsevier). The effect of hypertension in a rat with a stroke (hyperintense region in the T2-weighted image, (**d**) is detected as an apparent widespread activation (**e**) correlating to the change in blood pressure or an enhanced activation to forepaw stimulation (**f**) (modified from Tuor et al. 2007)

transduction processes at a cellular or subcellular level (Sosnovik and Weissleder 2007; Strijkers et al. 2007). Different approaches are being pursued to develop either targeted or activatable MR contrasts agents. *Targeted contrast agents* are directed to specific molecules or processes by conjugating the contrast agent with the appropriate ligand. The base component of such targeted contrasts have generally been selected from conventional nontargeted contrast agents, which form two basic classes: *paramagnetic agents* such as gadolinium, or those containing magnetic materials such as *iron oxide*. Another newer class of contrast agent comprises those containing exchangeable protons that resonate at a different frequency from bulk water and are detected directly using chemical exchange saturation transfer (CEST) imaging thereby forming a class of *CEST imaging agents*. In addition to targeted contrast agents, another approach has been to develop *activated or responsive probes* designed to image changes in the cell environment, changes in cell proteins, or to report gene activation (Gilad et al. 2007; Yoo and Pagel 2008). The change in MR signal can be achieved by linking the process of interest to a change in the contrast agent's physicochemical properties (Yoo and Pagel 2008). Much of the current development of molecular imaging contrast agents is still at the preclinical testing stage in phantoms, cell culture systems and animal models, as many of the steps in molecular MR contrast development are still being optimized to achieve an agent that is safe, has sufficient MR sensitivity, is target-accessible, and is specific for the disease or process under investigation. The brain presents a special challenge regarding the access of molecular contrast agents to their targets due to the blood–brain barrier preventing delivery of contrast across the cerebral vasculature.

As mentioned above, a major class of molecular imaging contrast agents are based on complexes that contain metal ion *paramagnetic agents* such as gadolinium (Gd). Because free metal ions are generally toxic, nontoxic complexes such as gadolinium diethylenetriaminepentaacetic acid (Gd-DTPA) have been developed and used clinically for years, e.g., for contrast-enhanced vascular and dynamic imaging (Caravan 2006; Lorusso et al. 2005; Strijkers et al. 2007; Weinmann et al. 2003). These paramagnetic agents have sites of water exchange that act catalytically to relax water protons to affect both T_1 and T_2 of the tissue, with a predominant effect on the T_1 relaxation of the local hydrogen atoms of water. This creates a hyperintense contrast on T_1-weighted images acquired using conventional T_1 pulse sequences. A major challenge in molecular imaging is to overcome the intrinsic low sensitivity of MR imaging, which requires the imaging of cellular processes at low concentrations (e.g., nanomolar amounts). Much effort in novel contrast design has therefore focused on improving the relaxivity or increasing the concentration of contrast delivered to the sites of interest. For paramagnetic agents, these approaches have included the formation of multiple gadolinium complexes or dendrimers or the incorporation of many thousands of gadolinium into amphiphilic lipid aggregates including micelles, lipid perfluoroemulsions or liposomes (Caravan 2006; Mulder et al. 2006). To achieve target specificity, these paramagnetic contrasts have been functionalized to provide sites of linkage with the ligands of interest. For example, lipid aggregates functionalized with biotin are coupled noncovalently using avidin–biotin complexes and biotinylated targeted ligands (Mulder et al. 2006). Covalent binding systems are also possible by functionalizing the nanoparticles with amide groups or maleimides with linkage to carboxylated or thiolated ligands, respectively. Incorporation of optical or fluorescent probes into lipid-based nanoparticles have produced multifunctional nanoparticles that are detectable with both MR and optical methods (Mulder et al. 2006). The ability to combine both in vivo MR and more sensitive optical methods in the development stage of novel contrast agents is particularly useful in preclinical testing in animal models to allow validation of the targeting of the agents, their specificity and localization.

A second major group of MR contrast agents for molecular MRI are based on the use of *superparamagnetic nanoparticles* with high magnetic susceptibility, such as iron oxide or other metals such as cobalt (Duguet et al. 2006; Gupta et al. 2007; McCarthy et al. 2007; Mulder et al. 2007; Thorek et al. 2006). In a magnetic field the net magnetic moment of iron oxide is several orders of magnitude greater than that of paramagnetic agents. This creates large microscopic field gradients for dephasing nearby protons, and results in a marked shortening of the transverse T_2 relaxation properties of the tissue. Gradient echo or T_2^* imaging is extremely sensitive for detecting the dephasing effect of these agents, but such sequences are also sensitive to artifacts produced by tissue and field inhomogeneities. Thus, at the very high fields often used for experimental imaging, T_2 rather than T_2^* maging is selected for optimal image quality.

Several steps and approaches have been taken in the development of the optimal superparamagnetic nanoparticles for molecular imaging (Gupta et al. 2007; McCarthy et al. 2007; Mulder et al. 2007). A first step in their synthesis is producing core particles of similar size and shape, where larger particles generally have greater magnetic effects and a different biodistribution than ultrasmall particles. The latter, which usually have a core $a < 10$ nm (<50 nm in diameter including their coating), more readily avoid immediate uptake by the liver and spleen, resulting in a relatively long plasma half-life (e.g., 20 h). Also, to improve their stability and to limit nonspecific cellular interactions and immune responses, the synthesis of iron oxide nanoparticles generally involves coating them with a hydrophilic/amphiphilic polymer such as polyethelyne glycol, polyvinyl alcohol or dextran, or with nonpolymeric stabilizers such as silica (Di et al. 2007; Gupta et al. 2007). The coatings can be synthesized to include carboxyl or amine groups on the surface, and these groups allow their derivatization with ligands such as antibodies or peptides/proteins to target the nanoparticles to cell-expressed surface receptors. The groups on the surface also allow the production of multifunctional nanoparticles by attaching fluorescent tags to produce probes that are then detectable with both optical and MR imaging methods (McCarthy et al. 2007; Mulder et al. 2007). The optimal superparamagnetic particle for MR imaging will depend on the biodistribution and targeting requirements of the intended application.

A third group of newer molecular imaging agents are based on CEST and *paramagnetic CEST (PARACEST)* agents (Woods et al. 2006). These contrast agents contain one or more pool of exchangeable protons with a sharply defined resonance frequency and a large chemical shift distinct from the bulk water pool. Image contrast is obtained by applying selective radiofrequency pulses to saturate the exchangeable protons, which transfer their magnetization to the free water upon exchange, leading to a drop in signal intensity and a negative contrast. Paramagnetic lanthanide chelates are examples of PARACEST agents that have a large chemical shift, resulting in faster exchange rates and better signal intensity changes than CEST agents containing exchangeable protons such as hydroxyl groups. One advantage of PARACEST contrast agents is that there is no image contrast without their presence, obviating the need for imaging pre- and post-contrast injection. Targeted PARACEST agents with high MR sensitivity are being developed using a number of approaches, including their entrapment within liposomes, or by attaching them to adenoviruses (Vasalatiy et al. 2008; Winter et al. 2006). PARACEST agents also hold promise for providing the possibility of targeted multispectral proton MRI to be performed (Aime et al. 2005), and for reporting physiological or metabolic information, as discussed next in greater detail.

MR responsive or activated contrast agents detect changes in the cellular environment that can include changes in concentrations or activities of proteins, enzymes and metabolites (Gilad et al. 2007; Yoo and Pagel 2008). The "response" is dependent on physicochemical phenomena such as accessibility of water, tumbling rate, and/or changes in local field inhomogeneities of the contrast agents, which change upon chemical reactions with the agents (e.g., gadolinium complexes). An example of a contrast agent that detects albumin is MS-325, which has an increase in tumbling time (and thereby a decrease in T_1 relaxation time) when this contrast agent binds to serum albumin. Many of the responsive agents are PARACEST agents which, when affected by the change in the cellular environment, undergo changes in proton exchange rate or MR frequency. A range of different PARACEST agents have been shown to detect changes in pH, temperature, various enzymes (such as caspase 3, when a peptidyl ligand of the agent is cleaved by this enzyme), various metabolites (such as L-lactate, which changes MR frequency when the PARACEST agent binds to this metabolite) or gases such as nitric oxide (Liu et al. 2007c; Yoo and Pagel 2008). Other substances imaged using responsive MR contrast agents include molecular imaging of oxygen, calcium and metal ions such as zinc. The development of MR reporter genes is also an active area of research that is a complex problem because imaging gene transcription using a reporter system analogous to optical reporters requires the encoding of an endogenous MR contrast agent, and substances such as ferritin have been considered (Gilad et al. 2007). Alternatively, a substance could be produced that when in contact with the MR contrast agent would produce an altered MR signal intensity. Major challenges include low sensitivity of detection, the need for substrates with associated undesirable pharmacokinetics, and difficult delivery of contrast at the site of gene expression. Nevertheless, there has been progress in the field, a recent example being the use of oligonucleotide sequences conjugated to superparamagnetic nanoparticles to target the expression of cerebral gene transcripts for cFOS (Liu et al. 2007a, b). Note, however, that such studies require delivery of these targeted contrasts directly into the ventricles of the brain. Despite these challenges, molecular imaging of gene expression can be expected to continue to be an exciting area for future applications of molecular MR imaging.

There are numerous *biomedical applications* of the molecular MR imaging contrast agents under development, and (as for the synthesis of the contrast agents) many of these have been reviewed elsewhere recently. The majority of the developments and applications for molecular imaging remain experimental with studies performed predominantly in animal models of disease. Promising applications for cardiovascular imaging include the imaging of vulnerable atherosclerotic plaques using magnetic nanoparticles or the targeted molecular imaging of fibrin and VCAM-1 (Briley-Saebo et al. 2007; Mulder et al. 2006; Sosnovik et al. 2007), as well as molecular imaging of stroke and the inflammatory process, including tracking of immune cells such as monocytes or macrophages and targeting of adhesion molecules on the endothelium (Barber et al. 2004; Heckl 2007; McCarthy et al. 2007; Mulder et al. 2006). Molecular imaging of cancer has also been extensively studied in animals, notable examples being the detection of angiogenesis using targets to integrins, and the imaging of specific tumor-related receptors such as those for human epidermal growth factor receptor (Mulder et al. 2006; Strijkers et al. 2007). Molecular imaging using magnetic nanoparticles will likely also prove a useful tool for monitoring stem cell migration and therapies (Magnitsky et al. 2007). In addition to detecting early stages of the disease process, nanoparticles may be designed to also deliver treatment in the future. Finally, future integration of molecular imaging into the process of drug development can be anticipated to facilitate the preclinical evaluation of targeted drug treatments.

Acknowledgments The author thanks D. Barry for the assistance provided in the preparation of the figures.

References

Aime S, Carrera C, Delli CD, Geninatti CS, Terreno E (2005) Tunable imaging of cells labeled with MRI-PARACEST agents. Angew Chem Int Ed Engl 44:1813–1815

Albayrak B, Samdani AF, Black PM (2004) Intra-operative magnetic resonance imaging in neurosurgery. Acta Neurochir (Wien) 146:543–556

Bagnato F, Ohayon JM, Ehrmantraut M, Chiu AW, Riva M, Ikonomidou VN (2006) Clinical and imaging metrics for monitoring disease progression in patients with multiple sclerosis. Expert Rev Neurother 6:599–612

Bammer R, Skare S, Newbould R, Liu C, Thijs V, Ropele S, Clayton DB, Krueger G, Moseley ME, Glover GH (2005) Foundations of advanced magnetic resonance imaging. NeuroRx 2:167–196

Barber PA, Foniok T, Kirk D, Buchan AM, Laurent S, Boutry S, Muller RN, Hoyte L, Tomanek B, Tuor UI (2004) MR molecular imaging of early endothelial activation in focal ischemia. Ann Neurol 56:116–120

Barbier EL, Liu L, Grillon E, Payen JF, Lebas JF, Segebarth C, Remy C (2005) Focal brain ischemia in rat: acute changes in brain tissue T1 reflect acute increase in brain tissue water content. NMR Biomed 18:499–506

Benveniste H, Blackband SJ (2006) Translational neuroscience and magnetic-resonance microscopy. Lancet Neurol 5:536–544

Benveniste H, Ma Y, Dhawan J, Gifford A, Smith SD, Feinstein I, Du C, Grant SC, Hof PR
(2007) Anatomical and functional phenotyping of mice models of Alzheimer's disease by MR
microscopy. Ann NY Acad Sci 1097:12–29

Blezer EL, Bauer J, Brok HP, Nicolay K, 't Hart BA (2007) Quantitative MRI–pathology correlations of brain white matter lesions developing in a non-human primate model of multiple
sclerosis. NMR Biomed 20:90–103

Bockhorst KH, Narayana PA, Liu R, Hobila-Vijjula P, Ramu J, Kamel M, Wosik J, Bockhorst T,
Hahn K, Hasan KM, Perez-Polo JR (2008) Early postnatal development of rat brain: In vivo
diffusion tensor imaging. J Neurosci Res 86(7):1520–1528

Borsook D, Becerra LR (2006) Breaking down the barriers: fMRI applications in pain, analgesia
and analgesics. Mol Pain 2:30

Borsook D, Becerra L (2007) Phenotyping central nervous system circuitry in chronic pain using
functional MRI: considerations and potential implications in the clinic. Curr Pain Headache
Rep 11:201–207

Borsook D, Becerra L, Hargreaves R (2006) A role for fMRI in optimizing CNS drug development.
Nat Rev Drug Discov 5:411–424

Boulby PA, Rugg-Gunn J (2003) T2: the transverse relaxation time. In: Tofts P (ed) Quantitative
MRI of the brain. Wiley, Chichester, pp 143–201

Briley-Saebo KC, Mulder WJ, Mani V, Hyafil F, Amirbekian V, Aguinaldo JG, Fisher EA, Fayad
ZA (2007) Magnetic resonance imaging of vulnerable atherosclerotic plaques: current imaging
strategies and molecular imaging probes. J Magn Reson Imaging 26:460–479

Caravan P (2006) Strategies for increasing the sensitivity of gadolinium based MRI contrast agents.
Chem Soc Rev 35:512–523

Chin CL, Pauly JR, Surber BW, Skoubis PD, McGaraughty S, Hradil VP, Luo Y, Cox BF, Fox
GB (2008) Pharmacological MRI in awake rats predicts selective binding of α_4, β_2 nicotinic
receptors. Synapse 62:159–168

Colonnese MT, Phillips MA, Constantine-Paton M, Kaila K, Jasanoff A (2008) Development of
hemodynamic responses and functional connectivity in rat somatosensory cortex. Nat Neurosci
11:72–79

Crosson B, McGregor K, Gopinath KS, Conway TW, Benjamin M, Chang YL, Moore AB, Raymer
AM, Briggs RW, Sherod MG, Wierenga CE, White KD (2007) Functional MRI of language
in aphasia: a review of the literature and the methodological challenges. Neuropsychol Rev
17:157–177

Dauguet J, Peled S, Berezovskii V, Delzescaux T, Warfield SK, Born R, Westin CF (2007) Comparison of fiber tracts derived from in-vivo DTI tractography with 3D histological neural tract
tracer reconstruction on a macaque brain. Neuroimage 37:530–538

de Zwart JA, van GP, Golay X, Ikonomidou VN, Duyn JH (2006) Accelerated parallel imaging for
functional imaging of the human brain. NMR Biomed 19:342–351

Di MM, Sadun C, Port M, Guilbert I, Couvreur P, Dubernet C (2007) Physicochemical characterization of ultrasmall superparamagnetic iron oxide particles (USPIO) for biomedical application
as MRI contrast agents. Int J Nanomedicine 2:609–622

Dijkhuizen RM (2006) Application of magnetic resonance imaging to study pathophysiology in
brain disease models. Methods Mol Med 124:251–278

Doty FD, Entzminger G, Kulkarni J, Pamarthy K, Staab JP (2007) Radio frequency coil technology
for small-animal MRI. NMR Biomed 20:304–325

Driehuys B, Nouls J, Badea A, Bucholz E, Ghaghada K, Petiet A, Hedlund LW (2008) Small
animal imaging with magnetic resonance microscopy. ILAR J 49:35–53

Dubois J, Benders M, Cachia A, Lazeyras F, HaVinh LR, Sizonenko SV, Borradori-Tolsa C,
Mangin JF, Huppi PS (2007) Mapping the early cortical folding process in the preterm newborn
brain. Cereb Cortex 18(6):1444–1454

Duguet E, Vasseur S, Mornet S, Devoisselle JM (2006) Magnetic nanoparticles and their
applications in medicine. Nanomed 1:157–168

Duong TQ (2006) Cerebral blood flow and BOLD fMRI responses to hypoxia in awake and
anesthetized rats. Brain Res 1135(1):186–194

Engelbrecht V, Rassek M, Preiss S, Wald C, Modder U (1998) Age-dependent changes in magnetization transfer contrast of white matter in the pediatric brain. AJNR Am J Neuroradiol
  19:1923–1929
Ferris CF, Febo M, Luo F, Schmidt K, Brevard M, Harder JA, Kulkarni P, Messenger T, King JA
  (2006) Functional magnetic resonance imaging in conscious animals: a new tool in behavioural
  neuroscience research. J Neuroendocrinol 18:307–318
Filippi M, Agosta F (2007) Magnetization transfer MRI in multiple sclerosis. J Neuroimaging
  17(Suppl 1):22S–26S
Fujita H (2007) New horizons in MR technology: RF coil designs and trends. Magn Reson Med
  Sci 6:29–42
Gilad AA, Winnard PT Jr, Van Zijl PC, Bulte JW (2007) Developing MR reporter genes: promises
  and pitfalls. NMR Biomed 20:275–290
Gochberg DF, Gore JC (2007) Quantitative magnetization transfer imaging via selective inversion
  recovery with short repetition times. Magn Reson Med 57:437–441
Gowland PA, Stevenson VL (2003) T1: The longitudinal relaxation time. In: Tofts P (ed)
  Quantitative MRI of the brain. Wiley, Chichester, pp 111–141
Gupta AK, Naregalkar RR, Vaidya VD, Gupta M (2007) Recent advances on surface engineering
  of magnetic iron oxide nanoparticles and their biomedical applications. Nanomed 2:23–39
Haku T, Miyasaka N, Kuroiwa T, Kubota T, Aso T (2006) Transient ADC change precedes persis-
  tent neuronal death in hypoxic-ischemic model in immature rats. Brain Res 19(1100):136–141
Heckl S (2007) Future contrast agents for molecular imaging in stroke. Curr Med Chem 14:1713–
  1728
Horsfield MA (2005) Magnetization transfer imaging in multiple sclerosis. J Neuroimaging
  15:58S–67S
Hua K, Zhang J, Wakana S, Jiang H, Li X, Reich DS, Calabresi PA, Pekar JJ, Van Zijl PC, Mori S
  (2008) Tract probability maps in stereotaxic spaces: analyses of white matter anatomy and
  tract-specific quantification. Neuroimage 39:336–347
Huppi PS, Dubois J (2006) Diffusion tensor imaging of brain development. Semin Fetal Neonatal
  Med 11:489–497
Huttunen JK, Grohn O, Penttonen M (2008) Coupling between simultaneously recorded BOLD
  response and neuronal activity in the rat somatosensory cortex. Neuroimage 39:775–785
Jack CR Jr, Marjanska M, Wengenack TM, Reyes DA, Curran GL, Lin J, Preboske GM, Poduslo
  JF, Garwood M (2007) Magnetic resonance imaging of Alzheimer's pathology in the brains of
  living transgenic mice: a new tool in Alzheimer's disease research. Neuroscientist 13:38–48
Jezzard P, Buxton RB (2006) The clinical potential of functional magnetic resonance imaging.
  J Magn Reson Imaging 23:787–793
Johnson GA, li-Sharief A, Badea A, Brandenburg J, Cofer G, Fubara B, Gewalt S, Hedlund
  LW, Upchurch L (2007) High-throughput morphologic phenotyping of the mouse brain with
  magnetic resonance histology. Neuroimage 37:82–89
Katscher U, Bornert P (2006) Parallel RF transmission in MRI. NMR Biomed 19:393–400
Laatsch L (2007) The use of functional MRI in traumatic brain injury diagnosis and treatment.
  Phys Med Rehabil Clin N Am 18:69–85, vi
Laufs H, Daunizeau J, Carmichael DW, Kleinschmidt A (2007) Recent advances in record-
  ing electrophysiological data simultaneously with magnetic resonance imaging. Neuroimage
  40(2):515–528
Laule C, Vavasour IM, Kolind SH, Li DK, Traboulsee TL, Moore GR, MacKay AL (2007)
  Magnetic resonance imaging of myelin. Neurotherapeutics 4:460–484
Liu CH, Huang S, Cui J, Kim YR, Farrar CT, Moskowitz MA, Rosen BR, Liu PK (2007a) MR
  contrast probes that trace gene transcripts for cerebral ischemia in live animals. FASEB J
  21:3004–3015
Liu CH, Kim YR, Ren JQ, Eichler F, Rosen BR, Liu PK (2007b) Imaging cerebral gene transcripts
  in live animals. J Neurosci 27:713–722
Liu G, Li Y, Pagel MD (2007c) Design and characterization of a new irreversible responsive
  PARACEST MRI contrast agent that detects nitric oxide. Magn Reson Med 58:1249–1256

Logothetis NK, Pauls J, Augath M, Trinath T, Oeltermann A (2001) Neurophysiological investigation of the basis of the fMRI signal. Nature 412:150–157

Lorusso V, Pascolo L, Fernetti C, Anelli PL, Uggeri F, Tiribelli C (2005) Magnetic resonance contrast agents: from the bench to the patient. Curr Pharm Des 11:4079–4098

Lowe AS, Barker GJ, Beech JS, Ireland MD, Williams SC (2008) A method for removing global effects in small-animal functional MRI. NMR Biomed 21:53–58

MacDonald CL, Dikranian K, Bayly P, Holtzman D, Brody D (2007) Diffusion tensor imaging reliably detects experimental traumatic axonal injury and indicates approximate time of injury. J Neurosci 27:11869–11876

Magnitsky S, Walton RM, Wolfe JH, Poptani H (2007) Magnetic resonance imaging as a tool for monitoring stem cell migration. Neurodegener Dis 4:314–321

Majcher K, Tomanek B, Tuor UI, Jasinski A, Foniok T, Rushforth D, Hess G (2007) Functional magnetic resonance imaging within the rat spinal cord following peripheral nerve injury. Neuroimage 38:669–676

Makris N, Caviness VS, Kennedy DN (2005) An introduction to MR imaging-based stroke morphometry. Neuroimaging Clin N Am 15:325–39, x

Masamoto K, Kim T, Fukuda M, Wang P, Kim SG (2007) Relationship between neural, vascular, and BOLD signals in isoflurane-anesthetized rat somatosensory cortex. Cereb Cortex 17:942–950

Matthews PM, Honey GD, Bullmore ET (2006) Applications of fMRI in translational medicine and clinical practice. Nat Rev Neurosci 7:732–744

May A, Matharu M (2007) New insights into migraine: application of functional and structural imaging. Curr Opin Neurol 20:306–309

McCarthy JR, Kelly KA, Sun EY, Weissleder R (2007) Targeted delivery of multifunctional magnetic nanoparticles. Nanomed 2:153–167

Mori S, Zhang J (2006) Principles of diffusion tensor imaging and its applications to basic neuroscience research. Neuron 51:527–539

Mori S, Zhang J, Bulte JW (2006) Magnetic resonance microscopy of mouse brain development. Methods Mol Med 124:129–147

Mulder WJ, Strijkers GJ, van Tilborg GA, Griffioen AW, Nicolay K (2006) Lipid-based nanoparticles for contrast-enhanced MRI and molecular imaging. NMR Biomed 19:142–164

Mulder WJ, Griffioen AW, Strijkers GJ, Cormode DP, Nicolay K, Fayad ZA (2007) Magnetic and fluorescent nanoparticles for multimodality imaging. Nanomed 2:307–324

Nair DG (2005) About being BOLD. Brain Res Brain Res Rev 50:229–243

Neeb H, Zilles K, Shah NJ (2006) Fully-automated detection of cerebral water content changes: study of age- and gender-related H_2O patterns with quantitative MRI. Neuroimage 29:910–922

Neema M, Stankiewicz J, Arora A, Guss ZD, Bakshi R (2007) MRI in multiple sclerosis: what's inside the toolbox? Neurotherapeutics 4:602–617

Negus SS, Vanderah TW, Brandt MR, Bilsky EJ, Becerra L, Borsook D (2006) Preclinical assessment of candidate analgesic drugs: recent advances and future challenges. J Pharmacol Exp Ther 319:507–514

Neil JJ (2008) Diffusion imaging concepts for clinicians. J Magn Reson Imaging 27:1–7

Nieman BJ, Bishop J, Dazai J, Bock NA, Lerch JP, Feintuch A, Chen XJ, Sled JG, Henkelman RM (2007) MR technology for biological studies in mice. NMR Biomed 20:291–303

Norris DG (2006) Principles of magnetic resonance assessment of brain function. J Magn Reson Imaging 23:794–807

Nucifora PG, Verma R, Lee SK, Melhem ER (2007) Diffusion-tensor MR imaging and tractography: exploring brain microstructure and connectivity. Radiology 245:367–384

Ohliger MA, Sodickson DK (2006) An introduction to coil array design for parallel MRI. NMR Biomed 19:300–315

Pelletier D, Garrison K, Henry R (2004) Measurement of whole-brain atrophy in multiple sclerosis. J Neuroimaging 14:11S–19S

Pirko I, Johnson AJ (2008) Neuroimaging of demyelination and remyeli-nation models. Curr Top Microbiol Immunol 318:241–266

Qiao M, Malisza KL, Del Bigio MR, Tuor UI (2001) Correlation of cerebral hypoxic-ischemic T2 changes with tissue alterations in water content and protein extravasation. Stroke 32:958–963

Qiao M, Latta P, Meng S, Tomanek B, Tuor UI (2004) Development of acute edema following cerebral hypoxia-ischemia in neonatal compared with juvenile rats using magnetic resonance imaging. Pediatr Res 55:101–106

Qiao M, Rushforth D, Wang R, Shaw RA, Tomanek B, Dunn JF, Tuor UI (2006) Blood-oxygen-level-dependent magnetic resonance signal and cerebral oxygenation responses to brain activation are enhanced by concurrent transient hypertension in rats. J Cereb Blood Flow Metab 27:1280–1289

Ritter P, Villringer A (2006) Simultaneous EEG-fMRI. Neurosci Biobehav Rev 30:823–838

Roberts TP, Mikulis D (2007) Neuro MR: principles. J Magn Reson Imaging 26:823–837

Schonberg T, Pianka P, Hendler T, Pasternak O, Assaf Y (2006) Characterization of displaced white matter by brain tumors using combined DTI and fMRI. Neuroimage 30:1100–1111

Sosnovik DE, Weissleder R (2007) Emerging concepts in molecular MRI. Curr Opin Biotechnol 18:4–10

Sosnovik DE, Nahrendorf M, Weissleder R (2007) Molecular magnetic resonance imaging in cardiovascular medicine. Circulation 115:2076–2086

Sotak CH (2004) Nuclear magnetic resonance (NMR) measurement of the apparent diffusion coefficient (ADC) of tissue water and its relationship to cell volume changes in pathological states. Neurochem Int 45:569–582

Sperling R (2007) Functional MRI studies of associative encoding in normal aging, mild cognitive impairment, and Alzheimer's disease. Ann N Y Acad Sci 1097:146–155

Staempfli P, Reischauer C, Jaermann T, Valavanis A, Kollias S, Boesiger P (2008) Combining fMRI and DTI: a framework for exploring the limits of fMRI-guided DTI fiber tracking and for verifying DTI-based fiber tractography results. Neuroimage 39:119–126

Strijkers GJ, Mulder WJ, van Tilborg GA, Nicolay K (2007) MRI contrast agents: current status and future perspectives. Anticancer Agents Med Chem 7:291–305

Talos IF, Mian AZ, Zou KH, Hsu L, Goldberg-Zimring D, Haker S, Bhagwat JG, Mulkern RV (2006) Magnetic resonance and the human brain: anatomy, function and metabolism. Cell Mol Life Sci 63:1106–1124

Thorek DL, Chen AK, Czupryna J, Tsourkas A (2006) Superparamagnetic iron oxide nanoparticle probes for molecular imaging. Ann Biomed Eng 34:23–38

Tofts P, Steens SCA, van Buchem MA (2003) MT: magnetization transfer. In: Tofts P (ed) Quantitative MRI of the brain. Wiley, Chichester, pp 257–298

Tuor UI, McKenzie E, Bascaramurty S, Campbell M, Foniok T, Latta P, Tomanek B (2001) Functional imaging of the early and late response to noxious formalin injection in rats and the potential role of blood pressure changes. Proc ISMRM 2001:1344

Tuor UI, Wang R, Zhao Z, Foniok T, Rushforth D, Wamsteeker JI, Qiao M (2007) Transient hypertension concurrent with forepaw stimulation enhances functional MRI responsiveness in infarct and peri-infarct regions. J Cereb Blood Flow Metab 27:1819–1829

Tuor UI, Meng S, Qiao M, Webster N, Crowley S, Foniok T (2008) Progression of magnetization transfer ratio changes following cerebral hypoxia-ischemia in neonatal rats: comparison of mild and moderate injury methods [abstract]. Proc ISMRM 2008:Abstract 64

Vasalatiy O, Gerard RD, Zhao P, Sun X, Sherry AD (2008) Labeling of Adenovirus Particles with PARACEST Agents. Bioconjug Chem 19:598–606

Wang R, Foniok T, Wamsteeker JI, Qiao M, Tomanek B, Vivanco RA, Tuor UI (2006) Transient blood pressure changes affect the functional magnetic resonance imaging detection of cerebral activation. Neuroimage 31:1–11

Ward NS (2006) The neural substrates of motor recovery after focal damage to the central nervous system. Arch Phys Med Rehabil 87:S30–S35

Weber R, Ramos-Cabrer P, Hoehn M (2006a) Present status of magnetic resonance imaging and spectroscopy in animal stroke models. J Cereb Blood Flow Metab 26:591–604

Weber R, Ramos-Cabrer P, Wiedermann D, van CN, Hoehn M (2006b) A fully noninvasive and robust experimental protocol for longitudinal fMRI studies in the rat. Neuroimage 29:1303–1310

Weber R, Ramos-Cabrer P, Justicia C, Wiedermann D, Strecker C, Sprenger C, Hoehn M (2008) Early prediction of functional recovery after experimental stroke: functional magnetic resonance imaging, electro-physiology, and behavioral testing in rats. J Neurosci 28:1022–1029

Weiller C, May A, Sach M, Buhmann C, Rijntjes M (2006) Role of functional imaging in neurological disorders. J Magn Reson Imaging 23:840–850

Weinmann HJ, Ebert W, Misselwitz B, Schmitt-Willich H (2003) Tissue-specific MR contrast agents. Eur J Radiol 46:33–44

Wheeler-Kingshott CAM, Barker GJ, Steens SCA, van Buchem MA (2003) D: the diffusion of water. In: Tofts P (ed) Quantitative MRI of the brain. Wiley, Chichester, pp 203–256

Whitwell JL, Jack CR Jr (2005) Comparisons between Alzheimer disease, frontotemporal lobar degeneration, and normal aging with brain mapping. Top Magn Reson Imaging 16:409–425

Winter PM, Cai K, Chen J, Adair CR, Kiefer GE, Athey PS, Gaffney PJ, Buff CE, Robertson JD, Caruthers SD, Wickline SA, Lanza GM (2006) Targeted PARACEST nanoparticle contrast agent for the detection of fibrin. Magn Reson Med 56:1384–1388

Woods M, Woessner DE, Sherry AD (2006) Paramagnetic lanthanide complexes as PARACEST agents for medical imaging. Chem Soc Rev 35:500–511

Wozniak JR, Lim KO (2006) Advances in white matter imaging: a review of in vivo magnetic resonance methodologies and their applicability to the study of development and aging. Neurosci Biobehav Rev 30:762–774

Xydis V, Astrakas L, Drougia A, Gassias D, Andronikou S, Argyropoulou M (2006) Myelination process in preterm subjects with periventricular leucomalacia assessed by magnetization transfer ratio. Pediatr Radiol 36:934–939

Yoo B, Pagel MD (2008) An overview of responsive MRI contrast agents for molecular imaging. Front Biosci 13:1733–1752

Chapter 3
Freehand 3D Ultrasound Calibration: A Review

Po-Wei Hsu, Richard W. Prager(✉), Andrew H. Gee, and Graham M. Treece

Abstract Freehand three-dimensional (3D) ultrasound is a technique for acquiring 3D ultrasound data by measuring the trajectory of a conventional 2D ultrasound probe as a clinician moves it across an object of interest.

The probe trajectory is measured by fixing some sort of position sensor onto it. The position sensor, however, can only measure its own trajectory, and a further six-degree-of-freedom transformation is required to map from the location and orientation of the position sensor to the location and orientation at which the ultrasound image is acquired. The process of determining this transformation is known as calibration. Accurate calibration is difficult to achieve and it is critical to the validity of the acquired data. This chapter describes the techniques that have been developed to solve this calibration problem and discusses their strengths and weaknesses.

Introduction

Three-dimensional (3D) ultrasound imaging is a medical imaging modality that allows the clinician to obtain a 3D model of the anatomy, possibly in real time (Nelson and Pretorius 1998; Fenster et al. 2001). Compared to other 3D modalities it has the advantages of being cheap, safe, quick and of giving fairly high resolution. However, ultrasound images are sometimes difficult to interpret because of artefacts generated in the acquisition process. Clinical applications of 3D ultrasound include obstetrics (Gonçalves et al. 2005), gynecology (Alcazar 2005), breast biopsy (Fenster et al. 2004b), cardiology (Fenster et al. 2004a), fetal cardiology (Yagel et al. 2007), neurosurgery (Unsgaard et al. 2006), radiology (Meeks et al. 2003) and surgery (Rygh et al. 2006).

Richard Prager
University of Cambridge, Department of Engineering, Trumpington Street, Cambridge CB2 1PZ, UK, e-mail: rwp@eng.cam.ac.uk

C.W. Sensen and B. Hallgrímsson (eds.), *Advanced Imaging in Biology and Medicine.*   47
© Springer-Verlag Berlin Heidelberg 2009

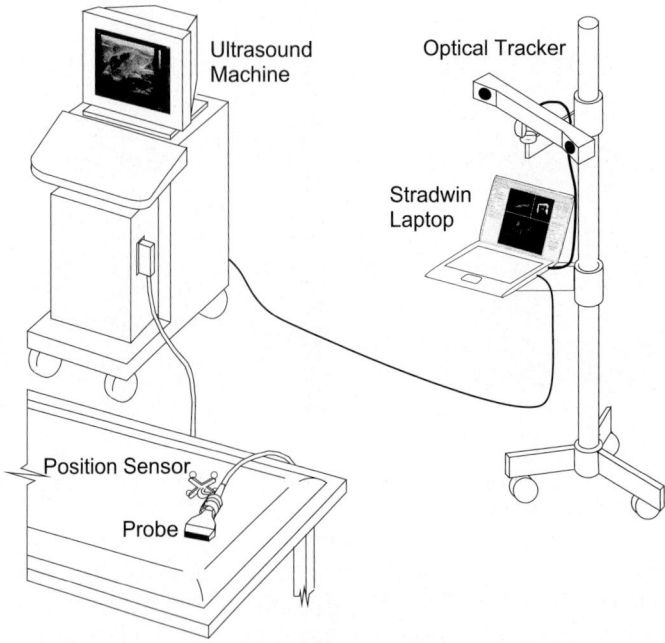

Fig. 3.1 General arrangement of a freehand 3D ultrasound system. Our freehand 3D ultrasound software is called *Stradwin*. In this diagram, Stradwin is run on a laptop which simultaneously records images from the ultrasound machine and the probe trajectory from the optical tracker.[1] The cameras in the optical tracker determine the position and orientation of the ultrasound probe by locating a set of balls that are mounted rigidly on it

Two main approaches are used to acquire the ultrasound data in 3D. Either the probe is held still and a fixed 3D volume is scanned, or the probe is moved manually over the volume of interest and a sequence of ultrasound images and corresponding probe positions are recorded. The second approach is called "freehand 3D ultrasound" (see Fig. 3.1) (Gee et al. 2003). This chapter is about this second approach and particularly focuses on ways of achieving effective tracking of the probe trajectory. First, however, we describe the motivation for using freehand 3D ultrasound data and illustrate some of its applications.

Why is Freehand 3D Ultrasound Useful?

Compared with a fixed 3D probe, freehand 3D ultrasound has the advantage that an arbitrarily large volume may be scanned. The data in this volume may also be accurately located in a fixed global coordinate system. However, there is the

---

[1] For example, the Polaris system produced by Northern Digital Inc.

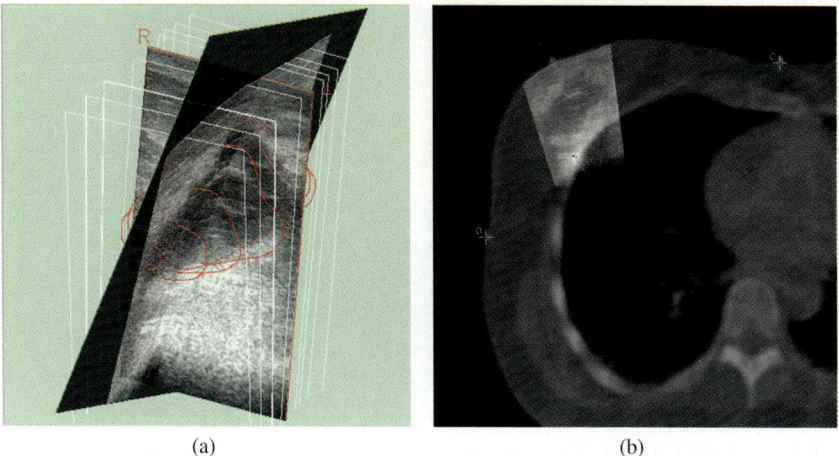

(a)   (b)

Fig. 3.2 Freehand 3D ultrasound used to assist with radiotherapy planning (Coles et al. 2007). The white "goal post" shapes in **a** show the positions of some typical B-scans in a freehand 3D dataset. The transducer array of the probe is located along the cross-bar of the goal-post shape. An example of one B-scan is shown, outlined in *red*. The *red contours* show where a clinician has outlined the position of a tumor bed (after breast-conserving surgery). The *green-edged plane* shows a slice through the data aligned with the coordinate system of a radiotherapy planning machine. In **b**, this slice is overlayed on data from a conventional X-ray CT radiotherapy planning system. Features identified by the clinician in the original B-scans (e.g., the red contours), can thus be used to improve the accuracy of the radiotherapy planning process

inconvenience of setting up the position-sensing equipment and maintaining a clear line of sight between the position sensor's cameras and the probe (see Fig. 3.1). Furthermore, the data acquired with a freehand system may be less regularly sampled than data from a fixed 3D probe.

Nevertheless, freehand 3D systems have their uses. They are particularly valuable when it is necessary to locate the subject in a fixed external coordinate system, as for example in radiotherapy planning. Figure 3.2 shows how freehand 3D ultrasound data may be incorporated into an existing X-ray CT-based radiotherapy planning system. It is also possible to perform the planning based on the ultrasound data alone.

The second group of applications arises when it is not possible to scan the object of interest in a single sweep of the probe. Figure 3.3 shows three surface models, each constructed from a separate freehand 3D sweep. Taken together, they provide a useful visualization of the shape of the liver, and an estimate of its volume that is accurate to within $\pm 5\%$. This accuracy is not dependent on the liver conforming to a prior model and is therefore robust to changes in shape brought about by pathology.

Figure 3.4 shows another example of multiple sweeps of the probe being used to enable the analysis of complex geometry: in this case the structure of a neonatal foot. The purpose of the imaging here is to provide a detailed description of the three-dimensional geometry of the unossified cartilage in the foot. This is useful in planning surgery to correct talipes equinovarus (club foot).

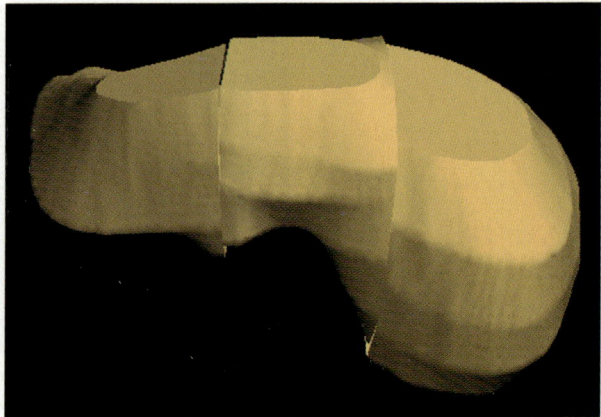

Fig. 3.3 Measuring the volume of a liver using a freehand three-dimensional ultrasound scan that passes across it three times to cover the whole shape (Treece et al. 2001). The liver is manually segmented in a number of the original B-scans, and the three independent surface models are constructed from the resulting contours

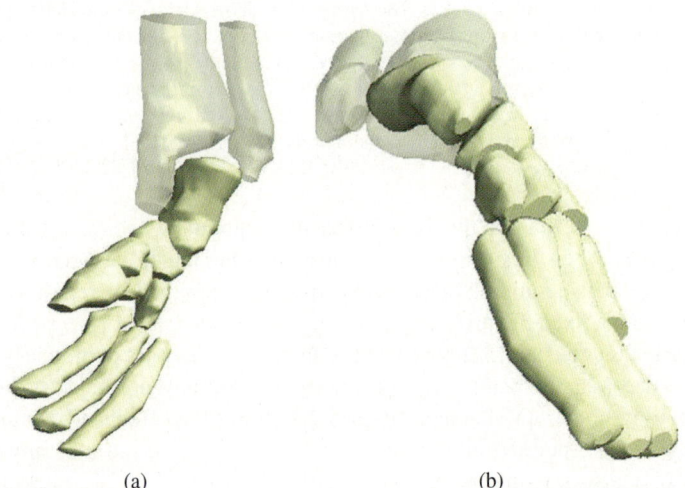

(a)   (b)

Fig. 3.4 Freehand 3D ultrasound study of a neonatal foot with talipes equinovarus (*club foot*) by Charlotte Cash et al. (2004). Parts **a** and **b** show two views of a surface model of the foot. This was constructed by manually segmenting the unossified cartilage in the original B-scans of several freehand 3D ultrasound sweeps

What is Probe Calibration and Why is it Important?

In freehand 3D ultrasound, the position sensor records the 3D location and orientation of the mobile part of the position sensor S relative to its stationary counterpart W, as shown in Fig. 3.5. It is therefore necessary to determine the position and orientation of the scan plane with respect to the electrical center of the position

Fig. 3.5 The coordinates associated with a freehand 3D ultrasound system

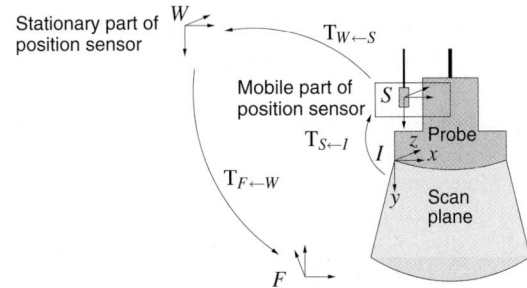

sensor. This transformation is determined through a process called probe calibration (Mercier et al. 2005).

Accurate calibration, or measurement of the position of the scan plane with respect to the position sensor, is vital for accurate freehand 3D ultrasound. The importance of this process is further heightened by that fact that the main applications of freehand 3D systems, as described above, require image features to be accurately located in 3D space. Radiotherapy requires points to be located correctly in an external coordinate system. Volume measurement or geometry analysis, such as the talipes example, require accurate relative positions. If a freehand 3D ultrasound system is incorrectly calibrated, it will produce dangerously plausible but misleading images and measurements.

This chapter provides a survey of approaches to the calibration of freehand 3D ultrasound systems. It attempts to provide an answer to the question "what is the best way to calibrate a system for a particular application?" We start by classifying each calibration technique according to its principles. This is followed by a discussion of the metrics used to assess calibration quality. Finally, we compare the calibration techniques, focusing on ease of use, speed of calibration and reliability. The comparison is performed based on our own experimental results, rather than figures quoted from previous papers, to eliminate factors caused by differences in 3D ultrasound systems and user expertise.

Probe Calibration

In this section we describe a mathematical framework for freehand 3D ultrasound calibration and then explain the calibration techniques found in the literature in terms of this framework.

A Mathematical Framework for Probe Calibration

The goal in freehand 3D ultrasound calibration is to determine the rigid-body transformation from the ultrasound B-scan to the electrical center of the position-sensing device that is clamped to the ultrasound probe. This fixed transformation $T_{S \leftarrow I}$

comprises six parameters—three translations in the direction of the x-, y- and z-axes, and the three rotations (azimuth, elevation and roll) about these axes. We have to deal with a number of coordinate systems, as shown in Fig. 3.5. There is the object being scanned F, the fixed part of the position sensor W (the cameras, the case of an optical sensor), the mobile part of the position sensor S, and the B-scan itself I. The stationary part of the position sensor is often called the world coordinate system, and the term "position sensor" is used to mean its mobile counterpart. We will also follow these conventions in this chapter. In general, a transformation involves both a rotation and a translation in 3D space. For brevity, we will use the notation $T_{B \leftarrow A}$ to mean a rotational transformation followed by a translation from the coordinate system A to coordinate system B.

Another issue before we can construct a volume in space is to determine the scales of the B-scans. A point $p^{I'} = (u,v,0)^t$ in a B-scan image, where u and v are the column and row indices, typically has units in pixels rather than in millimeters. A scaling factor $T_s = \begin{pmatrix} s_u & 0 & 0 \\ 0 & s_v & 0 \\ 0 & 0 & 0 \end{pmatrix}$, where s_u and s_v are the scales in millimetres per pixel, is necessary to change the unit of the point to metric units by $p^I = T_s p^{I'}$. In this chapter, we will use the notation p^A to denote the coordinates of a point p in the coordinate system A.

If both the calibration and the image scales are known, each point can be mapped to 3D space by:

$$p^F = T_{F \leftarrow W} T_{W \leftarrow S} T_{S \leftarrow I} T_s p^{I'}. \tag{3.1}$$

In the above equation, $T_{W \leftarrow S}$ can be read from the position sensor. The transformation from the world coordinate system to a phantom coordinate system $T_{F \leftarrow W}$ is in fact not necessary in 3D image analysis. Most of the time, it is nevertheless included for convenience. Should it be removed, all analysis on the resulting 3D image will remain correct. However, the anatomy may appear at an absurd orientation. We will see later in this chapter how the choice of $T_{F \leftarrow W}$ will help us to find the calibration parameters.

The 3D Localizer or "Pointer"

Before we start introducing the different methods to calibrate a probe, we briefly outline a device that is often used in modern probe calibration (Anagnostoudis and Jan 2005). This is a 3D localizer, often called a pointer or a stylus. Figure 3.6a shows one such stylus, consisting of a round shaft. On one end, it has position-sensing devices that can be tracked by the position tracking system; at the other end, it is sharpened to a point. The localizer can report the location of its tip in 3D space; hence we can get the location of any point in space by pointing the stylus at the target.

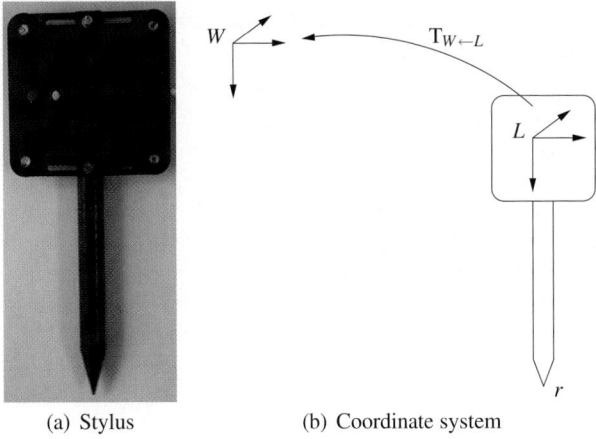

(a) Stylus   (b) Coordinate system

Fig. 3.6 A 3D localizer

Figure 3.6b shows the coordinate system involved when using a stylus. If the position of its tip r^L is known in the stylus's coordinate system L, then the position of the tip in 3D space is given by:

$$r^W = T_{W \leftarrow L} r^L, \tag{3.2}$$

where $T_{W \leftarrow L}$ is provided by the position sensor. The position of the tip r^L may be supplied by the manufacturer (Muratore and Galloway Jr. 2001). When this position is not known, it can be determined by a pointer calibration (Leotta et al. 1997).

During a pointer calibration, the stylus is rotated about its tip while the position sensor's readings are recorded. Since the tip of the stylus remains stationary throughout the pointer calibration process, its location r^W in 3D space is therefore fixed. We can then use (3.2) to solve for the position of the stylus tip, by minimizing

$$\sum_i \left| \overline{r_i^W} - T_{W \leftarrow L_i} r^L \right|, \tag{3.3}$$

where $|\cdot|$ denotes the usual Euclidean norm on $\mathbb{R}^3$ and $\overline{a_i}$ the mean of (a_i). We also used the notation r^L instead of r^{L_i} since r is invariant in every L_i.

The stylus is nevertheless prone to errors. These include errors from inaccurate tracking and pointer calibration errors. The accuracy of the pointer calibration is dependent on the size of stylus. Pointer calibrations typically have RMS errors between 0.6 and 0.9 mm, but errors up to 1.5 mm have been quoted (Hsu et al. 2007). The tracking error is dependent on the tracking system. A typical optical tracking system, such as the Polaris, has a tracking error of 0.35 mm. In general, a stylus has a positioning uncertainty of approximately 1 mm.

The stylus has become popular in probe calibration because of its ability to locate points in space. Such a stylus is often part of the package when purchasing the position sensor for a freehand ultrasound system, so it is available for probe calibration.

Point Phantom

A common approach used to perform probe calibration is to scan an object with known dimensions (a phantom). This phantom can be as simple as a point target. Indeed, this was one of the first phantoms (Detmer et al. 1994; State et al. 1994; Trobaugh et al. 1994) used for this purpose and continues to be used to this day (Barratt et al. 2006; Krupa 2006). Calibration with a point phantom can be divided into two classes, calibration with the aid of a stylus or without a stylus.

Point Phantom Without a Stylus

The point phantom can be formed by a pair of cross-wires (Detmer et al. 1994; Barry et al. 1997; Huang et al. 2005; Krupa 2006) or a spherical bead-like object (State et al. 1994; Leotta et al. 1997; Legget et al. 1998; Pagoulatos et al. 1999; Barratt et al. 2006). Trobaugh et al. (1994) and Meairs et al. (2000) imaged a phantom with multiple point targets one at a time, but their theory is no different to the case when only a single target is used. The point phantom p is scanned, and its location $p^{I} = (u, v, 0)^{t}$ segmented in the B-scan. Now, if we position the phantom coordinate system so that its origin coincides with the point phantom as shown in Fig. 3.7, then (3.1) becomes

$$T_{F \leftarrow W} T_{W \leftarrow S} T_{S \leftarrow I} T_s \begin{pmatrix} u \\ v \\ 0 \end{pmatrix} = \begin{pmatrix} 0 \\ 0 \\ 0 \end{pmatrix}. \qquad (3.4)$$

This is an equation with 11 unknowns—two scale factors, six calibration parameters and three parameters from $T_{F \leftarrow W}$. Only the three translations in $T_{F \leftarrow W}$ need to be determined, since we do not not care about the orientation of F; hence we can set the three rotations in $T_{F \leftarrow W}$ to arbitrary values, such as zeroes. If we capture N images of the point phantom from many directions and orientations, we can find these 11 unknowns by minimizing

$$f_{point1} = \sum_{i=1}^{N} \left| T_{F \leftarrow W} T_{W \leftarrow S_i} T_{S \leftarrow I} T_s p_i^{I} \right|, \qquad (3.5)$$

with the three rotations in $T_{F \leftarrow W}$ set to zero. This function can be minimized using iterative optimisation algorithms, such as the Levenberg–Marquardt algorithm (More 1977). After the calibration parameters and the scales have been found, the transformation $T_{F \leftarrow W}$ may be discarded and replaced with an alternative $T_{F \leftarrow W}$ that is convenient for visualization.

Fig. 3.7 The geometry of a point phantom

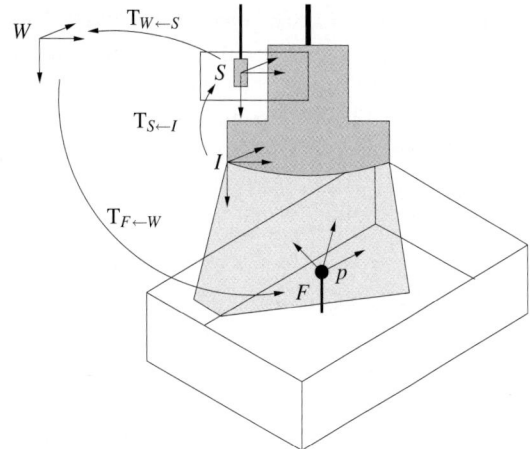

Point Phantom With a Stylus

When a stylus is available, we can find the position of the point phantom p^W in world space by pointing the stylus at the phantom. This approach was followed by Péria et al. (1995), Hartov et al. (1999), Amin et al. (2001) and Viswanathan et al. (2004). If the scales are unknown, the calibration can be solved by minimizing

$$f_{\text{point2}} = \sum_{i=1}^{N} \left| p^W - T_{W \leftarrow S_i} T_{S \leftarrow I} T_s p_i^{I'} \right|. \qquad (3.6)$$

There is little to be gained over (3.5), since the minimum of this function needs to be found by an iterative minimization algorithm. Viswanathan et al. (2004) implemented an alternative solution form used in robotics (Andreff et al. 2001) involving Kronecker products (Brewer 1978) to solve the calibration parameters and the image scales, but an iterative minimization algorithm is still required.

In some rare cases, the scales may be supplied by the manufacturer (Boctor et al. 2003) or by accessing the raw ultrasound signals (Hsu et al. 2006), but this requires special arrangements with the supplier. Otherwise, the scales can be obtained explicitly by using the distance measurement tool provided by the ultrasound machines (Hsu et al. 2006).

If the scales can be found (Péria et al. 1995), then the segmented image of the point phantom is known in millimeters: $p^I = T_s p^{I'}$. After the point has been located in world space by the stylus, it can be mapped to the sensor's coordinate system by the inverse of the position sensor readings, i.e., $p^S = T_{W \leftarrow S}^{-1} p^W$. This means that the point phantom is known in the two-coordinate system I and S, and we want to find a transformation $T_{S \leftarrow I}$ that best transforms $\{p^{I_i}\}$ to $\{p^{S_i}\}$. This can be found by minimizing

$$f_{\text{point3}} = \sum_{i=1}^{N} \left| T_{W \leftarrow S_i}^{-1} p^W - T_{S \leftarrow I} T_s p_i^{I'} \right|. \qquad (3.7)$$

Unlike the case with the previous function, the minimum of f_{point3} can be found in a closed form, provided that the point has been scanned at three noncollinear locations in the B-scans. Péria et al. (1995) scanned three distinct points, but this is not necessary. The most popular method used to find the minimum of f_{point3} is the singular value decomposition technique devised by Arun et al. (1987), and modified by Umeyama (1991). Eggert et al. (1997) detailed the strengths and weaknesses of the different solution forms.

Point Phantom Variants

There are three major difficulties when using the point phantom described above. Most importantly, the images of the phantom need to be segmented manually. Although some automatic algorithms may exist (Hsu et al. 2008b), the segmentation of isolated points in ultrasonic images is seldom reliable. This is evident from the fact that all of the abovementioned research groups who use a point target segmented their phantom manually. This makes the calibration process long and tiresome; it can take up to two hours depending on the number of points to be segmented. Secondly, it is very difficult to align the point phantom precisely with the scan plane. The finite thickness of the ultrasound beam makes the target visible in the B-scans even if the target is not precisely at the elevational center of the scan plane. This error can be up to several millimeters depending on the beam thickness and the ability of the user to align the scan plane with the phantom. Finally, the phantom also needs to be scanned from a sufficiently diverse range of positions, and its location spread throughout the B-scan images. This is to ensure the resulting system of constraints is not underdetermined with multiple solutions (Prager et al. 1998).

There are several phantoms that are designed to overcome the segmentation and alignment problems of the point phantom, while still being based on the same mathematical principles. Liu et al. (1998) scanned a pyramid transversely. The pyramid appears as a triangle of varying size in the B-scans, depending on where the pyramid is scanned. The side lengths of the triangle are used to find the precise intersection of the scan plane with the pyramid. The three points of intersection act as three distinct point targets.

Brendel et al. (2004) scanned a sphere with a known diameter. The center of the sphere acts as the virtual point phantom. The sphere appears as a circle in the B-scans and can be segmented automatically by using a Hough transform (Hough 1959). Alignment is ensured providing the circle has the correct diameter. However, the lack of good visual feedback in the B-scans means that alignment is difficult. Sauer et al. (2001) scanned five spheres and manually fitted their image to a circle with the corresponding diameter. Gooding et al. (2005) placed a cross-wire through the center of the sphere to ensure good alignment while maintaining automatic segmentation. Hsu et al. (2008b) scanned a phantom consisting of two cones joining at a circle. The center of this circle serves as the virtual point phantom, as shown in Fig. 3.8. Alignment of the scan plane with the circle is aided by the cones, so that

Fig. 3.8 An image of the cone
phantom

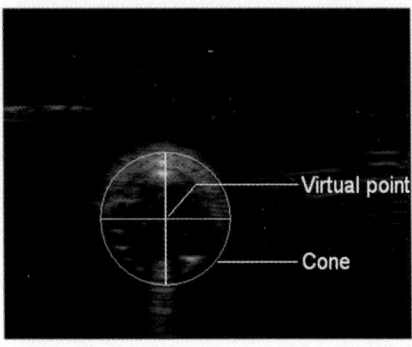

even a slight misalignment can be detected. The fact that circles can be segmented
automatically makes calibration simpler and quicker to perform.

Stylus

When a stylus is available, it is possible to perform calibration using just the stylus.
Instead of scanning a point phantom and finding its location with a stylus, the tip of
the stylus itself can be scanned. Muratore and Galloway Jr. (2001) were the first to
perform probe calibration with a stylus, and Zhang et al. (2006) also followed their
approach. The calibration process is almost identical to the one where a point target
is used. The tip of the stylus is scanned from many positions and orientations. This
places constraints on the calibration parameters. If the image scales are unknown,
a function similar to f_{point2} is minimized, the only difference being that the point
target is now free to move around in space. The function to be minimized is

$$f_{stylus} = \sum_{i=1}^{N} \left| T_{W \leftarrow L_i} r^L - T_{W \leftarrow S_i} T_{S \leftarrow I} T_S p_i^{I_i'} \right|, \tag{3.8}$$

where $T_{W \leftarrow L_i}$ is the location of the stylus in space.

    This technique is equivalent to a point phantom and is subject to most of its disad-
vantages. Hence alignment is a major source of error. Hsu et al. (2008b) designed a
Cambridge stylus with a thick shaft. This shaft is thinned at a point precisely 20 mm
above the tip of the stylus. Instead of locating the stylus's tip in the scan plane, this
thinned point is located. Any misalignment is detected visually.

    Khamene and Sauer (2005) solve the alignment problem by attaching a rod to
a position sensor, as shown in Fig. 3.9. Both ends of the rod are pointer-calibrated,
and their locations in space are therefore $T_{W \leftarrow L} r_1^L$ and $T_{W \leftarrow L} r_2^L$. The rod is scanned
at an arbitrary location, and the point of intersection segmented in the B-scan. This
point's location in world space is governed by (3.1), and lies on the line joining the

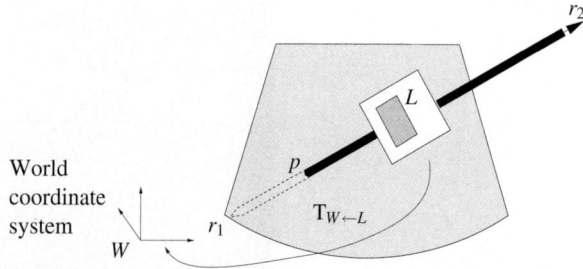

Fig. 3.9 The geometry of a rod stylus

two ends of the rod. The distance from the point p^W to the line segment $r_1^W r_2^W$ is

$$\frac{\left|\left(r_2^W - r_1^W\right) \times \left(r_1^W - p^W\right)\right|}{\left|r_2^W - r_1^W\right|}.$$

This distance must be zero, hence

$$\left|\left(T_{W \leftarrow L} r_2^L - T_{W \leftarrow L} r_1^L\right) \times \left(T_{W \leftarrow L} r_1^L - T_{W \leftarrow S} T_{S \leftarrow I} T_s p^{I'}\right)\right| = 0.$$

The $\times$ in the above equations denotes the cross product of two vectors in $\mathbb{R}^3$. Calibration can be obtained by minimizing

$$f_{\text{rod}} = \sum_{i=1}^{N} \left|\left(T_{W \leftarrow L_i} r_2^L - T_{W \leftarrow L_i} r_1^L\right) \times \left(T_{W \leftarrow L_i} r_1^L - T_{W \leftarrow S_i} T_{S \leftarrow I} T_s p_i^{I'}\right)\right|. \quad (3.9)$$

This is an equation with six unknowns—the six calibration parameters. Hsu et al. (2008b) pointed out that for a reliable optimization, the scales needs to be found explicitly and fixed before optimization.

Three-Wire Phantom

The three-wire phantom is solely used by Carr (1996). Instead of mounting a pair of cross-wires in a fluid, three mutually orthogonal wires are used. These three wires form the three principal axes of the phantom coordinate system, as shown in Fig. 3.10. Each wire is scanned along its length individually. Suppose that the wire defining the x-axis is being scanned; then the point on the wire that is being scanned must satisfy

$$T_{F \leftarrow W} T_{W \leftarrow S} T_{S \leftarrow I} T_s \begin{pmatrix} u \\ v \\ 0 \end{pmatrix} = \begin{pmatrix} p_x^F \\ p_y^F = 0 \\ p_z^F = 0 \end{pmatrix}. \quad (3.10)$$

Fig. 3.10 The geometry of a three-wire phantom

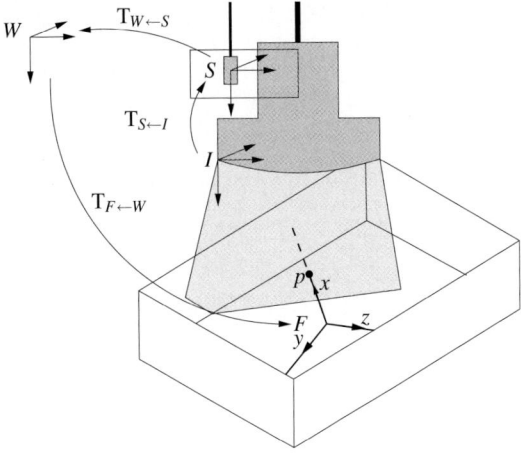

The y and z coordinates of p^F give rise to two equality constraints. If N_x, N_y and N_z points were recorded along the x-, y- and z-axes of the phantom coordinate system in that order, calibration can be solved by minimizing

$$f_{3\text{-wire}} = \sum_{i=1}^{N_x} \left(\left(p_{i_y}^F \right)^2 + \left(p_{i_z}^F \right)^2 \right) + \sum_{i=N_x+1}^{N_x+N_y} \left(\left(p_{i_x}^F \right)^2 + \left(p_{i_z}^F \right)^2 \right)$$

$$+ \sum_{i=N_x+N_y+1}^{N_x+N_y+N_z} \left(\left(p_{i_x}^F \right)^2 + \left(p_{i_y}^F \right)^2 \right). \tag{3.11}$$

This approach involves solving for 14 variables. These are the two scales, six calibration parameters and the six parameters that define the phantom coordinate system.

This technique does not require the user to align the scan plane with the phantom, and it potentially speeds up the calibration process. Segmentation remains slow, since manual intervention is required. The user also needs to keep track of which wire is being scanned. The phantom may need to be precision-manufactured to ensure that the wires are straight and orthogonal to each other.

Plane Phantom

Instead of scanning a point, it is possible to scan a plane. The design complexity of the plane varies from the floor of a container (Prager et al. 1998), a plexiglass plate (Rousseau et al. 2005), a nylon membrane (Langø 2000) to a precision-made Cambridge phantom (Prager et al. 1998) and its variants (Varandas et al. 2004; Ali and Logeswaran 2007).

Fig. 3.11 The geometry of a
plane phantom

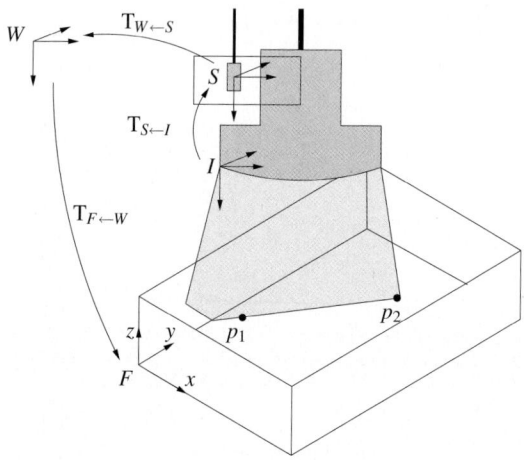

The plane appears as a straight line in the B-scans. If we align the phantom
coordinate system so that its xy-plane coincides with the plane phantom as shown
in Fig. 3.11, then every point on the line in the image must satisfy, by (3.1):

$$T_{F \leftarrow W} T_{W \leftarrow S} T_{S \leftarrow I} T_s \begin{pmatrix} u \\ v \\ 0 \end{pmatrix} = \begin{pmatrix} p_x^F \\ p_y^F \\ p_z^F = 0 \end{pmatrix}. \tag{3.12}$$

The equation for the z-coordinate of the phantom coordinate system is the required
constraint on the calibration parameters. For each segmented line, we get two independent constraints by choosing any two points on the line. Choosing more points
does not add any further information. The calibration parameters are solved by
minimizing

$$f_{\text{plane}} = \sum_{i=1}^{N} \left(\left(p_{1i_z}^F \right)^2 + \left(p_{2i_z}^F \right)^2 \right), \tag{3.13}$$

where N is the number of images of the plane. The above equation is a function of 11
variables—two scales, six calibration parameters and three parameters from $T_{F \leftarrow W}$.
These three parameters consist of two rotations and one translation. Since we only
require the xy-plane to coincide with the plane phantom, the two translations in the
plane and the rotation about a normal of the plane will be absent from the equation.

The plane technique is attractive because it enables an automatic segmentation
algorithm to be used, making the calibration process rapid to perform. The plane
appears as a straight line in the B-scans. There are several automatic algorithms for
segmenting a line, such as the Hough transform (Hough 1959) and wavelet-based
techniques (Kaspersen et al. 2001). Prager et al. (1998) implemented a simplified
version of the line detection algorithm by Clarke et al. (1996) and used the RANSAC
algorithm to reject outliers (Fischler and Bolles 1981).

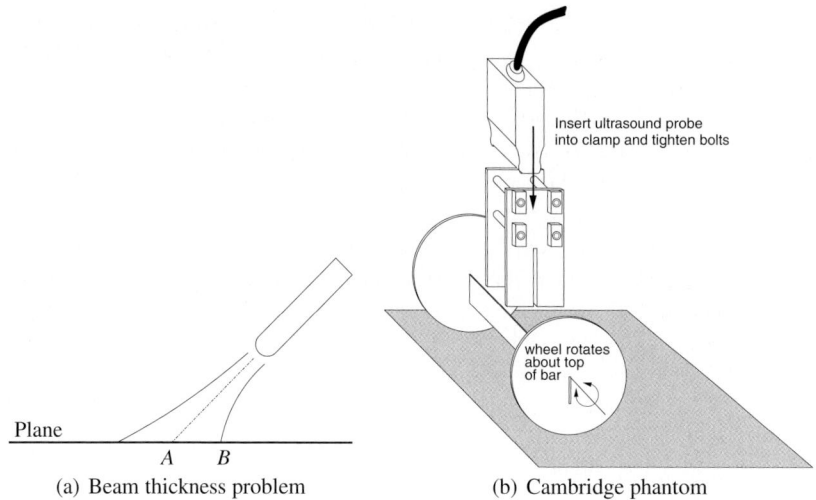

Fig. 3.12 The beam thickness problem and its solution

A major drawback of using this approach, similar to the case of a point phantom, is that the phantom needs to be scanned from a wide range of angles and positions (Prager et al. 1998). In particular, the user is required to scan the phantom obliquely, as shown in Fig. 3.12a. Due to the thick ultrasound beam, point B is encountered by the ultrasound pulse before point A. The echo from point B makes the plane appear at an incorrect position. The user is subsequently required to scan the plane at the same angle on both sides of the normal to limit this error (Prager et al. 1998). Furthermore, much of the ultrasound energy is reflected away from the plane. The echo received by the probe is therefore weak, making segmentation at these positions difficult.

It is possible to use a Cambridge phantom, as shown in Fig. 3.12b. The user is required to mount the probe onto the clamp, so that the scan plane is aligned with the slit of the clamp and hence with the brass bar. In order to assist the user in aligning the scan plane, a set of wedges (see Fig. 3.16c) can be placed on the brass bar. The user then aligns the scan plane with the wedges. In either case, aligning the scan plane with the brass bar may be difficult. The phantom is moved around in space by translating the phantom or rotating the wheels so that the phantom remains in contact with the floor of the container. Since the top of the brass bar joins the center of the wheels, it always remains at a fixed height above the floor. The top of the brass bar serves as a virtual plane that is scanned. The advantage of using the Cambridge phantom is that a strong and clear reflection is received from the brass bar, irrespective of the probe position. The user can scan the plane from different angles and still get a clear image. However, the user is still required to scan the phantom from a wide range of positions and angles. Calibrating with a plane phantom is therefore a skilled task and requires the user to be experienced. From our experience of supporting medical physicists and clinicians, an incorrect calibration is often obtained because the phantom has not been scanned from a sufficiently diverse set of positions and orientations. Although an eigenvalue metric exists to detect whether the system of equations is underconstrained (Hsu et al. 2006), the user is still required to be sufficiently trained.

Dandekar et al. (2005) used two parallel wires to mimic a plane phantom. The virtual plane is formed by the unique plane that passes through the two parallel wires. The idea is to scan the set of two wires; the points of intersection of the wires with the scan plane are chosen as the points p_1 and p_2 in Fig. 3.11. The phantom can be moved freely in the container so that both wires always intersect the scan plane. This phantom has the advantage that the beam thickness effect is minimized. When the plane is being scanned at an oblique angle, the plane no longer appears at an incorrect depth. The user therefore does not need to ensure that scans from the same angle to both sides of the normal are captured. However, the phantom needs to be precision-manufactured to ensure that the wires are parallel. Most importantly, the wires need to be manually segmented. This sacrifices the rapid segmentation advantage of the plane phantom, making calibration, once again, a time-consuming process. The user is still required to follow the same complex protocol and scan the phantom from a wide variety of positions and angles.

Two-Plane Phantom

Boctor et al. (2003) designed a phantom with a set of parallel wires forming two orthogonal planes. When the set of wires is being scanned, it appears as distinct dots in the shape of a cross. If we align the phantom coordinate system with the orthogonal planes as shown in Fig. 3.13, then a point p_h lying on the horizontal axis

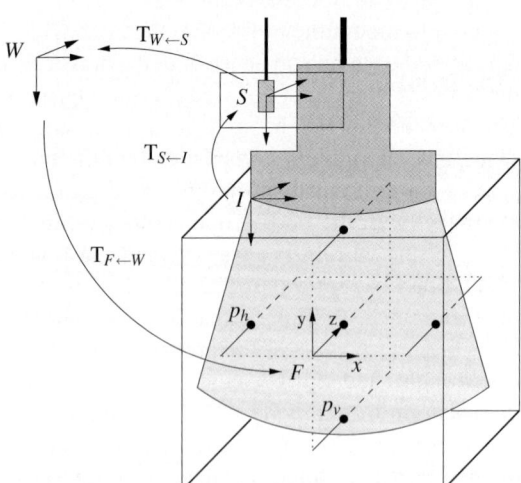

Fig. 3.13 The geometry of a two-plane phantom

of the cross lies on the xz-plane of the phantom coordinate system and must satisfy:

$$\begin{pmatrix} p_x^F \\ p_y^F = 0 \\ p_z^F \end{pmatrix} = T_{F \leftarrow W} T_{W \leftarrow S} T_{S \leftarrow I} T_S p_h^{I'}. \tag{3.14}$$

The y coordinate in the above equation is a constraint on the calibration parameter. A similar constraint can be obtained for each point p_v on the vertical axis of the cross. Suppose that N images of the phantom are captured, each consisting of M_h points on the horizontal axis and M_v points on the vertical axis of the cross; then the calibration parameter and the scales can be found by minimizing

$$f_{2\text{-plane}} = \sum_{i=1}^{N} \left(\sum_{j=1}^{M_h} \left(p_{i,hj_y}^F \right)^2 + \sum_{j=1}^{M_v} \left(p_{i,vj_x}^F \right)^2 \right), \tag{3.15}$$

where $p_{i,hj}$ and $p_{i,vj}$ denote the jth point on the horizontal and vertical axis of the cross in the ith image. This equation consists of 13 variables; only the translation in the z-axis of the phantom coordinate system can be arbitrary.

An advantage of the two-plane phantom is that the set of wires appear as a cross in the ultrasound image. This information can be used to automatically segment the wires. Just as in the case with a point and a plane phantom, the phantom needs to be scanned from a wide variety of positions to constrain the calibration parameters.

It may be possible to generalize this idea and scan the faces of a cube with the phantom coordinate system suitably defined. Points on each face of the cube need to satisfy the equation for that plane and this places a constraint on the calibration parameters. Calibration can be solved by minimizing a similar equation to f_{plane} and $f_{2\text{-plane}}$. However, nobody has yet applied this calibration technique in a freehand 3D ultrasound system.

Two-Dimensional Alignment Phantom

When calibration is performed using a point phantom with the aid of a stylus, with known scales, calibration only needs three noncollinear points to be positioned in the scan plane. If it is possible to align the scan plane with three such points at the same time, then even one frame is sufficient for calibration. Sato et al. (1998) was the first to use such a phantom. They scanned a thin board with three vertices, as shown in Fig. 3.14. The location of these vertices is determined using a stylus. The scan plane is then aligned with these vertices, and each vertex is segmented in the B-scan. Since the distance between each pair of vertices is known, and we can find their distance in pixels from the ultrasound images, the scale factors can easily be computed. The calibration parameters can be solved in a closed-form by minimizing a function similar to f_{point3}. If we have captured N images of a two-dimensional

Fig. 3.14 The coordinates of
a 2D alignment phantom

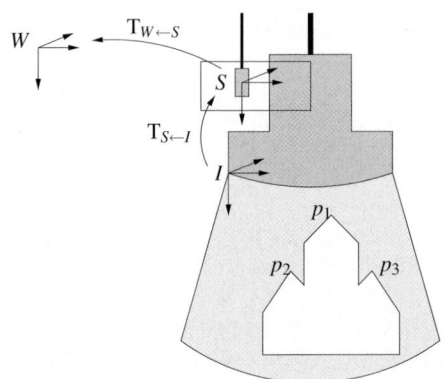

alignment phantom with M fiducial points, calibration is found by minimizing

$$f_{2D} = \sum_{i=1}^{N} \sum_{j=1}^{M} \left| T_{W \leftarrow S_i}^{-1} p_j^W - T_{S \leftarrow I} T_s p_j^{I_i'} \right|. \tag{3.16}$$

Several other groups use similar two-dimensional alignment phantoms with a variety of shapes and different numbers of fiducial points. Berg et al. (1999) aligned a jagged membrane with five corners and Welch et al. (2000) used an acrylic board with seven vertices. Beasley et al. (1999) scanned a ladder of wires with three weights fitted on the strings. Lindseth et al. (2003b) scanned a diagonal phantom, with the nine fiducial points formed by cross-wires. Leotta (2004) fitted 21 spherical beads on parallel wires at different axial depths. The main disadvantage of this phantom is the requirement to align the phantom with the fiducial points, which can be very difficult. An advantage is that only one frame ($N = 1$ in (3.16)) is theoretically needed for probe calibration, and a large number of fiducial points can be captured with just a few frames.

Z-Phantom

The Z-fiducial phantom was designed so that the user is not required to align the scan plane with the 2D phantom (Comeau et al. 1998, 2000). The phantom consists of a set of wires in a "Z" shape, as shown in Fig. 3.15. The end points of the "Z" wire configuration w_1, w_2, w_3 and w_4 can be found in space using a stylus. A typical Z-phantom may have up to 30 such "Z" configurations. Instead of pointing the stylus at each end of the wire, there are usually a number of fixed locations (divots) on the phantom. The Z-shaped wire configurations are precision-manufactured relative to these divots, and the positions of the divots are located in space by using a stylus (Gobbi et al. 1999; Pagoulatos et al. 2001; Boctor et al. 2004; Zhang et al. 2004; Hsu et al. 2008a). It may be possible to attach a position sensor to the Z-phantom (Lindseth et al. 2003b; Chen et al. 2006) and precision-manufacture the wires relative

Fig. 3.15 The geometry of a
Z-phantom

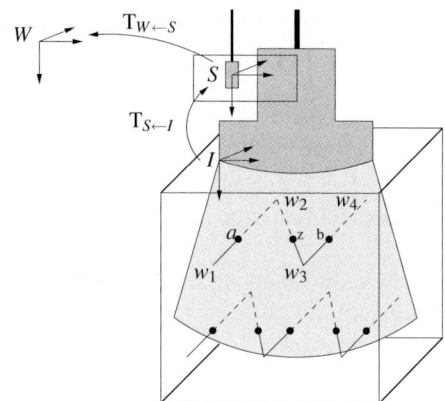

to the position sensor (Lindseth et al. 2003a). This requires the coordinates of the
position sensor to be known. Calibration then requires two objects to be tracked
simultaneously, but otherwise there is little difference between the two approaches.
In each case, the end-points of the wires can be found in space.

When the Z-phantom is scanned, the scan plane intersects the wire $w_1w_2w_3w_4$
at a, z and b. These points are segmented in the ultrasound images. Assuming the
image scales are known, the distances $|z - b|$ and $|a - b|$ can be measured off the
B-scan images. The location of z is given by:

$$
\begin{aligned}
z^W &= w_3^W + \frac{|z - w_3|}{|w_2 - w_3|} \left(w_2^W - w_3^W \right) \\
&= w_3^W + \frac{|z - b|}{|a - b|} \left(w_2^W - w_3^W \right),
\end{aligned}
\tag{3.17}
$$

since $\triangle aw_2z$ and $\triangle bw_3z$ are similar. If N images of the phantom are captured,
each consisting of M Z-fiducials, then the calibration parameters can be found by
minimizing

$$
f_{\text{Z-phantom}} = \sum_{i=1}^{N} \sum_{j=1}^{M} \left| T_{W \leftarrow S_i}^{-1} z_{ij}^W - T_{S \leftarrow I} T_s z_{ij}^{I_i'} \right|,
\tag{3.18}
$$

where z_{ij} is the jth Z-fiducial in the ith frame. This function differs slightly from f_{2D}
since the Z-fiducials are at different positions depending on the scan plane, while 2D
alignment phantoms are fixed in space.

The Z-phantom has the advantage that it does not require the alignment of the
scan plane with the phantom. It also maintains other advantages of a 2D alignment
phantom, e.g., only one frame is needed for calibration. However, the scale factors can no longer be measured off the B-scan images, and need to be found using
other approaches, as described in the section earlier in this chapter entitled "Point
Phantom With a Stylus."

It may be possible to segment the wires automatically. Chen et al. (2006) simplified their phantom to just two "Z" wire configurations. Their segmentation algorithm

involves finding two sets of parallel wires. Hsu et al. (2008a) mounted a membrane on top of their phantom, which is treated as a plane and can be segmented automatically. The wires are at known locations below the membrane, and this information is used to find the wires. This allows calibration to be completed in just a few seconds.

Mechanical Instrument

Gee et al. (2005) built a mechanical instrument that performs probe calibration by calibrating the position sensor and the scan plane separately to the gantry on the instrument. Since the two calibrations are independent, once the position sensor is calibrated, the depth and zoom settings can be changed and only the scan plane needs to be recalibrated each time. This is achieved by using a specialized probe holder. Both the position sensor and the ultrasound probe are attached to the probe holder. The probe holder is positioned onto the phantom during calibration, and removed when the calibration is complete.

The phantom's coordinate system is defined by its gantry G, where the probe holder H is mounted repeatably at a fixed location, as shown in Fig. 3.16a. The transformation $T_{H \leftarrow G}$ is therefore constant and determined by the geometry. The position sensor is mounted onto the probe holder at a fixed location as well. The transformation $T_{S \leftarrow H}$ is therefore also fixed. In order to find this transformation, the probe holder is placed into the sensor's volume while the position sensor is attached. A stylus is then used to locate fixed landmarks on the probe holder and record the corresponding locations in the sensor's coordinate system. Since this part of the calibration process is independent of the probe, replacing the probe or changing any of the ultrasound settings will not affect the position of the sensor relative to the gantry.

In order to calibrate the scan plane, the user is required to align the scan plane with a 2D phantom by adjusting a set of micrometers. The 2D phantom consists of two parallel wires, with three sets of wedges p_1, p_2 and p_3 mounted on these wires at known locations. The coordinate system of the wires R is defined so that its origin coincides with p_1, as shown in Fig. 3.16b. Once the wires and these wedges are aligned with the scan plane, the image scales are found from the known distance

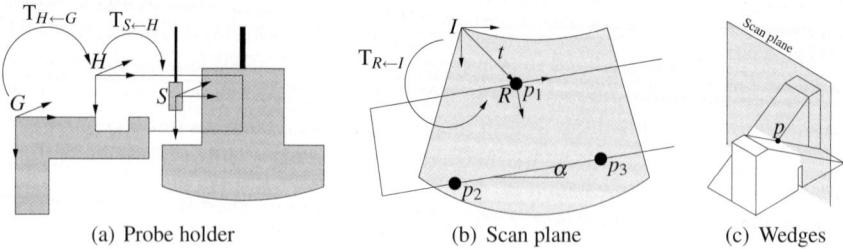

(a) Probe holder   (b) Scan plane   (c) Wedges

Fig. 3.16 Geometry of the mechanical device for calibration

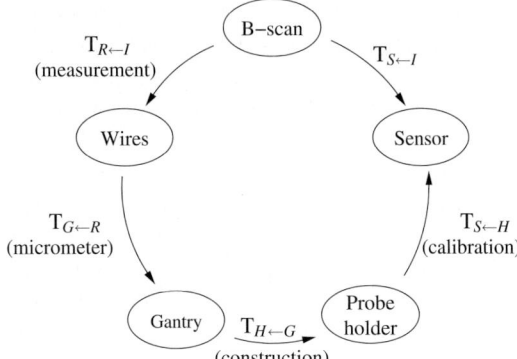

Fig. 3.17 Principle of the mechanical device for calibration

between these wedges, as in the case with other 2D phantoms. If we rely on the user to ensure that p_2 is to the left of p_3, the transformation $T_{R \leftarrow I}$ only has three degrees of freedom—two translations t and a rotation α. The translation is found from the location of p_1 in the B-scan, and the rotation is found from the orientation of p_2 and p_3.

The three sets of wedges also help in aligning the scan plane, rather than merely placing landmarks on the two wires. A set of wedges is shown in Fig. 3.16c. It consists of two triangular blocks. When the scan plane is aligned perfectly with the two wedges, a symmetrical reflection will be obtained in the ultrasound image. The surface of the wedges are roughened to ensure a strong reflection. This visual feedback allows the 2D plane to be aligned with the scan plane to a high degree of accuracy.

Now, once the two calibrations are complete, the transformation that relates the wires to the gantry $T_{G \leftarrow R}$ is simply read off the micrometers. Calibration is found as a series of transformations mapping from the B-scan to the wires, then to the gantry, the probe holder, and finally to the position sensor, as shown in Fig. 3.17. Calibration is therefore given by

$$T_{S \leftarrow I} = T_{S \leftarrow H} T_{H \leftarrow G} T_{G \leftarrow R} T_{R \leftarrow I}. \tag{3.19}$$

Image Registration

Another technique used to calibrate a probe is image registration. When a point phantom is used for probe calibration, the point is scanned from different positions and orientations. The 3D image of the point can be constructed by using an assumed calibration and image scales. An iterative optimization algorithm is implemented to find the calibration and scales so that the constructed image best fits the model. Here, the *best fit* is measured by the amount of variation of the reconstructed point.

Once the best fit has been found, the required calibration is the corresponding values that result in the least variation of the reconstructed point. This idea is used in other phantoms as well, each one using a different measure to define what the best fit is. For example, the plane phantom requires the reconstructed points to lie on a plane. Thus, the best fit is measured by the deviation of the reconstructed points from a particular plane. What these techniques have in common is that particular points of the phantom are selected, and the best fit is measured as a function of the deviation of these points from their ideal location.

Blackall et al. (2000) built a gelatin phantom with tissue-mimicking properties. The geometric model of the phantom is acquired by a magnetic resonance (MR) scan. The phantom is scanned with a freehand 3D ultrasound system. A 3D image of the phantom can be reconstructed by using an assumed calibration and (3.1). An iterative optimization algorithm is implemented to find the calibration and the image scales where the reconstructed image best fits the MR model. The similarity measure between two 3D volumes A and B is given by their mutual information (Studholme et al. 1999):

$$I(A,B) = \frac{H(A)+H(B)}{H(A,B)}, \tag{3.20}$$

where $H(A)$ and $H(B)$ denote the marginal entropies of the images and $H(A,B)$ represents their joint entropy.

This technique is dependent on the image quality of the phantom and the similarity measure used. The impact of choosing another similarity measure (Pluim et al. 2003; Zitová and Flusser 2003) is unknown.

3D Probe Calibration

Although it is not the main focus of this chapter to investigate calibrations for a 3D probe (a mechanically swept or a 2D array probe), we mention in passing that all of the techniques that are used to calibrate a 2D probe are equally valid for the calibration of 3D probes. In fact, the exact same phantoms have been used, such as the point phantom (Sawada et al. 2004; Poon and Rohling 2007) and the Z-phantom (Bouchet et al. 2001). The mathematical principles remain the same. However, since a 3D probe is used, a 3D image of the phantom is obtained. This is useful for segmenting the phantom. Lange and Eulenstein (2002) and Hastenteufel et al. (2003) used an image registration technique. Poon and Rohling (2005) provided a detailed discussion comparing calibrations using the various phantoms, including a three-plane phantom that has not been used to calibrate conventional 2D probes.

Calibration Quality Assessment

Probe calibration is a critical component of every freehand 3D ultrasound system, and its quality has a direct impact on the performance of the imaging system. It is therefore crucial to quantify the accuracy achievable with each calibration technique. Unfortunately, there has not been an agreed standard for assessing calibration quality. As a result, every research group may assess calibration quality differently, depending on what is available and convenient. Comparing calibration qualities between different research groups is therefore not straightforward. The quoted figures need to be interpreted on an individual basis; for example, some may quote standard deviation and others may quote the 95% confidence interval. Nevertheless, we may classify all quality measures broadly into two classes, namely precision and accuracy.

Precision

One of the first measures used was formulated by Detmer et al. (1994) and used by various other research groups (Leotta et al. 1997; Prager et al. 1998; Blackall et al. 2000; Meairs et al. 2000; Muratore and Galloway Jr. 2001; Brendel et al. 2004; Dandekar et al. 2005). Now commonly named the reconstruction precision (RP), this measure is calculated by scanning a point phantom p from different positions and orientations. The point phantom is segmented in the B-scans and reconstructed in 3D space by using (3.1). If N images of the point are captured, we get a cloud of N points spread in world space. Reconstruction precision is measured by the spread of this cloud of points, i.e.,

$$\mu_{RP1} = \frac{1}{N} \sum_{i=1}^{N} \left| T_{W \leftarrow S_i} T_{S \leftarrow I} T_S p_i^{I_i'} - \overline{p_i^W} \right|. \tag{3.21}$$

This equation can be generalized to include multiple calibrations:

$$\mu_{RP2} = \frac{1}{MN} \sum_{i=1}^{N} \sum_{j=1}^{M} \left| T_{W \leftarrow S_i} T_{S \leftarrow I_j} T_s p_i^{I_i'} - \overline{p_{ij}^W} \right|, \tag{3.22}$$

where N is the number of images of the point phantom and M the number of calibrations.

Reconstruction precision measures the point reconstruction precision of the entire system, rather than calibration itself. This is dependent on a lot of factors, such as position sensor error, alignment error and segmentation error. Nevertheless, it is not unrelated to calibration. If calibration is far from correct, and the point phantom has been scanned from a sufficiently diverse set of positions, then every image of the point will be mapped to an incorrect location, resulting in a huge spread and subsequently a large reconstruction error. A good reconstruction precision is therefore necessary, but unfortunately not sufficient, for a good calibration.

An alternative measure, based on the same idea as reconstruction precision, is called calibration reproducibility (CR) (Prager et al. 1998). Calibration reproducibility measures the variability in the reconstructed position of points in the B-scan. Suppose that only a single frame of the phantom is captured, and that N calibrations were performed. We can then map the single point in space by using the different calibrations. Now, it is not necessary to reconstruct the point in world space, since the transformation $T_{W \leftarrow S}$ is independent of calibration. Equivalently, the reconstruction can be done in the sensor's coordinate system. This removes position-sensing variations. Furthermore, the point phantom image itself is unnecessary. Imaging a point introduces alignment and segmentation errors. Instead, we may conveniently assume that a point has been imaged without scanning such a point physically, and that we have perfectly aligned and segmented its location $p^{I'}$ on the B-scan. Calibration reproducibility is computed as follows:

$$\mu_{CR} = \frac{1}{N} \sum_{i=1}^{N} \left| T_{S \leftarrow Ii} T_S p^{I'} - \overline{p^{S_i}} \right|. \tag{3.23}$$

Clearly, the measure for calibration reproducibility is not just dependent on the calibrations $(T_{S \leftarrow Ii})$, but also on the point $p^{I'}$. When Prager et al. (1998) first used this measure, they chose $p^{I'}$ to be the center of the image. Many research groups also gave the variation at the center of the image (Meairs et al. 2000; Lindseth et al. 2003b). Pagoulatos et al. (2001) quoted variations for multiple points down the middle of the image. When there is an error in the calibration (of one of the rotational parameters, say), often the scan plane is incorrect by a rotation about some axis near the center of the image. This means that points near the center of the image are still roughly correct, but errors measured by points towards the edges are more visible. Therefore, many papers in recent years quote calibration reproducibility for a point at a corner of the image (Blackall et al. 2000; Rousseau et al. 2005), points along the left and right edges of the image (Leotta 2004) and the four corners of the image (Treece et al. 2003; Gee et al. 2005; Hsu et al. 2006). Leotta (2004) and the Cambridge group (Treece et al. 2003; Gee et al. 2005; Hsu et al. 2006) also quoted the spread in the center of the image. Brendel et al. (2004) gave the maximum variation of every point in the B-scan. Calibration reproducibility is a measure solely based on calibration, and does not incorporate errors like the position sensor or human skills such as alignment and segmentation. For this reason, calibration reproducibility has started to become the norm when precision is measured. In some papers, precision is simply defined as calibration reproducibility and referred to as "precision."

Some research groups give the variation of the six calibration parameters (Amin et al. 2001; Boctor et al. 2003, 2004; Viswanathan et al. 2004). Other research groups give the variation of the three calibration translations and each entry in the rotational transformation matrix (Pagoulatos et al. 2001; Leotta 2004); this is not appropriate, since these values are not independent. In any case, interpreting these results is difficult since it is the variation due to the combination of these six parameters that is useful.

Accuracy

Precision measures the spread of a point in some coordinate system. This does not measure calibration accuracy, as there may be a systematic error. In fact, it is almost impossible to measure calibration accuracy, since the true calibration is unknown. If there was a technique that was able to give us the exact error, then this technique could be used to find the calibration parameters in the first place. Gee et al. (2005) measured their accuracy by considering the error in each component of their instrument. However, they had to assume that the scan plane can be aligned with their phantom without a systematic bias.

Many research groups quote accuracy for the entire freehand 3D ultrasound system. The calibration accuracy can then be deduced or inferred from the system accuracy, with a careful quantization of every error source in the system evaluation (Lindseth et al. 2002). In fact, this is the ultimate accuracy that is important to a clinician, who is interested in the performance of the system, rather than some individual component. However, in such an environment, the accuracy of interest would be the in vivo accuracy. This is again difficult to assess. The reasons for this are that it is difficult and inconvenient to scan a live patient in the laboratory, and that the shape of the real anatomical structure is unknown. This is why the ultrasound system was built in the first place. Some research groups produce in vivo images in their papers (Meairs et al. 2000; Ali and Logeswaran 2007), but merely as examples of images constructed by their system. As a result, accuracy experiments are often performed on artificial phantoms in a well-controlled environment. Note that there are many papers on freehand 3D ultrasound systems as a whole. Although these papers may include probe calibration, their goal is to evaluate the accuracy of their system, rather than the calibration. We have thus excluded these accuracy assessments in this chapter. Their methods will favor clinical quantities, such as volume and in vivo images.

In vitro accuracy is nevertheless very different to in vivo accuracy. First, the image of the phantom usually has a better quality than that in in vivo images. Unlike the speckle in in vivo images, phantom images have a clear border and segmentation is usually more accurate. For this reason, Treece et al. (2003) scanned a tissue-mimicking phantom when assessing the accuracy of their system. Scanning in vivo is also subject to tissue deformation due to probe pressure (Treece et al. 2002). Furthermore, sound travels at different speeds as it passes through the various tissue layers, which does not occur in in vitro experiments. For a given system, the in vitro accuracy is generally better than the in vivo accuracy. Nevertheless, in vitro accuracy defines what can be achieved with such a system in an ideal environment.

Point Reconstruction Accuracy

Point reconstruction accuracy (PRA) is probably the most objective measure of accuracy. However, it is only recently, with the increased use of the stylus, that this technique has became widely used. A point p is scanned and its location reconstructed in 3D space. The 3D location of the point phantom is usually verified by the stylus (Blackall et al. 2000; Muratore and Galloway Jr. 2001; Pagoulatos et al. 2001). The only exception is Lindseth et al. (2003b), who precision-manufactured their point phantom relative to the position sensor. Point reconstruction accuracy is given by the discrepancy between the reconstructed image and the stylus reading, i.e.,

$$\mu_{\text{PRA}} = p^W - T_{W \leftarrow S} T_{S \leftarrow I} T_s p^{I'}. \tag{3.24}$$

This is a measurement of system accuracy, and includes errors from every component of the system. The main error is due to manual misalignment of the scan plane with the point phantom used for accuracy assessment. As described before, when calibrating with a point phantom, manual alignment is difficult due to the thick beam width. There are other sources of error, such as segmentation error and position-sensor error, and these should not be neglected. Of course, for better measurement, the point should be scanned from different positions and at different locations in the B-scan. A large number of images should be captured and the results averaged.

It is important that the image of the point phantom is scanned at different locations in the B-scans. This is because, if the calibration was performed by capturing a series of images incorrectly in one region of the B-scan, then calibration would be most accurate for points near the same region of the B-scan. If the image of the point phantom used for accuracy assessment is again captured at the same region, the measured accuracy will appear to be higher than the true accuracy. In order to find the true calibration accuracy, the point phantom needs to be imaged at different locations throughout the B-scan.

Note that it is bad practice to use the same phantom that was used for calibration to assess its accuracy, especially when the location of the fiducial point is dependent on phantom construction (Liu et al. 1998; Chen et al. 2006). This means that point reconstruction accuracy is not very appropriate to assess a calibration performed using a point target if the same phantom and the same algorithm are used. This is because such errors will cause an offset in the calibration if there is a flaw in the construction of the phantom. The same error will occur during accuracy assessment, and will remain unnoticed.

Distance Accuracy

Before the stylus was developed, enabling the evaluation of point reconstruction accuracy, many groups assessed accuracy by measuring the distances between objects (Leotta et al. 1997; Prager et al. 1998; Blackall et al. 2000; Boctor et al. 2003; Lindseth et al. 2003b; Leotta 2004; Dandekar et al. 2005; Hsu et al. 2006;

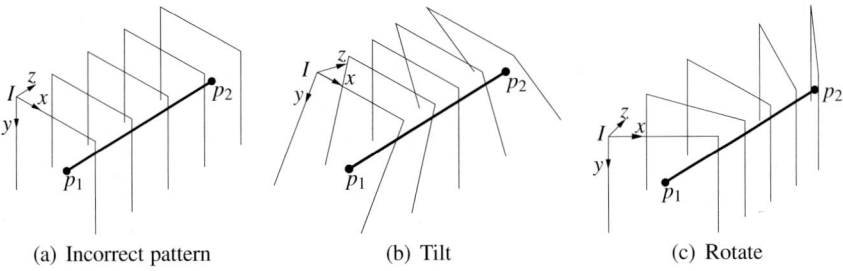

(a) Incorrect pattern   (b) Tilt   (c) Rotate

Fig. 3.18 The different types of scanning pattern during accuracy assessment

Krupa 2006). This technique is popular because the experiment is easy to set up. A phantom is manufactured with distinct landmarks. Even though the exact locations of these landmarks are unknown in 3D space, the distances between the landmarks are known. This means that when the phantom is scanned and its image recon-structed in 3D space, we can compute the distances between the landmarks and see whether each computed distance is correct. The error measure is

$$\mu_{\text{Distance}} = \left| p_1^W - p_2^W \right| - \left| T_{W \leftarrow S_1} T_{S \leftarrow I} T_s p_1^{I'} - T_{W \leftarrow S_2} T_{S \leftarrow I} T_s p_2^{I'} \right|. \qquad (3.25)$$

The idea behind this measure is that if a line is scanned with an incorrect calibra-tion, the image of the line in 3D space should be distorted. However, this depends on the way in which the phantom is scanned. Very often, when assessing the accu-racy by distance measurement, a single sweep of the phantom is performed in one direction, as shown in Fig. 3.18a. Accuracy assessment performed in this way is incorrect. If the calibration is wrong, then the whole line will be incorrect in the same way. Each point will be offset by the same value and the reconstructed image will appear to be correct. What the user ends up assessing is the resolution of the ultrasound system. It is therefore not surprising that many research groups quote a high distance measurement accuracy.

In order to successfully detect an incorrect calibration, the line should be scanned by tilting or rotating the probe in different directions, as shown in Fig. 3.18b, c. This ensures that calibration errors will map different points on the line in different directions. The line image will be a distorted curve for incorrect calibrations, and the distance between the two end-points will be incorrect.

Volume Measurement

Some researchers produced a phantom with a known volume (Rousseau et al. 2005; Dandekar et al. 2005). The phantom is scanned and reconstructed in world space. The volume of the phantom can be calculated from their 3D imaging system. The computed volume is then compared with the known volume, and the difference quoted.

The advantage of this measure is that it gives the user an expected error for volume measurements. Often in diagnosis, the clinician is concerned about the volume of some anatomy, rather than its absolute location. Hence errors in volume measurements are important. As in the case for distance accuracy, the position of the phantom may be incorrect. Also, the volume of such an object may be correct even if the calibration is incorrect, unless the phantom has been scanned with the probe rotated in some direction.

Comparison

It is very difficult to compare results quoted from different research groups because of the differences in each measure. Treece et al. (2003) analyzed these differences and made an attempt to compare the results from different research groups. However, even for calibration reproducibility, which does not contain user-induced errors other than errors from calibration itself, it is difficult to compare results across different groups. Different calibrations are probably performed at different depth settings. Furthermore, the size of the cropped B-scan is probably different since a different ultrasound machine is used. Point reconstruction accuracy is highly dependent on the point phantom that is imaged. This has a direct impact on the ability to align the scan plane with the phantom accurately. Distance and volume measurements are highly dependent on the scan motion, which in turn is solely dependent on the user. Even so, many papers fail to describe the probe motion when performing such an assessment. This means that distance or volume measurement results are unlikely to be meaningful. Due to these differences, it has become common practice to give multiple accuracy measures (Lindseth et al. 2003b).

Choosing a Phantom

In this section, we will try to answer the question "what is the best way to calibrate a freehand 3D ultrasound system for a particular application?" At first glance, the question may seem trivial to answer: one should simply choose the most accurate technique. However, this accuracy is dependent on many factors, such as the probe type, user skill and calibration time.

Accuracy

Table 3.1 shows the precision (CR) and accuracy (PRA) achievable with the different phantoms. These results are from our research group, using the same precision and accuracy measures on the same accuracy assessment phantom and with similar ultrasound settings when performing the calibrations. The figures in this table are

Table 3.1 Precision and accuracy (in mm) of calibrations performed by our research group

Phantom	Probe	Depth (cm)	Precision (CR)		Accuracy (PRA)	
			Center	Mean	Center	Mean
Point (cone) (Hsu et al. 2008b)	Linear	3	0.27	0.59	1.86	1.77
Stylus (spherical) (Hsu et al. 2008b)	Linear	3	0.31	0.44	3.07	3.63
Stylus (Cambridge) (Hsu et al. 2008b)	Linear	3	0.45	0.61	1.52	2.18
Plane	Linear	3	0.39	0.57	2.46	2.28
Cambridge phantom	Linear	3	0.83	0.88	1.56	1.67
Mechanical instrument (Gee et al. 2005)	Linear	6	0.15	0.19	–	–
Mechanical instrument (Gee et al. 2005)	Curvilinear	12	0.24	0.44	–	–
Z-phantom (Hsu et al. 2008a)	Curvilinear	8	0.47	0.78	–	–
	Curvilinear	15	1.07	1.54	–	–

therefore directly comparable. Where a citation is missing in the first column, we have performed calibrations with the same ultrasound machine and settings as the ones used by Hsu et al. (2008b), so that the results are comparable.

For precision, we have given the variation at the center of the B-scan as well as the mean of the variations at the four corners and the center of the B-scan. The PRA is computed by scanning the tip of a 1.5 mm-thick wire (Hsu et al. 2008b). We scanned the wire tip at five different regions in the B-scans—near the four corners and the center of the B-scan. The probe is rotated through a full revolution about the lateral axis at six different positions. Five images of the wire are taken at each probe position and in each region of the B-scan. The PRA for images captured near the center of the B-scan as well as the mean of every point captured are given in the table.

From the table, it can be seen that the calibration performed using the Cambridge phantom is the most accurate, closely followed by the cone phantom. The Cambridge stylus and the plane phantom produce suboptimal accuracies, and the spherical stylus produces the worst accuracy. The best precision is obtained by calibrating with the mechanical instrument designed by Gee et al. (2005). The Cambridge phantom is least precise when calibrating at 3 cm.

Table 3.2 shows the quoted precision reported by the various research groups. We have not included accuracy measurements for reasons outlined in the previous section. Each group uses different phantoms to assess their point reconstruction accuracy, and such accuracies are therefore dependent on the phantom used. From the table it can be seen that all of the values for precision are of the same order. Precision increases as the depth setting becomes shallower. This is exactly what we would expect.

Table 3.2 Precision (in mm) of calibrations performed by the various research groups

Phantom	Probe	Depth (cm)	Precision (CR)		
			Center	Corner	Mean
Point (Meairs et al. 2000)	Linear	3.5	1.81	–	–
Point (Lindseth et al. 2003b)	Linear	8	0.62	–	–
Point (sphere) (Hsu et al. 2007)	Linear	3	0.31	0.47	0.44
Point (cone) (Hsu et al. 2007)	Linear	3	0.27	0.67	0.59
Plane (Rousseau et al. 2003)	Linear	–	–	0.89	–
Plane (Rousseau et al. 2005)	Sector	–	–	2.75	–
Plane	Linear	3	0.39	0.61	0.57
Cambridge phantom	Linear	3	0.83	0.89	0.88
2D alignment (Lindseth et al. 2003b)	Linear	8	0.44	–	–
2D alignment (Leotta 2004)	Sector	10	0.67	1.32	1.19
Z-phantom (Lindseth et al. 2003b)	Linear	8	0.63	–	–
Z-phantom (Hsu et al. 2008a)	Curvilinear	8	0.47	0.86	0.78
Z-phantom (Hsu et al. 2008a)	Curvilinear	15	1.07	1.66	1.54
Image registration (Blackall et al. 2000)	Linear	4	–	1.05	–

Calibration Factors

There are several factors that should be taken into account when choosing a particular phantom. The most important factors, other than the precision and accuracy requirements, are the type of probe, the difficulty of the calibration procedure and the calibration time.

Probe Type

There is a large difference between calibrating a linear and a curvilinear probe. Curvilinear probes usually have a lower frequency and are used for imaging at a higher depth setting. It is generally less accurate to calibrate a probe at a higher depth setting, since the image degrades away from the focus and constraining a larger image is more difficult. Despite all these effects, it is still very different to calibrate a linear and a curvilinear probe, even at the same depth. Some phantoms, such as the point phantom, may be equally suitable for calibrating both a linear and a curvilinear probe. On the other hand, 2D alignment phantoms are more suitable for a curvilinear probe, and the Cambridge phantom is more suitable for calibrating a linear probe. A 2D alignment phantom, particularly the Z-phantom, requires getting as many fiducials as possible into the same B-scan frame. Although it is theoretically possible to scan just a part of the phantom repeatedly with a linear probe, this defeats the purpose of using such a phantom. On the other hand, using a plane phantom to calibrate a curvilinear probe may be difficult. If calibration is not performed in a solution where sound travels at a speed similar to that in soft tissue, the distortions will cause the plane to appear as a curve, and not a line. The image needs to be rectified for accurate segmentation. A simple solution is to calibrate in hot water to match the sound speed in water and in soft tissue. If the Cambridge phantom is used to calibrated a probe at a high depth setting, the reverberation due to the clamp may degrade the image so badly that the brass bar is undetectable.

Ease of Use

The point phantom is difficult to use in the sense that the user needs to align the scan plane with the phantom. This requires a certain amount of skill and experience. If a stylus is available, the phantom only needs to be scanned at three noncollinear locations in the B-scans. Therefore, a novice user will perform a slightly worse calibration, but probably not much worse than an expert. The 2D alignment phantoms require precise alignment of the whole phantom with the scan plane. This may be difficult to achieve for a novice user. The user will probably take a very long time to complete the task, but as in the case of the point phantom with a stylus, the calibration should be fairly reliable. In contrast, the Z-phantom does not need any alignment. Not much skill or experience is required to calibrate a probe. The accuracy achieved by an expert and a beginner should be similar. On the other hand, a point and a plane phantom are difficult to use. It is crucial to scan the phantom from a wide variety of positions. An inexperienced user usually neglects one or more of the required probe motions, leading to an incorrect calibration. The Cambridge phantom requires the user to mount the probe accurately, which is also a skilled task.

Calibration Time

The time needed for calibration is dependent on the image quality and segmentation. Images of a point phantom often need to be segmented manually. Nevertheless, automatic segmentation algorithms have been implemented. The automatic segmentation of the plane phantom makes it attractive to use, as the time is shortened considerably. The Z-phantom can calibrate a probe in seconds, outperforming every other phantom in this sense.

Phantom Comparison

We want to answer the question "which phantom is most suitable for probe calibration?" We have listed and discussed the factors that should be taken into account when choosing such a phantom. Based on these factors, we will now compare the strengths and weaknesses of the four major phantoms, namely: point phantom, stylus, plane phantom and Z-phantom, as well as their variants.

The two variants of the point phantom are the cone phantom and the spherical stylus used by Hsu et al. (2008b). Although the cone phantom is physically not a point phantom, it is based on the mathematical principle of a point phantom and has been classified as such in this review. From our perspective, this phantom shows what can be achieved when the point can be aligned and segmented accurately. We will use the results from the spherical stylus to represent a typical point phantom. This phantom is subject to typical alignment and segmentation problems associated with point phantoms. The only advantage is that the point phantom can be moved around, which is unlikely to be a huge advantage. The stylus to be compared will be the Cambridge stylus. This shows what a stylus can achieve with a good alignment. The two variants of the plane phantom are the Cambridge phantom and a plexiglass plate.

In this chapter, we will disregard some phantoms used by individual groups in the comparison. These phantoms include the three-wire phantom, the two-plane phantom, the ordinary 2D alignment phantom and the mechanical instrument. The main problem with the three-wire phantom is that a large number of frames is necessary for an accurate calibration. Manual segmentation is also required. Due to these drawbacks, this phantom has not been used in the last decade. The two-plane phantom works on the same principle as a plane phantom, and can be classified and compared as such. For a 2D alignment phantom, it is difficult to align the scan plane with the whole 2D phantom. It is probably easier to scan individual points one-by-one on such a phantom. For this reason, the 2D alignment phantom is inferior to the point phantom. The mechanical instrument is expensive to manufacture, making it uneconomical to purchase for a freehand 3D ultrasound system. Also, the position sensor needs to be mounted at a fixed position relative to the phantom. This means that either a specific probe holder needs to be used, or the probe holder needs to be calibrated as well. Neither of these approaches offers a straightforward solution.

Table 3.3 ranks the six phantoms according to the different factors that are deemed important for calibration. For each factor, the phantoms are ranked from 1 to 6, where 6 is given to the least suitable phantom. The table is drawn up based on our experience with the phantoms.

From the table, we see that the Z-phantom is the easiest to use. Calibration can be completed within seconds. The calibration performed by a novice should be reliable and have a similar accuracy to that obtained by an expert. However, the precision and accuracy achievable by the phantom is among the worst of all the available phantoms. In contrast, the plane phantoms are difficult to use. The user needs to be sufficiently trained in order to use a plane phantom. The accuracy of a Cambridge phantom is nevertheless the best among the available phantoms. The Cambridge phantom also becomes very easy to use if the user is sufficiently skilled. The plexiglass plate also achieves moderate accuracy and is simple to make. The cone phantom is also very accurate. Not much training is required to use this phantom. However, the phantom needs to be aligned manually by the user, and the segmentation requires human intervention to mark the search region. The Cambridge stylus

Table 3.3 Probe calibration factors for the different phantoms

Factor	Point		Stylus	Plane		2D alignment
	Sphere	Cone	Cambridge	Plate	Cambridge	Z-phantom
Precision	1	2	2	2	6^a	5
Accuracy	5	1	3	3	1	5
Easy to use (novice)	4	2	2	4	4	1
Easy to use (expert)	5	2	2	5	2	1
Segmentation	6	5	2	2	1	2
Speed	6	3	2	4	4	1
Reliability	4	2	2	4	4	1
Phantom simplicity	1	4	3	1	4	4
Linear probe	✓	✓	✓	✓	✓	$✗^b$
Curvilinear probe	✓	✓	✓	✓	$✗^b$	✓

[a] This is based on the precision at 3 cm. The precision is better when calibrating at a higher depth
[b] It is possible, but difficult, to use these phantoms to calibrate the corresponding probe

and the point target lie in the middle. They are not particularly simple nor very difficult to use, and can produce calibrations in a reasonable time. Automatic segmentation is also possible with good image quality. The Cambridge stylus produces better accuracy with a slightly more complicated design. Most phantoms are suitable for calibrating both a linear and a curvilinear probe. The Z-phantom may be more suitable for a curvilinear probe, so that a large number of fiducials can be captured in the same frame, enabling very rapid calibration. The Cambridge phantom is not suitable for a curvilinear probe at high depth since the reverberation effect from the clamp corrupts the image so badly that the plane cannot be detected.

Conclusion

In this chapter, we have classified all of the phantoms used for freehand 3D ultrasound calibration by their mathematical principles. The strengths and weaknesses of each phantom were discussed. The different measures used to assess the calibration quality were analyzed and the accuracy of each phantom quantified. In the end, we pointed out the situations where a particular phantom may be more suitable than the others. Unfortunately, there is no single phantom that outperforms the rest. The Cambridge phantom and the cone phantom are the most accurate. The Cambridge phantom is the most difficult to use for a novice user, but easy to use for an expert. The Z-phantom is easiest to use and produces a calibration within seconds, but its accuracy remains poor. The other phantoms lie between these extremes, offering moderate accuracy, ease of use and phantom complexity.

References

Alcazar JL (2005) Three-dimensional ultrasound in gynecology: current status and future perspectives. Curr Women's Health Rev 1:1–14

Ali A, Logeswaran R (2007) A visual probe localization and calibration system for cost-effective computer-aided 3D ultrasound. Comp Biol Med 37:1141–1147

Amin DV, Kanade T, Jaramaz B, DiGioia III AM, Nikou C, LaBarca RS, Moody Jr JE (2001) Calibration method for determining the physical location of the ultrasound image plane. In: Proceedings of the Fourth International Conference on Medical Image Computing and Computer-Assisted Intervention (Lecture Notes in Computer Science 2208). Springer, Berlin, pp 940–947

Anagnostoudis A, Jan J (2005) Use of an electromagnetic calibrated pointer in 3D freehand ultrasound calibration. In: Proc Radioelektronika, Brno, Czech Republic, 3–4 May 2005

Andreff N, Horaud R, Espiau B (2001) Robot hand-eye calibration using structure from motion. Int J Robot Res 20(3):228–248

Arun KS, Huang TS, Blostein SD (1987) Least-squares fitting of two 3-D point sets. IEEE Trans Pattern Anal Mach Intell 9(5):698–700

Barratt DC, Penney GP, Chan CSK, Slomczykowski CM, Carter TJ, Edwards PJ, Hawkes DJ (2006) Self-calibrating 3D-ultrasound-based bone registration for minimally invasive orthopedic surgery. IEEE Trans Med Imag 25(3):312–323

Barry CD, Allott CP, John NW, Mellor PM, Arundel PA, Thomson DS, Waterton JC (1997) Three-dimensional freehand ultrasound: image reconstruction and volume analysis. Ultrasound Med Biol 23(8):1209–1224

Beasley RA, Stefansic JD, Herline AJ, Guttierez L, Galloway Jr RL (1999) Registration of ultrasound images. Proc SPIE Med Imag 3658:125–132

Berg S, Torp H, Martens D, Steen E, Samstad S, Høivik I, Olstad B (1999) Dynamic three-dimensional freehand echocardiography using raw digital ultrasound data. Ultrasound Med Biol 25(5):745–753

Blackall JM, Rueckert D, Maurer Jr CR, Penney GP, Hill DLG, Hawkes DJ (2000) An image registration approach to automated calibration for freehand 3D ultrasound. In: Proceedings of the Third International Conference on Medical Image Computing and Computer-Assisted Intervention (Lecture Notes in Computer Science 1935. Springer, Berlin, pp 462–471

Boctor EM, Jain A, Choti MA, Taylor RH, Fichtinger G (2003) A rapid calibration method for registration and 3D tracking of ultrasound images using spatial localizer. Proc SPIE 5035:521–532

Boctor EM, Viswanathan A, Choti M, Taylor RH, Fichtinger G, Hager G (2004) A novel closed form solution for ultrasound calibration. In: IEEE Int Symp Biomedical Imaging: Nano to Macro, Arlington, VA, 15–18 April 2004, 1:527–530

Bouchet LG, Meeks SL, Goodchild G, Bova FJ, Buatti JM, Friedman, WA (2001) Calibration of three-dimensional ultrasound images for image-guided radiation therapy. Phys Med Biol 46:559–577

Brendel B, Winter S, Ermert H (2004) A simple and accurate calibration method for 3D freehand ultrasound. Biomed Tech 49:872–873

Brewer J (1978) Kronecker products and matrix calculus in system theory. IEEE Trans Circuit Syst 25(9):772–781

Carr JC (1996) Surface reconstruction in 3D medical imaging. Ph.D. thesis, University of Canterbury, Christchurch, New Zealand

Cash C, Berman L, Treece G, Gee A, Prager R (2004) Three-dimensional reconstructions of the normal and abnormal neonatal foot using high-resolution freehand 3D ultrasound. In: Proceedings of the Radiological Society of North America (RSNA 2004), Chicago, IL, 28 Nov–3 Dec 2004

Chen TK, Abolmaesumi P, Thurston AD, Ellis RE (2006) Automated 3D freehand ultrasound calibration with real-time accuracy control. In: Proceedings of the Ninth International Conference on Medical Image Computing and Computer-Assisted Intervention (Lecture Notes in Computer Science 4190). Springer, Berlin, pp 899–906

Clarke JC, Carlsson S, Zisserman A (1996) Detecting and tracking linear features efficiently. In: Proceedings of the British Machine Vision Conference 1996, Edinburgh. British Machine Vision Association, Malvern, UK, pp 415–424

Coles CE, Cash CJC, Treece GM, Miller FNAC, Hoole ACF, Gee AH, Prager RW, Sinnatamby R, Britton P, Wilkinson JS, Purushotham AD, Burnet NG (2007) High definition three-dimensional ultrasound to localise the tumour bed: a breast radiotherapy planning study. Radiother Oncol 84(3):233–241

Comeau RM, Fenster A, Peters TM (1998) Integrated MR and ultrasound imaging for improved image guidance in neurosurgery. Proc SPIE 3338:747–754

Comeau RM, Sadikot AF, Fenster A, Peters TM (2000) Intraoperative ultrasound for guidance and tissue correction in image-guided neurosurgery. Med Phys 27(4):787–800

Dandekar S, Li Y, Molloy J, Hossack J (2005) A phantom with reduced complexity for spatial 3-D ultrasound calibration. Ultrasound Med Biol 31(8):1083–1093

Detmer PR, Bashein G, Hodges T, Beach KW, Filer EP, Burns DH, Stradness Jr DE (1994) 3D ultrasonic image feature localization based on magnetic scanhead tracking: in vitro calibration and validation. Ultrasound Med Biol 20(9):923–936

Eggert DW, Lorusso A, Fisher RB (1997) Estimating 3-D rigid body transformations: a comparison of four major algorithms. Mach Vision Appl 9:272–290

Fenster A, Downey DB, Cardinal HN (2001) Three-dimensional ultrasound imaging. Phys Med Biol 46:R67–R99

Fenster A, Landry A, Downey DB, Hegele RA, Spence JD (2004a) 3D ultrasound imaging of the carotid arteries. Curr Drug Targets—Cardiovasc Hematol Disord 4(2):161–175

Fenster A, Surry KJM, Mills GR, Downey DB (2004b) 3D ultrasound guided breast biopsy system. Ultrasonics 42:769–774

Fischler MA, Bolles RC (1981) Random sample consensus: a paradigm for model fitting with applications to image analysis and automated cartography. Commun ACM 24(6):381–395

Gee AH, Prager RW, Treece GH, Berman LH (2003) Engineering a freehand 3D ultrasound system. Pattern Recognit Lett 24:757–777

Gee AH, Houghton NE, Treece GM, Prager RW (2005) A mechanical instrument for 3D ultrasound probe calibration. Ultrasound Med Biol 31(4):505–518

Gobbi DG, Comeau RM, Peters TM (1999) Ultrasound probe tracking for real-time ultrasound/MRI overlay and visualization of brain shift. In: Proceedings of the Second International Conference on Medical Image Computing and Computer-Assisted Intervention (Lecture Notes in Computer Science 1679). Springer, Berlin, pp 920–927

Gonçalves LF, Lee W, Espinoza J, Romero R (2005) Three- and 4-dimensional ultrasound in obstetric practice. Does it help? J Ultrasound Med 24:1599–1624

Gooding MJ, Kennedy SH, Noble JA (2005) Temporal calibration of freehand three-dimensional ultrasound using image alignment. Ultrasound Med Biol 31(7):919–927

Hartov A, Eisner SD, Roberts DW, Paulsen KD, Platenik LA, Miga MI (1999) Error analysis for a free-hand three-dimensional ultrasound system for neuronavigation. Neurosurg Focus 6(3):5

Hastenteufel M, Mottl-Link S, Wolf I, de Simone R, Meinzer H-P (2003) A method for the calibration of 3D ultrasound transducers. Proc SPIE 5029:231–238

Hough PVC (1959) Machine analysis bubble chamber pictures. In: International Conference on High Energy Accelerators and Instrumentation. CERN, Geneva, pp 554–556

Hsu P-W, Prager RW, Gee AH, Treece GM (2006) Rapid, easy and reliable calibration for freehand 3D ultrasound. Ultrasound Med Biol 32(6):823–835

Hsu P-W, Prager RW, Houghton NE, Gee AH, Treece GM (2007) Accurate fiducial location for freehand 3D ultrasound calibration. Proc SPIE 6513:15

Hsu P-W, Prager RW, Gee AH, Treece GM (2008a) Real-time freehand 3D ultrasound calibration. Ultrasound Med Biol 34(2):239–251

Hsu P-W, Treece GM, Prager RW, Houghton NE, Gee AH (2008b) Comparison of freehand 3D ultrasound calibration techniques using a stylus. Ultrasound Med Biol 34(10):1610–1621

Huang QH, Zheng YP, Lu MH, Chi ZR (2005) Development of a portable 3D ultrasound imaging system for musculosketetal tissues. Ultrasonics 43:153–163

Kaspersen JH, Langø T, Lindseth F (2001) Wavelet-based edge detection in ultrasound images. Ultrasound Med Biol 27(1):89–99

Khamene A, Sauer F (2005) A novel phantom-less spatial and temporal ultrasound calibration method. In: Proceedings of the Eighth International Conference on Medical Image Computing and Computer-Assisted Intervention (Lecture Notes in Computer Science 3750). Springer, Berlin, pp 65–72

Krupa A (2006) Automatic calibration of a robotized 3D ultrasound imaging system by visual servoing. In: Proc 2006 IEEE Int Conf on Robotics and Automation, Orlando, FL, 15–19 May 2006, pp 4136–4141

Lange T, Eulenstein S (2002) Calibration of swept-volume 3-D ultrasound. In: Proceedings of Medical Image Understanding and Analysis, Portsmouth, UK, 22–23 July 2002, 3:29–32

Langø T (2000) Ultrasound guided surgery: image processing and navigation. Ph.D. thesis, Norwegian University of Science and Technology, Trondheim, Norway

Legget ME, Leotta DF, Bolson EL, McDonald JA, Martin RW, Li X-N, Otto CM, Sheehan FH (1998) System for quantitative three-dimensional echocardiography of the left ventricle based on a magnetic-field position and orientation sensing system. IEEE Trans Biomed Eng 45(4):494–504

Leotta DF (2004) An efficient calibration method for freehand 3-D ultrasound imaging systems. Ultrasound Med Biol 30(7):999–1008

Leotta DF, Detmer PR, Martin RW (1997) Performance of a miniature magnetic position sensor for three-dimensional ultrasound imaging. Ultrasound Med Biol 23(4):597–609

Lindseth F, Langø T, Bang J (2002) Accuracy evaluation of a 3D ultrasound-based neuronavigation system. Comput Aided Surg 7:197–222

Lindseth F, Bang J, Langø T (2003a) A robust and automatic method for evaluating accuracy in 3-D ultrasound-based navigation. Ultrasound Med Biol 29(10):1439–1452

Lindseth F, Tangen GA, Langø T, Bang J (2003b) Probe calibration for freehand 3-D ultrasound. Ultrasound Med Biol 29(11):1607–1623

Liu J, Gao X, Zhang Z, Gao S, Zhou J (1998) A new calibration method in 3D ultrasonic imaging system. In: Proc 20th Annu Int Conf of IEEE Eng Med Biol Soc, Hong Kong, 29 Oct–1 Nov 1998, 20:839–841

Meairs S, Beyer J, Hennerici M (2000) Reconstruction and visualization of irregularly sampled three- and four-dimensional ultrasound data for cerebrovascular applications. Ultrasound Med Biol 26(2):263–272

Meeks SL, Buatti JM, Bouchet LG, Bova FJ, Ryken TC, Pennington EC, Anderson KM, Friedman WA (2003) Ultrasound-guided extracranial radiosurgery: technique and application. Int J Radiat Oncol Biol Phys 55(4):1092–1101

Mercier L, Langø T, Lindseth F, Collins DL (2005) A review of calibration techniques for freehand 3-D ultrasound systems. Ultrasound Med Biol 31(4):449–471

More JJ (1977) The Levenberg–Marquardt algorithm: implementation and theory. In: Numerical analysis (Lecture Notes in Mathematics 630). Springer, Berlin, pp 105–116

Muratore DM, Galloway RL Jr (2001) Beam calibration without a phantom for creating a 3-D freehand ultrasound system. Ultrasound Med Biol 27(11):1557–1566

Nelson TR, Pretorius DH (1998) Three-dimensional ultrasound imaging. Ultrasound Med Biol 24(9):1243–1270

Pagoulatos N, Edwards WS, Haynor DR, Kim Y (1999) Interactive 3-D registration of ultrasound and magnetic resonance images based on a magnetic position sensor. IEEE Trans Inf Technol Biomed 3(4):278–288

Pagoulatos N, Haynor DR, Kim Y (2001) A fast calibration method for 3-D tracking of ultrasound images using a spatial localizer. Ultrasound Med Biol 27(9):1219–1229

Péria O, Chevalier L, François-Joubert A, Caravel J-P, Dalsoglio S, Lavallée S, Cinquin P (1995) Using a 3D position sensor for registration of SPECT and US images of the kidney. In: Proc

First Int Conf on Computer Vision, Virtual Reality and Robotics in Medicine (Lecture Notes in Computer Science 905), Nice, France, 3–6 April 1995, pp 23–29

Pluim JPW, Maintz JBA, Viergever MA (2003) Mutual-information-based registration of medical images: a survey. IEEE Trans Med Imag 22(8):986–1004

Poon TC, Rohling RN (2005) Comparison of calibration methods for spatial tracking of a 3-D ultrasound probe. Ultrasound Med Biol 31:1095–1108

Poon TC, Rohling RN (2007) Tracking a 3-D ultrasound probe with constantly visible fiducials. Ultrasound Med Biol 33(1):152–157

Prager RW, Rohling RN, Gee AH, Berman L (1998) Rapid calibration for 3-D freehand ultrasound. Ultrasound Med Biol 24(6):855–869

Rousseau F, Hellier P, Barillot C (2003) Robust and automatic calibration method for 3D freehand ultrasound. In: Proceedings of the Sixth International Conference on Medical Image Computing and Computer-Assisted Intervention (Lecture Notes in Computer Science 2879). Springer, Berlin, pp 440–448

Rousseau F, Hellier P, Barillot C (2005) Confhusius: A robust and fully automatic calibration method for 3D freehand ultrasound. Med Image Anal 9(1): 25–38

Rygh OM, Nagelhus Hernes TA, Lindseth F, Selbekk T, Brostrup T, Müller TB (2006) Intra-operative navigated 3-dimensional ultrasound angiography in tumor surgery. Surg Neurol 66(6):581–592

Sato Y, Nakamoto M, Tamaki Y, Sasama T, Sakita I, Nakajima Y, Monden M, Tamura S (1998) Image guidance of breast cancer surgery using 3-D ultrasound images and augmented reality visualization. IEEE Trans Med Imag 17(5):681–693

Sauer F, Khamene A, Bascle B, Schinunang L, Wenzel F, Vogt S (2001) Augmented reality visualization of ultrasound images: system description, calibration, and features. In: Proc IEEE ACM Int Symp on Augmented Reality, New York, 29–30 Oct 2001, pp 30–39

Sawada A, Yoda K, Kokubo M, Kunieda T, Nagata Y, Hiraoka M (2004) A technique for noninvasive respiratory gated radiation treatment system based on a real time 3D ultrasound image correlation: a phantom study. Med Phys 31(2):245–250

State A, Chen DT, Tector C, Brandt A, Chen H, Ohbuchi R, Bajura M, Fuchs H (1994) Case study: observing a volume rendered fetus within a pregnant patient. In: Proceedings of the Conference on Visualization '94, IEEE Visualization. IEEE Computer Society Press, Los Alamitos, CA, pp 364–368

Studholme C, Hill DLG, Hawkes DJ (1999) An overlap invariant entropy measure of 3D medical image alignment. Pattern Recognit 32:71–86

Treece GM, Prager RW, Gee AH, Berman L (2001) 3D ultrasound measurement of large organ volume. Med Image Anal 5(1):41–54

Treece GM, Prager RW, Gee AH, Berman L (2002) Correction of probe pressure artifacts in freehand 3D ultrasound. Med Image Anal 6:199–214

Treece GM, Gee AH, Prager RW, Cash CJC, Berman LH (2003) High-definition freehand 3-D ultrasound. Ultrasound Med Biol 29(4):529–546

Trobaugh JW, Richard WD, Smith KR, Bucholz RD (1994) Frameless stereotactic ultrasonography: method and applications. Comput Med Imag Graph 18(4):235–246

Umeyama S (1991) Least-squares estimation of transformation parameters between two point patterns. IEEE Trans Pattern Anal Mach Intell 13(4):376–380

Unsgaard G, Rygh OM, Selbekk T, Müller TB, Kolstad F, Lindseth F, Nagelhus Hernes TA (2006) Intra-operative 3D ultrasound in neurosurgery. Acta Neurochir 148(3):235–253

Varandas J, Baptista P, Santos J, Martins R, Dias J (2004) VOLUS—a visualization system for 3D ultrasound data. Ultrasonics 42:689–694

Viswanathan A, Boctor EM, Taylor RH, Hager G, Fichtinger G (2004) Immediate ultrasound calibration with three poses and minimal image processing. In: Proceedings of the Fourth International Conference on Medical Image Computing and Computer-Assisted Intervention (Lecture Notes in Computer Science 3217). Springer, Berlin, pp 446–454

Welch JN, Johnson JA, Bax MR, Badr R, Shahidi R (2000) A real-time freehand 3D ultrasound system for image-guided surgery. IEEE Ultrasonics Symp 2:1601–1604

Yagel S, Cohen SM, Shapiro I, Valsky DV (2007) 3D and 4D ultrasound in fetal cardiac scanning: a new look at the fetal heart. Ultrasound Obstet Gynecol 29(1):81–95

Zhang WY, Rohling RN, Pai DK (2004) Surface extraction with a three-dimensional freehand ultrasound system. Ultrasound Med Biol 30(11):1461–1473

Zhang H, Banovac F, White A, Cleary K (2006) Freehand 3D ultrasound calibration using an electromagnerically tracked needle. Proc SPIE Med Imag 6141:775–783

Zitová B, Flusser J (2003) Image registratioin methods: a survey. Image Vision Comput 21:977–1000

Chapter 4
Laser Scanning: 3D Analysis of Biological Surfaces

Matthew W. Tocheri

Abstract This chapter introduces laser scanning to students and researchers who are interested in using this three-dimensional (3D) acquisition method for biological research. Laser scanning is yet another tool for transforming biological structures into 3D models that contain useful geometric and topological information. Current laser scanning technology makes it relatively straightforward to acquire 3D data for visualization purposes. However, there are many additional challenges that are necessary to overcome if one is interested in collecting and analyzing data from their laser-scanned 3D models. In this chapter, I review some basic concepts, including what laser scanning is, reasons for using laser scanning in biological research, how to choose a laser scanner, and how to use a laser scanner to acquire 3D data, and I provide some examples of what to do with 3D data after they have been acquired.

Introduction

As a teenager, most of my time was spent studying geometry, physics, and economics in an applied context. By that I mean I could invariably be found playing pool and snooker rather than attending class. I tried to supplement my educational experiences at the local pool rooms by reading as many how-to-play-pool books as possible. I always remember that while there were many books available, they all seemed to be missing one important chapter—how to actually play! Sure these books told you about the rules, necessary equipment, how to hold the cue, and how to stand, but they never described the thought processes that are necessary to approach even the most basic situations that occur again and again at the table. Instead, they always assumed you already knew what you were supposed to be thinking about. The reason I can say this is because I became a reasonably proficient

M.W. Tocheri
Human Origins Program, Department of Anthropology, National Museum of Natural History, Smithsonian Institution, Washington DC 20013-7012, USA, e-mail: tocherim@si.edu

player by learning the hard way what I should be thinking about at different times. Although my billiard career is long extinct, I have never forgotten that learning the thinking process behind any desired task is often critical for any long-term success.

Laser scanning is a lot like pool. It looks easy, almost magical when performed well, and it even has a certain "cool factor" surrounding it. In reality, however, it is extremely challenging, involving constant problem-solving, troubleshooting, and creative and critical thinking. Since 2001, this particular type of imaging technology has formed an integral part of my research program (Tocheri 2007; Tocheri et al. 2003, 2005, 2007). Over the years, I have given countless demonstrations to other interested students and researchers as well as general advice on how to incorporate laser scanning into their research. However, the high costs of equipment and software have precluded many from being able to do so. Even with access to a laser scanner, the combination of a steep learning curve and constant troubleshooting still often results in research efforts grinding to a halt despite initially high levels of expectation and excitement.

I hope this chapter helps reverse this unfortunate trend. Recent technological breakthroughs have resulted in laser scanners that cost only a fraction of what they did a few years ago. Given this new affordability, there is no doubt that more students and researchers are going to give laser scanning a try. To ensure that these newly acquired pieces of equipment collect 3D data rather than dust, my primary goal in this chapter is to convey some of the creative and critical thinking skills that are necessary to successfully incorporate laser scanning into an active research program. To accomplish this, I have focused the chapter around five main questions that continually arise every time I demonstrate the laser scanning process to interested students and researchers. By presenting information about laser scanning in this manner, I hope the reader is better prepared to solve the common problems (and their many variations) that they will encounter as the initial "cool factor" wears off and they are forced to face the many challenges that laser scanning will bring to their research.

What is Laser Scanning?

Laser scanning is simply a semiautomated method of 3D data capture. In other words, it is a method for generating a numerical description of an object. A straightforward way to describe an object numerically is to construct an array of coordinate values for points that lie on the object's surface (Bernardini and Rushmeier 2002). In principle, you could do this by hand by measuring how far away different locations on a surface are from a single designated location in space. The chosen location would represent the origin of a 3D coordinate system and you would convert your depth measurements into x, y, and z coordinates for each measured point on the surface.

Laser scanners accomplish these tasks for you by projecting a laser beam onto an object (Fig. 4.1). As the laser contacts and moves across the surface of the object,

 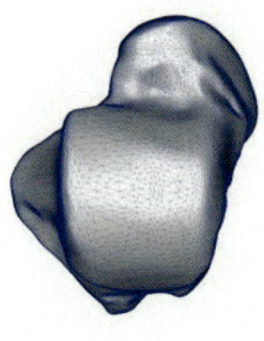

Fig. 4.1 By projecting a laser onto the surface of an object, depth information is recorded at particular intervals along the surface through a sensor (typically a CCD camera), which tracks the reflectivity of the laser on the surface (*left*). The depth values are converted into 3D point coordinates, which are used to generate a polygonal mesh that is a numerical description of the object's surface (*right*)

depth information is recorded at particular intervals along the surface through a sensor (typically a CCD camera), which tracks the reflectivity of the laser on the surface (Bernardini and Rushmeier 2002; Kappelman 1998; Zollikofer and Ponce de Leon 2005). Most often, the positions and orientations of the laser and the sensor are calibrated via triangulation. Therefore, the acquired depth information is converted into 3D point values (*x*, *y*, *z*) using the scanner's coordinate system. The result is a 3D point cloud of data that represents a collection of geometrical information about various locations on the object's surface. Generally, neighboring points are also connected by an edge (i.e., a straight line). Together, the edges form polygons (usually either triangles or quadrilaterals), resulting in a 3D mesh structure (Figs. 4.1 and 4.2). The resulting polygonal mesh is a numerical description that contains geometric information (*x*, *y*, *z* values) and topological information (how the points are connected) (Bernardini and Rushmeier 2002).

By changing the position of the object and repeating the scanning process, additional scans of the object's surface are acquired. Using portions of surface overlap between multiple scans, these are then "stitched" together to form the 3D model of the object (Bernardini and Rushmeier 2002; Kappelman 1998). The task of stitching scans together may be performed manually or automatically, depending on the specifics of the scanning software. Most often, the stitching process involves first registering or aligning the multiple scans together. After the scans are aligned, they are then merged together to form a single polygonal mesh. Laser scanners are typically sold with accompanying software that enables you to perform the alignment and merging steps relatively easily.

Although every type of laser scanner operates differently (along with the accompanying software), the overall process is the same. Before describing in more detail how one goes about using a laser scanner to generate 3D numerical descriptions (hereafter termed "3D models") of real-world objects, I first want to discuss why

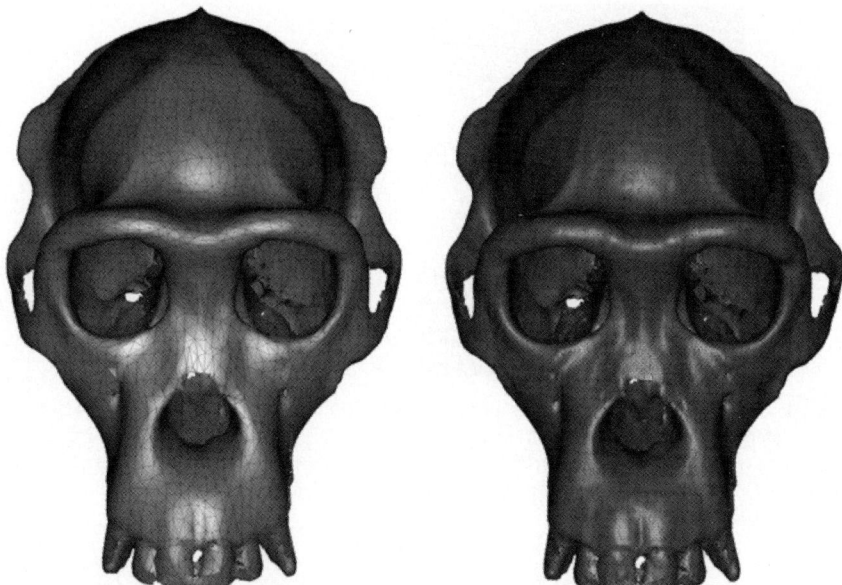

Fig. 4.2 Polygonal meshes are a combination of geometry (x, y, z values) and topology (edges that connect neighboring coordinates). Denser point clouds result in more surface detail; compare the mesh of a chimpanzee cranium on the *left* (15,000 triangles) to the one on the *right* (400,000 triangles)

a biological or medical researcher would want to use laser scanning to acquire 3D models.

Why Use Laser Scanning?

3D models can be acquired using many different kinds of technology, many of which are described in the other chapters of this book. Laser scanners are specifically used to acquire 3D models that contain surface information only, with no information about the internal properties of the object (Bernardini and Rushmeier 2002; Kappelman 1998; Zollikofer and Ponce de Leon 2005). Therefore, laser scanners should be used when the data of interest relate to the surfaces or external shape properties of an object (e.g., Ungar 2004; Dennis et al. 2004; Tocheri et al. 2003, 2005, 2007). While it is obvious that if the data of interest relate to the trabecular structure of bone then laser scanning is not the appropriate 3D acquisition method, there are other scenarios that are less obvious that deserve attention.

The "cool factor" of laser scanning results in many wanting to use the technology almost for the sake of using it. My general advice is if you want to take linear measurements then use calipers to measure the actual object or if you want acquire

3D landmarks for geometric morphometrics then use a point digitizer, again on the actual object. I say this because laser scanning results in a *model* of the actual object rather than an exact replica of it. In other words, a model is an approximation of an object, and its accuracy in representing the object is dependent on a variety of factors, including the equipment, software, algorithms, and expertise used to generate it. Therefore, a good rule of thumb is to use laser scanning when you want to measure some aspect of the object that is easier or more practical to do using a model of the object rather than measuring the object itself. It is true that you can take linear measurements and landmarks from a laser-scanned 3D model, but two factors argue against doing so if it is the sole reason you are using a laser scanner. First, you end up with a measure of the model that you could have easily acquired directly from the actual object. This may result in a loss of precision of the measurement, since it is being acquired indirectly from an approximation of the object. Second and most important, laser scanning simply takes longer. If laser scanning resulted in a complete 3D model in less time than it takes to acquire the caliper measurements or a set of landmarks, then any loss of precision could be better justified. But the fact remains that the laser scanners that are currently the most widely available to biological researchers simply do not produce measurable 3D models instantaneously. Instead, laser scanning often requires anywhere between at least 20 min to several hours to acquire a complete 3D model of an object. Until laser scanning technology can cost-effectively and rapidly produce 3D models that can be easily validated as practical replicas of their real-world counterparts, I would argue that it is a better strategy to try and use the current technology more appropriately.

Appropriate uses of laser scanning should directly involve or focus on surfaces, such as any type of surface visualization or quantification. Surface visualization benefits from the ability to manipulate the model on a computer screen, making it extremely easy to view the model from any angle without any fear of dropping or mishandling the actual object. In addition, most 3D software applications allow for complete digital control of lighting and other conditions that significantly enhance the visualization experience. These advantages justify the use of the 3D model despite the fact that it is an approximation of the actual object. Whether used for research or educational purposes, visualization of laser-scanned 3D biological surface data is a powerful tool.

Surface quantification involves an attempt to measure some property of the surface. A simple and straightforward example is the calculation of surface area. Using a laser-scanned 3D model to calculate the surface area of an object (or a portion thereof) is appropriate because the surface area of the actual object cannot be quantified in any standard manner but the surface area of the model is easily computed (exceptions include objects of standard geometric shape for which there are formulae for calculating surface area). Inherent in the surface area measurement of the model is the notion that it is an approximation of the "real" surface area, but since the actual object cannot be measured directly an approximation of the measure is appropriate. The same basic principle is true when measuring angles between surfaces, surface curvatures, and surface cross-sections to name a few—they all are approximate measures of surface properties made possible by the model.

Whether laser scanning is used for surface visualization, quantification, or both, an additional benefit is the ability to archive surface data (e.g., Rowe and Razdan 2003). Laser-scanned 3D data are not just an approximation of some real surface; they are an approximation of some real surface at a particular moment in time. Therefore, laser scanning automatically results in a digital archive of 3D surfaces and objects. Such data can be used if the actual object is no longer accessible (e.g., to take additional measurements) or it can be used to compare how an object's surface properties have changed over time if the object is laser scanned more than once (e.g., before and after plastic surgery; Byrne and Garcia 2007).

Choosing a Laser Scanner

There are hundreds of makes and models of laser scanners available today. Deciding which laser scanner is best suited for particular biological research is challenging. I do not wish to endorse any particular laser scanner over another; however, there are several key factors that I think should be considered if you are choosing a laser scanner for biological research purposes. These key factors include object size, auto-alignment capability and performance, 3D model acquisition speed, and portability.

Unfortunately, there is no one-size-fits-all laser scanner available. Instead, laser scanners are invariably configured to handle a particular range of object sizes. You want to purchase a scanner that is optimized for the sizes of object that you intend to scan the most. Laser scanners are configured to have maximum precision within a field of view that is a particular distance from the object (some scanners may have multiple fields of view). Therefore, you want to ensure that your object fits within a field of view that is configured for maximum precision. If the objects you want to scan are bigger than the optimized field of view, then the scanner is not appropriate for your research needs.

You also need to consider whether the objects you want to scan are too small for the field of view of the scanner. This will depend on how many points the scanner is capable of sampling as the laser moves across the surface of the object. Remember that scanners will on average sample the same number of points per millimeter within its entire field of view. Therefore, the smaller the object is in comparison to the field of view, the smaller the number of points sampled. If the field of view is about the size of a page (e.g., $8.5'' \times 11''$), then objects the size of a quarter are probably not going to scan very well, simply because not enough points on its surface are being sampled.

One of the most important factors to consider is the ability of accompanying software to align consecutive scans together. The basic idea of a laser scanner is to automatically sample a large number of points from a surface so that you do not have to spend hours or days manually digitizing each point by hand. However, the majority of laser scanners do not have the ability to automatically align consecutive scans together. Instead, the user has to manually align scan after scan by selecting common points on each surface and then waiting for an algorithm to complete its best mathematical guess at the alignment. This is a huge issue for biological research, where (in most cases) we do not want to invest hours or days generating a 3D model for each specimen we want to include in our analyses. My advice is that if the laser scanner does not come equipped with additional hardware and software that performs automatic alignment of consecutive scans robustly and accurately, then it is most likely not useful for any type of biological research that requires reasonable sample sizes—it will simply require too much time to generate enough 3D models for the research. Ideally, the laser scanner should come equipped with an attachable turntable and software that includes an auto-alignment feature. The best algorithms for auto-alignment take into account how far away the turntable is from the laser and sensor and how many degrees the turntable moves in-between consecutive scans. Incorporating such information enables consecutive scans to be aligned and merged with maximum precision.

Auto-alignment capability directly leads to the next important factor: 3D model acquisition speed. Notice that the emphasis is on 3D model acquisition, not individual scan acquisition. Almost all laser scanners can acquire 3D data reasonably quickly and efficiently. However, the first question you need to ask yourself is: what is it that you are going to visualize or measure? If you are going to use individual scans then scanning speed is all you need to worry about. But if you want to visualize or measure a 3D model, which is generated from a combination of individual scans, then you need to reliably estimate how long it will take you to acquire such a model. This includes scanning time plus alignment, merging, and any additional post-scanning procedures necessary to acquire the 3D model. Remember that if it takes you longer to first make a copy of the object you want to measure rather than measuring it directly, then how are you justifying making a copy in the first place? Determine how long it takes you to acquire one 3D model and then multiply this by how many models you estimate you will need to reach your research objectives. You may well realize that the amount of time you require just to generate the sample of 3D models far exceeds the time you have available. For example, it took an average of 30 min to acquire each of the 1,250 3D bone models used in my dissertation research sample (Tocheri 2007): roughly 625 h of scanning and post-processing, equivalent to about four months of focused effort (e.g., 8 h/day, 5 days/week). The bottom line is to not allow the "cool factor" of laser scanning to cloud your judgment regarding how much time and effort is needed to acquire a reasonable sample of 3D models for you research. Of course, I think it is worth investing the time and effort if the research question justifies it; but I would be lying if I told you that I had not seen many research projects fall apart because they did not realistically take into account this important factor.

Finally, portability is an additional factor to consider. This includes whether it is the laser scanner, the objects, or both that are portable. You need to consider how you will get the laser scanner, the computer that runs it, and the objects you want to scan together in the same room. Unfortunately, as is the case with most manufactured products, portability typically means you are getting less but paying more. Portable laser scanners are generally restricted in some capacity in comparison to their nonportable equivalents. The same is true of laptop computers, which are often necessary to run portable laser scanners. These factors need to be weighed against whether the 3D models you will acquire are sufficient for your research objectives.

There may be additional factors to consider depending specifically on the research objectives, such as the ability to capture texture information. For the most part, however, the four discussed here should be sufficient as a starting guide for narrowing down the wide variety of available choices. New laser scanners with improved capabilities become available every year, and they continue to become more affordable. When making your decision about which laser scanner to use, always remember that it is only a tool to help you reach your research objectives. Therefore, stay focused on issues relating to the size, quality, and speed of the 3D models it generates, and factor in the required time and effort to build up the required study sample.

How to Laser Scan

A good rule of thumb when laser scanning is to be patient, flexible, and willing to experiment. Keep in mind that laser scanning is like a form of 3D digital sculpture, wherein you as the sculptor are attempting to acquire a digital representation of the object you are scanning. You must be flexible and willing to experiment in everything from how you place the object in relation to the laser and the sensor, how you edit (clean) each scan, to how you perform the alignment and merging. Patience, flexibility, and experimentation will enable you to become comfortable using many different laser scanners and 3D software packages and will ultimately result in final 3D models that are more accurate digital representations of the objects you scan.

The first step involves some imagination and decision-making. You must decide how you are going to place your object in relation to the laser and the sensor. My general advice is to pretend that you are the object; can you, metaphorically speaking (i.e., do not actually try to look directly at the laser), "see" the laser and the sensor simultaneously? Remember that you must be able to "see" the laser (typically straight in front of you) and the sensor in order for the depth information to be captured. Any portion of the object's surface which is not in the line-of-sight of both the laser and the sensor will not be captured, resulting in holes in the digital surface reconstruction (Bernardini and Rushmeier 2002; Zollikofer and Ponce de Leon 2005). Additional consideration must be given to the surface topography. Biological surfaces tend to be quite complex with lots of curves and indentations. Because of the triangulation procedure used by the laser and sensor to capture surface information, portions of the surface that are more perpendicular to the direction of the laser will be captured with higher precision (less error) (Kappelman 1998). Therefore, the further portions of the surface are from being perpendicular to the laser, the more erroneous the resulting surface reconstruction. To avoid this problem, you need to consider how to best position your object so that you maximize the amount of surface that is close to perpendicular to the laser. Remember that, as you take multiple scans from different orientations, different portions of the surface will be more perpendicular to the laser. This means that data acquired from one scan poorly because of its position in relation to the laser may be acquired more precisely from another scan. This is where a good imagination comes in handy. Imagine that you are digitally reconstructing the object by creating different pieces of the object (like pieces to a 3D jigsaw puzzle). Each scan results in an additional piece. You want to think about how you are going to create each piece such that you maximize the amount of high-precision data within each piece. Understandably, this decision-making process is not always easy, but as you gain experience in scanning differently shaped objects, you will notice improvements in how long it takes you to scan an object as well as the accuracy of the resultant 3D model.

A good strategy involves positioning the object such that you maximize the amount of surface that is captured perpendicular to the laser while minimizing the amount of surface that will not be captured. Using a scanner that has an attachable turntable is advantageous because multiple scans can be acquired, each at a slightly different orientation. For example, if we decide on eight scans, then after the first scan is complete the turntable rotates 45° (360/8) and begins the second scan. After scanning is complete, it is a good idea to inspect each scan and delete any data that do not belong to the object. For instance, if you used plasticine or modeling clay to stabilize the object, or if you notice any data that appears odd, such as curled edges or proportionately large triangles, then you will want to delete it from the model.

After cleaning the scans, each of them now consists of a collection of coordinates (x, y, z values) and topological information (how the coordinates are connected). Consecutive scans share a certain amount of scanned surface with each other. This overlapping surface information is what is used to align the scans with one another. Many 3D software packages include the ability to select points on the surface that are shared by two scans. An algorithm, usually some variant of the ICP (iterative closest point) algorithm (Besl and McKay 1992), is then used to align the scans with one another. This process is continued until all the scans are aligned. Alternatively, laser scanners that utilize turntables are advantageous in that if the software knows how far away the turntable is from the scanner then the different scans can be aligned with one another automatically—an important time-saving feature that also often reduces errors that result from manually aligning multiple scans with one another.

After initial alignment, reinspect the scanned data for any inconsistencies that are not apparent on the actual object and edit the individual scans accordingly. Once you are satisfied with the aligned set of scans, you can merge the scans together. Merging involves using an algorithm to compute a final nonoverlapping mesh to represent the entire scanned surface both geometrically and topologically. Merging algorithms typically generate an average surface based on overlapping scanned areas. Now you reposition the object and go through the same steps in order to generate a second mesh that includes areas that are missing from the first mesh and vice versa. These two meshes are then aligned and merged as above, resulting in a 3D model of the object.

If there are still large areas of surface missing from your model, then you need to take additional scans following the same steps as above in order to fill in these "holes" in the surface mesh. Alternatively, you may choose to simply fill small holes in the mesh using a hole-filling algorithm. After you have deleted any unwanted polygons that do not correspond to the object's surface, you may wish to make some general improvements to the mesh overall such as global smoothing. Always keep in mind that your acquired 3D model is an approximation of the actual object. Do not fall into the trap of thinking that the initially acquired point cloud and surface mesh data are necessarily the most precise numerical representation of the actual object. The precision of the 3D data is dependent not only on the particular laser scanner you use, but also the software and algorithms that are used during the various steps that are necessary to generate a final model, as well as your levels of expertise in performing each step. Additional modeling procedures such as hole-filling and smoothing will often result in a 3D model that is better for visualization and analytical purposes, and may in fact also be a more accurate numerical representation of the actual object. For example, if the sampling error of the depth measurements from the surface is truly random, then the acquired points will tend to always fall around the actual surface rather than directly on it. This will result in a modeled surface that appears rougher (or "noisier") than the actual surface. By applying a smoothing algorithm to the acquired point data, the sampling error is averaged out, resulting in a modeled surface that more accurately represents the actual surface.

When unsure, however, run experiments to empirically determine how any scanning or modeling procedures you are using are affecting the 3D model and the measurements being derived from it. For instance, scan the same object multiple times and calculate the same measurement from the resulting 3D models. You will immediately discover how precise your selected measurement is relative to your acquisition and measuring procedures, and you can use this information to determine whether the measurement will be useful in the selected comparative context. The comparative context is an important distinction that separates the use of laser scanning for typical biological research rather than for reverse engineering purposes. In the latter, the goal is often to produce a replica that is as close as possible to the original; therefore, the laser scanned 3D data must be simultaneously precise and accurate. However, because variation is pervasive in biology, accuracy in measurement is often more critical than precision. For example, in biological research it is often less important to determine precisely that the area of a specific surface of one individual is exactly $10.03 \pm 0.01 \, \text{cm}^2$, and more important to determine that the mean areas of a surface differ significantly between two or more groups. In other words, measurement errors should be random and proportionately small enough not to have a significant effect on the overall accuracy of the results.

Using Laser Scanned 3D Data

As if choosing an appropriate laser scanner and then figuring out the best scanning protocol were not challenging enough, the biggest hurdle facing individuals who want to incorporate laser scanning into their research is working out what to do after they have acquired their 3D models. Just as there are a variety of laser scanners available on the market, there are also countless commercial software programs and applications to work with laser-scanned 3D data. Unfortunately, there is no one-size-fits-all category for 3D software either. Therefore, for those interested in biological research, most often the only solution involves serious compromise.

Commercial 3D software is invariably geared toward the reverse engineering or computer animation markets. While many of these are extremely powerful programs, none have been designed specifically with the biological researcher in mind. This is not surprising given the fact that Hollywood production and automobile manufacturing companies have considerably bigger operating budgets than your typical biological researcher. How should the researcher face this challenge? My general advice on this matter is to become familiar with as many 3D programs as possible. In other words, try to diversify as much as you can. Note also that I use the word "familiar" rather than "expert." By familiar I mean that you are capable of opening and saving a 3D file within the software environment, and that you are comfortable navigating through some of the specific options available in that environment. You will soon discover that every program has different capabilities and some are better suited for certain tasks over others. As an example, some researchers have been very successful in using geographic information system (GIS) software to analyze their laser-scanned 3D models (Ungar and Williamson 2000; M'Kirera and Ungar 2003; Ungar and M'Kirera 2003; Ungar 2004). For your own research needs, do not be surprised if you find yourself using multiple programs. In fact, mixing and matching to best suit your research needs will enable you to capitalize on the strengths of each program while avoiding particular weaknesses.

In order to use various 3D programs, you will need to become familiar with multiple 3D file formats and their corresponding three-letter extensions (e.g., .stl, .obj, etc.). Do not let the barrage of file formats available intimidate you. In most cases, these different file formats are simply different ways of organizing the various components of a 3D model in a textual form. Remember that a 3D model is a numerical description of an object. Exactly how this numerical description is written down as textual data corresponds to a file format (Fig. 4.3). A basic 3D model consists of x, y, z coordinate data only. Thus, written as text, such a 3D model is simply a list of the numerical values for each coordinate. More complex 3D models must include text that describes additional model information such as faces, normals, and sometimes even texture (Fig. 4.4). As you may imagine, there is more than one way of listing all of this information as text, and different 3D file formats represent different ways of accomplishing this task. Sometimes the only major difference between two types of 3D file formats is how information is listed in the header, which is simply text that appears above the actual numerical description.

It is important to recognize exactly what model information each 3D file format contains, so that you understand what data are lost when you convert from one format to another. For example, if you take any polygonal 3D file format (e.g., .stl, .obj, .ply, etc.) and save the file as a point cloud or vertex only file, then you will lose all of the information except for the x, y, z coordinate values. If you do not know how a particular file format is organized, try opening the file using a simple text editor, such as Notepad or Wordpad, or search the Internet for documentation on the specific file structure. Knowing the structure of the file format allows you to either

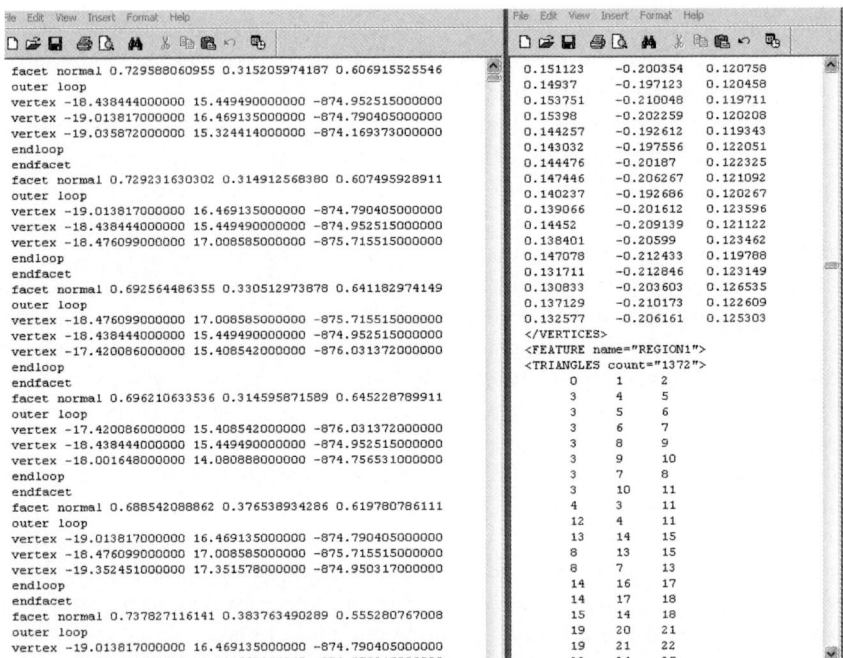

Fig. 4.3 Two examples of how 3D data is stored as a text file. On the *left*, each triangle is described by listing each vertex (i.e., point) that belongs to it as well as its normal vector. Note that each vertex belongs to more than one triangle. On the *right*, the file consists of an ordered list of vertices followed by a list of the vertices that together form triangles. Note that this structure also indicates which triangles belong to a given region of the 3D model by listing the triangles within a "FEATURE" name

write your own analytical routines using programs such as MATLAB, or to better communicate exactly what you are trying to accomplish using your 3D models to programmers you collaborate with. The bottom line is that it is a good idea to always have a clear understanding how the raw 3D data of your models directly relate to the metrics you are interested in quantifying.

Putting All the Steps Together

There are now many examples of laser scanning being utilized in biological research, including forensics (Park et al. 2006), medicine (Byrne and Garcia 2007; Da Silveira et al. 2003; Hennessy et al. 2005; Wettstein et al. 2006), and physical anthropology (Aiello et al. 1998; M'Kirera and Ungar 2003; Tocheri et al. 2003, 2005, 2007; Ungar 2004; Ungar and Williamson 2000; Ungar and M'Kirera 2003). A majority of the studies that have been published or presented at meetings thus far, however, still principally deal with the potential applications of laser scanning to their respective biological discipline. I think it is safe to say that laser scanning is a useful

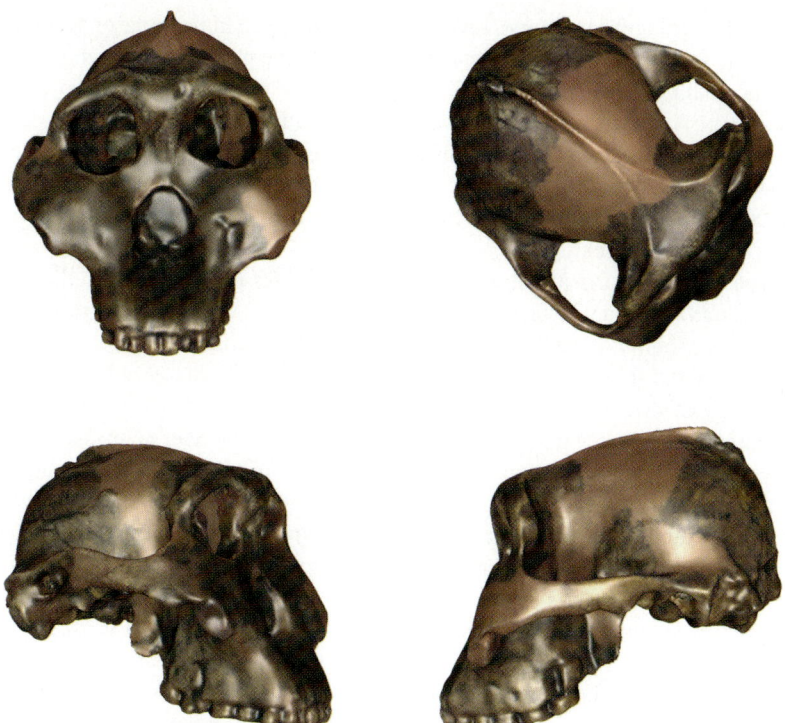

Fig. 4.4 Many laser scanners are also capable of acquiring texture information in addition to the geometry of the object. The texture data are superimposed onto the geometry, enhancing the visualization experience. Shown here are four views of laser scans of a cast of OH 5, a fossil cranium belonging to the extinct hominin species *Paranthropus boisei*. Note that the texture information provides additional surface detail (e.g., the lighter brown areas indicate which parts of the fossil were reconstructed)

technology that has incredible potential for different areas of biological research, but the onus falls on interested students and researchers to more fully develop the ways in which they use this technology to answer specific research questions. Since the primary readers of this chapter probably have little or no experience working with laser scanners and laser-scanned data, it is probably useful to provide a brief overview of my own experiences with using laser scanning in biological research.

My first introduction to laser scanning occurred in 2001 when I was a first year graduate student in physical anthropology at Arizona State University (ASU). I attended a public seminar that summarized how researchers from a variety of disciplines at ASU were using methods such as laser scanning to acquire 3D data for research purposes. I was immediately struck by the idea of transforming real world objects into digital information that could then be visualized and analyzed using computers. To make a long story short, I arranged to see one of these laser scanners in action and very quickly made the decision that I wanted to incorporate this technology into my graduate student research.

As a physical anthropologist, I was initially interested in using laser scanning to acquire 3D models of human skeletal material with the goal of developing methods to better quantify biological variation. This interest soon expanded into a larger evolutionary framework wherein I wanted to quantitatively compare the skeletal morphology between humans and our closest living relatives, the great apes. Such comparative data could then be used to evaluate the morphology of fossil hominids (extinct members of the human–great ape family Hominidae) in both a functional and evolutionary context.

I began to brainstorm about what laser scanning was giving me (i.e., 3D coordinate data and resulting surface representation) and how I could use it to calculate metrics that would capture shape differences. I now recognize this brainstorming stage as a critical step that everyone must perform if they want to use laser scanning successfully in their research. It is imperative that you figure out exactly how the data you want to collect from your laser scans relate to the data your laser scanner gives you. Remember that your 3D model is a numerical representation and that any metric you try to quantify from it is some function of the numbers behind the model. For example, if your laser scanner gives you a triangular mesh, then you can quantify the surface area of the mesh (or any portion thereof) by summing the areas of each included triangle.

For my dissertation research, I investigated shape differences in wrist anatomy between humans, great apes, and fossil hominins (Tocheri 2007). As part of this research, I laser scanned 1,250 hand and wrist bones from more than 300 individuals. After I had acquired my study sample of 3D models, my goal was to quantify two aspects of wrist bone shape that could be measured because of the numerical properties of each model. The two selected metrics were the relative areas of each articular and nonarticular surface, and the angles between articular surfaces.

In order to generate these two metrics, I first needed to identify which points belonged to each articular and nonarticular area. There are many commercial 3D programs that allow you to select and label particular regions of a model, and there are also particular 3D file formats (e.g., .obj) that will retain region information within the file structure (see Fig. 4.3). In other words, I was able to transform each initial 3D model into a segmented 3D model. The former consisted of an unorganized list of x, y, and z coordinate values and the triangles to which each coordinate belonged, whereas the latter added to which articular or nonarticular region each coordinate and triangle belonged.

Using these segmented 3D models, surface areas of the mesh were quantified by summing the areas of each included triangle. I calculated relative areas by dividing each articular or nonarticular area by the surface area of the entire bone, resulting in scale-free shape ratios (Jungers et al. 1995; Mosimann and James 1979). To quantify the angle between two articular surfaces, a least-squares plane was fit to each articular surface by performing a principal components analysis on the coordinates of each surface. Each angle was calculated as $180°$ minus the inverse cosine of the dot product of the normal vectors of the two respective least-squares planes (note the eigenvector associated with the smallest eigenvalue is the normal vector of each least-squares plane). Notice that both of these metrics relate specifically to the surface information acquired by laser scanning and are easily quantified because of the numerical properties of a 3D model.

Using these two metrics, I was able to statistically demonstrate major quantitative shape differences in wrist bone morphology between humans and great apes (Tocheri 2007; see also Tocheri et al. 2003, 2005). Moreover, these multivariate analyses demonstrated that Neandertals and early humans show wrist bone shapes that are characteristic of modern humans, whereas earlier hominins, such as species of *Australopithecus* and early *Homo*, show wrist bone shapes characteristic of African apes (Tocheri 2007). After my dissertation, this comparative dataset became a critical component of a collaborative study on the wrist remains of *Homo floresiensis* (Tocheri et al. 2007)—the so-called "hobbits" of hominin evolution.

One of the potentials of incorporating laser scanning in biological research involves the ability to combine a statistical evaluation of biological shape properties along with simple yet extremely informative visualization techniques for sharing the analytical results. It is this particular potential that I and my collaborators used to demonstrate that LB1, the holotype specimen of *Homo floresiensis*, retains wrist morphology that is primitive for the African ape–human clade (Fig. 4.5). For example, Fig. 4.5 shows our comparison of the trapezoid, the wrist bone situated directly proximal to the index finger. Each row corresponds to a particular view of trapezoid

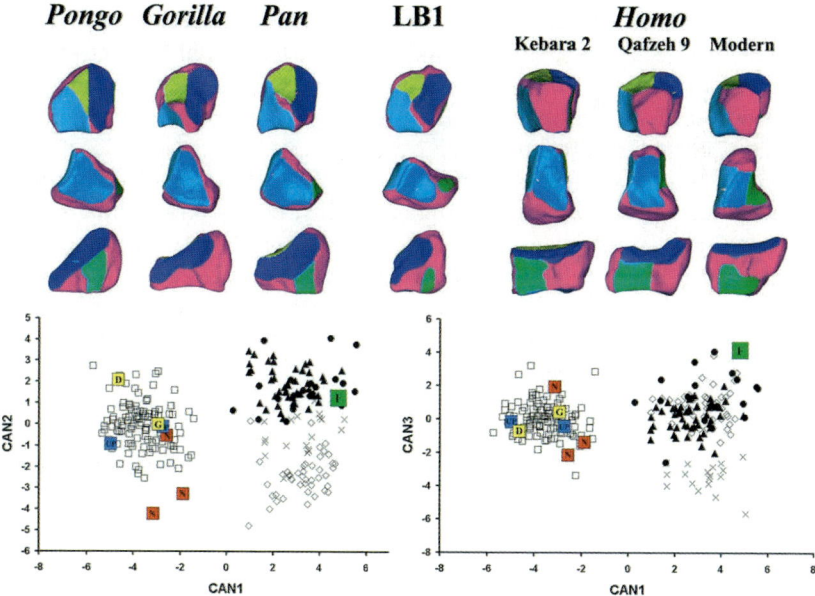

Fig. 4.5 Results of a comparative analysis of trapezoid morphology in modern and fossil hominids using laser-scanned 3D models. *Above*: each row corresponds to a particular view of trapezoid anatomy obtained using the segmented 3D models that illustrate the different articular and nonarticular surfaces that were quantified in the analysis. *Below*: the results of the statistical analysis are summarized in canonical plots generated from a quantitative comparison of the relative areas and angles (see Tocheri et al. 2007 for the full caption)

anatomy using laser-scanned and segmented 3D models that illustrate the different articular and nonarticular surfaces that were quantified in the analysis. Below the visual comparison of the segmented 3D models, the results of the multivariate statistical analysis are summarized in canonical plots generated from the quantitative comparison of the relative areas and angles. Without laser scanning and the ability to work with and analyze the resulting 3D models, a figure such as this, which effectively and succinctly summarizes years of research, would not be possible. The surface visualization informs the statistical analysis and vice versa.

This is only a single example of what can be done with laser-scanned models in biological research, but I hope it helps illustrate the main theme of this chapter: if you want to incorporate laser scanning into your research, be prepared to think creatively and critically at every stage of the process, from initial data acquisition to analysis and presentation of the results.

Summary and Conclusions

In this chapter, I have reviewed some basic concepts, including what laser scanning is, why it should be used for biological research, how to choose a laser scanner, how to use a laser scanner to acquire 3D data, and what to do with the 3D data after they have been acquired. In my experience, these are the kinds of questions that students and researchers most often ask when they are first introduced to laser scanning. My answers to these questions should be used as a rough introductory guide to help interested students and researchers make decisions relating to laser scanning and their short and long-term research goals.

There are many practical challenges surrounding laser scanning that relate to equipment and software availability and capability, but these can often be overcome with imaginative and creative solutions. However, do not make the mistake of thinking that laser scanning will solve all of your research objectives. Always keep in mind that laser scanning, like pool, often looks a lot easier than it is. Unless you are willing to dedicate a significant portion of your research time and effort to solving the many curves that laser scanning will throw at you, then it will probably not serve the purpose you hope it will. If, on the other hand, you are willing to put in the necessary time and effort it takes to respond to the many challenges involved in incorporating laser scanning into biological research, then it can be highly rewarding and effective method for accomplishing your research objectives.

References

Aiello L, Wood B, Key C, Wood C (1998) Laser scanning and paleoanthropology: an example from Olduvai Gorge, Tanzania. In: Strasser E, Fleagle J, Rosenberger A, McHenry H (eds) Primate locomotion. Plenum, New York, pp 223–236

Bernardini F, Rushmeier H (2002) The 3D model acquisition pipeline. Comput Graph Forum 21:149–172

Besl P, McKay N (1992) A method for registration of 3-D shapes. IEEE Trans Pattern Anal Machine Intell (PAMI) 14:239–256

Byrne PJ, Garcia JR (2007) Autogenous nasal tip reconstruction of complex defects: a structural approach employing rapid prototyping. Arch Facial Plastic Surg 9:358–364

Da Silveira AC, Daw JL, Kusnoto B, Evans C, Cohen M (2003) Craniofacial applications of three-dimensional laser surface scanning. J Craniofacial Surg 14:449–456

Dennis JC, Ungar PS, Teaford MF, Glander KE (2004) Dental topography and molar wear in *Alouatta palliata* from Costa Rica. Am J Phys Anthropol 125:152–161

Hennessy RJ, McLearie S, Kinsella A, Waddington JL (2005) Facial surface analysis by 3D laser scanning and geometric morphometrics in relation to sexual dimorphism in cerebral-craniofacial morphogenesis and cognitive function. J Anat 207:283–295

Jungers WL, Falsetti AB, Wall CE (1995) Shape, relative size, and size-adjustments in morphometrics. Yearbook Phys Anthropol 38:137–161

Kappelman J (1998) Advances in three-dimensional data acquisition and analysis. In: Strasser E, Fleagle J, Rosenberger A, McHenry H (eds) Primate locomotion. Plenum, New York, pp 205–222

M'Kirera F, Ungar PS (2003) Occlusal relief changes with molar wear in *Pan troglodytes troglodytes* and *Gorilla gorilla gorilla*. Am J Primatol 60:31–42

Mosimann JE, James FC (1979) New statistical methods for allometry with application to Florida red-winged blackbirds. Evolution 33:444–459

Park HK, Chung JW, Kho HS (2006) Use of hand-held laser scanning in the assessment of craniometry. Forensic Sci Int 160:200–206

Rowe J, Razdan A (2003) A prototype digital library for 3D collections: tools to capture, model, analyze, and query complex 3D data. Museums and the Web 2003, Charlotte, NC (see http://www.archimuse.com/mw2003/papers/rowe/rowe.html)

Tocheri MW (2007) Three-dimensional riddles of the radial wrist: derived carpal and carpometacarpal joint morphology in the Genus *Homo* and the implications for understanding the evolution of stone tool-related behaviors in hominins. Ph.D. dissertation, Arizona State University, Tempe, AZ

Tocheri MW, Marzke MW, Liu D, Bae M, Jones GP, Williams RC, Razdan A (2003) Functional capabilities of modern and fossil hominid hands: 3D analysis of trapezia. Am J Phys Anthropol 122:101–112

Tocheri MW, Razdan A, Williams RC, Marzke MW (2005) A 3D quantitative comparison of trapezium and trapezoid relative articular and nonarticular surface areas in modern humans and great apes. J Human Evol 49:570–586

Tocheri MW, Orr CM, Larson SG, Sutikna T, Jatmiko E, Saptomo EW, Due RA, Djubiantono T, Morwood MJ, Jungers WL (2007) The primitive wrist of *Homo floresiensis* and its implications for hominin evolution. Science 317:1743–1745

Ungar PS (2004) Dental topography and diets of *Australopithecus afarensis* and early *Homo*. J Hum Evol 46:605–622

Ungar PS, M'Kirera F (2003) A solution to the worn tooth conundrum in primate functional anatomy. Proc Natl Acad Sci USA 100:3874–3877

Ungar P, Williamson M (2000) Exploring the effects of tooth wear on functional morphology: a preliminary study using dental topographic analysis. Palaeontologia Electronica 3 (see http://palaeo-electronica.org/2000_1/gorilla/issue1_00.htm)

Wettstein R, Kalbermatten DF, Rieger UM, Schumacher R, Dagorov P, Pierer G (2006) Laser surface scanning analysis in reconstructive rhytidectomy. Aesthetic Plastic Surg 30:637–640

Zollikofer CPE, Ponce de Leon MS (2005) Virtual reconstruction: a primer in computer-assisted paleontology and biomedicine. Wiley, Hoboken, NJ

Chapter 5
Optical Coherence Tomography: Technique and Applications

J.B. Thomsen, B. Sander, M. Mogensen, L. Thrane, T.M. Jørgensen, G.B.E. Jemec, and P.E. Andersen(✉)

Abstract Optical coherence tomography (OCT) is a noninvasive optical imaging modality providing real-time video rate images in two and three dimensions of biological tissues with micrometer resolution. OCT fills the gap between ultrasound and confocal microscopy, since it has a higher resolution than ultrasound and a higher penetration than confocal microcopy. Functional extensions are also possible, i.e., flow, birefringence or spectroscopic measurements with high spatial resolution. In ophthalmology, OCT is accepted as a clinical standard for diagnosing and monitoring the treatment of a number of retinal diseases. The potential of OCT in many other applications is currently being explored, such as in developmental biology, skin cancer diagnostics, vulnerable plaque detection in cardiology, esophageal diagnostics and a number of other applications within oncology.

Introduction

Optical coherence tomography (OCT) is a noninvasive optical imaging technique that has developed rapidly since it was first realized in 1991 (Huang et al. 1991). At present, OCT is commercially available and accepted as a clinical standard within ophthalmology for the diagnosis of retinal diseases. Emerging applications within biology and medicine are currently being explored by many research groups worldwide. Various intense research efforts aimed at technical improvements regarding imaging speed, resolution, image quality and functional capabilities are currently underway.

Optical coherence tomography is often characterized as the optical analog to ultrasound, where light is used instead of sound to probe the sample and map the variation of reflected light as a function of depth. OCT is capable of providing

P. E. Andersen

Department of Photonics Engineering, Technical University of Denmark, Frederiksborgvej 399, Roskilde, DK-4000, Denmark, e-mail: peter.andersen@fotonik.dtu.dk real-time video rate images with a resolution of typically better than $\sim$10–20 μm, and imaging with a resolution of 1 μm or less has been demonstrated (Drexler 2004; Hartl et al. 2001; Povazay et al. 2002). The penetration depth is highly tissue-dependent and is typically limited to a few millimeters. The combination of high resolution and relatively high imaging depth places OCT in a regime of its own, filling the gap between ultrasound and confocal microscopy.

The chapter starts with an introduction to the principle of OCT, which is followed by an overview of different applications. The technical part is aimed at providing an understanding of the feasibilities and limitations of OCT. Following the introduction, a number of important applications are described. The first one is ophthalmology, which is currently the most clinically successful application of OCT. The next application concerns dermatology, with an emphasis on skin cancer diagnostics. Because skin is a highly scattering medium, the interpretation of OCT images of skin is a challenging task. Finally, other applications of OCT, such as in developmental biology and endoscopically in medicine, are briefly discussed.

The Principle of Optical Coherence Tomography

Optical coherence tomography is based on the interference of light, which is related to the coherence properties of light. In contrast to a laser, which radiates almost monochromatic light (one color), OCT employs a broadband light source that emits polychromatic light (several colors). In this section, the principle of operation of OCT is explained in relation to the first implementation, so-called time-domain OCT (TD-OCT), as well as a more effective scanning scheme known as Fourier-domain OCT (FD-OCT), introduced in 1995, which has improved imaging speed and sensitivity. A discussion of resolution, choice of wavelength and functional OCT will then follow.

Time-Domain OCT (TD-OCT)

The principle of OCT can be explained by referring to Fig. 5.1 (Bouma and Tearney 2002; Brezinski 2006; Fercher et al. 2003). A Michelson interferometer, as shown in Fig. 5.1a, can be used to measure the ability of light to interfere with itself, i.e., its ability to amplify or blur itself ("constructive" and "destructive" interference, respectively). Light is split into two paths using a beam splitter (a half-transparent mirror). The light directed against the mirrors is reflected, recombined at the beam splitter, and detected. Interference between the two reflections is possible only when the path lengths of the two arms are matched within the so-called coherence length of the light source. The coherence length is determined by the spectral width of the light; a broad optical spectrum corresponds to a short coherence length, and a narrow optical spectrum to a long coherence length. When a light source with a

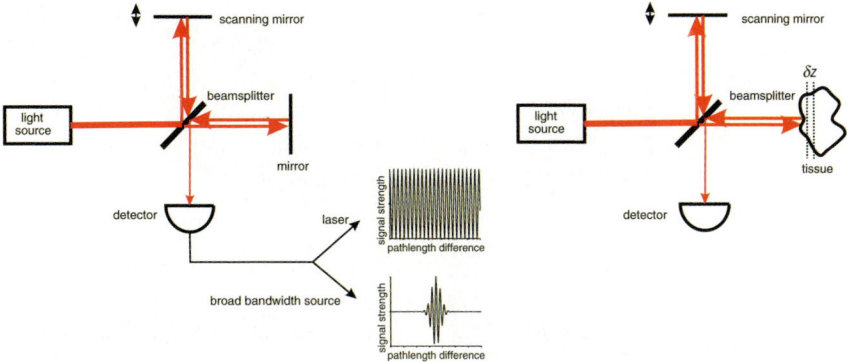

Fig. 5.1 (**a**) Michelson interferometer. (**b**) Michelson interferometer with the fixed mirror replaced by a sample

large coherence length is used, interference arises for even very large differences in path length. Correspondingly, when a source with a small coherence length is used, interference only arises when the two path lengths are matched within the coherence length of the light. It is exactly this effect that is used in OCT to distinguish signals from different depths in the sample. The axial resolution is set by the coherence length, with a small coherence length corresponding to high axial resolution.

Consider one of the mirrors in the Michelson interferometer interchanged with a biological sample as shown in Fig. 5.1b. In this case, every different position of the scanning mirror yields a signal from a different thin slice in the sample. In other words, it is possible to determine the depth from where the reflection originates. The thickness δ_z of the slice that contributes to the signal, see Fig. 5.1b, is equal to the depth resolution of the system and inversely proportional to the bandwidth of the light source (see also the section "Resolution and Sensitivity"). The mechanism for selecting signal from a specific depth is also referred to as coherence gating. By moving the scanning mirror, the coherence gate successively selects interference signals from different depths. This produces a depth scan recording, also referred to as an A-scan. The depth scanning range is limited by the mirror displacement. The transverse resolution is determined by the spot size, which can be changed with the focusing optics. It is important to point out that the transverse resolution and the depth resolution are independent, in contrast to the situation in, for example, microscopy (see also "Resolution and Sensitivity").

Two-dimensional data is obtained by moving the beam across the sample and acquiring data (B-scan). By translating the beam in two directions over a surface area it is possible to acquire three-dimensional data. The time it takes to acquire a B-scan image is set by the time it takes to acquire an A-scan. A B-scan consisting of n A-scans is acquired in the time $n \times t$, where t denotes the time needed to acquire an A-scan. In general, two- and three-dimensional data can be acquired in real time; see the section "Fourier-Domain OCT (FD-OCT)." Obtaining an image parallel to the surface at a certain depth in the sample is known as *en face* imaging.

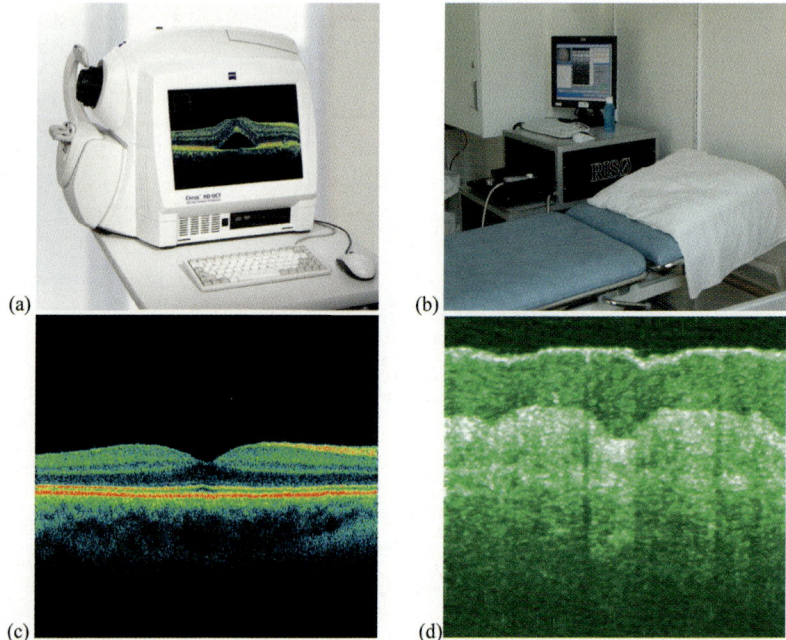

Fig. 5.2 (**a**) Zeiss Cirrus OCT system tailored for ophthalmology (courtesy of Carl Zeiss Meditec AG). (**b**) Prototype OCT system for use in a clinical environment. (**c**) OCT image of the healthy retina in a human eye showing distinct retinal layers. (**d**) OCT image of human skin (*palmar aspect, thumb*) with a clear visible boundary between the stratum corneum (*dark upper layer*) and the living part of the epidermis, and a visible change from a high-intensity band (*epidermis*) to a darker area (*dermis*)

The interference signal is amplified, filtered to improve the signal-to-noise ratio, and then digitized and transferred to a computer. The reflection strength is extracted from the digital signal and mapped using either a grayscale or color palette. Free-space optical beam propagation (as sketched in Fig. 5.1) is possible, even though optical fiber-based systems are more convenient for clinical use. Figure 5.2a shows a commercial OCT system for ophthalmologic diagnostics (Zeiss), Fig. 5.2b shows a fiber-based prototype OCT system in a clinical environment, Fig. 5.2c an OCT image of the human retina, and Fig. 5.2d depicts an OCT image acquired on human skin.

Fourier-Domain OCT (FD-OCT)

In the time-domain scheme, an A-scan is acquired by successively changing the depth of the coherence gate, recording interference at each depth. A more efficient scheme for acquiring interference data was suggested and first realized in 1995 (Fercher et al. 1995), which involves detecting the interference from all depths in the sample simultaneously. This method is referred to as Fourier-domain OCT (FD-OCT) and can be further divided into spectrograph-based systems, denoted spectral-domain OCT (SD-OCT), and swept-source OCT (SS-OCT), which employs a scanning laser and single detector (note that there is unfortunately no naming convention, so FD-OCT and SD-OCT are frequently interchanged). The light source used for SD-OCT is the same as that used for TD-OCT (broadband), but the reference mirror is fixed. A spectrograph is used for detection at the interferometer output, i.e., light is dispersed using a grating for example, or alternatively it is split into single wavelengths, and every component is detected with a separate detector (array detector). Signal originating from a given depth in the sample corresponds to an oscillation in the array detector with a specific frequency. In other words, the depth information is now frequency-encoded at the array detector. By performing a frequency analysis, the detected signals from all depths can be divided into the depth-dependent signal. The depth scanning range is basically limited by the resolution of the spectrograph, which is closely related to the number of elements on the array detector. The frequency at the detector array increases with depth. Because the detector samples the signal in equal wavelength intervals $\delta\lambda$, the corresponding wave number interval $\delta k \propto k^2 \delta\lambda$ increases with wavenumber or equivalently with imaging depth. This uneven sampling results in a degradation of the resolution with depth if the detector signal is used directly. Nonlinear scaling algorithms are needed to correct for this effect (Brezinski 2006).

In SS-OCT, the broad-bandwidth light source is exchanged for a laser that can scan over a range of wavelengths. When only a single wavelength is shone onto the sample at a time, only a single detector is needed. By changing the laser wavelength with time or equivalently changing the wavelength by a constant rate, the frequency of the detected signal only depends on the path length difference between sample and reference arm, i.e., the sample depth. Therefore, a frequency analysis of the detector signal gives the reflection strength as a function of depth. The depth scanning range is determined by the line width of the scanning laser, because this determines the coherence length of the light. Comparing the two FD-OCT schemes, SD-OCT can be regarded as the superposition of many single-wavelength interferometers operating in parallel, whereas SS-OCT operates with one wavelength at a time. The advantage of FD-OCT systems is their simultaneous detection of interference from all depths, which can be used to increase the speed or sensitivity of the technique (Choma et al. 2003). However, there is some controversy regarding this issue. A recent analysis concluded that TD-OCT theoretically has the highest sensitivity, followed by SS-OCT and SD-OCT (Liu and Brezinski 2007).

While time-domain OCT is limited by the speed of the scanning mirror, FD-OCT is limited by the read-out speed of the detector array or the laser sweep rate, respectively. The highest A-scan rate achieved is currently 370 kHz, which was obtained with a swept source (Huber et al. 2005). This A-scan rate is far beyond those of time-domain systems, which typically operate at on the order of 10 kHz. With an A-scan rate of 370 kHz, a B-scan image consisting of 500 A-scans can be acquired in about 1 ms, corresponding to a frame rate of roughly 1,000 Hz, i.e., 1,000 images per second. Therefore, the focus is now mainly on FD-OCT systems. The fast imaging rate allows 3D data to be acquired with limited motion artifacts, which has been used in ophthalmology for example to visualize a macular hole in 3D (Hangai et al. 2007).

Resolution and Sensitivity

As mentioned above, the depth resolution is dependent on the bandwidth of the light source. Assuming a Gaussian spectrum, the depth resolution δ_z is defined as $\delta_z = 0.44\lambda^2/(n\Delta\lambda)$, with λ denoting the center wavelength of the light source, n the index of refraction of the sample, and $\Delta\lambda$ the spectral bandwidth of the source (full width at half maximum). This equation is valid for both time- and spectral-domain systems, even though the number of elements on the array detector in practice can limit the resolution in SD-OCT. When the depth resolution is a few micrometers or less, the regime is denoted *ultrahigh resolution OCT* (UHR OCT), and such resolutions have been obtained in several OCT implementations (Drexler 2004).

The transverse resolution δx is determined by the spot size of the optical beam $\delta_x = 4\lambda f/(\pi d)$, where f is the focal length of the lens used for focusing and d is the diameter of the beam at the lens (Hecht 1998). As a unique property compared to for example confocal microscopy, the transverse and depth resolutions are independent (depth resolution depends on the bandwidth of the source; transverse resolution depends only on focusing optics). This property makes imaging with a high depth resolution possible with a relative large working distance; which is necessary for example when imaging the human retina. The spot size—and therefore the transverse resolution—varies with depth on a scale known as the depth of focus $z_d = \pi\delta x^2/2\lambda_0$. From this formula, it is clear that a higher transverse resolution results in a smaller depth of focus. Thus, a higher transverse resolution is associated with a faster degradation of transverse resolution with imaging depth. In other words, there is a trade-off between transverse resolution and the maximum imaging depth in OCT. Therefore a moderate transverse resolution is often used, allowing a reasonable maximum imaging depth. In addition to this inevitable degradation of transverse resolution with depth, light scattering also causes degradation with depth, which is a well-known effect in microscopy. The trade-off between imaging depth and transverse resolution can be overcome by overlapping the coherence gate and the position of the spot during a depth scan, in so-called dynamic focusing. However, this is only possible within the TD-OCT regime because FD-OCT acquires all points in an A-scan at the same time. Using an axicon lens, it is possible to maintain a small spot over a larger depth, which to some extent overcomes the problem (Leitgeb et al. 2006). When using a high transverse resolution, imaging in a plane parallel to the surface is convenient, since then only slow dynamic focusing is needed. This is known as *en face* scanning and, when combined with high transverse resolution, is referred to as optical coherence microscopy (OCM). This is analogous to confocal microscopy. Notice, however, that fluorescence, which is often the main contributor to the signal in confocal microscopy, is not registered using OCT because fluorescent light is incoherent. The advantage of OCM compared to confocal microscopy is an increased imaging depth, because multiply scattered out-of-focus light is partly filtered out using the interference principle in OCT (Izatt et al. 1996).

The sensitivity of an OCT system is a measure of the weakest signals that can be detected relative to the input, and it therefore relates to the maximum imaging depth. There is a trade-off between imaging speed and sensitivity. Most OCT systems achieve sensitivity in the range of 90–100 dB, which means that signals as small as 10^{-9}–10^{-10} of the input can be detected.

In Fig. 5.3, the resolution and penetration of OCT and OCM are sketched together with those of confocal microscopy and ultrasound. Confocal microscopy has superior resolution but very limited penetration, whereas ultrasound has high penetration but coarser resolution. OCT and OCM fills the gap between ultrasound and confocal microscopy, having a greater penetration than confocal microscopy and a higher resolution than ultrasound. Furthermore, OCT does not need physical contact with the sample. Upon comparing OCT to other imaging modalities, their lack of resolution compared to OCT is striking (consider magnetic resonance imaging and X-ray imaging for instance). Another advantage of OCT is the ability to produce very small probes, which is highly relevant to endoscopic use, where a fiber-optic implementation is also feasible.

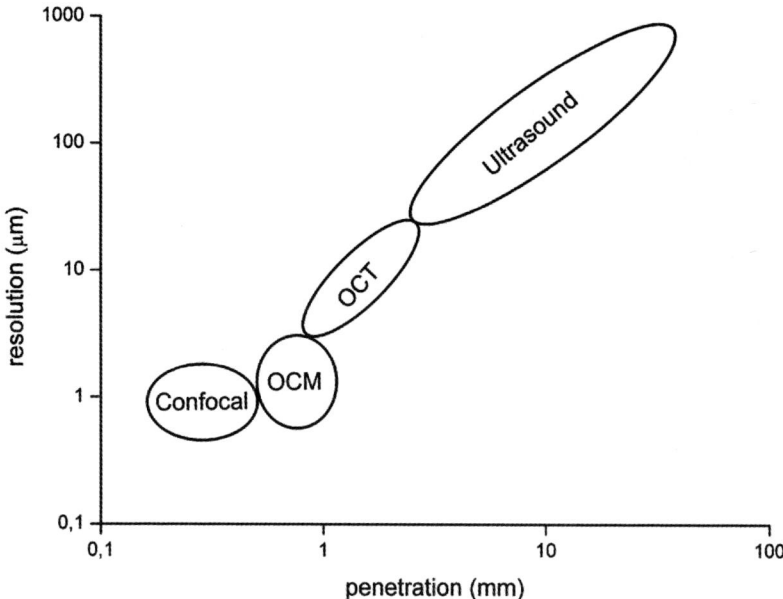

Fig. 5.3 OCT and OCM compared to ultrasound and confocal microscopy regarding resolution and penetration (note the log scale on both axes)

Choice of Light Source

As already mentioned, the light source for OCT needs to emit a broad spectrum, because this is equivalent to a short coherence length and thus a high depth resolution. Optical power, spectral shape and noise characteristics are also important parameters (Brezinski 2006; Fercher et al. 2003).

Commercial turnkey sources are available, but only with a bandwidth that allows resolution on the order of $\sim$10 µm. Sources with larger bandwidths are desirable and achievable on a research basis, i.e., using more complicated and expensive setups that are not yet suitable for clinical use. As an example, a femtosecond laser pumping a photonic crystal fiber has been used to achieve a resolution of less than 1 µm (Povazay et al. 2002). Note that, in this context, the same light sources can be used for both TD-OCT and SD-OCT. A scanning laser is employed for SS-OCT, which is also commercially available with a scanning range corresponding to a depth resolution of $\sim$10 µm and a limited speed compared to state-of-the-art sources developed in the research area. Compact and low-cost broadband sources (including scanning lasers) are currently a highly active research area.

The choice of wavelength has an important influence on the imaging depth and depends on the application. When light is propagating through tissue it is attenuated by scattering and absorption. Light scattering decreases with wavelength, which is an argument for choosing longer wavelengths. However, the absorption of light due to the water content can be considerably higher at longer wavelengths, and so there is a trade-off when choosing the wavelength. Another issue is dispersion, which is the difference in the velocity of light for different wavelengths, which results in a degradation in the depth resolution. In the case of water, for example, the dispersion is zero for a wavelength of about 1,050 nm (Hillman and Sampson 2005). To image the retina, center wavelengths in the range of 800–1,100 nm are desirable for minimizing water absorption and dispersion in the outer part (Drexler et al. 1998). In highly scattering tissue, such as skin, center wavelengths in the range of 1,300 nm are typically employed, lowering scattering losses.

Functional OCT

Up to this point we have only considered the information represented by the amplitude of backscattered light, also known as the structural OCT image. Functional OCT adds additional information extracted from the measured signal. This includes flow measurements (Doppler OCT), birefringence detection (polarization-sensitive OCT) or spectroscopic information (spectroscopic OCT). Functional OCT can provide valuable knowledge of the sample that is not present in the structural OCT image.

Doppler OCT (DOCT)

The first system implementations acquired the velocity of the scatterers by analyzing a single A-scan (Chen et al. 1997; Izatt et al. 1997). Later, another method was introduced that acquired more A-scans in the same transverse position. Upon comparing these A-scans, the velocity of the scatterers can be found by finding the displacement of the signal assuming that the time between two A-scans is known. This method is known as sequential A-scan processing, and the advantage it has over the former technique is an increased imaging speed, which allows real-time Doppler imaging while retaining the capability of sub-mm s^{-1} velocity sensitivity (Zhao et al. 2000).

Flow information is also obtainable using laser Doppler imaging, but unlike DOCT there is no depth resolution (Stucker et al. 2002). Using Doppler OCT (DOCT) it is, for example, possible to rapidly visualize the embryonic vascular system (Mariampillai et al. 2007). Even though it is the velocity parallel to the beam direction that is measured, an absolute velocity determination without prior knowledge of the flow direction has been achieved by analyzing the backscattered light spectrum (Wu 2004). The flow information can be mapped in the same way as structural OCT images using a grayscale palette, but it is typically shown in colors. More commonly, however, DOCT is superimposed on the structural OCT image, such that the flow information is shown in color and the structure in grayscale, making it easy to assess flow relative to structure. Blood flow in the retinal vessels in the human retina has been demonstrated (Yazdanfar et al. 2000). DOCT is expected to be important for the diagnosis of a number of retinal diseases including glaucoma and diabetic retinopathy, but research in this area is still at an early stage (van Velthoven et al. 2007).

Polarization-Sensitive OCT (PS-OCT)

In general, light can be described as a transverse wave that allows two possible independent directions for the plane of vibration. The direction of the vibrational plane is known as the polarization plane (Hecht 1998). The velocity of light in a medium is usually independent of the polarization. However, in some materials the velocity depends on the polarization; these are referred to as "birefringent" materials, or are said to exhibit birefringence (two indices of refraction). In biological tissues, it is usually highly organized tissue (for example, collagen fibrils) that exhibits birefringence. The net effect of the propagation of light through a birefringent material is a change in the direction of polarization with propagation distance. Polarization-sensitive OCT (PS-OCT), first demonstrated with free-space optics in 1992, is capable of measuring this change (Hee et al. 1992). Fiber-based polarization-sensitive systems were later demonstrated (Saxer et al. 2000). PS-OCT can be used to identify tissue that is birefringent and as a diagnostic tool for diseases that break up the ordered structure, resulting in a loss of birefringence. In Brezinski

(2006), examples and a more detailed interpretation of PS-OCT images is given. In ophthalmology it has been suggested that PS-OCT could be used for a more precise determination of the retinal nerve fiber layer thickness, which is important for glaucoma diagnosis (Cense et al. 2004).

Spectroscopic OCT

Light absorption at specific wavelengths is a signature of the sample composition. The use of a broadband light source in OCT makes absorption profile measurements of the sample possible, which is denoted spectroscopic OCT (Morgener et al. 2000). Spectroscopic OCT combines spectroscopy with high-resolution imaging (enabling mapping of the chemical structure), and becomes increasing relevant with broader-bandwidth sources since it is more likely to coincide with certain relevant or desired sample absorption spectra. The technique has been used to assess the oxygen saturation level in blood (Faber et al. 2005).

Because OCT typically operates in the infrared region, elastic scattering is actually more prevalent than absorption. Imaging separately with two different wavelengths is another spectroscopic approach; this technique probes wavelength-dependent scattering. A differential image can be constructed with increased contrast from these separate images (Spoler et al. 2007).

Optical Coherence Tomography in Ophthalmology

OCT was first applied in ophthalmology, and its use has expanded rapidly since the introduction of the first commercial system in 1996. Eye diseases are common and early diagnosis is important in many cases in order to avoid visual decline, making high-resolution imaging relevant. Today OCT is widely used clinically because it provides *in vivo* images of the retinal layers with higher resolution than possible with any other technique. In order to image the retina and the optic nerve, penetration through the outer $\sim$0.5 mm cornea is necessary, which requires a relatively weak light focus (a long working distance), resulting in moderate transverse resolution. With confocal microscopy, a weak focus inevitable leads to low depth resolution. On the other hand, when OCT is used, the transverse and depth resolutions are independent (see "Time-Domain OCT (TD-OCT)", and imaging with moderate transverse resolution does not affect the depth resolution, since this is determined by the bandwidth of the light source. Therefore, OCT is capable of imaging the retina *in vivo* with high depth resolution. Furthermore, no physical contact with the eye is needed, in contrast to (for example) ultrasound.

Due to the commercialization and use of OCT as a clinical standard, this section focuses primarily on the use of OCT in the clinic. The ongoing technical development of OCT for ophthalmology is described in the last part of this section.

Commercial OCT Systems for Ophthalmology

The first commercial OCT system that was applied to ophthalmology was introduced to the market in 1996 (Carl Zeiss Meditec), and subsequent generations of systems have improved resolution, speed and image handling software (see Fig. 5.2a for the newest Zeiss Cirrus OCT system). More than 9,000 systems have been sold by Zeiss (http://www.zeiss.ca), and more manufacturers have entered the market. OCT is currently used in larger clinics, and it is spreading to smaller clinics as well. Because the cornea mainly consists of water, a wavelength of 800 nm has been preferred in order to minimize absorption (see "Choice of Light Source"). The axial resolution is typically 5–10 μm, sufficient to distinguish different layers in the retina.

Retinal structures inaccessible with any other techniques can be detected with OCT, which is the reason for its success in ophthalmology. OCT is the clinical standard for a number of retinal diseases, and is used on a daily basis in the clinic for the diagnosis, monitoring and treatment control of, e.g., macular holes, age-related macular degeneration, glaucoma and diabetes. A recent review with a more complete description of ophthalmic OCT can be found in van Velthoven et al. (2007).

Routine Examinations of the Retina Using OCT

Visualization of the layered structure of the retina is possible, as illustrated by the OCT image of a healthy retina in Fig. 5.4. The fundus image of the retina shows the fovea in the middle—the place with maximum visual resolution or power that is used for example when reading. The right part of the image contains the optic nerve, with nerve fibers and blood vessels running in and out of the retina. The white bar corresponds to the position of the OCT scan shown in Fig. 5.4b (6 mm, 512 A-scans). The innermost layer of the retina is the nerve fiber layer (top Fig. 5.4), where

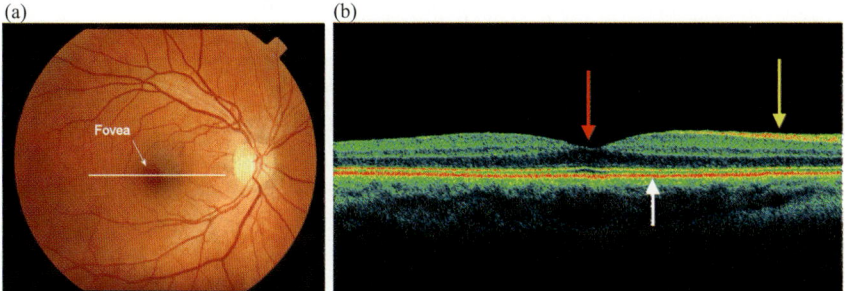

Fig. 5.4 (**a–b**) Images of a normal retina. (**a**) Fundus image with the *white line* indicating the position of the OCT scan. (**b**) OCT image corresponding to the position of the white line in (**a**) (6 mm wide). *Red arrow*, position of fovea; *yellow arrow*, nerve fiber layer; *white arrow*, retinal pigment epithelium

nerve fibers carrying the signal from the photoreceptors combine in a layer linked to the optic nerve. The fovea contains the largest density of photoreceptors and is also the thinnest part of the retina, being approximately 160 μm thick. The nerve fiber layer is highly reflective (red and white in Fig. 5.4b), while the photoreceptors are less reflective (black and blue). In the outer part of the retina, the hyperreflecting retinal pigment epithelium is seen anterior to the underlying Bruch's membrane and choroidea. Reproducible thickness measurements of the retinal nerve fiber layer can be obtained from a range of 2D OCT images (Schuman et al. 1996). The thickness of the nerve fiber layer is likely to be an important parameter in the early diagnosis of glaucoma; currently diagnosis of glaucoma is only possible when visual damage has happened (Medeiros et al. 2005). Furthermore, retinal thickness measurements and the visualization of intraretinal and subretinal fluid are relevant to the diagnosis and follow-up of age-related macular degeneration, a disease where an effective treatment has been introduced worldwide over the last few years (Hee et al. 1996). Macular holes developing around the fovea are also easily visualized using OCT, which is useful because the ophthalmologist might not be able to diagnose a small macular hole using standard methods (Puliafito et al. 1995).

Macular edema is caused by protein and fluid deposition and is often a consequence of diabetes. It may reduce visual acuity, particularly when the edema is close to the fovea. In Fig. 5.5, images from a diabetic patient are shown. In the case of diabetes, the blood vessels become fragile, and the first sign is small bulges in the blood vessels in the retina. Bleeding from vessels and leakage of plasma occurs at a later stage of the disease. Because plasma contains proteins and lipids, depositions of these are often noticed. This is seen as the white spots on the fundus image in Fig. 5.5 and as high reflecting areas in the OCT image. For the case shown in Fig. 5.5, the leakage close to the fovea is very limited and the corresponding lipid deposits on the fundus image are relatively faint, while the OCT image clearly shows the highly reflecting deposit and an underlying shadow. The evaluation of possible edema from fundus images or direct observation of the eye requires stereoscopic techniques and is very difficult. This can be judged by examining the OCT image instead. If edema is present, some of the fragile blood vessels can be closed by laser treatment.

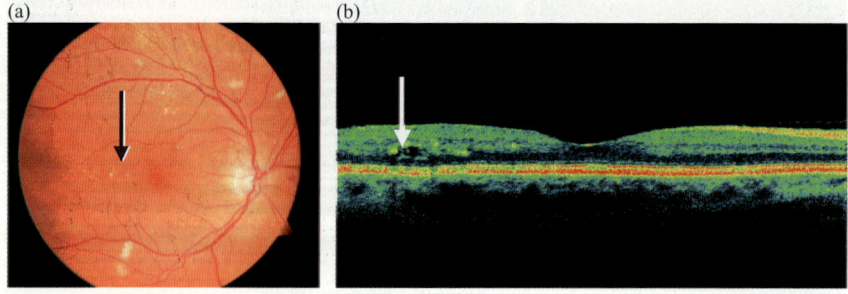

Fig. 5.5 (a–b) Images of the retina belonging to a diabetic patient. (**a**) Fundus image with the *arrow* indicating faint protein precipitation close to the fovea. (**b**) OCT image of the retina where the protein precipitation can easily be seen (*white arrow*)

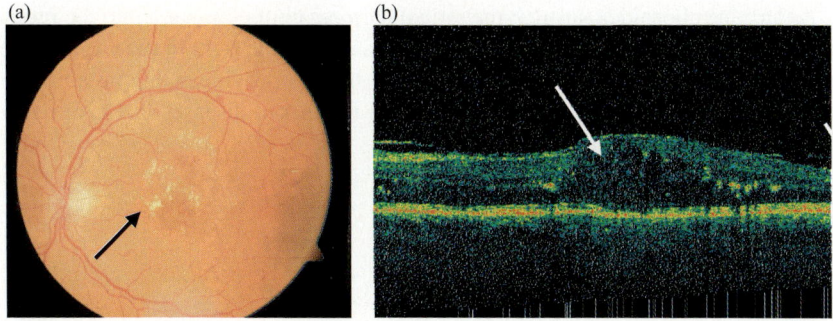

Fig. 5.6 (**a–b**) Images of the eye from a patient with a large edema. (**a**) Fundus image showing the deposition of proteins as white spots (*black arrow*). (**b**) Corresponding OCT image which clearly shows the edema (*white arrow*). The quality of the OCT image is reduced due to increased scattering and absorption in the thickened retina

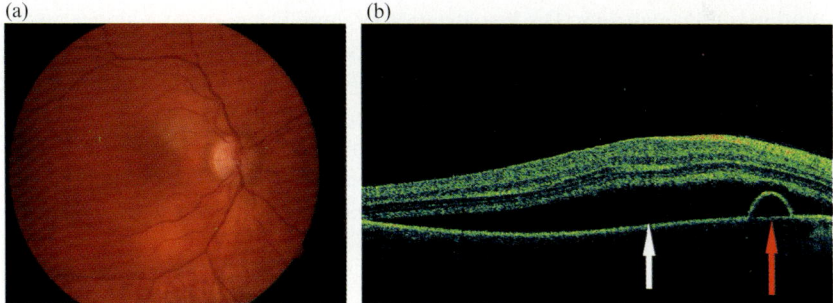

Fig. 5.7 (**a–b**) Images from a patient with serous chorioretinopathy leading to a large, distinct fluid accumulation, both subretinal (i.e., just anterior to the pigment epithelium) and a smaller, dome-shaped, fluid accumulation posterior to the pigment epithelium. (**a**) Fundus image. (**b**) Corresponding OCT image clearly showing the presence of fluid accumulation anterior to the pigment epithelium (*white arrow*) and posterior to the pigment epithelium (*red arrow*)

The images in Fig. 5.6 are acquired on a patient with a large edema in the retina, and the visual power is degraded to a level that makes reading difficult. Many protein depositions can be seen around the fovea (white spots) in the image. It is difficult even for trained medical doctors to judge whether a thickening/edema is present and if laser treatment is necessary. On the other hand, the OCT technique is capable of visualizing the edema, and the retinal thickness before and after treatment is easily assessed. When comparing new treatment techniques, it is important to quantify their effects for optimization and comparison. This is possible using OCT.

Figure 5.7 shows another example of edema in the retina. In this case, the patient is diagnosed with serous chorioretinopathy, a disease which seems to be related to stress. In the fundus image in Fig. 5.7a it is very difficult to spot the slightly difference in the red color of the retina. On the other hand, the edema is easily seen in the OCT image, as indicated by the arrow in Fig. 5.6b. The fluid is seeping from the underlying choroidea through the retinal pigment epithelium and into the retina. Moreover, there is also fluid under the retina, as seen to the right in the OCT image.

The cases shown here illustrate how OCT images provide a unique ability to detect the fluid accumulation associated with a number of diseases that give rise to completely different morphological changes.

Future Directions

OCT has just started its second decade as a commercial instrument for ophthalmo-logic diagnostics and treatment control. Its rapid development is a sign of its clinical relevance. This subsection deals with advances in OCT that can be expected to find their way to the clinic during the second decade of this century.

When identifying retinal layers and determining layer thicknesses, a higher depth resolution is advantageous. The visualization of more layers using a higher resolu-tion has been demonstrated, but the light source setup needed is not yet suitable for widespread clinical use (Ko et al. 2004). With the development of broadband light sources suited to clinical use, it is expected that higher depth resolution will be available in future commercial OCT systems. Attempts to improve the trans-verse resolution are complicated by aberrations introduced by the cornea. The use of adaptive optics can compensate for these aberrations to a certain degree (Hermann et al. 2004; Zhang et al. 2005; Zawadzki et al. 2005).

Due to eye motion, faster image acquisition is preferable and can be achieved using FD-OCT systems. The clinical advantages of using faster systems are not yet fully apparent, since FD-OCT systems have only just been released commercially. Fast systems makes 3D imaging possible, provide a more complete characteriza-tion of the retina, and can yield better diagnostics and treatment control (Povazay et al. 2007).

Traditionally a wavelength of 800 nm has been used to image the retina due to the low absorption by water in this region. However, to increase the penetration, it is advantageous to use a slightly longer wavelength (about 1,050 nm), thereby reducing the scattering (see "Choice of Light Source"). Moreover, the dispersion is lower in this region, resulting in better preservation of the depth resolution. An improved penetration of 200 μm has been demonstrated at a center wavelength of 1,040 nm (Unterhuber et al. 2005).

The presence of speckle as a result of interference of light reflected from closely spaced scatterers can reduce the contrast significantly and effectively blur the image. Different techniques to suppress speckle noise, resulting in a better signal-to-noise ratio and better delineation of retinal layers, have been demonstrated (Sander et al. 2005; Jorgensen et al. 2007).

Polarization-sensitive OCT has recently been applied for retinal imaging (Cense et al. 2002). The retinal nerve fiber layer exhibits birefringence, and a better estimate of the thickness can be made by using PS-OCT to delineate the borders. Further-more, it is known that (for example) glaucoma causes nerve fiber layer damage, so PS-OCT could potentially be used as a diagnostic tool for these diseases. Retinal blood flow is an important parameter in the characterization of a number of diseases, such as diabetes and glaucoma, and has been assessed with high spatial resolution using Doppler OCT (Yazdanfar et al. 2000; White et al. 2003). Finally, changes in the amount of backscattered light from the dark-adapted retina caused by light stimulation have been demonstrated, thus improving our understanding of retinal physiology and pathology (Bizheva et al. 2006).

Optical Coherence Tomography in Dermatology

Skin abnormalities can generally be identified by the naked eye. However, simple visual inspection is highly dependent on operator skills and does not always allow for high diagnostic accuracy. Furthermore, visual inspection only considers the surface of the skin. For this reason, biopsy with subsequent histopathological analysis is the reference standard for confirming clinical diagnosis and examining deeper skin layers. However, biopsies can be time-consuming to perform, they are invasive, and they have potential complications, which make it important to investigate whether a noninvasive technology such as OCT can be used as a diagnostic tool in dermatology.

Besides diagnosis, OCT may be potentially useful as a noninvasive monitoring tool during treatment. Such monitoring would allow for more precise individual adjustment of topical or systematic nonsurgical therapy. The main focus of OCT in the area of dermatology is skin cancer, although OCT has also been studied in relation to photodamage, burns, and inflammatory diseases such as psoriasis and eczema.

Diagnosis of Skin Cancer Using OCT

The two major skin cancer types are malignant melanoma (MM) and nonmelanoma skin cancer (NMSC). MM is less prevalent than NMSC, but MM exhibits high mortality and increasing incidence [in 2003, according to data from the US Surveillance, Epidemiology, and End Results (SEER) registry, it was estimated that 54,200 Americans were diagnosed with melanoma]. NMSC is more common than MM. In the US, over a million people are diagnosed with NMSC each year, and its incidence appears to be rising (Neville et al. 2007). Although the mortality is low, depending on the type of NMSC, the morbidity can be high due to local tissue destruction and subsequent complications or tumor recurrence.

The clinical standard for the diagnosis of skin cancer is biopsy, which is invasive. In addition, biopsies are usually taken as 2–4 mm punch biopsies and can therefore potentially introduce sampling errors if taken in suboptimal places. Furthermore, the handling and processing of the biopsy takes time resulting in treatment delays and

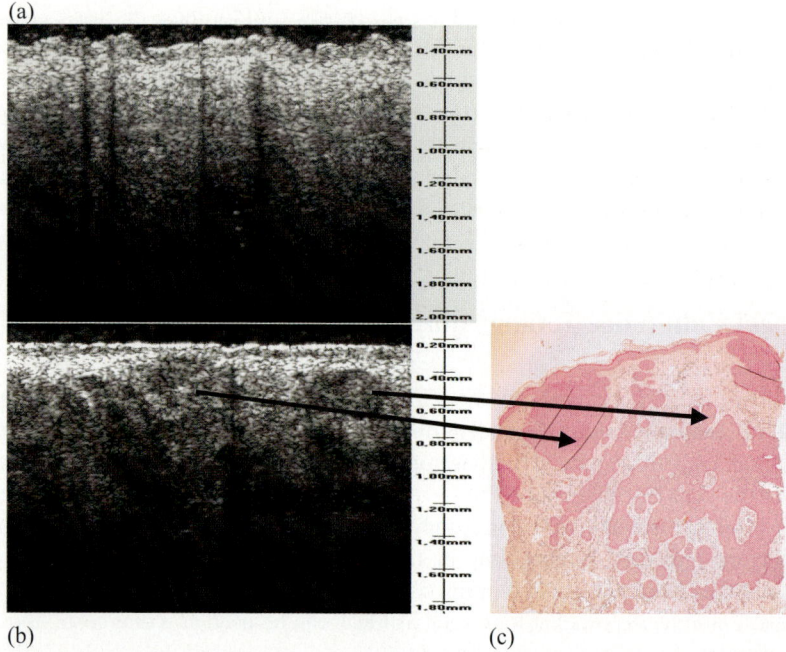

(a)

(b)   (c)

Fig. 5.8 (**a**) OCT image of normal skin. (**b**) OCT image of a basal cell carcinoma. (**c**) Corresponding histology (HE stain, magnification ×40). A clear structural difference between normal skin and the basal cell carcinoma is seen. The main features of the basaloid carcinoma cell islands in the histological image can also be retrieved in the OCT image (*arrows*)

higher costs. Noninvasive *in vivo* imaging techniques such as OCT may therefore be valuable as adjunct diagnostic tools in the clinic (Welzel 2001). It has been speculated that the imaging depth of OCT may also be potentially useful for identifying the extent of the lesions. A knowledge of the borders of the skin cancer lesion is also relevant when choosing the optimal treatment.

The typical layered structure of normal skin is shown in Fig. 5.8a. A clear boundary between the stratum corneum and the living part of the epidermis is usually seen in glabrous skin. The epidermis can be distinguished from the dermis in OCT images (Mogensen 2007a). Because OCT is typically not capable of resolving single cells, the diagnosis must rely on a change in the skin structure, such as a breaking up of tissue layers, general disordered structure, a change in the amount of backscattered light, etc., compared to normal skin (Welzel 2001; Steiner et al. 2003; Mogensen 2007a).

Malignant melanomas have been investigated in a study comparing dermoscopic observations with histology and OCT (Giorgi et al. 2005). Morphological correlation was reported in six out of ten cases. It was not possible to differentiate benign and malignant lesions according to the authors. Improved resolution allowing single-cell identification was suggested in order to make reliable diagnosis. The

OCT system used in the study did not provide a resolution that was high enough for this to be achieved. Another group included 75 patients in a study of MM and benign melanocytic nevi, and reported a loss of normal skin architecture in MM compared to benign lesions (Gambichler et al. 2007). Characteristic morphological OCT features in both nevi and MM were demonstrated and confirmed by histology. Interestingly, the presence of vascularity was not limited to the MM lesions (note that the identification of vessels was performed using structural OCT images and not Doppler OCT images). The specific diagnostic accuracy of OCT was not calculated in this study.

A number of studies of the diagnostic potential of OCT in NMSC have also been conducted. In contrast to potentially reducing mortality with early MM diagnosis, the main advantage for NMSC would be the potential to reduce the number of biopsies. Figure 5.8 shows an OCT image of a basal cell carcinoma (BCC), the most common type of NMSC, together with an OCT image of normal skin adjacent to the lesion. A difference in structure between normal skin and the skin cancer is demonstrated in the OCT image of the BCC, which shows a more generally disordered tissue. Furthermore, the main structures in the histology of the BCC are also seen in the OCT image.

In a clinical setting, the task is to differentiate between benign and malignant lesions. This is much more challenging, because benign lesions often show the same structure as malignant ones in the OCT images. Furthermore, variations with age, skin type, anatomical site, etc., must be taken into account (Mogensen 2007a; Gambichler et al. 2006a, b). Many studies have, however, reported a high correlation between OCT images and histology, suggesting that OCT can be used to recognize NMSC lesions (Bechara et al. 2004; Olmedo et al. 2006, 2007; Gambichler et al. 2006c). An element of subjectivity was introduced in the evaluation, because decisions were based on visual inspection of the OCT images in most of the cited studies. Thus it has been suggested that machine learning tools may aid and improve the diagnostic accuracy of OCT. The potential of automatic feature extraction using nonparametric machine learning algorithms was demonstrated in a pilot study concerning the possibility of distinguishing basal cell carcinoma (the most prevalent NMSC type) from actinic keratosis (Jørgensen et al. 2008). Another study, including more than 100 patients, aimed at distinguishing between sun-damaged skin, actinic keratosis and normal skin (Korde et al. 2007). A horizontal edge detection technique was employed to measure the presence of layered structure in epidermis. The edge detection was automated and resulted in about 70% correct classification. The presence of a dark band in epidermis evaluated by the naked eye gave a correct classification rate of about 85%.

Functional OCT has also been suggested to improve the accuracy. The use of PS-OCT images has been investigated for NMSC (Strasswimmer et al. 2004). The highly organized collagen fibers in dermis result in birefringence. The ordered structure breaks down in skin cancer lesions. Therefore, it was suggested that it should be possible to distinguish normal skin and basal cell carcinoma using PS-OCT, and this was confirmed by the two cases studied (Strasswimmer et al. 2004). In addition to tissue changes caused by invasive growth, the neovascularization of tumors is an important biological and morphological feature of malignancy. Doppler OCT is another possible approach to the diagnosis of malignant lesions, because increased vascularity in malignant skin tumors would be expected. Doppler OCT data from clinical studies of skin tumors are not available at present. Diagnosis of skin cancer lesions has already been attempted through the use of laser Doppler imaging, but this method gives no depth resolution, which could be important for determining the borders of the lesion (Stucker et al. 2002).

Multiphoton imaging is an emerging method that may provide information about the chemical structure of tissues *in vivo* (Lin et al. 2006). Multiphoton imaging has shown promising results in skin cancer diagnosis, but the methods appear to be restricted by a limited penetration depth (Lin et al. 2006). OCT and the multiphoton technique can be combined into unified system such that two types of images are acquired at the same time. One approach may therefore be to use the multiphoton image to determine whether a lesion is benign or malignant and the OCT image to determine the thickness of the lesion. This requires that the OCT image carries information from the deeper region about the transition from tumor tissue to normal tissue. Using Raman spectroscopy, the change in chemical composition between BCC and the surrounding normal skin can be detected with high accuracy (Nijssen et al. 2002). However, Raman spectroscopy does not provide depth-resolved information like OCT, and so delineation of tumor borders based on depth is not possible.

Emerging Applications in Dermatology

The organized collagen structures in skin break down when exposed to high temperatures. Because the collagen structures can be detected using PS-OCT, it is possible to measure the depth and extent of a burn, which are crucial parameters when deciding which treatment is necessary. PS-OCT may also be used to monitor the healing process. A significant difference in birefringence between burns and unaffected skin has been reported (Pierce et al. 2004).

The common skin disease psoriasis is challenging to treat, and there are currently no precise instruments for evaluating the effect of psoriasis treatment, which is essential for improving treatment possibilities. The use of OCT to monitor changes during a treatment period has been suggested (Welzel et al. 2003).

Another study used OCT to measure the thickness of human nails (Mogensen et al. 2007b). A more detailed visualization of the nail morphology compared to 20 MHz ultrasound imaging was reported, and thickness was measured with higher reproducibility (Mogensen et al. 2007b). Visualization of the small tunnel in the skin excavated by cutaneous larva migrans (a tropical disease) was possible using OCT, and this result suggests that OCT could be helpful in the diagnosis of skin infestations (Morsy et al. 2007a). The pigment color in a tattoo was also clearly visible when using OCT (Morsy et al. 2007b).

Other Applications in Biology and Medicine

Besides the applications mentioned above, OCT has been applied in several other areas. In this section, some of these will be described briefly, with emphasis placed on cardiology, oncology and developmental biology. This is not a complete list of feasible applications, but rather they highlight the areas where OCT is applied or where the potential of OCT is currently being investigated.

The skin has an easy accessible surface that can be investigated using OCT. Moving to the inside of the body, a number of so-called hollow organs—such as the cardiovascular system, gastrointestinal tract, bladder, etc.—also contain accessible surfaces that can be reached using an endoscope. On these epithelial surfaces, tools for detecting neoplastic changes are needed in order to perform early diagnostics. Therefore, it is important to investigate if OCT is capable of detecting precancerous and neoplastic lesions.

In general, a feasible application must take advantage of some of the unique properties of OCT, such as the relatively high penetration combined with micrometer resolution, the fact that physical contact with the sample is not needed, or the ability to produce small probes for endoscopic applications. In cardiology, gastroenterology, gynecology, urology and respiratory medicine, it is essential to use a small probe that can be integrated with an endoscope. In developmental biology, it is the micrometer resolution combined with millimeter penetration and noninvasiveness that makes OCT attractive. Moreover, the noninvasiveness of the technique makes it possible to track the progression over time, which is of high value in many applications.

Cardiology: Detection of Vulnerable Plaque

Cardiologic diseases are the leading cause of death worldwide; for example, according to the World Health Organization, 16.7 million people died from cardiovascular diseases in 2003, corresponding to about 30% of all deaths (http://www.who.int). A vulnerable plaque in the arteries is the most prevalent condition leading to myocardial infarctions. Therefore, the ability to detect this type of plaque is important. Early diagnosis is essential for prognosis and makes high-resolution imaging of the cardiovascular system highly relevant.

A vulnerable plaque is characterized by a $\sim 50\,\mu\mathrm{m}$ fibrous cap overlying a lipid pool. Previously, ultrasound imaging was used to characterize plaque lesions, but the resolution of this technique is not sufficient to resolve structures and make a reliable diagnosis (Jang et al. 2002; Low et al. 2006). OCT provides a much higher resolution than ultrasound, making more detailed characterization of the morphology possible (Tearney et al. 2003, 2006). In Fig. 5.9, a comparison of fibrous coronary plaque imaged with OCT and ultrasound is shown (rotational scan). It is evident that OCT provides a much more detailed image with better delineation of layers than ultrasound. Ex vivo measurements demonstrated an ability to distinguish

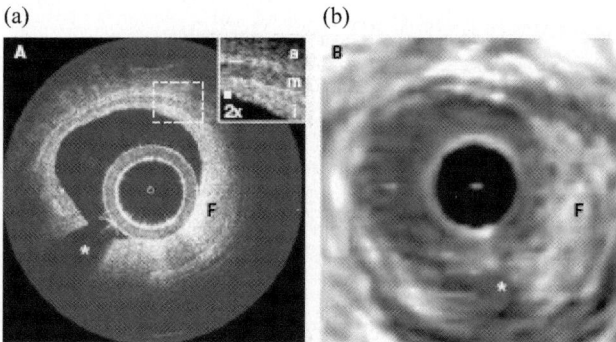

Fig. 5.9 (**a–b**) Comparing fibrous coronary plaque imaged with OCT and ultrasound. (**a**) OCT image. (**b**) Corresponding 30 MHz ultrasound image. i, Intima with intimal hyperplasia; m, media; a, adventitia; f, fibrous plaque; *, guidewire artifact. (Figure 5.2 from Jang et al. 2002; reprinted here with permission)

different plaque types (Levitz et al. 2004; Yabushita et al. 2002). For fibrous plaque, the sensitivity and specificity were 71–79% and 97–98%, respectively (Yabushita et al. 2002). For fibrocalcific plaque they were 95–96% and 97%, and for lipid-rich plaque they were 90–94% and 90–92% (Yabushita et al. 2002). *In vivo* characterization of plaque has also been reported, and is currently an issue of intense research (Brezinski 2007; Jang et al. 2005). Whether the use of PS-OCT can assist further in identifying vulnerable plaque has also been investigated (Giattina et al. 2006). Due to a loss of organized collagen in these lesions detectable with PS-OCT, it may be possible to diagnose with an even higher accuracy than achieved by structural OCT alone.

Oncology

Apart from OCT imaging of skin cancer (as described in "Optical Coherence Tomography in Dermatology"), a number of other applications in oncology have been explored. This includes attempts to detect precancerous states and neoplastic changes in the esophagus, bladder, lung, breast and brain. Early detection of cancer is important in order to maximize the survival rate. Therefore, it is important to investigate whether noninvasive and high-resolution imaging with high diagnostic accuracy can be provided by OCT. Due to the limited penetration, OCT imaging of inner organs must be performed either during open surgery or via an endoscope. During surgery, OCT can potentially be used to guide and help the surgeon to delineate tumor borders, and to guide surgery. Examination of the esophagus is performed for the diagnosis of cancer and pre-cancer conditions. Usually the esophagus is visually inspected through an imaging endoscope, but it is difficult to detect neoplastic changes using this approach. Excisional biopsies suffer from sampling errors and have a number of disadvantages when performed in the esophagus, such as bleeding. High-frequency endoscopic ultrasound has also been used to image the esophagus, but it lacks the resolution to detect cancer with high accuracy (Waxman et al. 2006). The same drawbacks apply to X-ray and magnetic resonance imaging. Because OCT provides a much higher resolution than the imaging techniques mentioned above, it is important to investigate whether OCT can improve diagnostic accuracy and detect cancer at an earlier stage, which is essential for prognosis.

Barrett's esophagus is a condition that can progress to cancer, and once diagnosed, a screening is regularly performed in order to monitor its development. Barrett's esophagus has been imaged with endoscopic ultrahigh-resolution OCT and compared with histology (Chen et al. 2007). A difference between normal esophagus vs. Barrett's esophagus and esophageal cancer was reported, and it was concluded that OCT is capable of imaging the fine structures in esophageal tissue. Another study investigating this subject reported 78% accuracy in the detection of dysplasia in patients with Barrett's esophagus (Isenberg et al. 2005).

Bladder cancer is traditionally diagnosed using cystoscopy, i.e., visual inspection through an endoscope. This method gives information about the surface of the bladder, and therefore diagnosis of subsurface tumors is not possible. Detection of neoplastic changes is not possible either. OCT has recently been used to examine the bladder walls (Jesser et al. 1999; Zagaynova et al. 2002). In healthy bladder tissue, the mucosa, submucosa and muscularis layers were visible in the OCT images, whereas the break-up of this structure in invasive carcinoma was reported. A study including 87 areas in 24 patients reported a sensitivity and a specificity of 100% and 89%, respectively (Manyak et al. 2005).

Optical coherence tomography has also been applied to image the bronchial airways (Whiteman et al. 2006; Yang et al. 2004). A high correlation between OCT images and histology was reported, and the authors suggest a potential application in lung cancer diagnostics. The use of OCT for breast cancer diagnosis has been suggested (Boppart et al. 2004; Luo et al. 2005). Furthermore, OCT can be used as monitoring tool to assess tumor margins during breast cancer surgery (Boppart et al. 2004).

Developmental Biology

Within developmental biology, OCT has been used to visualize many different developmental processes in many different small-animal models for more than a decade. It is not possible to cover all of the applications here, but a recent rather detailed review of OCT in developmental biology can be found in Boppart et al. (1996).

OCT was introduced to developmental biology in 1996 by Boppart et al. (1996). One of the most promising and fascinating applications of OCT within developmental biology is related to the study of human heart development. The heart is the first organ to *form* and *function* in vertebrates and undergoes simultaneous *structural* and *functional* maturation as it transforms in a dynamic process from a straight tube to

Fig. 5.10 3D OCT scan of
a 56-hour-old (HH stage 15)
chick embryo heart of a CNC-
ablated embryo. *Cutaway*
reveals further internal structural details. *O*, Outflow limb;
i, inflow limb; v, presumptive
ventricle. Bar = 0.250 mm.
(Figure 2E from Yelbuz
et al., 2002; reprinted here
with permission)

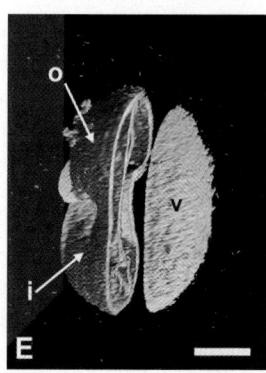

a four-chambered heart (Keller 1998). Malformations of the heart are among the most common birth defects, and are the leading cause of birth defect-related deaths (Hoffman 1995). However, the cause of most heart defects remains unknown. In order to enhance our understanding of normal and abnormal heart development (as studied in different small-animal models), there is a need for a real-time, noninvasive imaging technique with micrometer resolution and millimeter penetration depth. It has been found that OCT is well suited to anatomic and functional imaging of the developing cardiovascular system, as has been demonstrated by various groups in recent years for different species (Boppart et al. 1997; Yelbuz et al. 2002; Choma et al. 2006; Jenkins et al. 2007; Luo et al. 2006; Mariampillai et al. 2007). As an example, Fig. 5.10 shows a 3D OCT scan of a 56-hour-old (HH-stage 15) chick embryo heart of a CNC-ablated embryo (Yelbuz et al. 2002).

Among the most recent studies in this field is an investigation of one of the most critical but poorly understood processes during cardiovascular development: the establishment of a functioning coronary artery (CA) system. Using OCT, the first *in vivo* recordings of developing CAs in chick embryo hearts have been obtained (Norozi et al. 2008). Again, this demonstrates the use and strength of OCT to help understand the complex processes studied in developmental biology.

Summary

In conclusion, OCT is a noninvasive optical imaging technique that provides real-time video rate micrometer-resolution images with millimeter penetration, thereby filling the gap between ultrasound and confocal microscopy. Moreover, functional extensions that enable flow detection (Doppler OCT), birefringence detection (polarization-sensitive OCT) and extraction of spectroscopic information (spectroscopic OCT) are possible. Technical development has resulted in faster systems, higher resolution and better image quality.

Regarding the applications of OCT, the technique is already established as a clinical standard in ophthalmology and has proven invaluable in diagnostics and treatment control of many eye diseases. Most OCT systems used in the clinic today are relatively slow compared to state-of-the-art research OCT systems, although faster FD-OCT systems are now emerging in the clinic. It is therefore believed that not all of the advantages of higher speed have become apparent yet. Furthermore, the OCT systems used in the clinic do not have functional OCT extensions. Preliminary research results show a remarkable improvement in diagnostic information upon using higher resolution, faster data acquisition (3D data), and functional OCT. Therefore, it is expected that major improvements in diagnostics are feasible in the coming decade of commercial OCT.

OCT possesses many useful features for skin cancer diagnostics; it has been studied clinically, although convincing studies reporting high-accuracy diagnosis are sparse. It seems that a different approach is needed to make it a diagnostic tool in the clinic. Cellular resolution would make it possible to see how the cells are arranged in the skin, which is one parameter used for diagnosis by histology. Also, confocal microscopy studies have reported a diagnostic accuracy of higher than 95% (Gerger et al. 2006). Therefore, a much higher resolution is probably required to make OCT a strong imaging modality for skin cancer diagnostics. This requires either broader-bandwidth light sources (better depth resolution) or tighter focus (better transverse resolution). The latter is conveniently implemented in an *en face* mode, like confocal microscopy (optical coherence microscopy, see "Resolution and Sensitivity"). In this context, it is important to emphasize that the depth and transverse resolution degrades with depth in highly scattering tissue. If high-resolution images are necessary to make an accurate diagnosis, it is probably only possible to generate useful images for diagnostics to a depth of less than the 1–2 mm penetration depth.

In cardiology there is a need for the detection of vulnerable plaques, and OCT has already shown promising results in this respect. Within oncology, a wide variety of lesions have been examined, including esophagus, bladder, lung, breast and brain, with promising results. Finally, in developmental biology, OCT has proven its value by providing *in vivo* micrometer-resolution images that promote understanding of the complex processes studied in this field.

References

Bechara FG, Gambichler T, Stucker M et al. (2004) Histomorphologic correlation with routine histology and optical coherence tomography. Skin Res Technol 10:169–173

Bizheva K, Pflug R, Hermann B et al. (2006) Optophysiology: Depth-resolved probing of retinal physiology with functional ultrahigh-resolution optical coherence tomography. Proc Natl Acad Sci USA 103:5066–5071

Boppart SA, Brezinsky ME, Bouma BE et al. (1996) Investigation of developing embryonic morphology using optical coherence tomography. Dev Biol 177:54–63

Boppart SA, Tearney GJ, Bouma BE et al. (1997) Noninvasive assessment of the developing Xenopus cardiovascular system using optical coherence tomography. Proc Natl Acad Sci USA 94:4256–4261

Boppart SA, Luo W, Marks DL et al. (2004) Optical coherence tomography: feasibility for basic research and image-guided surgery of breast cancer. Breast Cancer Res Treat 84:85–97

Bouma BE, Tearney GJ (2002) Handbook of optical coherence tomography. Marcel Dekker, New York

Brezinski ME (2006) Optical coherence tomography: principles and applications. Elsevier, Amsterdam

Brezinski ME (2007) Applications of optical coherence tomography to cardiac and musculoskeletal diseases: bench to bedside? J Biomed Opt 12:051705

Cense B, Chen TC, Park BH et al. (2002) In vivo depth-resolved birefringence measurements of the human retinal nerve fiber layer by polarization-sensitive optical coherence tomography. Opt Lett 27:1610–1612

Cense B, Chen TC, Park BH et al. (2004) Thickness and birefringence of healthy retinal nerve fiber layer tissue measured with polarization-sensitive optical coherence tomography. Inv Opth Vis Sci 45:2606–2612

Chen ZP, Milner TE, Dave D et al. (1997) Optical Doppler tomographic imaging of fluid flow velocity in highly scattering media. Opt Lett 22:64–66

Chen Y, Aguirre AD, Hsiung PL et al. (2007) Ultrahigh resolution optical coherence tomography of Barrett's esophagus: preliminary descriptive clinical study correlating images with histology. Endoscopy 39:599–605

Choma MA, Sarunic MV, Yang CH et al. (2003) Sensitivity advantage of swept source and Fourier domain optical coherence tomography. Opt Express 11:2183–2189

Choma MA, Izatt SD, Wessels RJ et al. (2006) Images in cardiovascular medicine: in vivo imaging of the adult *Drosophila melanogaster* heart with real-time optical coherence tomography. Circulation 114:e35–e36

Drexler W (2004) Ultrahigh-resolution optical coherence tomography. J Biomed Opt 9:47–74

Drexler W, Hitzenberger CK, Baumgartner A et al. (1998) Investigation of dispersion effects in ocular media by multiple wavelength partial coherence interferometry. Exp Eye Res 66:25–33

Faber DJ, Mik EG, Aalders MCG et al. (2005) Toward assessment of blood oxygen saturation by spectroscopic optical coherence tomography. Opt Lett 30:1015–1017

Fercher AF, Hitzenberger CK, Kamp G et al. (1995) Measurement of intraocular distances by backscattering spectral interferometry. Opt Commun 117:43–48

Fercher AF, Drexler W, Hitzenberger CK et al. (2003) Optical coherence tomography: principles and applications. Rep Prog Phys 66:239–303

Gambichler T, Matip R, Moussa G et al. (2006a) In vivo data of epidermal thickness evaluated by optical coherence tomography: effects of age, gender, skin type, and anatomic site. J Derm Sci 44:145–152

Gambichler T, Huyn J, Tomi NS et al. (2006b) A comparative pilot study on ultraviolet-induced skin changes assessed by noninvasive imaging techniques in vivo. J Photochem Photobiol 82:1103–1107

Gambichler T, Orlikov A, Vasa R et al. (2006c) In vivo optical coherence tomography of basal cell carcinoma. J Dermatol Sci 45:167–173

Gambichler T, Regeniter P, Bechara FG et al. (2007) Characterization of benign and malignant melano-cytic skin lesions using optical coherence tomography in vivo. J Am Acad Dermatol 57:629–637

Gerger A, Koller S, Weger W et al. (2006) Sensitivity and specificity of confocal laser-scanning microscopy for in vivo diagnosis of malignant skin tumors. Cancer 107:193–200

Giattina SD, Courtney BK, Herz PR et al. (2006) Assessment of coronary plaque collagen with polarization sensitive optical coherence tomography (PS-OCT). Int J Cardiol 107:400–409

Giorgi GV, de Stante M, Massi D et al. (2005) Possible histopathologic correlates of dermoscopic features in pigmented melanocytic lesions identified by means of optical coherence tomography. Exp Dermatol 14:56–59

Hangai M, Jima Y, Gotoh N et al. (2007) Three-dimensional imaging of macular holes with high-speed optical coherence tomography. Ophthalmology 114:763–773

Hartl I, Li XD, Chudoba C et al. (2001) Ultrahigh-resolution optical coherence tomography using continuum generation in air-silica microstructure optical fiber. Opt Lett 26:608–610

Hecht E (1998) Optics. Addison Wesley, Reading, MA

Hee MR, Huang D, Swanson EA et al. (1992) Polarization-sensitive low-coherence reflectometer for birefringence characterization and ranging. J Opt Soc Am B 9:903–908

Hee MR, Baumal CR, Puliafito CA et al. (1996) Optical coherence tomography of age-related macular degeneration and choroidal neovascularization. Ophthalmology 103:1260–1270

Hermann B, Fernandez EJ, Unterhuber AB et al. (2004) Adaptive-optics ultrahigh-resolution optical coherence tomography. Opt Lett 29:2142–2144

Hillman TR, Sampson DD (2005) The effect of water dispersion and absorption on axial resolution in ultrahigh-resolution optical coherence tomography. Opt Express 13:1860–1874

Hoffman JI (1995) Incidence of congenital heart disease, I: postnatal incidence. Pediatr Cardiol 16:103–113

Huang D, Swanson EA, Lin CP et al. (1991) Optical coherence tomography. Science 254:1178–1181

Huber R, Wojtkowski M, Fujimoto JG et al. (2005) Three-dimensional and C-mode OCT imaging with a compact, frequency swept laser source at 1300 nm. Opt Express 13:10523–10538

Isenberg G, Sivak MV, Chak A et al. (2005) Accuracy of endoscopic optical coherence tomography in the detection of dysplasia in Barrett's esophagus: a prospective, double-blinded study. Gastrointest Endos 62:825–831

Izatt JA, Kulkarni MD, Wang H-W et al. (1996) Optical coherence tomography and microscopy in gastrointestinal tissues. IEEE J Sel Top Quant Electr 4:1017–1028

Izatt JA, Kulkami MD, Yazdanfar S et al. (1997) In vivo bidirectional color Doppler flow imaging picoliter blood volumes using optical coherence tomography. Opt Lett 22:1439–1441

Jang IK, Bouma BE, Kang DH et al. (2002) Visualization of coronary atherosclerotic plaques in patients using optical coherence tomography: Comparison with intravascular ultra-sound. J Am Coll Cardiol 39:604–609

Jang IK, Tearney GJ, MacNeill B et al. (2005) In vivo characterization of coronary atherosclerotic plaque by use of optical coherence tomography. Circulation 111:1551–1555

Jenkins MW, Adler DC, Gargesha M et al. (2007) Ultrahigh-speed optical coherence tomography imaging and visualization of the embryonic avian heart using a buffered Fourier domain mode locked laser. Opt Express 15:6251–6267

Jesser CA, Boppart SA, Pitris C et al. (1999) High resolution imaging of transitional cell carcinoma with optical coherence tomography: feasibility for the evaluation of bladder pathology. Br J Radiol 72:1170–1176

Jørgensen TM, Thomadsen J, Christensen U et al. (2007) Enhancing the signal-to-noise ratio in ophthalmic optical coherence tomography by image registration: method and clinical examples. J Biomed Opt 12:041208

Jørgensen TM, Tycho A, Mogensen M et al. (2008) Machine-learning classification of non-melanoma skin cancers from image features obtained by optical coherence tomography. Skin Res Technol 14:364–369

Keller BB (1998) Embryonic cardiovascular function, coupling, and maturation: a species view. In: Burggren WW, Keller BB (eds) Development of cardiovascular systems. Cambridge Univ Press, Cambridge

Ko TH, Fujimoto JG, Duker JS et al. (2004) Comparison of ultrahigh- and standard-resolution optical coherence tomography for imaging macular hole pathology and repair. Ophthalmology 111:2033–2043

Korde VR, Bonnema GT, Xu W et al. (2007) Using optical coherence tomography to evaluate skin sun damage and precancer. Las Surg Med 39:687–695

Leitgeb RA, Villiger M, Bachmann AH et al. (2006) Fourier domain optical coherence microscopy with extended focus depth. Opt Lett 31:2450–2452

Levitz D, Thrane L, Frosz MH et al. (2004) Determination of optical scattering properties of highly-scattering media in optical coherence tomography images. Opt Express 12:249–259

Lin SJ, Jee SH, Kuo CJ et al. (2006) Discrimination of basal cell carcinoma from normal dermal stroma by quantitative multiphoton imaging. Opt Lett 31:2756–2758

Liu B, Brezinski ME (2007) Theoretical and practical considerations on detection performance of time domain, Fourier domain, and swept source optical coherence tomography. J Biomed Opt 12:044007

Low AF, Tearney GJ, Bouma BE et al. (2006) Technology insight: optical coherence tomography: current status and future development. Nat Clin Pract Cardiovasc Med 3:154–162

Luo W, Nguyen FT, Zysk AM et al. (2005) Optical biopsy of lymph node morphology using optical coherence tomography. Technol Cancer Res Treat 4:539–547

Luo W, Marks DL, Ralston TS et al. (2006) Three-dimensional optical coherence tomography of the embryonic murine cardiovascular system. J Biomed Opt 11:021014

Manyak MJ, Gladkova ND, Makari JH et al. (2005) Evaluation of superficial bladder transitional-cell carcinoma by optical coherence tomography. J Endourol 19:570–574

Mariampillai A, Standish BA, Munce NR et al. (2007) Doppler optical cardiogram gated 2D color flow imaging at 1000 fps and 4D in vivo visualization of embryonic heart at 45 fps on a swept source OCT system. Opt Express 15:1627–1638

Medeiros FA, Zangwill LM, Bowd C et al. (2005) Evaluation of retinal nerve fiber layer, optic nerve head, and macular thickness measurements for glaucoma detection using optical coherence tomography. Am J Ophthal 139:44–55

Mogensen M et al. (2007a) Morphology and epidermal thickness of normal skin imaged by optical coherence tomography. Dermatology 217:14–20

Mogensen M, Thomsen JB, Skovgaard LT et al. (2007b) Nail thickness measurements using optical coherence tomography and 20 MHz ultrasonography. Br J Derm 157:894–900

Morgener U, Drexler W, Kärtner FX et al. (2000) Spectroscopic optical coherence tomography. Opt Lett 25:111–113

Morsy H, Mogensen M, Thomsen J et al. (2007a) Imaging of cutaneous larva migrans by optical coherence tomography. Travel Med Infect Dis 5:243–246

Morsy H, Mogensen M, Thrane L et al. (2007b) Imaging of intradermal tattoos by optical coherence tomography. Skin Res Technol 13:444–448

Neville JA, Welch E, Leffell DJ (2007) Management of nonmelanoma skin cancer in 2007. Nat Clin Pract Oncol 4:462–469

Nijssen A, Schut TCB, Heule F et al. (2002) Discriminating Basal cell carcinoma from its surrounding tissue by Raman spectroscopy. J Invest Derm 119:64–69

Norozi K, Thrane L, Männer J et al. (2008) In vivo visualisation of coronary artery development by high-resolution optical coherence tomography. Heart 94:130

Olmedo JM, Warschaw KE, Schmitt JM et al. (2006) Optical coherence tomography for the characterization of basal cell carcinoma in vivo: A pilot study. J Am Acad Derm 55:408–412

Olmedo JM, Warschaw KE, Schmitt JM et al. (2007) Correlation of thickness of basal cell carcinoma by optical coherence tomography in vivo and routine histologic findings: A pilot study. Dermatol Surg 33:421–426

Pierce MC, Sheridan RL, Park BH et al. (2004) Collagen denaturation can be quantified in burned human skin using polarization-sensitive optical coherence tomography. Burns 30:511–517

Povazay B, Bizheva K, Unterhuber A et al. (2002) Submicrometer axial resolution optical coherence tomography. Opt Lett 27:1800–1802

Povazay B, Hofer B, Hermann B et al. (2007) Minimum distance mapping using three-dimensional optical coherence tomography for glaucoma diagnosis. J Biomed Opt 12:041204

Puliafito CA, Hee MR, Lin CP et al. (1995) Imaging of macular diseases with optical coherence tomography. Ophthalmology 102:217–229

Sander B, Larsen M, Thrane L et al. (2005) Enhanced optical coherence tomography imaging by multiple scan averaging. Br J Ophthal 89:207–212

Saxer CE, de Boer JF, Park BH et al. (2000) High-speed fiber-based polarization-sensitive optical coherence tomography of in vivo human skin. Opt Lett 25:1355–1357

Schuman JS, PedutKloizman T, Hertzmark E et al. (1996) Reproducibility of nerve fiber layer thickness measurements using optical coherence tomography. Ophthalmology 103:1889–1898

Spoler F, Kray S, Grychtol P et al. (2007) Simultaneous dual-band ultra-high resolution optical coherence tomography. Opt Express 15:10832–10841

Steiner R, Kunzi-Rapp K, Scharfetter-Kochaner K (2003) Optical coherence tomography: clinical applications in dermatology. Med Laser Appl 18:249–259

Strasswimmer J, Pierce MC, Park BH et al. (2004) Polarization-sensitive optical coherence tomography of invasive basal cell carcinoma. J Biomed Opt 9:292–298

Stucker M, Esser M, Hoffmann M et al. (2002) High-resolution laser Doppler perfusion imaging aids in differentiating benign and malignant melanocytic skin tumors. Acta Derm 82:25–29

Tearney GJ, Yabushita H, Houser SL et al. (2003) Quantification of macrophage content in atherosclerotic plaques by optical coherence tomography. Circulation 107:113–119

Tearney GJ, Jang IK, Bouma BE (2006) Optical coherence tomography for imaging the vulnerable plaque. J Biomed Opt 11:021002

Unterhuber A, Povazay B, Hermann B et al. (2005) In vivo retinal optical coherence tomography at 1040 nm-enhanced penetration into the choroid. Opt Express 13:3252–3258

van Velthoven MEJ, Faber DJ, Verbraak FD et al. (2007) Recent developments in optical coherence tomography for imaging the retina. Prog Retinal Eye Res 26:57–77

Waxman I, Raju GS, Critchlow J et al. (2006) High-frequency probe ultrasonography has limited accuracy for detecting invasive adenocarcinoma in patients with Barrett's esophagus and high-grade dysplasia or intramucosal carcinoma: a case series. Am J Gastroent 101:1773–1779

Welzel J (2001) Optical coherence tomography in dermatology: a review. Skin Res Technol 7:1–9

Welzel J, Bruhns M, Wolff HH (2003) Optical coherence tomography in contact dermatitis and psoriasis. Arch Dermatol Res 295:50–55

White BR, Pierce MC, Nassif N et al. (2003) In vivo dynamic human retinal blood flow imaging using ultra-high-speed spectral domain optical Doppler tomography. Opt Express 11:3490–3497

Whiteman SC, Yang Y, van Pittius DG et al. (2006) Optical coherence tomography: Real-time imaging of bronchial airways microstructure and detection of inflammatory/neoplastic morphologic changes. Clin Cancer Res. 12:813–818

Wu L (2004) Simultaneous measurement of flow velocity and Doppler angle by the use of Doppler optical coherence tomography. Opt Lasers Eng 42:303–313

Yabushita H, Bourna BE, Houser SL et al. (2002) Characterization of human atherosclerosis by optical coherence tomography. Circulation 106:1640–1645

Yang Y, Whiteman S, van Pittius DG et al. (2004) Use of optical coherence tomography in delineating airways microstructure: comparison of OCT images to histopathological sections. Phys Med Biol 49:1247–1255

Yazdanfar S, Rollins AM, Izatt JA (2000) Imaging and velocimetry of the human retinal circulation with color Doppler optical coherence tomography. Opt Lett 25:1448–1450

Yelbuz TM, Choma MA, Thrane L et al. (2002) Optical coherence tomography: a new high-resolution imaging technology to study cardiac development in chick embryos. Circulation 106:2771–2774

Zagaynova EV, Streltsova OS, Gladkova ND et al. (2002) In vivo optical coherence tomography feasibility for bladder disease. J Urol 167:1492–1496

Zawadzki RJ, Jones SM, Olivier SS et al. (2005) Adaptive-optics optical coherence tomography for high-resolution and high-speed 3D retinal in-vivo imaging. Opt Express 13:8532–8546

Zhang Y, Rha JT, Jonnal RS et al. (2005) Adaptive optics parallel spectral domain optical coherence tomography for imaging the living retina. Opt Express 13:4792–4811

Zhao YH, Chen ZP, Saxer C et al. (2000) Phase-resolved optical coherence tomography and optical Doppler tomography for imaging blood flow in human skin with fast scanning speed and high velocity sensitivity. Opt Lett 25:114–116

Chapter 6
Mass Spectrometry-Based Tissue Imaging[*]

Carol E. Parker, Derek Smith, Detlev Suckau, and Christoph H. Borchers(✉)

Abstract In MALDI (matrix-assisted laser desorption/ionization) mass spectrometry, the sample consists of a thin film of proteins or peptides that has been cocrystallized with a matrix selected to "match" the frequency of a UV laser. The laser vaporizes and ionizes the sample, which is then mass-analyzed, typically in a time-of-flight (TOF) mass analyzer. Since the footprint of the laser is small, and the sample is a solid rather than a solution, it is easy to see how this led to the idea of "rastering" the laser across the sample to form a molecular image.

After about ten years of development, MALDI imaging has finally come of age. This is partly due to newer MALDI-MS instrumentation that is capable of higher mass accuracy and resolution, as well as the development of MALDI-MS/MS for gas-phase sequencing. Several commercially-available sprayer/spotters have recently been developed which can produce a uniform coating of matrix on the sample. These sprayer/spotters can also be used to deposit enzyme solutions on targeted areas so that differentially-localized proteins can be identified.

This chapter describes some of the recent work in MALDI imaging, as well as some of the clinical applications of this technique. Finally, a new technique is described (MRM MALDI) which allows the quantitation of differentially-localized proteins on the basis of their peptide MS/MS spectra.

Introduction

The idea of creating a molecular image of a tissue for the purpose of differentiating between diseased and healthy tissue, as well as for biomarker discovery, has been a "holy grail" for mass spectrometrists for over ten years (Caprioli

C.H. Borchers

Department of Biochemistry & Microbiology, University of Victoria—Genome British Columbia Protein Centre, University of Victoria, 3101-4464 Markham Street, Vancouver Island Technology Park, Victoria, BC, Canada V8Z 7X8, e-mail: christoph@proteincentre.com

[*]Dedicated to our colleague and friend Viorel Mocanu, 2/23/69–4/8/08

C.W. Sensen and B. Hallgrímsson (eds.), *Advanced Imaging in Biology and Medicine.*   131
© Springer-Verlag Berlin Heidelberg 2009

et al. 1997b), and direct MALDI ("matrix-assisted laser desorption/ionization") analysis of bacteria has been performed since 1996 (Claydon et al. 1996; Holland et al. 1996; Krishnamurthy et al. 1996; Parker et al. 1996).

Clearly, the development of MALDI, a mass spectrometric technique where a laser is used to vaporize and ionize proteins and peptides that have been cocrystallized with matrix on a conductive target, was a major breakthrough for tissue imaging. This technology opened up the possibility of the direct analysis of tissue slices, but there were still many challenges to be overcome, including problems in uniformly applying the matrix solution to the tissue slices, minimizing analyte "spreading" (which would lead to low-resolution molecular images), automating the rastering across the sample during data acquisition, and reconstructing the acquired spectra into images.

These difficulties have now largely been overcome, and new instruments and software for molecular imaging by MALDI-MS (sometimes called imaging mass spectrometry, IMS) are now commercially available. Current software packages can produce false-color images, with colors corresponding to the location and/or abundances of specific masses from target peptides or proteins. These images thus reveal the spatial distributions of specific masses in the tissue samples (Caprioli 2004a, b, 2005, 2006, 2007; Caprioli et al. 1997a, b; Reyzer and Caldwell et al. 2006; Meistermann et al. 2006; Caldwell and Caprioli 2005; Chaurand et al. 2002, 2003, 2004a–e, 2005, 2006a, b, 2007; Groseclose et al. 2007; Khatib-Shahidi et al. 2006; Cornett et al. 2007; McLean et al. 2007; Crecelius et al. 2005; Caprioli and Chaurand 2001; Chaurand et al. 2002, 2001; Todd et al. 2001; Stoeckli et al. 2001, 1999). Multivariate statistical treatment has been applied to protein IMS for diagnostic applications (Schwamborn et al. 2007) and MALDI imaging techniques have been further applied to metabolite detection and phospholipids. In this chapter, however, we will focus on MALDI imaging for proteins.

Sample Preparation

Unlike normal tissue processing for histological studies, the preferred method of sample preparation is the sectioning of frozen tissue into thin slices of approx. 10–15 μm thickness. If possible, samples for MALDI imaging based on direct protein detection should not be formalin-treated, since formalin induces irreversible crosslinks by adding methylene groups between free amines. This reduces the protein signal intensities (Groseclose et al. 2005) and leads to the formation of adducts which interfere with the determination of the molecular weight (Redeker et al. 1998; Rubakhin et al. 2003).

However, the existence of large repositories of formalin-fixed tissue in hospitals causes continued interest in the utilization of these specimens and the development of new preparation approaches (Lemaire et al. 2007b) (Fig. 6.1).

Sample preparation for MALDI imaging is similar to that used in conventional light microscopy, except that the tissue has to be kept cold during processing. Instead

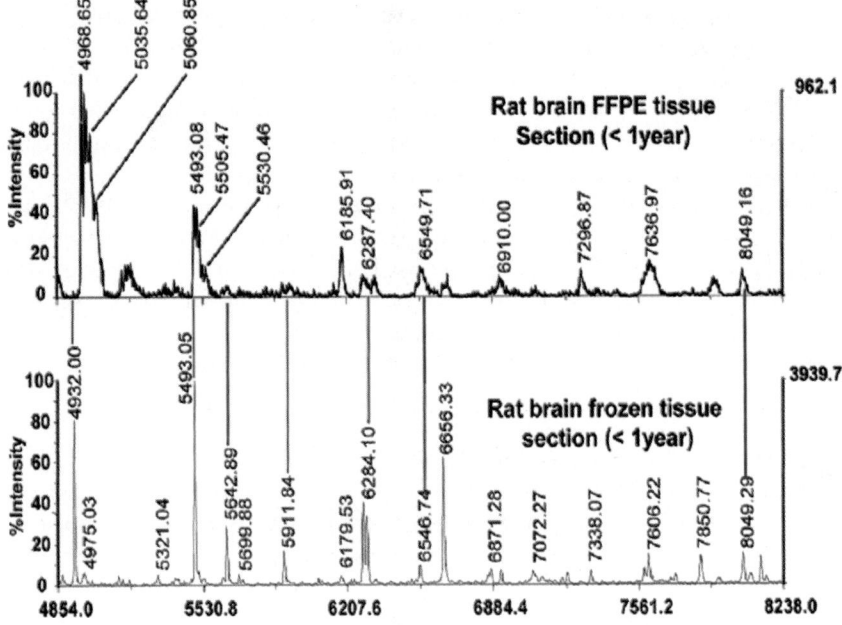

Fig. 6.1 Comparison of MALDI mass spectra from the direct analysis of a <1-year-old FFPE and fresh frozen rat brain tissues (obtained in the linear positive mode and recorded in the same region with sinapinic acid as matrix; each experiment was conducted five times) (Lemaire et al. 2007b) (figure and figure legend reprinted from Lemaire et al. 2007b with permission)

of being placed on a typical microscope glass slide, a ∼5–15 μm-thick slice of tissue is placed on a transparent, electrically conductive glass slide for MALDI analysis. For optimal imaging of mass spectra, neither paraffin embedding nor the use of OCT (optimal cutting temperature) polymer is employed (McDonnell and Heeren 2007). Instead, the frozen tissue is placed at −10 to −20°C onto the stage and sections are deposited onto cold plates, which are quickly warmed (thaw mounting) (Schwartz et al. 2003). Typically, thaw-mounted sections are washed with 70% ethanol (2 × 30 s) to remove salts and blood. This yields increased signal intensity and reduced ion suppression (Schwartz et al. 2003). To avoid degradation of the sample, these processes must be performed *immediately* before the matrix is coated onto the sample (Cornett et al. 2007). Once prepared, the samples are reasonably stable for several days under vacuum (Caprioli et al. 1997b). Another common technique is cryogenic slicing of the tissue, the 10 μm-thick slice being placed directly on the conductive MALDI target (Deininger et al. 2008). For best alignment of the images, hematoxylin-eosin (H&E) staining should follow the MALDI imaging analysis and be performed on the same section (Schwamborn et al. 2007; Deininger et al. 2008). The work done on consecutive sections has shown that it is difficult to unequivocally assign MALDI image features in highly-resolved microscopic images in detailed histomorphological work.

These have included electrospraying a solution of MALDI matrix (usually DHB, 2,5-dihydroxybenzoic acid) onto the tissue (Caprioli et al. 1997b), or pipetting matrix solution onto the tissue, or dipping the tissue into a matrix solution (Kutz et al. 2004). These techniques have now evolved into several commercial instruments, one of which, the Bruker ImagePrep^{TM}, uses vibrational vaporization-based spraying of the MALDI matrix over the tissue sample (Schuerenberg et al. 2007), while the other instrument, the Shimadzu Chemical Inkjet Printer "ChIP^{TM}," uses inkjet printer technology to deposit small droplets of picoliter volumes in a designated pattern on the tissue (Shimma et al. 2006). A third instrument, based on piezoelectric deposition, has been developed in the Caprioli laboratory (Aerni et al. 2006), and is available through LabCyte.

Inkjet spotters produce reproducible spots ($\sim$100 pL volume) that dry to form matrix crystals $\sim$100 μm in diameter, providing routine MALDI image resolutions in the 100–500 μm range (Cornett et al. 2007). One major advantage is the ability to go back over the same sample, deposit an enzyme solution on the same spots (to an accuracy of within 5 μm), allowing the identification of the proteins present on the basis of the molecular weights of the intact proteins, their peptide molecular weights and MS/MS-based peptide sequencing. In addition, because four different solutions can be used with the ChIP^{TM}, other chemistries or enzymes can also be used, allowing more flexibility in the tissue treatment prior to mass spectrometry. This instrument can spot an array of spots (at a spatial resolution corresponding to 250 μm spacing) covering an entire tissue section, up to a maximum of 120×70 mm (the size of MALDI targets).

The Bruker ImagePrep^{TM} device (see http://www.bdal.de/uploads/media/tech18_imageprep.pdf) produces a mist of matrix droplets with average diameter of $\sim$20 μm using a piezoelectric nebulizer, providing a dense layer of matrix crystals of about the same size. This eliminates the bottleneck for high-resolution MALDI imaging; 20 μm MALDI images can be obtained that allow very detailed matching with microscopic images for histological evaluation. All matrices and enzyme solutions can be sprayed in this device with a free choice of solvent composition in a robust manner, i.e., without any clogging of the piezoelectric nozzles. The spray process occurs under conditions of full automation in cycles of spraying, incubation and drying; it is controlled by a scattering light sensor that monitors the degree of wetness and the thickness of the matrix layer. This permits reproducible preparation for high-resolution imaging and, at the same time, it ensures sufficiently intense incubation of the small droplets in a controlled environment, providing efficient analyte extraction from the tissue. As the matrix is coated uniformly across the glass slides, the achievable MALDI image resolution depends essentially on the MALDI acquisition regime, i.e., the spot size and scanning raster geometry. Typically, spectral qualities comparable to those yielded by spotters or manual preparation are achieved at an image resolution (pixel size) of 20–50 μm.

Commercial matrix-application instruments work with mass spectrometers and tissue-imaging software from different vendors, and an image spotter/sprayer can be selected depending on the analytical requirements defined by the specific application.

MALDI-MS Imaging from Tissue

Tissue imaging by MALDI mass spectrometry has seen tremendous growth recently due to advances in sample preparation protocols, deposition technology, MALDI instrumentation and commercially available image analysis software. This software allows the correlation of results from the mass spectral data to specific locations of the tissue. There has been a recent dramatic increase in the use of MALDI imaging for tissue profiling, for example for determining cell types in cancerous tissue (Fig. 6.2, right panel, and Fig. 6.3). Also, novel approaches are being developed to increase the throughput of the technology for biomarker analysis for potential use in a clinical setting. A screening study of lung biopsy samples (Fig. 6.2, left panel) was done in approximately a single day from the time the samples were received (Groseclose et al. 2005).

Correlation of MALDI Images with Histology

Mass spectrometric requirements for high-resolution imaging are the ability to adapt the laser spot size to the desired pixel size (typically 10–100 μm) and an acquisition speed (200–1,000 Hz) and lifetime ($\sim 10^9$ shot) of the solid state laser that permit data acquisition in a reasonable timeframe. The ability to prepare and acquire high-resolution mass images triggered attempts to precisely match these with the corresponding microscopic images in order to establish unequivocal molecular correlations with histological features such as cancer *in situ* vs. invasive tumors, etc. (Fuchs et al. 2008).

Traditionally, MALDI and histological images (H&E staining, immunohistochemistry) were acquired on consecutive sections. This enabled straight parallel tissue preparation and matching of images at resolutions of 200–500 μm. The increased resolution of MALDI images in recent years has triggered the development of approaches that enable high-accuracy image matching, basically by using the same section for dual readout in the mass spectrometer and histostaining. A first approach described initial staining by a number of dyes (e.g., methylene blue and cresyl violet) that are compatible with subsequent MALDI analysis (Chaurand et al. 2004b). More recently, reversed staining analysis was developed in order to avoid the compatibility requirement between the staining method and the MALDI process (Crecelius et al. 2005; Schwamborn et al. 2007). This approach allows precision matching of 20 μm MALDI images with microscopic images that would not work if consecutive sections were used (Figs. 6.4, 6.5). It is extremely difficult to arrange successive sections on the glass slides such that their images accurately overlap across several mm. In addition, given a thickness of 10–15 μm, and the existence of histomorphological features that are not perpendicular to the surfaces of the sections, successive sections typically contain features that do not match at the 50 μm scale.

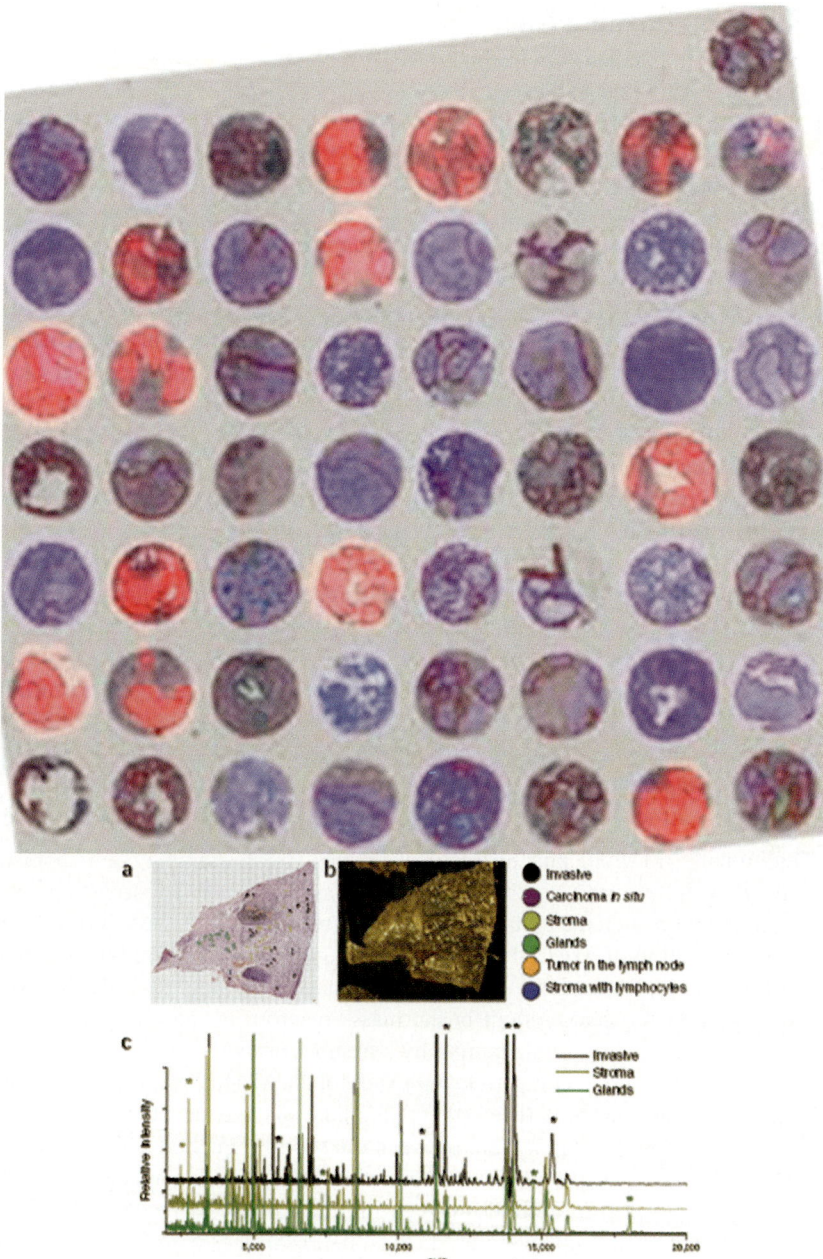

Fig. 6.2 MALDI molecular imaging. *Left*: Biomarker peptides found for two types of non-small cell lung cancers from human biopsies. (cancer 1, *blue*: *m/z* 2,854.34; cancer 2, *red*: *m/z* 1,410.81) (Groseclose et al. 2005) The workflow for this process took ~1 day from the time the tissue sample was received. *Right*: Histology-directed molecular profiling. Cell-specific peptides are marked (*) (Cornett et al. 2007) (figure and figure legends from Groseclose et al. 2005 and Cornett et al. 2007; reprinted with permission)

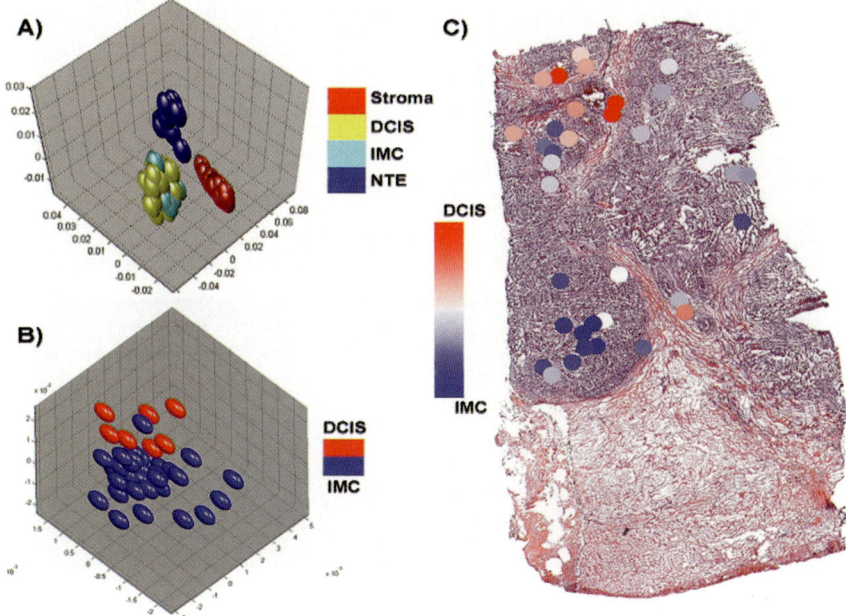

Fig. 6.3 (a–c) MALDI molecular imaging—profiling of breast cancer tissue. **(a)** Results of an unsupervised classification of the profiles of specific cell types acquired from one breast tumor section as determined by multidimensional scaling of results from an unsupervised classification of whole spectra. Colors indicate the histopathology of the profile site. **(b)** Spatial plot representing the similarity in the profiles of ductal carcinoma *in situ* (DCIS) and invasive mammary cancer (IMC), as determined by multidimensional scaling of the top-ranked markers identified by supervised classification. Each profile is colored according to the histopathology of the profile site. **(c)** H&E-stained section with annotation marks colored to represent the results of classification analysis. A gradient color scale derived from the supervised classification indicating the degree of similarity to DCIS or IMC characteristics is used (Cornett et al. 2006) (figure and figure legend from Cornett et al. (2006), reprinted with permission)

Biomarker Discovery and Identification

MALDI imaging can be considered to be a type of direct mixture analysis by MALDI. As such, the same problems that plague other direct mixture analysis techniques can be expected to occur here as well. The first is suppression effects—i.e., high-abundance proteins will suppress the ionization and detection of low-abundance proteins (Perkins et al. 1993; Billeci and Stults 1993).

Second, although a different pattern of masses may be observed in, for example, cancerous tissue vs. noncancerous tissue, the actual *identification* of the differentially expressed analyte may be challenging, especially for proteins. In general, it is not possible to identify a protein based only on its molecular weight. The identification of proteins that were detected by tissue imaging typically requires their extraction from tissue followed by classical protein separation approaches such as

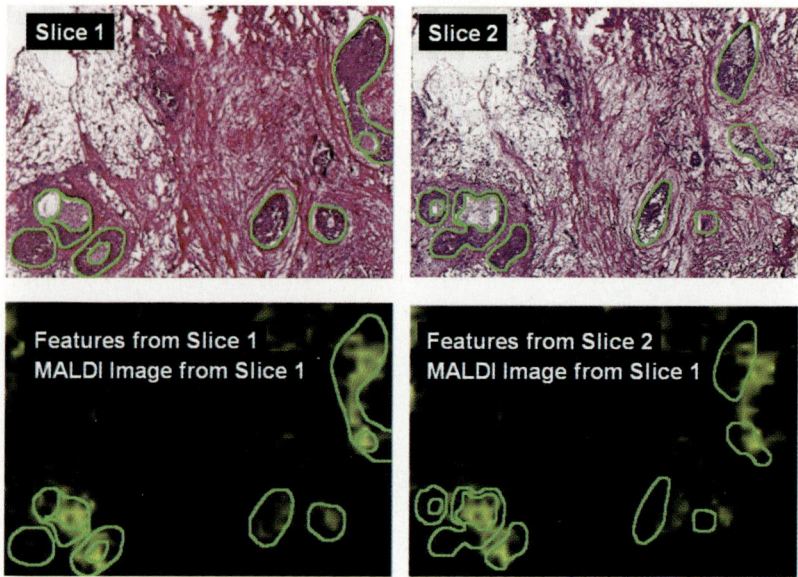

Fig. 6.4 Consecutive slice problem: two consecutive slices of human breast cancer H&E stains are shown (*top*) with specific histological structures marked in green. A MALDI image from slice 1 (*bottom*) is overlapped with the marked features on the same slice (*left*) and the consecutive slice (*right*), showing several feature mismatches (adapted from Deininger et al. 2008 with permission)

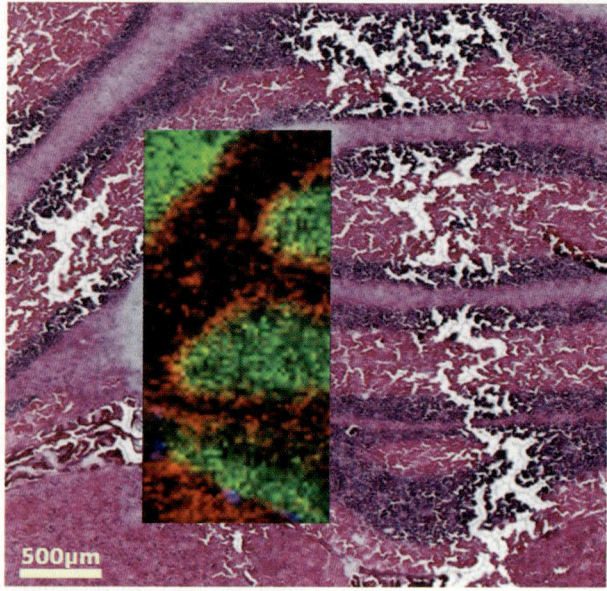

Fig. 6.5 Image of rat cerebellum. MALDI matrix sinapinic acid was applied with the ImagePrep™ and the image (25 µm pixel size, false color insert) was acquired prior to H&E staining using a 20 µm focus size of a smartbeam™ laser (Holle et al. 2006). Both images were aligned with the Bruker FlexImaging™ software (adapted from Holle et al. 2006 with permission)

proteolytic digestion followed by MALDI-MS/MS or LC-ESI-MS/MS (for methods used in our laboratories, see Parker et al. 2005).

Instruments that are capable of high-sensitivity high-mass-accuracy MALDI-MS/MS peptide sequencing are already commercially available. This instrumentation has already been used for the "bottom-up" identification of proteins digested on-target from the tissue slices. A solution of, for example, trypsin can be sprayed or "printed" on the tissue slice (Groseclose et al. 2007). Instead of simply showing patterns of different masses, it is now possible to identify the differentially expressed proteins based on "standard" peptide-based protein identification techniques (Holle et al. 2006). Unless the target protein is particularly abundant, peptide mass fingerprinting (Perkins et al. 1993) based on MALDI-MS may not be successful, since there are likely to be multiple proteins present in a digested "spot." MALDI-MS/MS-based sequence-tag protein identification techniques have now been used on tissue slices, allowing the creation of a true "protein map" (Groseclose et al. 2007). This has been a major breakthrough, facilitating biological interpretation of the results.

Another reason to use a bottom-up imaging approach is that it allows the detection of small peptides from formalin-treated samples that do not produce intact proteins. These peptides can be sequenced and identified, allowing the imaging of archived formalin-treated tissue samples (Groseclose et al. 2005).

However, technological challenges still remain, including suppression effects on the peptide level. Using a "bottom-up" approach, the complexity of the tissue eluate is increased from perhaps 3,000 proteins to 100,000 peptides, which by far exceeds the resolving capabilities of the TOF mass analyzer and the ionization capabilities of the MALDI process. MALDI-FTICR, however, could enable not only the detection but also the identification of molecules in tissues, including metabolites, because of its high mass accuracy and high-energy fragmentation options. This would be particularly beneficial for analyzing larger molecules.

MALDI from Membranes

The same imaging technology can be used for protein identification from PVDF membranes (Nadler et al. 2004). This means that proteins can be identified from Western blots, after on-gel digestion with trypsin or another enzyme. Various applications of this technique have included the localization of phosphotyrosine-containing proteins by probing the Western blot with anti-pY antibodies (Nakanishi et al. 2007) (Fig. 6.6). Tyrosine phosphorylation has been implicated in the signaling pathways for various cancers, including androgen-independent prostate cancer (Mahajan et al. 2007). Phosphotyrosine-containing proteins can then be identified by peptide mass fingerprinting and MALDI-MS/MS sequence determination. Since the process of protein identification is based on peptide identification, this system could be useful for the identification of pY-containing peptides.

Another promising application would be the analysis of phosphopeptides from nitrocellulose membranes. MALDI analysis of peptides from nitrocellulose

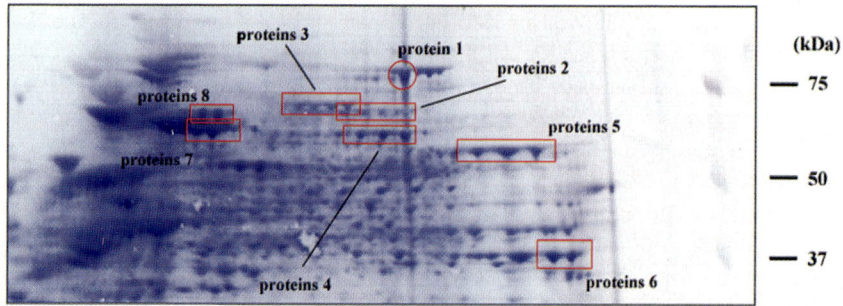

Fig. 6.6 Phosphoproteins detected and identified by peptide mass fingerprinting after on-membrane pY detection and MALDI-MS analysis (Nakanishi et al. 2007) (figure and figure legend from Nakanishi et al. 2007, reprinted with permission)

membranes has already been demonstrated (Caprioli et al. 1997b; Dufresne-Martin et al. 2005), and this technique could be adapted for the detection of nitrocellulose-separated phosphopeptides, probably by substituting a "cooler" MALDI matrix (e.g., DHB).

The multienzyme capabilities of the Shimadzu inkjet printing approach also allow the study of glycoproteins, since combinations of enzymes, such as trypsin and deglycosylating enzymes (e.g., PNGase F) can be used on the same PVDF spot (Kimura et al. 2007). However, the effectiveness and applicability of these methods remain to be demonstrated.

Quantitative Imaging of Targeted Proteins in Tissues

As described above, MALDI imaging has recently gained popularity, allowing researchers to determine the spatial distribution and relative abundance of proteins or other molecules in thin tissue sections. Targeted analysis of proteins is, in fact, possible if immune-specific amplification steps are included in the experimental design. The "tag-mass" technique allows the detection and localization of specific analytes in a tissue slice by *in situ* modification of the target analyte (Lemaire et al. 2007a; Thiery et al. 2007). The tag-modified probe binds to the analyte on the MALDI target. When the tag is released during the MALDI process, an image of a specific target analyte is produced (Lemaire et al. 2007a).

For example, in the "tag-mass" approach, tagged antibodies can be used to detect target proteins. Thus, rather than detecting whatever molecules happen to be present and above the detection threshold in the desorption pixel, photocleavable, charge-stabilized mass tags are attached to antibodies to the target proteins. "Staining" of histological sections is carried out in a similar way to common immune-histochemical procedures, with chemiluminescent or fluorescent detection using all of the antibodies of a multiplex simultaneously. Mass tags with discrete masses are released from their respective antibodies by a laser pulse without added matrix. After scanning, mass spectrometry-based images are created for the mass of each tag. This permits the multiplexed targeted analysis of proteins below the detection threshold of direct MALDI-MS imaging. The lack of MALDI matrix in the mass tag approach eliminates the charge competition problem that limits the analysis to a few (say 100–500) molecular species that can be visualized, permitting the targeted detection of the labeled proteins.

A major limitation of MALDI imaging, however, is still that it provides a *qualitative* picture of analyte distribution, but—because of suppression effects and because of compound-dependent ionization efficiencies—*quantitative* measurements are usually not possible without the use of labeled standards. The Applied Biosystems 4000 QTrap mass spectrometer equipped with a prototype vacuum MALDI source has the potential to overcome this limitation and should be able to provide data for both the detection and the absolute quantitation of targeted proteins, peptides, and drug metabolites in tissues.

Using this novel MS instrumentation, the Borchers group has pioneered a new technique that couples the ability to perform multiple reaction monitoring (MRM) analyses on a triple quadrupole mass spectrometer with the spatial resolving power made possible with a MALDI source.

MRM is a technique that uses multiple stages of mass spectrometry to achieve high levels of sensitivity and specificity in the quantitation of proteins from a complex mixture. Although MRM assays have been widely used for many years in pharmaceutical companies to quantify small molecules in biological specimens, their applicability to the quantitation of peptides has only recently been demonstrated (Lange et al. 2008). The greatest advantages of this technique for MALDI imaging are increased specificity through the precise selection of a targeted mass of peptides, metabolites and other compounds in the first quadrupole, and a specific fragment ion mass in the third quadrupole of these molecules. Analytes can now be quantitated more accurately through the addition of stable isotope-labeled internal standards, and sensitivity is increased due to the lack of interference with competing ions. Preliminary studies were undertaken using a mixture of peptides spotted onto rat liver slices and coated with CHCA, and these have shown the feasibility of detecting and quantifying targeted proteins at spatial resolution. Custom software is being written in the Borchers group to plot the data as a three-dimensional image (Figs. 6.7 and 6.8).

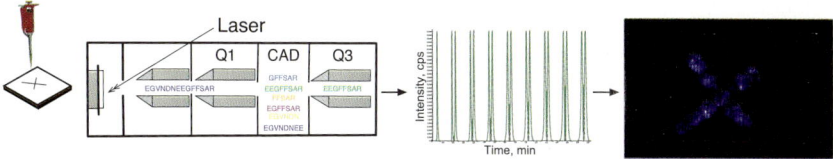

Fig. 6.7 Schematic of the MRM MALDI imaging process. Glu-fibrinopeptide $(1\,pmol\,\mu L^{-1})$ is spotted onto a MALDI target plate and coated with CHCA matrix. The sample plate is then rastered under the laser and Q1 selects the parent ion and Q3 selects the peptide specific fragment ion. The data file is then converted into a three-dimensional image with in-house software

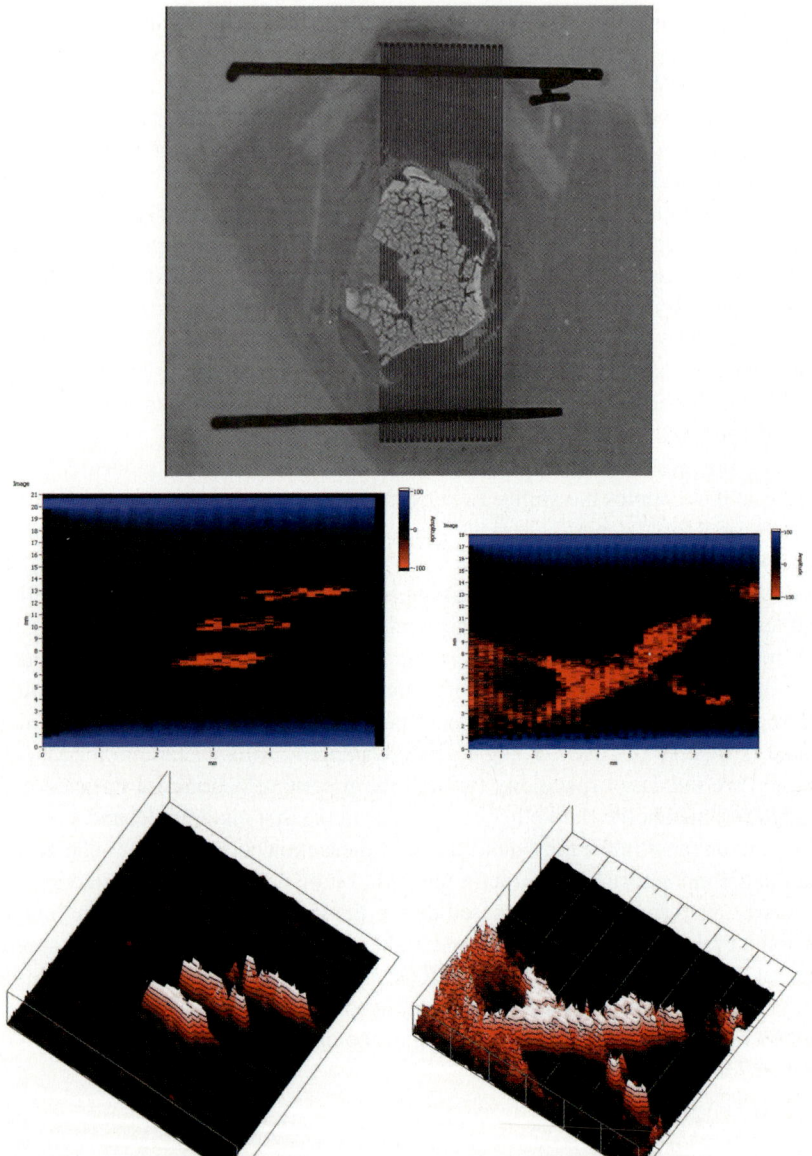

Fig. 6.8 Peptide spotted onto 12 μm-thick slice of mouse liver. Tissue slices were prepared by cryostat and excess OCT was removed by hand. Glu-fibrinopeptide (100 pmol μL^{-1}) was applied to tissue as either three strokes or in an X using a fine artist's paintbrush. The tissue was then coated 10× with α-cyano-4-hydroxy-cinnaminic acid using an airbrush sprayer. Data was acquired on an Applied Biosystems QTrap 4000 mass spectrometer fitted with a prototype vacuum MALDI source. One-dimensional data files were converted to two- and three-dimensional image files using custom software

To date, MRM MALDI imaging has been limited to relatively small ions such as peptides and drug metabolites because of the limited mass range of the quadrupoles (typically about 3,000 Da) and the low collision energies used, which precludes fragmentation of larger peptides.

Using newly-developed on-plate tryptic digestion protocols developed by the Caprioli group (Groseclose et al. 2007), it is now possible to enzymatically digest proteins while they remain fixed in the two-dimensional space of the tissue slice. The resulting tryptic peptides can then be measured in MRM mode on a triple quadrupole mass spectrometer and, with the addition of stable isotope-labeled peptide analogs, the absolute abundance of proteins can be determined.

Conclusion

It is impossible to overstate the contribution that Richard Caprioli has made to the field of MALDI imaging. New mass spectrometric instrumentation has provided many researchers, including pathologists and histologists, with several platforms for rugged and reliable MALDI imaging, and an ever-increasing number of other researchers are making exciting contributions to this field. This, in turn, has led to new commercially available matrix-deposition devices, including ultrasonic sprayers and "printers," further advancing this area of research and its applications. When one considers the recent developments in qualitative and quantitative MALDI imaging, this technology truly seems poised to make the leap from the mass spectrometry research laboratory first to the clinical research laboratory and then to the diagnostic clinic.

Acknowledgements This book chapter was supported by funding from Genome BC, Genome Canada (CB & DS) and by an anonymous donor (CP).

References

Aerni H-R, Cornett DS, Caprioli RM (2006) Automated acoustic matrix deposition for MALDI sample preparation. Anal Chem 78(3):827–834
Billeci TM, Stults JT (1993) Tryptic mapping of recombinant proteins by matrix-assisted laser desorption/ionization mass spectrometry. Anal Chem 65(13):1709–1716
Caldwell RL, Caprioli RM (2005) Tissue profiling and imaging by matrix-assisted laser desorption ionization mass spectrometry. Proteomics Cancer Res :107–116
Caldwell RL, Gonzalez A, Oppenheimer SR, Schwartz HS, Caprioli RM (2006) Molecular assessment of the tumor protein microenvironment using imaging mass spectrometry. Cancer Genomics Proteomics 3(4):279–288
Caprioli RM (2004a) Imaging of proteins in tissue sections using mass spectrometry as a discovery tool. Zhipu Xuebao 25(Suppl):119–120
Caprioli RM (2004b) Direct imaging and profiling of proteins in tissues using mass spectrometry to aid diagnosis and treatment of disease and to identify therapeutic targets. Abstracts from the

56th Southeast Regional Meeting of the American Chemical Society, Research Triangle Park, NC, 10–13 Nov. 2004, GEN-211

Caprioli RM (2005) Direct imaging and profiling of proteins in tissues using mass spectrometry to aid diagnosis and treatment of disease and to identify therapeutic targets. Abstracts of Papers from the 229th ACS National Meeting, San Diego, CA, 13–17 March 2005, ANYL-385

Caprioli RM (2006) In situ molecular imaging of proteins in tissues using mass spectrometry: A new tool for biological and clinical research. Abstracts of Papers from the 231st ACS National Meeting, Atlanta, GA, 26–30 March 2006, ANYL-276

Caprioli RM, Chaurand P (2001) Profiling and imaging of peptides and proteins in biological tissues using MALDI MS. Adv Mass Spectrom 15:3–18

Caprioli RM, Farmer TB, Zhang H, Stoeckli M (1997a) Molecular imaging of biological samples by MALDI MS. Book of Abstracts from the 214th ACS National Meeting, Las Vegas, NV, 7–11 Sept. 1997, ANYL-113

Caprioli RM, Farmer TB, Gile J (1997b) Molecular imaging of biological samples: localization of peptides and proteins using MALDI-TOF MS. Anal Chem 69(23):4751–4760

Chaurand P, Caprioli RM (2001) Direct tissue analysis for profiling and imaging of proteins. Abstracts of Papers from the 222nd ACS National Meeting, Chicago, IL, 26–30 Aug. 2001, TOXI-003

Chaurand P, Caprioli RM (2002) Direct profiling and imaging of peptides and proteins from mammalian cells and tissue sections by mass spectrometry. Electrophoresis 23(18):3125–3135

Chaurand P, Schwartz SA, Caprioli RM (2002) Imaging mass spectrometry: a new tool to investigate the spatial organization of peptides and proteins in mammalian tissue sections. Curr Opin Chem Biol 6(5):676–681

Chaurand P, Fouchecourt S, DaGue BB, Xu, BJ, Reyzer ML, Orgebin-crist M-C, Caprioli RM (2003) Profiling and imaging proteins in the mouse epididymis by imaging mass spectrometry. Proteomics 3(11):2221–2239

Chaurand P, Sanders ME, Jensen RA, Caprioli RM (2004a) Proteomics in diagnostic pathology: profiling and imaging proteins directly in tissue sections. Am J Pathol 165(4):1057–1068

Chaurand P, Schwartz SA, Caprioli RM (2004b) Profiling and imaging proteins in tissue sections by MS. Anal Chem 76(5):86A–93A

Chaurand P, Schwartz SA, Caprioli RM (2004c) Assessing protein patterns in disease using imaging mass spectrometry. J Proteome Res 3(2):245–252

Chaurand P, Schwartz SA, Billheimer D, Xu BJ, Crecelius A, Caprioli RM (2004d) Integrating histology and imaging mass spectrometry. Anal Chem 76(4):1145–1155

Chaurand P, Hayn G, Matter U, Caprioli RM (2004e) Exploring the potential of cryodetectors for the detection of matrix-assisted laser desorption/ionization produced ions: application to profiling and imaging mass spectrometry. Zhipu Xuebao 25(suppl):205–206,216

Chaurand P, Schwartz S, Reyzer M, Caprioli R (2005) Imaging mass spectrometry: principles and potentials. Toxicol Pathol 33(1):92–101

Chaurand P, Cornett DS, Caprioli RM (2006a) Molecular imaging of thin mammalian tissue sections by mass spectrometry. Curr Opin Biotechnol 17(4):431–436

Chaurand P, Norris JL, Cornett DS, Mobley JA, Caprioli RM (2006b) New developments in profiling and imaging of proteins from tissue sections by MALDI mass spectrometry. J Proteome Res 5(11):2889–2900

Chaurand P, Schriver KE, Caprioli RM (2007) Instrument design and characterization for high resolution MALDI-MS imaging of tissue sections. J Mass Spectrom 42(4):476–489

Claydon MA, Davey SN, Edwards-Jones V, Gordon DB (1996) The rapid identification of intact microorganisms using mass spectrometry. Nat Biotechnol 14:1584–1586

Cornett DS, Mobley JA, Dias EC, Andersson M, Arteaga CL, Sanders ME, Caprioli RM (2006) A novel histology-directed strategy for MALDI-MS tissue profiling that improves throughput and cellular specificity in human breast cancer. Mol Cell Proteomics 5(10):1975–2198

Cornett DS, Reyzer ML, Chaurand P, Caprioli RM (2007) MALDI imaging mass spectrometry: molecular snapshots of biochemical systems. Nat Methods 4(10):828–833

Crecelius AC, Cornett DS, Caprioli RM, Williams B, Dawant BM, Bodenheimer B (2005) Three-dimensional visualization of protein expression in mouse brain structures using imaging mass spectrometry. J Am Soc Mass Spectrom 16(7):1093–1099

Deininger S-O, Krieg R, Rauser S, Walch A (2008) Advances in molecular histology with the MALDI molecular imager. Bruker Appl Note MT-89:1–6

Dufresne-Martin G, Lemay J-F, Lavigne P, Klarskov K (2005) Peptide mass fingerprinting by matrix-assisted laser desorption ionization mass spectrometry of proteins detected by immunostaining on nitrocellulose. Proteomics 5:55–66

Fuchs B, Schiller J, Suess R, Zscharnack M, Bader A, Mueller P, Schuerenberg M, Becker M, Suckau D (2008) Analysis of stem cell lipids by offline HPTLC-MALDI-TOF MS. Anal Bioanal Chem. Aug 5, Epub ahead of print

Groseclose MR, Chaurand P, Wirth PS, Massion PP, Caprioli RM (2005) Investigating formalin fixed paraffin embedded tissue specimens by imaging MALDI mass spectrometry: application to lung cancer tissue microarrays. Presented at HUPO 4, 29 Aug.–1 Sept. 2005, Munich, Germany

Groseclose MR, Andersson M, Hardesty WM, Caprioli RM (2007) Identification of proteins directly from tissue: in situ tryptic digestions coupled with imaging mass spectrometry. J Mass Spectrom 42(2):254–262

Holland RD, Wilkes JG, Rafii F, Sutherland JB, Persons CC, Voorhees KJ, Lay JOJ (1996) Rapid identification of intact whole bacteria based on spectral patterns using matrix-assisted laser desorption/ionization with time-of-flight mass spectrometry. Rapid Commun Mass Spectrom 10:1227–1232

Holle A, Haase A, Kayser M, Höhndorf J (2006) Optimizing UV laser focus profiles for improved MALDI performance. J Mass Spectrom 41(6):705–716

Khatib-Shahidi S, Andersson M, Herman JL, Gillespie TA, Caprioli RM (2006) Direct molecular analysis of whole-body animal tissue sections by imaging MALDI mass spectrometry. Anal Chem 78(18):6448–6456

Kimura S, Kameyama A, Nakaya S, Ito H, Narimatsu H (2007) Direct on-membrane glyco-proteomic approach using MALDI-TOF mass spectrometry coupled with microdispensing of multiple enzymes. J Proteome Res 6(7):2488–2494

Krishnamurthy T, Ross PL, Rajamani U (1996) Detection of pathogenic and nonpathogenic bacteria by matrix-assisted laser desorption/ionization time-of-flight mass spectrometry. Rapid Commun Mass Spectrom 10:883–888

Kutz KK, Schmidt JJ, Li L (2004) In situ tissue analysis of neuropeptides by MALDI FTMS in-cell accumulation. Anal Chem 76(19):5630–5640

Lange V, Malmström JA, Didion J, King NL, Johansson, BP, Schäfer J, Rameseder J, Wong CH, Deutsch EW, Brusniak MY, Bühlmann P, Björck L, Domon B, Aebersold R (2008) Targeted quantitative analysis of *Streptococcus pyogenes* virulence factors by multiple reaction monitoring. Mol Cell Proteomics 7(8):1489–1500

Lemaire R, Stauber J, Wisztorski M, Van Camp C, Desmons A, Deschamps M, Proess G, Rudlof I, Woods AS, Day R, Salzet M, Fournier I (2007a) Tag-mass: specific molecular imaging of transcriptome and proteome by mass spectrometry based on photocleavable tag. J Proteome Res 6(6):2057–2067

Lemaire R, Desmons A, Tabet JC, Day R, Salzet M, Fournier I (2007b) Direct analysis and MALDI imaging of formalin-fixed, paraffin-embedded tissue sections. J Proteome Res 6(4):1295–1305

Mahajan NP, Liu Y, Majumder S, Warren MR, Parker CE, Mohler JL, Earp HS III, Whang YE (2007) Activated Cdc42-associated kinase Ack1 promotes prostate cancer progression via androgen receptor tyrosine phosphorylation. Proc Natl Acad Sci USA 104(20):8438–8443

McDonnell LA, Heeren RMA (2007) Imaging mass spectrometry. Mass Spectrom Rev 26:606–643

McLean JA, Ridenour WB, Caprioli RM (2007) Profiling and imaging of tissues by imaging ion mobility-mass spectrometry. J Mass Spectrom 42(8):1099–1105

Meistermann H, Norris JL, Aerni H-R, Cornett DS, Friedlein A, Erskine AR, Augustin A, De Vera Mudry MC, Ruepp S, Suter L, Langen H, Caprioli RM, Ducret A (2006) Biomarker discovery by imaging mass spectrometry: transthyretin is a biomarker for gentamicin-induced nephrotoxicity in rat. Mol Cell Proteomics 5(10):1876–1886

Nadler TK, Wagenfeld BG, Huang Y, Lotti RJ, Parker KC, Vella GJ (2004) Electronic Western blot of matrix-assisted laser desorption/ionization mass spectrometric-identified polypeptides from parallel processed gel-separated proteins. Anal Biochem 332:337–348

Nakanishi T, Ando E, Furuta M, Tsunasawa S, Nishimura O (2007) Direct on-membrane peptide mass fingerprinting with MALDI–MS of tyrosine-phosphorylated proteins detected by immunostaining. J Chromatogr B 846:24–29

Parker CE, Papac DI, Tomer KB (1996) Monitoring cleavage of fusion proteins by matrix-assisted laser desorption ionization/mass spectrometry: recombinant HIV-1IIIB p26. Anal Biochem 239(1):25–34

Parker CE, Warren MR, Loiselle DR, Dicheva NN, Scarlett CO, Borchers CH (2005) Identification of components of protein complexes, vol. 301. Humana, Patterson, NJ, pp 117–151

Perkins JR, Smith B, Gallagher RT, Jones DS, Davis SC, Hoffman AD, Tomer KB (1993) Application of electrospray mass spectrometry and matrix-assisted laser desorption ionization time-of-flight mass spectrometry for molecular weight assignment of peptides in complex mixtures. J Am Soc Mass Spectrom 4(8):670–84

Redeker V, Toullec JY, Vinh J, Rossier J, Soyez D (1998) Combination of peptide profiling by matrix-assisted laser desorption/ionization time-of-flight mass spectrometry and immunodetection on single glands or cells. Anal Chem 70(9):1805–1811

Reyzer ML, Caprioli RM (2005) MS imaging: new technology provides new opportunities. In: Korfmacher WA (ed) Using mass spectrometry for drug metabolism studies. CRC, Boca Raton, FL, pp 305–328

Reyzer ML, Caprioli RM (2007) MALDI-MS-based imaging of small molecules and proteins in tissues. Curr Opin Chem Biol 11(1):29–35

Rubakhin SS, Greenough WT, Sweedler JV (2003) Spatial profiling with MALDI MS: distribution of neuropeptides within single neurons. Anal Chem 75(20):5374–5380

Schuerenberg M, Luebbert C, Deininger S, Mueller R, Suckau D (2007) A new preparation technique for MALDI tissue imaging. Presented at the ASMS Workshop on Imaging Mass Spectrometry, Sanibel, FL, 19–22 Jan. 2007

Schwamborn K, Krieg RCMR, Jakse G, Knuechel R, Wellmann A (2007) Identifying prostate carcinoma by MALDI imaging. Int J Mol Med 20:155–159

Schwartz SA, Reyzer ML, Caprioli RM (2003) Direct tissue analysis using matrix-assisted laser desorption/ionization mass spectrometry: practical aspects of sample preparation. J Mass Spectrom 38(7):699–708

Shimma S, Furuta M, Ichimura K, Yoshida Y, Setou M (2006) Direct MS/MS analysis in mammalian tissue sections using MALDI-QIT-TOFMS and chemical inkjet technology. Surf Interface Anal 38:1712–1714

Stoeckli M, Farmer TB, Caprioli RM (1999) Automated mass spectrometry imaging with a matrix-assisted laser desorption ionization time-of-flight instrument. J Am Soc Mass Spectrom 10(1):67–71

Stoeckli M, Chaurand P, Hallahan DE, Caprioli RM (2001) Imaging mass spectrometry: A new technology for the analysis of protein expression in mammalian tissues. Nat Med (New York) 7(4):493–496

Thiery G, Shchepinov MS, Southern EM, Audebourg A, Audard V, Terris B, Gut IG (2007) Multiplex target protein imaging in tissue sections by mass spectrometry—TAMSIM. Rapid Commun Mass Spectrom 21:823–829

Todd PJ, Schaaff TG, Chaurand P, Caprioli RM (2001) Organic ion imaging of biological tissue with secondary ion mass spectrometry and matrix-assisted laser desorption/ionization. J Mass Spectrom 36(4):355–369

Chapter 7
Imaging and Evaluating Live Tissues at the Microscopic Level

John Robert Matyas

Abstract There are many ways in which images can be acquired, including the transmission, reflection, and emission of light, radiation, sound, magnetism, heat and other sources of energy. Some methods are more or less suitable for imaging living tissues and cells. Images, while informative and useful for generating hypotheses, may be quantified to actually test hypotheses. Various methods of image quantification are discussed, including pixel counting and stereology. Multimodal imaging, and multidimensional combinations of images with intensity information in the domains of genomics, proteomics, and metabolomics pave the way to a fuller understanding of biology and medicine.

Introduction

Early studies of animal and plant biology involved the direct examination of living tissues. Such curiosity was further piqued with improvements in the quality of optical lenses, which enabled ever-closer glimpses into the microscopic world. Still, the word "microscopic" conjures images in the minds of many students of thinly sliced fixed tissues mounted on glass slides that can readily be viewed by light microscopy. The utility of paraffin sectioning to modern biology and medicine is also natural. Spurred on by the industrial revolution, advances in organic chemistry (particularly the development of natural and synthetic textile dyes) as well as the production of high-quality steel knives and glass lenses opened up vast new opportunities for biologists and medical scientists to explore the fundamental organization of tissues and cells. These advances served the twentieth century well; with the advent of the twenty-first century, further advances in physics, engineering, and computing have promoted a renaissance in live tissue imaging, which now can be done at spatial and

J.R. Matyas
Faculty of Veterinary Medicine, 3330 Hospital Drive NW, Calgary, Alberta, Canada T2N 4N1, e-mail: jmatyas@ucalgary.ca

C.W. Sensen and B. Hallgrímsson (eds.), *Advanced Imaging in Biology and Medicine*.
© Springer-Verlag Berlin Heidelberg 2009   147

temporal resolutions never before imagined. While this chapter describes a variety of techniques in microscopy and histomorphometry that are suited to the study of fixed and embedded tissues, it focuses especially on those techniques that can be used to an advantage when imaging and evaluating living tissues and cells at the microscopic level.

Forming and Acquiring an Image of Live Tissue

One of the simplest, most informative, and underutilized methods for imaging living organisms, organs, tissues, and even cells is a hand lens. Much can be learned about the form and organization of both simple organisms and complex organs simply by looking with a magnifying glass, a dissecting stereomicroscope, or the inverted ocular lens of a compound microscope. I have one of my undergraduate professors at Cornell University to thank for instilling this habit into me before beginning a dissection (indeed, this course was graded partly on how well the dissection was illustrated, a strong incentive for high-quality inspection). This habit was later reinforced during my training in an anatomical pathology laboratory, where eyeballing each glass slide at arm's length preceded mounting the slide on the microscope. Important architectural features otherwise obscured by a limited field of view or depth of focus can often be discerned with the naked eye or a hand lens. Applying vital stains such as methylene blue to the surface of tissues can reveal valuable information about the two and three-dimensional arrangement of cells within tissues. Additional textural features can be enhanced at low magnification through the use of monochromatic, polarized, or oblique illumination techniques (Fig. 7.1).

Light Microscopy

Most biological research laboratories are equipped with a light microscope, designed in either the upright (fixed objectives above a moving stage) or inverted (moving objectives below a fixed stage) configuration. Either microscope design can be used for examining living or fixed tissues, though those of the inverted design are typically supplied with long working distance objectives that are well suited to focusing on tissues and cells through the bottom of tissue culture plasticware. Whereas microscopes are ordered with a standard configuration, all research microscopes can be customized with a wide assortment of special lenses and filters that can substantially improve performance. For critical work with living tissues, it may desirable to invest in lenses designed for water immersion, and in illumination systems that enhance contrast without histochemical staining. Phase-contrast illumination is the most common method of enhancing the contrast of unstained tissues and cells, though differential interference contrast ("Nomarski") and Rheinberg illumination are two different methods of enhancing the contrast of both stained and unstained tissues and cells. Polarized light microscopy is also highly effective for studying the

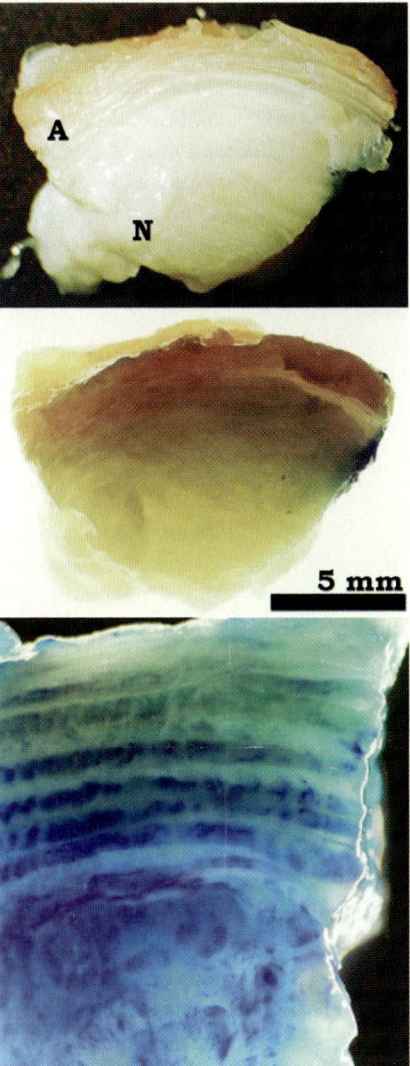

Fig. 7.1 Quarter-round segment of calf tail intervertebral disc cut in the axial plane. *Top panel*: unstained reflected polarized light; *middle panel*: unstained transmitted light. Note the texture of the collagen fibers in the annulus (*A*) compared to the more gelatinous nucleus pulposus (*N*). *Bottom panel*: a higher magnification using oblique illumination, showing the disc surface stained with vital dye (1% aqueous methylene blue). Note the additional contrast of the lamellae; the *smallest blue dots* are individual cells. Images taken with Zeiss Stemi stereomicroscope (2–×)

presence and distribution of birefringent macromolecules and substances, including fibrillar elements of the cytoskeleton (e.g., Inoué and Oldenbourg 1998) and the extracelluar matrix. Indeed, these illumination methods not only enhance the visualization of biological structure, they can also be used to varying degrees to quantify tissue characteristics (see below).

Most students remember from high school biology that images in a typical (upright) microscope are formed from light rays that pass through a biological sample, are collected by the objective lens, and are then focused into the eyepiece of the microscope and in turn to the eye. This configuration, wherein light passes through the specimen, is known as trans-illumination. It is also possible to form images of specimens by collecting light (or sound) reflected off a specimen, or by detecting various signals emitted from a specimen. The following sections describe the methods of microscopic imaging in transmission, reflection and emission. At the time of writing, many superb Internet resources for microscopy are available, several hosted by microscopy manufacturers and suppliers (e.g., http://www.olympusmicro.com, http://www.microscopyu.com, http://www.zeiss.com, http://www.leica-microsystems.com, http://micro.magnet.fsu.edu/primer/index.html).

Transmission

As mentioned above, transmitting light through a specimen is the most typical method of microscopy used in most biology laboratories. To allow light to pass through it and reveal its structure, the specimen must be sufficiently thin, or else light scattered within the tissue or at its surface will obscure the structural details. Routine histological sections range from 5 to 10μm, which is usually sufficient to resolve cellular detail—in particular the clear distinction between nuclear and cytoplasmic elements. This is true for both living tissues and those that are fixed and stained. While many tissues require sectioning to make visible their internal structure, others are thin enough to be examined directly. The abdominal mesentery or cremaster muscle, for example, transferred from the abdominal or scrotal cavity of an animal to a stage of a microscope, is available for directly studying the dynamics of blood and endothelial cells in the capillaries of the gut (e.g., Cara and Kubes 2003). As mentioned above, the contrast of unstained tissues (living or not) can be enhanced further by polarized light, differential interference contrast (e.g., Fig. 7.2), and phase contrast. Dark-field illumination, and its magnificent color variant, Rheinberg illumination, can also be used to great advantage with small living organisms, and is particularly informative for highlighting silver grains in emulsion autoradiographs.

The principle of passing visible light through a specimen can be extended to other forms of radiation, including electron beam radiation. Such is the case for transmission electron microscopy (TEM), wherein electrons are boiled from a filament and accelerated and focused through a very thin section (e.g., 50–100nm). As this method typically involves epoxy embedding and a high-vacuum chamber, living tissue is not readily studied by TEM. Nonetheless, it is possible to pass other forms of electromagnetic radiation through living specimens, including lasers (Tromberg et al. 2000), X-rays (see Chap. 10 in this volume), and synchrotron irradiation (Miao et al. 2004), all of which can reveal very fine details in living specimens. Depending on the energy and wavelength, and the thickness of the specimens, the

Fig. 7.2 High-magnification photomicrographs of the meniscus fibrocartilage of the guinea pig knee joint. *Top panel*: differential interference contrast (DIC) illumination; note the texture evoked at the edges of both collagen fibers and cells. *Bottom panel*: full polarization with DIC. Note that the collagen fibers appear either bright (+45°) or dark (−45°) depending on their orientation relative to the plane of polarization. Fixed, paraffin-embedded tissues stained with H&E are shown

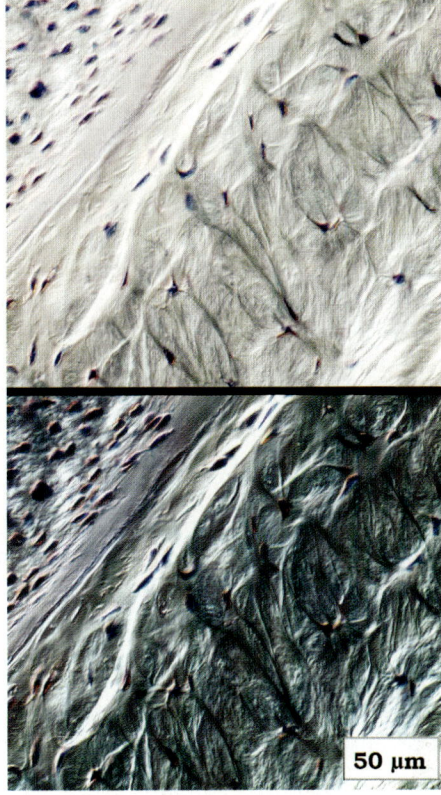

50 μm

transmission of radiation is retarded and scattered to varying degrees. While this scattering achieves the desired effect, the interpretation and quantification of this scattering can be less than straightforward.

Reflection

While typical of macrophotography, wherein a camera lens is mounted on a bellows extension or a special macro lens is used, light reflected off the surface of an object forms an image, whether on film or on a digital sensor. Indeed, the last decade saw a revolution in the resolution, sensitivity, and color quality of consumer digital cameras that have been a tremendous boost in terms of both efficiency and sensitivity to biology and medical laboratories. While digital cameras are eminently practical, there still are special uses for film, including capturing specimens illuminated with ultraviolet and infrared wavelengths, though digital cameras can be modified to capture these as well. Monochromatic light can be used to advantage for specimen photography in reflection (e.g., Cutignola and Bullough 1991). Fresh biological specimens are often moist and, depending on the lighting, specular reflection from wet surfaces can be troublesome to image. This problem can be minimized by careful lighting, in particular through the use of polarizing filters.

It has been possible for some time to optically measure specimen thickness or roughness using interference objectives built specially for microscopy, though these methods are exploited primarily in materials science. Recently, a new class of optical devices has been commercialized to measure surface height with nanometer resolution (optical profilometry). These devices serve both the fields of material science and biology, and can accurately record surface roughness and texture (profilometry, dimensional metrology, microtopography) of living tissues, including human skin. Some of these devices use laser illumination, while others employ white light and take advantage of chromatic aberration phenomena (which is beyond the scope of this text) to make these measurements. Such measurements can be especially useful for evaluating surface geometry and wear; these are measurements that were previously limited mainly to specimens fixed and coated for examination by scanning electron microscopy (SEM). Only very recently have so-called environmental SEMs (ESEM) become commercially available for studying unfixed, hydrated biological specimens. The ESEM requires very steep pressure gradients to be maintained between the electron beam column ($\sim 10^{-9}$ torr) and the sample chamber (~ 10 torr of water vapor). Like older SEMs, ESEM form images of biological specimens by collecting secondary electrons reflected off their surfaces to reach the detector.

Laser light reflected off the surface of living tissue can also be used to measure blood flow. Laser Doppler and, more recently, laser speckle imaging can be used to form images of blood moving in surface vessels, and can be used to determine blood flow (e.g., Forrester et al. 2004).

A much more widespread use of reflected light illumination is found in fluorescence microscopy. Whereas it is possible to image tissues with transmitted fluorescent light, it is much more common to collect reflected fluorescent light for microscopy. This configuration is known as epifluorescence microscopy, which can be performed on either an upright or inverted microscope. Epifluorescence typically implements a dichroic mirror capable of reflecting the shorter excitation wavelength and transmitting the longer emission wavelength of fluorescent light. This configuration permits both illumination through, and collection of emitted light by, the objective lens. As virtually none of the excitation light fluorescence re-enters the objective, the background is extremely low. Therefore, even though only a small fraction of fluorescence emission is collected, this method is very sensitive due to the high signal-to-noise ratio.

So-called wide-field fluorescence microscopy typically uses a mercury or xenon light source, which emit a wide spectrum of light. To maximize the signal-to-noise ratio in these systems, only a fraction of this spectrum is transmitted to the specimen due to the use of an appropriate band-pass excitation filter. Confocal scanning fluorescence microscopy (CSLM) follows essentially the same light path as used for widefield fluorescence, but a laser is used as an excitation light source. A powerful laser light source is advantageous in CSLM because its wavelength is concentrated at one or more narrow "lines" (e.g., 488 nm), and because a very bright light source

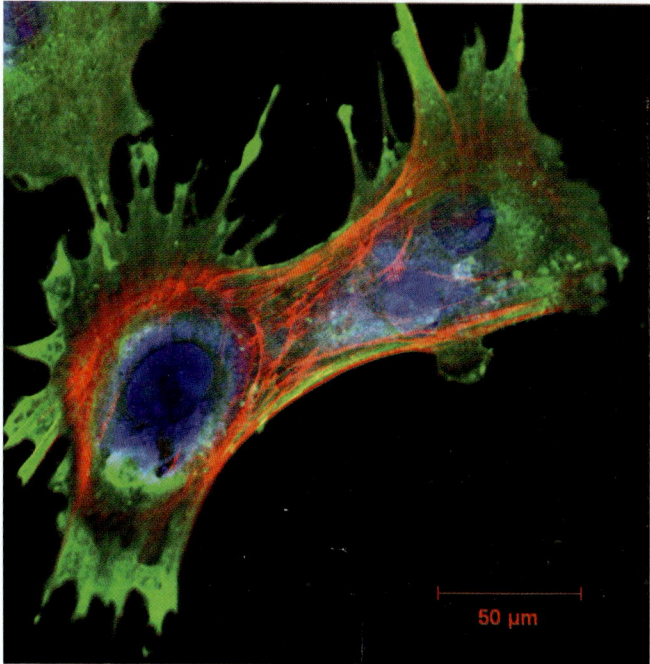

Fig. 7.3 Multichannel laser scanning confocal micrograph of mouse embryonic stem cell in culture (day 15). Immunohistochemical staining for Oct-4 (*green channel*), and histochemical staining for actin cytoskeleton (phalloidin: *red channel*) and nucleus (Toto3: *blue channel*). Section thickness = 1 μm (courtesy of Dr. Roman Krawetz)

is needed to illuminate the photodetector, which is positioned behind a pinhole aperture to minimize out-of-focus light. By mechanically scanning the pinhole before the detector, the CSLM can produce thin, serial, optical sections while eliminating out-of-focus glare, thus improving resolution and clarity compared to ordinary microscopes (Fig. 7.3).

Historically, choosing the optimal fluorophores for labeling and the appropriate excitation and emission filters for detection was a painstaking process. More recently, a large number of fluorophores have been developed, which can be conjugated to affinity reagents (antibodies, biotin, etc.) that can subsequently be excited at the wavelengths that are most commonly available in systems using laser illumination (e.g., CSLM and flow cytometers). As with microscopy, there are now outstanding Internet reference sources for fluorophores (one particularly rich source is published online by Molecular Probes at: http://probes.invitrogen.com/handbook/). Some newer developments in nanotechnology, specifically quantum dot nanocrystal semiconductors, solve some of the problems associated with using traditional fluorophores (stability, photobleaching, quenching). Still, quantum dots also have limitations in some live tissue applications, as their size often precludes efficient clearance from the circulation in the kidneys, and their long-term toxicity is uncertain (Iga et al. 2007).

With rapid improvements in camera sensors, microscope optics and fluorochrome design, it is now possible to use fluorescent probes in the near-infrared part of the spectrum (~700 nm). This is especially useful for live-cell imaging, as the energy of long excitation wavelengths is low compared to shorter wavelengths. Hence, working with fluorophores that are excited in the near-infrared spectrum is less damaging to living cells (e.g., George 2005). Morever, working with longer wavelengths reduces scattering and avoids the autofluorescence of many naturally occurring molecules, which might complicate work with fluorescence in the visible spectrum. This approach has considerable utility for in vivo work, as the absorption of light in the near-infrared range (700–1,000 nm) is low, and so light can penetrate deeply into the tissues, i.e., up to several centimeters.

Atomic force microscopy (AFM) is a physical method that can be used to generate a height image using a scanning cantilever beam to which a tip of varying geometry and size is attached. The position of the tip in space is recorded from the reflection of a laser off the cantilever onto a sensor. As the spring constant of the cantilever is known and can be calibrated, the mechanical properties of a specimen can be interrogated, as well as the specimen height. Depending on the probe tip, various dimensions of a tissue or molecule can be interrogated, often with quite different results (e.g., Stolz et al. 2004).

Although humans are best equipped to see their world by reflected light, some species are specialized to sense their environment by reflected sounds (e.g., bats, whales). It is not surprising that acoustic microscopy, particularly scanning acoustic microscopy, has found a niche in the imaging of materials and tissues using reflected sounds. After being focused on a tissue, sound may be scattered, transmitted, or absorbed. Reflected and transmitted sounds can be collected and measured to form an image based on "contrast" due to geometric and material characteristics of the tissue. Varying the transmitter frequency (ranging from 1 Hz to 1 GHz) influences the resolution of the image. The mechanics of the cytoskeleton, for example, have been explored using acoustic microscopy (e.g., Bereiter-Hahn and Lüers 1998).

Emission

Radioactive probes emitting beta or gamma radiation can be detected by film or various sensors, so images of tissues or cells (living or dead) can be acquired. Historically, autoradiography has been an invaluable tool in histology because of its inherent sensitivity (Fig. 7.4). With the development of solid-state detectors, gamma scintigraphy and single photon emission computed tomography (SPECT) have become commonplace in clinical nuclear medicine. Specialty clinics with access to more expensive equipment can detect particles emitting positrons by positron emission tomography (PET). Images obtained from nuclear scintigraphy, which reflect physiological processes, are powerful when combined with anatomical imaging techniques like CT scanning (e.g., Boss et al. 2008). Such multimodal imaging is a very capable exploratory tool, whether in the clinic or in the laboratory.

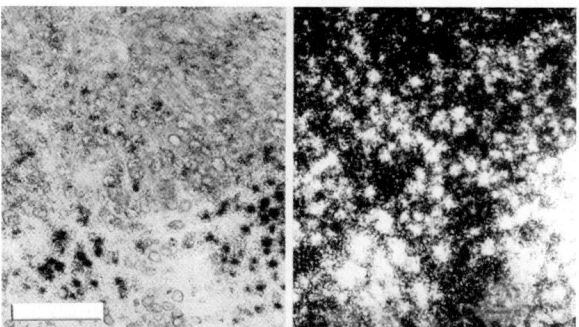

Fig. 7.4 Autoradiograph of type II procollagen in situ hybridization of fibrocartilage. Oligonu-cleotide probe labeled with ^{35}S-dATP. *Left panel*: bright-field image showing silver grains developed in liquid emulsion overlay of a frozen histological section counterstained with hematoxylin. *Right panel*: dark-field image of the same field. Bar $= 100\,\mu$m

Bioluminescent imaging is an indirect technique based on the expression of firefly luciferase, an enzyme that when conjugated to affinity reagent, or when expressed by recombinant proteins transfected into cells or transplanted into animals, radiates light. When the firefly luciferase gene is expressed and the enzyme luciferase is produced, the presence of its substrate luciferin, which can either be injected or inhaled, produces light in living cells and tissues within an organism. While useful to a remarkable number of invertebrates, insects and plants for startling predators or attracting mates, bioluminescence is particularly useful to biologists when studying temporal expression in living cells. In addition to substrate, bioluminescence depends on ATP energy and molecular oxygen for the production of light, which depends on the concentration of these nutrients. Indeed, its ATP dependency has made luciferase the tool of choice for ATP measurement in cells (e.g., Jouaville et al. 1999). Moreover, sensitive detectors have enabled the real-time monitoring of tumors by luciferase activity in vivo (e.g., Miretti et al. 2008).

Unlike the exogenous introduction of radioactivity and bioluminescent proteins, the chemical structures of tissues themselves can serve as a source of information that can be assembled into an image. Two such methods are Fourier transform infrared (FTIR) microscopy (e.g., Petibois et al. 2007) and second harmonic generation (e.g., Williams et al. 2005). While beyond the limits of this chapter, these types of emerging methods spell good fortune to those exploring the molecular structure of tissues. There can be little doubt that advances in these techniques will facilitate interrogation of the complex molecular dynamics of living cells and tissues.

Magnetic resonance imaging (MRI) involves placing a specimen or organism within a high magnetic field, which induces the alignment of protons in biological tissue. The alignment of protons can be disturbed by a radiofrequency pulse, and the time (in milliseconds) that it takes for the protons to realign with the magnetic field can be the source of image information, as protons in different chemical environments behave somewhat differently. Hence, the protons in fat, muscle, cartilage, and brain can be distinguished under different pulse sequences. This is the (oversimplified) basis for clinical MRI in ∼1–3 T magnets, as well as for very high-field, high-resolution MR imaging (MR microscopy) in 7–15 T magnets. With pixel resolutions on the order of tens of microns, such devices can produce remarkably clear images, particularly of soft tissues (rich in protons) (Fig. 7.5).

Evaluating Images

Living or dead, big or small, once acquired (unless taken purely for aesthetic reasons), images taken for studies of biology or medicine are destined to be evaluated. Most commonly, images are assessed qualitatively without rigorous measurement, which is not to say they remain unscrutinized. Indeed, the human mind is such an efficient analyzer of complex images that meaningful patterns can often be readily discerned. (This feat has not yet been reproduced successfully by computers.) This practiced art form remains invaluable to patients and medical practitioners alike. Histopathology, while a reasonable source of information about which to generate hypotheses, is a rather poor environment in which to test hypotheses. The experimental pathologist, however, requires evaluation tools to do exactly that.

Dealing with Analog Data

The last century saw an explosion in the science and application of photographic methods for recording medical and biological images. Yet, before the end of the first decade of this new millennium, for all practical purposes, emulsion-based photomicroscopy and radiography will be extinct. Nostalgia notwithstanding, a haunting question remains of how one should go about quantifying images formed on film. Optical densitometry was once widely practiced as a method for measuring the density of an emulsion, though it is a tedious manual process. Few today would have the courage or patience for such an ordeal since digitization has become so widely available. Still, converting an analog film image into a digital image can be problematic if poorly done, as the fidelity of the conversion depends on many factors, including the quality of the film image (of primary importance), the resolution and the dynamic range of the scanner, the quality and intensity of light, and the quality of the output device (i.e., computer monitor). Anyone who has heard an original LP record that has been transferred poorly onto a compact disc or DVD understands that converting images from analog to digital, just like maintaining audio fidelity, is a challenging task.

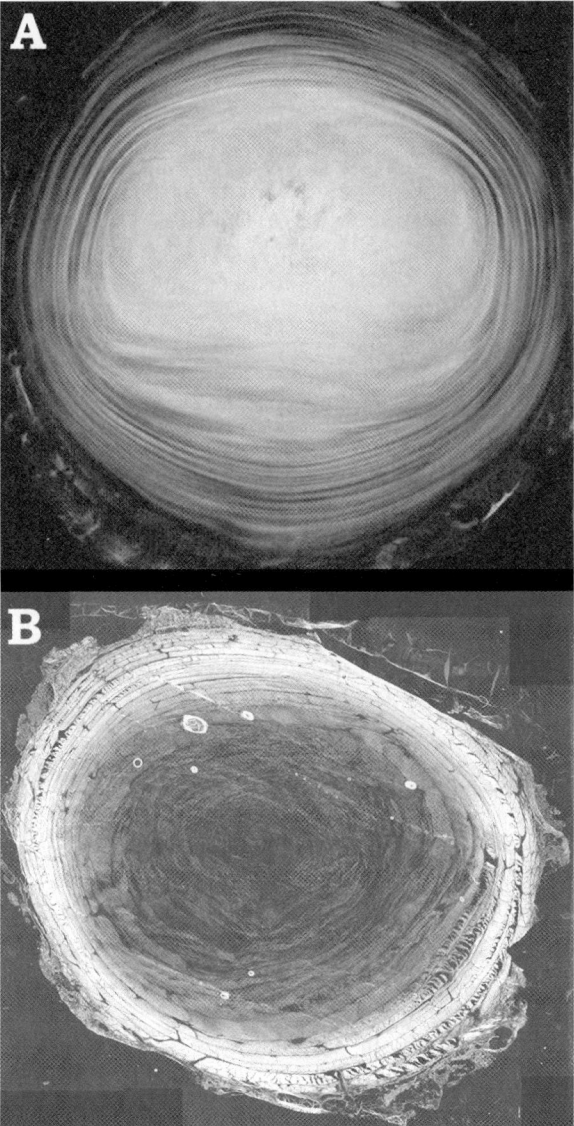

Fig. 7.5 (a–b) Axial section through bovine caudal intervertebral disc. **(a)** High-field (9.4 T) MR image (T2-weighted; 1 mm thick). Note the collagenous lamellae in the outer annulus fibrosus, and the high signal intensity in the water-rich central nucleus pulposus. **(b)** Histological section of bovine disc in the same orientation illuminated with circularly polarized light. Note the high signal (high retardance of polarized light) in the collagenous annulus. Circularly polarized light does not have the same orientation dependence as plane-polarized light (image taken with the LC-Polscope system, CRI, Woburn, MA, USA)

Types of Data

Not all data are created equal, and different data types are not necessarily inter-changeable for the purpose of testing hypotheses using inferential statistics. It is important to know what type(s) of data are being analyzed and what statistical methods are most appropriate for handling these data. Continuous, discrete, and ratio scale data can all be regarded as "quantitative," whereas ordinal data are often regarded as "semiquantitative." Nominal data is regarded as "qualitative." All data types can contribute to a histomorphometric evaluation, and "semiquan-titative" grading scales are particularly common in histopathology. When ordinal data are entered inappropriately into a parametric statistical analysis, they may void the underlying assumption of normally distributed continuous variables. As a rule, qualitative data are useful for hypothesis generation; quantitative data are especially useful for hypothesis testing.

Different types of data are presented in Table 7.1.

Accuracy and Precision

Quantitative measurements reveal three attributes about a structure or group of struc-tures: the numerical quantity, the unit of measure, and a measure of uncertainty. The expression of uncertainty informs the reader of the care taken in the measurement and in calibrating the measuring instrument, though unless explicitly stated it is rea-sonable to assume that the former presumes the latter. The accuracy of a "ruler" is verified by how near the ruler value is to the true value (typically by calibrating to a standard), as well as the "fineness" of the divisions on the ruler. In practice, the calibration of a measurement tool in microscopy, be it an eyepiece graticule or a digital rule, is commonly performed with a calibrated "stage micrometer." Once calibrated, a scaling factor is calculated. While less of a problem nowadays, early computer monitors had relatively large pixels that were typically rectangular, so calibrating a digital ruler using the computer monitor required care to ensure that a sufficiently large number of pixels were included as well as that the calibration is done in both the X and Y directions. This sometimes leads to awkward measurement

Table 7.1 Definitions and examples of different types of data

Type	Defined	Example
Continuous	Values between any two other values	Distance
Discrete	Intermediate values not possible	Integer
Ratio scale	Constant interval size and true zero	Height:weight ratio
Interval scale	Constant interval; zero is arbitrary	Linear (temperature); circular (time)
Ordinal	Ordered; different/undefined interval size	Lesion grade (–, +, ++, +++)
Nominal	Attribute	Eye color

Table 7.2 Examples of pixel calculations

Microns	Pixels	Pixels/micrometer	Micrometer/pixel
250	250	1	1
25	250	10	0.1
250	25	0.1	10
24	251	10.45833333	0.09561753

units, such as microns/pixel, that transform length from a continuous to a discrete variable. As arithmetic division is used to calculate the scale factor, it is important to keep in mind the number of true significant figures. The number of significant digits used when reporting a quantity (e.g., 5 vs. 5.000 mm) reflects the precision of the measurement (i.e., its repeatability).

It is very important to remember that pixel dimensions should be given in units that reflect the resolution of the system (Table 7.2). The dangers of division are obvious when calculating "pixels per micrometer," which is always mathematically possible, but is often physically impossible (ponder the meaning of 0.1 pixel). Moreover, "micrometers per pixel" can be calculated to many decimal places (e.g., row four). Yet, this calculated value is well below the optical resolution limit of even the best light microscopes (~0.2 μm), so reporting measurements using these units unduly promotes the system's resolution.

As a rule, no directly measured linear dimension or value that is calculated from a linear dimension should be reported with an implied accuracy superior to the resolution limit of the measuring device. In the case of row 2, for example, two discrete pixels on the screen are insufficient to discriminate two objects within the sample, so this is at least a twofold "empty" *digital* magnification. While two discrete pixels exist, two objects cannot be discriminated by light microscopy.

Evaluating in 2D

In all biological systems, anatomy and physiology operate in three dimensions, but in all but the smallest of organisms, access to three-dimensional structure and function is visibly obscured. Hence, traditional methods of biological sampling used to clarify internal structures almost always involve some form of slicing. The problem with slicing is the loss of the spatial relationships, thus confusing the interpretation of three-dimensional organization and content. Indeed, the central problems with biological morphometry are that structures are three-dimensional, geometrically imperfect, and embedded within other opaque structures. Particularly when working in transmission, cutting thin histological sections allows the passage of light with minimal scattering, which makes otherwise opaque three-dimensional structures visible for viewing in two dimensions. It is also possible to make slices based on contrast imaging (e.g., CT, MRI, PET, ultrasound, and optical), and all of these methods reduce 3D information to 2D data. Only with recent advances in computing technologies has it become possible to view data in 3D.

For most practical purposes, a histological section of a typical thickness of 4–6 μm can essentially be considered and treated as a two-dimensional plane of intersection with an organ. When a plane intersects a three-dimensional object, at least three types of measurements can be made directly on such a two-dimensional surface, including: area (square units), length (one dimension), and profile count (zero dimensions). Regardless of how 2D images are formed from 3D objects, the measurements of the sampled object depends on its size, shape, and orientation.

The evaluation of structural characteristics on a planar surface is known as planimetry, which may be acquired either manually or digitally, and typically yields measures of area, perimeter, and size (e.g., caliper and Feret diameter). While it is possible to use a digitizing tablet mouse to physically trace an object, it can be as efficient to use a "map wheel" to make measurements on printed photomicrographs. It is now far more commonplace to use a computer to assist in the tracing process of digitized photomicrographs, which typically involves some form of image analysis. Setting an intensity threshold and binarizing a digital image assigns pixels to either black or white, which simplifies the digital counting of contiguous pixels of one or both types. Summing the distances between adjacent pixels on the boundary of an object yields the perimeter; summing the area of pixels inside or outside an object yields an area. Counting the number of discrete objects in a field yields a profile count. Such pixel-based planimetry has the built-in inaccuracy of using a two-dimensional measuring unit (the pixel), so a rule must be devised for counting pixels on the outside or inside of the boundary. There is another complexity with using pixel counting methods on grayscale bitmap images that are rotated: a straight horizontal or vertical line is a row of identical pixels at 0°, but it is commonly interpolated as a set of steps when rotated to 45° (Fig. 7.6). Such rotation and interpolation influences measures of area and perimeter, which can be measured for defined geometries (for a hollow square, the area can change by more than 15%), but is more difficult to determine for irregular objects, which also use pixel steps to delineate the boundaries of curves.

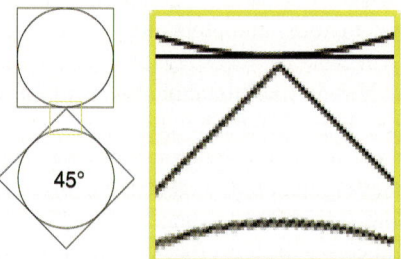

Fig. 7.6 Bitmap image of a circle embedded in a square. *Left-hand panel*: original and rotated (45°) figures. The yellow square in the left panel is "magnified" in the *center panel*, showing the pixellation of curves and rotated lines. The yellow square in the center panel is further magnified in the *right panel* to show how geometrical figures are represented on a computer screen by pixels. Note that the steps needed to form a continuous curve have two dimensions, as well as different intensities

Fig. 7.7 Array of points cast randomly on three different objects. As the same number of points (defined as the intersections of the crosses, *small arrow*) falls on each object, they are estimated to have the same area using this spacing of points

There is another approach that can be used to make estimates of two-dimensional parameters that does not employ pixel counting. This approach quantifies the number of intersections of points and lines with objects in the plane. It has long been known, for example, that the number of regularly spaced (dimensionless) points is an estimate of the area of a planar object. This is equivalent to counting the number of tiles on the floor to estimate its area. For a square tile measuring one foot on a side, counting the number of tiles gives an estimate of the area in square feet; if we switch focus from the tile to the intersection of four tiles, counting the intersection *points* is equivalent to counting the tiles and is an estimate of area (Fig. 7.7). (For anyone needing to specify lab space on a grant application, this is an efficient alternative to using a tape measure.) Similarly, counting the number of intersections between a boundary and an array of line probes (of known length) yields an estimate of perimeter. As points are dimensionless and lines have but one-dimension, measuring area and boundary length with these probes avoids the built-in inaccuracies of measuring with pixels.

While profile number can be counted directly on a plane, as mentioned above, this number depends on the size, shape, and orientation of the object in three-dimensional space. Indeed, the probability of a plane intersecting an object is a function of its volume: larger objects are more likely to be hit than smaller objects. For example, let's imagine a room filled with equal numbers of basketballs, tennis balls, and ping-pong balls that are distributed evenly in this volume. Imagine the room is cut at random by a plane. It is obvious that fewer ping-pong balls will be hit than basketballs. This well-understood phenomenon is the basis for the poorly acknowledged volume-weighted bias of number (Fig. 7.8). This bias is inherent in

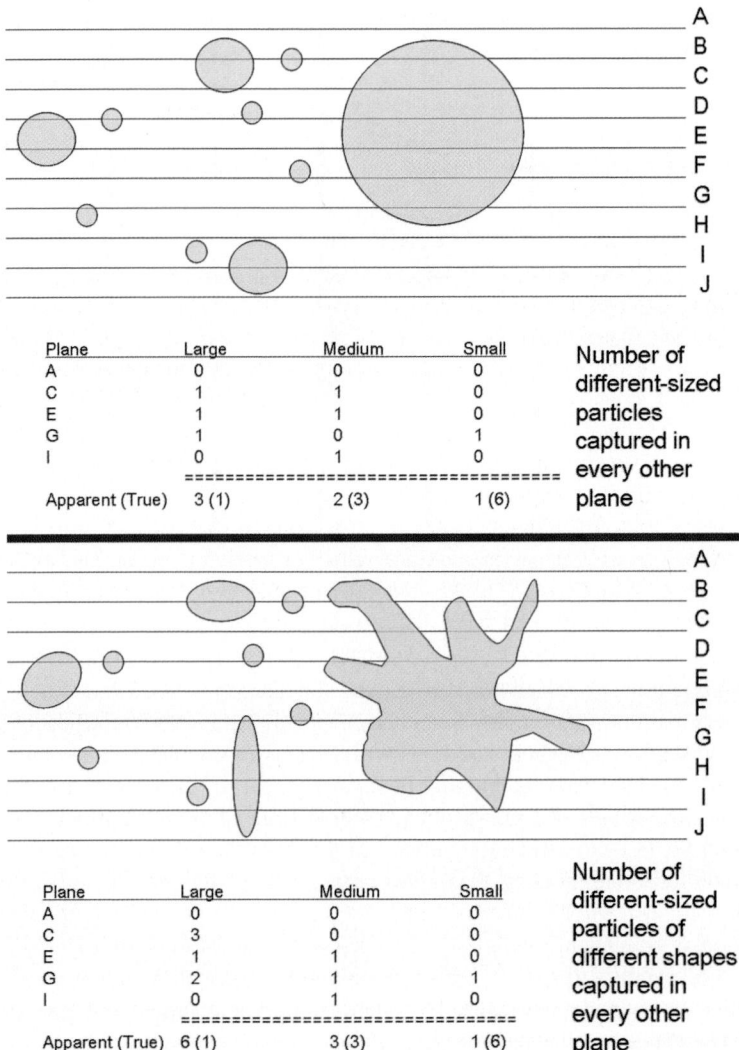

Fig. 7.8 Illustration showing in two dimensions the problem of counting objects in three dimensions. *Top panel*: embedded spheres of different sizes sectioned from top to bottom using a systematic random sample of sectioning planes (*a–j*). Note that the large spheres appear in slices more often than the smaller spheres. *Bottom panel*: the same number of embedded objects of varying sizes, orientations, and geometries cut by the same sectioning planes as above. Note that the apparent number of objects differs from both the true value (in parentheses) and from the values obtained with spheres. The disector method involves the evaluation of two slices; an object is only counted when it enters or leaves the volume created by the pair of slices. Hence, an object is counted only once, in proportion to its number rather than volume

clinical and experimental pathology when only one representative section, rather than a stack of serial sections, is examined.

Special Cases of Evaluation in 2D

Some techniques of image formation, notably the transmission of ionizing radiation or light, create an image by projection. For example, routine planar radiography and dual-photon absorptiometry (a measure of bone density) generate two-dimensional projections of three-dimensional objects. Such projected images may be binary (no transmission) or may have varying intensities depending on the density of the tissue being imaged. In the case of a projected image formed on film, the density of silver grains is proportional to the amount of transmitted light, which is inversely proportional to the density of the tissue (radiation is retarded more by dense tissue than by less dense tissue). The density of silver grains formed in an emulsion depends on various factors, including the concentration of silver halide on the emulsion, the exposure time, the energy of the radiation, the density of the tissue and the chemical development of the exposed film. It is worth noting that film emulsions all have characteristic curves that have linear and nonlinear response regions, which need to be understood when using film density to quantify projection images (an outstanding discussion of how and why film responds to light was given by photographer Ansel Adams 1981). Similarly, digital images are formed on sensitive detector elements that have a wide response range, though these will still have ranges of linearity and nonlinearity that are often difficult to identify for most users. Whether the image is formed on film or a digital sensor, experimentation to define the working range of a system is generally worth the investment of extra time and energy.

Nevertheless, the evaluation of projected images, in addition to planimetry, typically involves densitometry. At very high magnification, it is possible to actually count individual silver grains on film emulsions, an approach still practiced when using liquid emulsion-based autoradiography, which remains a highly sensitive technique for evaluating the emission of photons from tissues labeled with radioactive precursors. Analog densitometry has quickly been replaced by digital densitometry, wherein each pixel is assigned a grayscale value of intensity/density. It is now quite straightforward to generate histograms of intensity in images or regions of interest in projection images. Other common measures include maximum, minimum, mean, median, and mode for individual gray values or ranges of gray values. When working outside the linear response range, densitometric evaluation is compromised. For example, a film or sensor that is overexposed or underexposed clips the data and introduces bias to the analysis due to the nonlinear relationship between density and exposure.

A rather unique method of physically sampling the surface of a live tissue employs a two-dimensional porous membrane substrate (Matyas et al. 1995). This "tissue transfer" or "tissue print" technique can be used to sample, for observation and evaluation, the surface structures and architectures of many types of soft tissues, e.g., liver and kidney. Moreover, tissue transfers of living cells can be maintained in culture and thus offer the opportunity to evaluate the metabolism of a 2D lawn of cells, still in the context of their tissue organization. Evaluating such two-dimensional lawns of cells is a straightforward application of the methods described above.

Just as film density offers an added dimension in 2D analysis, so too does metabolism, whether measured in living or fixed tissues. Hence, immunohistochemistry, enzyme and lectin histochemistry, and in situ hybridization histochemistry can be quantified as a layer of intensity and mapped onto a planar surface. This emerging capability, particularly when extended into 3D, has an exciting future as a metric of tissue physiology in an era of proteomics, genomics, and metabolomics.

Evaluating Tissues in 3D: The Escape from Flatland

As stated above, the essential problems of histomorphometry have long hindered the evaluation of tissue structure and function in three dimensions. Nonetheless, seminal work in the 1950s led to developments in the area known as stereology that made estimates of 3D parameters from 2D sections possible (Weibel 1979). Common stereological probes include an assortment of points, lines, surfaces, and volumes. As we have seen, it is impractical to directly probe 3D objects with lines or points, so these probes are typically placed onto section planes. Points and lines can be easily implemented by overlaying a grid onto a photomicrograph or computer display.

As early as the seventeenth century, it was known that the volume of an object can be estimated as the product of the slice thickness and the sum of the cross-sectional areas of the slices (Cavalieri 1966). While unbiased for volume, estimates for other metrics often required assumptions, particularly about object geometry, that were difficult or impossible to verify and hence could be biased. The evolution of stereology included concomitant developments in histomorphometry, statistics, and stochastic geometry in the 1970s and 80s (e.g., Gundersen et al. 1988) that allowed many assumptions to be cast aside and allowed the implementation of unbiased methods.

The fundamental premise of the stereological approach is that 2D sections contain quantitative information about 3D structures only in a statistical sense, and when applying an appropriate set of sampling rules and estimators, it is possible to make efficient, unbiased estimates of 3D quantities from 2D images. This so-called design-based approach uses some form of randomization in sampling, such that there is an equal probability of sampling an object in 3D space. When sampled appropriately, there is an equivalence of measures across dimensions: the proportion of points counted equates to the areal fraction and to the volume fraction of the objects of interest in a volume of interest. This equivalence is given by the shorthand:

$$P_P = A_A = V_V.$$

Save for volume and number, most stereological estimators including 3D surface area and length require some kind of isotropy, but all biological structures have some anisotropy. Hence, the area of a curved surface of arbitrary shape and orientation in space is proportional to the mean number of intersections between a surface and a straight line probe of known length *when placed uniformly and isotropically at random in space*. Whereas an unbiased estimate of volume can be calculated from profile area, an unbiased estimate of surface area cannot necessarily be calculated directly from profile perimeter *unless the lines have an isotropic distribution in 3D space—not just in the section plane*. Hence, a number of sampling strategies have been devised for generating isotropic, uniform random (IUR) section planes in 3D (e.g., the "Orientor;" Mattfeldt 1990). Once IUR sections are available, stereological line probes (or grids) of specified dimensions can then be used to make unbiased estimates of surface area or length by counting intersections. However, as histopathologists often view tissues in particular orientations, it is very easy to lose track of tissue architecture on IUR sections. A novel method has been developed to produce isotropic test lines while maintaining a preferred direction of the section plane: vertical sections (Baddeley et al. 1986).

A vertical section is defined as being perpendicular to a given horizontal reference plane which defines the orientation of the section. The principle of creating vertical sections is to regain the degree of freedom given up when choosing a preferred orientation, in order to re-enable unbiased estimation of surface area. A classic application of the vertical section method is skin, which is best recognized in the surface-to-deep orientation. Laying a piece of skin on the lab bench naturally defines the horizontal orientation. The lost degree of freedom is regained by rotating the sample at random about the vertical axis. In practice, this involves sticking a pin through the surface of the skin and rotating about this axis before sectioning parallel to the vertical direction. The four requirements for making vertical sections include: (1) identifying or generating a vertical axis; (2) identifying the vertical axis on each 2D section; (3) relative to the horizontal plane, vertical sections must have a random position and orientation (i.e., must be rotated); and (4) the test line used has a weight proportional to the sine of the angle between the test line and the vertical (cycloid test lines). While a fuller explanation of the implementation of vertical sections is beyond the scope of this chapter, the relative cost of implementing this sampling design reaps the tangible benefit of unbiased measures of surface area in 3D. Given the number of important physiological processes that are dependent on surface area (e.g., gas exchange, fluid absorption, etc.), such an effort can be worthwhile, particularly when surfaces are unavailable by 3D methods (see below). Obtaining an unbiased estimate of the length of a thin tubular structure on vertical section has been described, wherein intersections of the tubular structure with the cycloid line probe are counted (Gokhale 1990; Gokhale 1993). This method holds special significance when evaluating vessel or neuron length in biology and medicine.

Unlike the metric properties of volume, area, and length, number is what is known as a topological property of a structure. Number is a centrally important measure of biological form, be it the number of cells, the number of some other functional element (e.g., glomerulus), or the number of nodes in a network (e.g., bone or lung). Yet, obtaining an unbiased estimate of object number from a single thin section is impossible. An elegantly simple solution to the problem of evaluating object number was described by Sterio (1984), wherein a (3D) volume probe is used to measure the 0-D parameter of cardinality (number). The stereological tool known as the disector probe is implemented using two (or more) physical, optical, or contrast sections, respectively. The disector operates simply by counting but the objects entering or leaving a volume (rather than the objects in a section). Practically speaking, cells in a known volume (disector) of tissue can be counted using thick sections ($>10\,\mu m$), a high numerical aperture objective lens (with a thin optical depth of focus), and a digitally encoded depth gauge (to monitor the precise movement of the microscope stage). Nuclei are counted when they come into focus within a defined counting frame (of known area) as the microscope stage is lowered a known distance (to define a known volume); this method is therefore known as the optical disector (Fig. 7.8). The disector method is efficient as nuclei are generally stained in high contrast compared to the cell cytoplasm or membrane, and (with few exceptions) represent a unique feature within the cell. Being smaller than a cell, the nucleus can be captured in a smaller disector volume (i.e., a thinner slice). Larger objects that are only rarely observed entering or leaving a thin section, such as glomeruli ($>100\,\mu m$ in diameter), are generally best evaluated on two physical sections separated by larger intervals, and this method is known as the physical disector. An obvious requirement of the disector is that the object of interest must be recognized in order to be counted, but once recognized, the disector is an unbiased estimate of object number in a defined volume.

Counting Efficiency in Stereology

Many people are under the impression that point counts, line intersections, and disector counts must be extensive before an accurate estimate can be obtained. Yet, this is often not the case. Indeed, most of the error in typical biological experiments is associated with the highest level of sampling (e.g., the organism), and the contribution of stereological measurement error to the overall variance is typically very low. Indeed, a classic stereological paper subtitled "Do more less well" explains how increasing "n" is generally more efficient at lowering variation, and hence at discerning a treatment effect, rather than increasing the number of counts or measurements (Gundersen et al. 1981).

Computers and Stereology

Whereas stereology can be practiced with very inexpensive equipment, various commercial computer systems have been developed and marketed for this purpose. Computerization has increased the repertoire of probes readily available to the average user, and some systems have implemented probes that were impractical to implement manually (e.g., virtual planes). While planimetry and image analysis software are commonplace, software that specifically implements stereological sampling probes is still a specialty market (e.g., http://www.stereology.info/, http://www.visiopharm.com/, http://www.disector.com/, http://www.mcid.co.uk/software/stereology/), though some microscopy software vendors are increasingly aware of client needs in this area. It is worth noting that stereological principles can be readily applied to digital data, including 3D datasets. It is also worth noting that whereas scale and dimensions may influence the performance of pixel and voxel counting methods, stereological methods can be applied equally well to dimensions from nanometers to light years.

Evaluating Three-Dimensional Images

Advancements in imaging and computer power have enabled the acquisition and presentation of true 3D datasets. These digital datasets are formed from pixels that are extended into three-dimensional voxels that have an intensity. Voxels may be cubic (isometric) or are more typically rectangular prisms when the resolution or spatial depth differs in one dimension compared to the other two (typical of confocal microscopy and MRI image stacks). As voxels are the building blocks of 3D volumes, the internal structures of 3D datasets can be interrogated with and evaluated on digital slices, as shown in other chapters of this volume. As voxel counting is a straightforward extension of pixel counting, it carries forward the same efficiencies as well as built-in inaccuracies when measuring boundary lengths, areas, and volumes.

Sampling and Efficiency in Stereology

One of the basic principles of stereology is unbiasedness, which stems in part from using stereological probes, though it also depends in good part on the method used to sample the object of interest. Sampling can be exhaustive, but need not necessarily be exhausting. Indeed, in arriving at a stereological estimate, sampling can contain elements of both randomness and uniformity, as in IUR sections and vertical sections. Considerable emphasis has been placed on efficiency in stereological sampling, which is pleasing for both theoretical and practical reasons. Random sampling, while theoretically acceptable, is highly inefficient, meaning that it may take more sampling to yield an accurate estimate. Isotropic uniform random (IUR) sampling reduces or eliminates bias and enables the use of systematic random sampling (SRS), in which a random start is followed by parallel slices at regular intervals, improving the efficiency of sampling, particularly for whole objects (such as the brain for instance).

Fractionator sampling is one of the most powerful sampling tools in stereology (Gundersen et al. 1988). The fractionator is a very important and powerful tool in the stereology of histological sections, as it is immune from dimensional distortions associated with histological processing. Simply put, an unbiased estimate of a parameter (number, volume, etc.) in a whole object can be made from a known fraction of random subsamples. For example, the number of jellybeans in a large jar can be estimated by dividing the jellybeans into 100 piles (the more similar the size of the piles, the more precise the estimate). With a random start, the piles are sub-sampled with a known fraction (e.g., every tenth pile). The estimate of the total number of jellybeans in the jar is the number of jellybeans counted in the subsamples divided by the sampling fraction (here one-tenth). In an experiment involving microscopy, fractionator sampling can be done at the level of an organ (e.g., one-tenth), tissue block (e.g., one-fifth), stack of sections (one-hundredth), microscopic fields (one-twenty-fifth), and even volume of a thick section (one-third). An unbiased estimate of the total is given by the value of the measured parameter divided by the product of all the subsampling fractions:

$$(0.1)(0.2)(0.01)(0.04)(0.33) = 0.00000264;$$
$$\text{Est } X = x/0.00000264$$

Although systematic random sampling has been a staple method in stereology for sampling an object with known probability, which ensures an unbiased result, such a sampling strategy it is not an absolute necessity. The development of digitized microscopy stages and digital photomicroscopy enables the practical implementation of a method wherein the sampling probability is unequal in size; rather it is proportional to the size or intensity of the objects. This method is the "proportionator" (Gardi et al. 2008), which can improve sampling efficiency significantly as compared to systematic random sampling.

Clinically, the area of pathology that has seen the most activity in histomorphometry is in bone, where the form of the bone relates to its mechanical function and metabolic health. Hence, commercial computerized systems were in clinical use in the late 1970s (planimetry), and these have now advanced to the level that they can perform highly sophisticated measurements and analyses based on 3D μCT datasets (see Chap. 10). Yet even in contemporary bone histomorphometry, the resolution of μCT remains insufficient to detect bone surfaces containing osteoclasts with certainty, or to count the number of cells, which must still be done on 2D histological sections by planimetry or stereology. Yet, in the laboratory, neuroscientists have been most active in implementing stereology in experimental studies.

The Escape from 3D

As biological processes occur in three-dimensional space over time, and many biological signals are governed by the rate of change in activity or concentration (as opposed to absolute concentrations), it would be valuable to evaluate biological processes in four or more dimensions. Initial attempts in this direction have already begun (e.g., Guillaud et al. 1997 2004; Kamalov et al. 2005), and the future is bright for combining the activities of cells and tissues in three dimensions with data on the genome, proteome, and metabolome. Indeed, there has been an explosion of such activity in the emerging area of Imaging Informatics (e.g., http://www.siimweb.org/). This topic will become increasingly important over time to our quest to understand medical and biological phenomena.

References

Adams A (1981) The negative. The new Ansel Adams photography series, vol. 2. New York Graphic Society, Little Brown and Company, Boston, MA

Baddeley AJ, Gundersen HJ, Cruz-Orive LM (1986) Estimation of surface area from vertical sections. J Microsc 142:259–276

Bereiter-Hahn J, Lüers H (1998) Subcellular tension fields and mechanical resistance of the lamella front related to the direction of locomotion. Cell Biochem Biophys 29:1085–1091

Boss DS, Olmos RV, Sinaasappel M, Beijnen JH, Schellens JH (2008) Application of PET/CT in the development of novel anticancer drugs. Oncologist 13:25–38

Cara DC, Kubes P (2003) Intravital microscopy as a tool for studying recruitment and chemotaxis. In: D'Ambrosio D, Sinigaglia F (eds) Cell migration in inflammation and immunity (Methods in Molecular Biology, vol. 239). Humana, Clifton, UK

Cavalieri B. Geometria Indivivisibilibus Continuorum. Bononi: Typis Clementis Ferronij, 1635. Reprinted as Geometria degli Indivisibili. Torino: Unione Tipografico-Editrice Torinese, 1966

Cutignola L, Bullough PG (1991) Photographic reproduction of anatomic specimens using ultraviolet illumination. Am J Surg Pathol 15:1096–1099

Forrester KR, Tulip J, Leonard C, Stewart C, Bray RC (2004) A laser speckle imaging technique for measuring tissue perfusion. IEEE Trans Biomed Eng 51:2074–2084

Gardi JE, Nyengard JR, Gundersen HJG (2008) Automatic sampling for unbiased and efficient stereological estimation using the proportionator in biological studies. J Microsc 230:108–120

George N (2005) Microscopes expand near-infrared applications. Biophotonics March:1–4

Gokhale AM (1990) Unbiased estimation of curve length in 3D using vertical slices. J Microsc 159:133–141

Gokhale AM (1993) Estimation of length density Lv from vertical sections of unknown thickness. J Microsc 170:3–8

Guillaud M, Matthews JB, Harrison A, MacAulay C, Skov K (1997) A novel image cytometric method for quantitation of immunohistochemical staining of cytoplasmic antigens. Analyt Cell Pathol 14:87–99

Guillaud M, Cox D, Malpica A, Staerkel G, Matisic J, van Nickirk D, Adler-Storthz K, Poulin N, Follen M, MacAulay C (2004) Quantitative histopathological analysis of cervical intra-epithelial neoplasia sections: Methodological issues. Cell Oncology 26:31–43

Gundersen HJ et al. (1981) Optimizing sampling efficiency of stereological studies in biology: or "do more less well." J Microsc 121:65–73

Gundersen HJ et al. (1988) The new stereological tools: disector, fractionator, nucleator and point sampled intercepts and their use in pathological research and diagnosis. APMIS. 96:857–881

Iga AM, Robertson JH, Winslet MC, Seifalian AM (2007) Clinical potential of quantum dots. J Biomed Biotechnol. 2007:76087

Inoué S, Oldenbourg R (1998) Microtubule dynamics in mitotic spindle displayed by polarized light microscopy. Mol Biol Cell 9:1603–1607

Jouaville LS, Pinton P, Bastianutto C, Rutter GA, Rizzuto R (1999) Regulation of mitochondrial ATP synthesis by calcium: Evidence for a long-term metabolic priming. Proc Natl Acad Sci USA 96:13807–13812

Kamalov R, Guillaud M, Haskins D, Harrison A, Kemp R, Chiu D, Follen M, MacAulay C (2005) A Java application for tissue section image analysis. Comput Methods Programs Biomed 77:99–113

Matyas JR, Benediktsson H, Rattner JB (1995) Transferring and culturing an architecturally intact layer of cells from animal tissue on membrane substrates. Biotechniques 19:540–544

Miao J, Chapman HN, Kirz J, Sayre D, Hodgson KO (2004) Taking X-ray diffraction to the limit: macromolecular structures from femtosecond X-ray pulses and diffraction microscopy of cells with synchrotron radiation. Annu Rev Biophys Biomol Struct 33:157–176

Miretti S, Roato I, Taulli R, Ponzetto C, Cilli M et al. (2008) A mouse model of pulmonary metastasis from spontaneous osteosarcoma monitored in vivo by Luciferase imaging. PLoS ONE 3(3):e1828. doi:10.1371/journal.pone.0001828

Mattfeldt T, Mall G, Gharehbaghi H, Moller P (1990) Estimation of surface area and length with the orientor. J. Microscopy 159:301–317

Petibois C, Drogat B, Bikfalvi A, Déléris G, Moenner M (2007) Histological mapping of biochemical changes in solid tumors by FT-IR spectral imaging. FEBS Lett 581:5469–74

Sterio DC (1984) The unbiased estimation of number and sizes of arbitrary particles using the disector. J Microsc 134:127–136

Stolz M, Raiteri R, Daniels AU, VanLandingham MR, Baschong W, Aebi U (2004) Dynamic elastic modulus of porcine articular cartilage determined at two different levels of tissue organization by indentation-type atomic force microscopy. Biophys J 86:3269–3283

Tromberg BJ, Shah N, Lanning R, Cerussi A, Espinoza J, Pham T, Svaasand L, Butler J (2000) Non-invasive in vivo characterization of breast tumors using photon migration spectroscopy. Neoplasia 2:26–40

Weibel ER (1979) Stereological methods, vol. 1: practical methods for biological morphology. Academic, London

Williams RM, Zipfel WR, Webb WW (2005) Interpreting second-harmonic generation images of collagen I fibrils. Biophys J 88:1377–1386

Chapter 8
Ultrastructure Imaging: Imaging and Probing the Structure and Molecular Make-Up of Cells and Tissues

Matthias Amrein

"New directions in science are launched by new tools much more often than by new concepts"

- Freeman J. Dyson (Princeton)

Abstract The past few years have seen dramatic improvements in life science microscopy. This chapter points out important recent developments that will probably affect many researchers in this field. Super-resolution fluorescence light microscopy allows for ultrastructure imaging of living cells. In electron tomograms, macromolecular complexes can be directly identified by shape without the need for labeling. Atomic force microscopes image isolated elements of the ultrastructure, including membranes and macromolecules, at submolecular resolution. This technology also makes molecular interactions directly accessible by measuring binding forces. In the chapter, new concepts are discussed in relation to current mainstream techniques. We will end with a brief discussion of novel concepts that could allow the molecular compositions of cells and tissues to be mapped in an unbiased, systematic way.

Introduction

Cells are organized into compartments and functional units down to the macromolecular level. The frameworks of these functional and structural units are either complexes containing only proteins, protein complexes that also include nucleic acids, or protein complexes that also include lipids. The recently completed map of the human genome as well as the systematic mapping of the proteins expressed

M. Amrein
Microscopy and Imaging Facility, Department of Cell Biology and Anatomy, Faculty of Medicine, University of Calgary, 3330 Hospital Drive N.W., Calgary, Alberta, T2N 4N1 Canada,
e-mail: mamrein@ucalgary.ca

C.W. Sensen and B. Hallgrímsson (eds.), *Advanced Imaging in Biology and Medicine.*   171
© Springer-Verlag Berlin Heidelberg 2009

in tissues and cells that this initiated (proteomics) are part of a concerted effort to rapidly advance our understanding of the functions of macromolecular functional units and the cell. However, proteomics only reveals the basic inventory of a cell, and this inventory is insufficient to explain the function of each element and the orchestration of the components. Attempting to understand life without observing the structures behind the functions is inconceivable. At the cellular and subcellular level, microscopy plays this important role by putting the molecular elements into their structural context. The pace of discovery of these elements has increased substantially via proteomics and this has also substantially increased the need for more and more sophisticated microscopy in the recent years.

Microscopy has existed for more than 500 years, and started with the simple magnifing glass. Work with the magnifying glass reached its peak in the 1600 s with the work of Anton von Leeuwenhoek, who was able to see single-celled animals and even bacteria. However, simple magnifiers had severe limitations in terms of illumination, image distortions, and resolution. The following sections describe how these limitations were gradually overcome by the introduction of increasingly ingenious "compound" light microscopes, thus achieving our first insights into the ultrastructure of a cell, or by exchanging the light—part of the electromagnetic spectrum—for a beam of electrons, thus reaching submolecular resolution.

Until very recently, the resolving power of the light microscope appeared to be invariably limited by diffraction. However, this barrier has now been surmounted to a degree, allowing elements of the ultrastructure of a cell to be well resolved. For now, the one dominant remaining limitation of light microscopy remains the need for labeling. Electron microscopy too has seen a dramatic change due to the implementation of tomographic data acquisition over the past few years. The resolving power of these microscopes is limited to the level where large molecular complexes are recognizable by shape. Higher resolutions appear to be invariably excluded because of the damage caused to the sample by the illuminating beam.

Not all microscopes depend on the emission of radiation and the subsequent recording of it at a distance from the sample using lenses. The atomic force microscope (AFM) is part of a family of microscopes that rely on an effect that is only present when a physical probe is in the immediate proximity of the sample, and this will be discussed below as well. It sometimes offers better resolution (or a better "signal-to-noise" ratio) than an electron microscope. More importantly, it can be used as a nanorobot to perform local experiments on cells or single molecules.

This chapter is not intended to be a history of microscopy, nor an exhaustive technical review. It points out important recent developments in life science microscopy that will probably affect many of those in this segment of the research community. The new concepts are described from the perspective of current mainstream techniques. We will end with a brief discussion of novel concepts that might allow the molecular compositions of cells and tissues to be mapped in an unbiased, systematic way.

Recent Developments in Light Microscopy

Recent developments have extended the resolution of light microscopy into the subcellular and ultrastructure range. Most of the improvements relate to fluorescent light microscopy and include a large and rapidly increasing set of fluorescent dyes and ingenious new means to improve the resolution and sensitivity at which the fluorescently labeled structures within a cell or tissue can be detected. Some of the most prominent of these developments will be discussed below. However, not everything can be fluorescently labeled, and improvements have also occurred in the visualization of unstained cellular structures by regular transmission light microscopy of unlabeled samples. One such development, hollow cone illumination by a "cardioid annular condenser" from CytoViva, is addressed below. First, we will briefly introduce some of the basic principles of light microscopy, an understanding of which is needed in order to comprehend the recent improvements in this field.

Basic principles of light microscopy (Fig. 8.1): The objective (together with a tube lens inside the microscope) produces a magnified image of the specimen inside the optical path leading up to the eyepiece or a camera. The image is then seen after being further magnified through the eyepiece or projected onto the light-sensitive chip of a camera. The objective is the most prominent optical component of the microscope. Its basic function is to collect the light passing through the specimen and then to project an accurate, magnified image of the specimen. Objectives have become increasingly sophisticated in terms of corrections for image distortions. A single magnifying lens focuses light of different wavelengths (colors) into different focal spots (chromatic aberration), and also focuses light entering the lens far from the optical axis into a different focal spot than light traveling through the central region of the lens (spherical aberration). Furthermore, not all of the specimen seen through the lens is in focus at the same time; when the center is in focus the periphery is not, and vice versa (field curvature). Corrections are indispensable when attempting to achieve the highest possible resolution.

The most basic correction is established by using "achromatic" objectives, which are corrected for chromatic aberration at two wavelengths (red and blue) and for spherical aberration at the color green but not for field curvature. Planapochromatic objectives yield the highest level of correction by eliminating chromatic aberration and spherical aberration almost entirely. They also produce "flat" images (without field curvature).

The resolving power of even the best-corrected objective is "diffraction limited" irrespective of whether fluorescence light or regular transmission light microscopy is being performed. As a consequence, the useful magnification is limited, and any additional magnification does not result in a more detailed image. When light from a specimen passes through the objective and is reconstituted as the image, the small features of the specimen appear in the image as small disks (Airy disks) rather than as points because of diffraction. Equally, sharp edges in the specimen appear blurred in the image. The function that describes the spreading of a point or edge in the specimen into a disk or blurred edge in the image is called the point spread function (PSF) and depends notably on the objective but also on other optical elements of the

CCD camera

projection lens

intermediate image

tube lens

objective
sample
condensor

light source

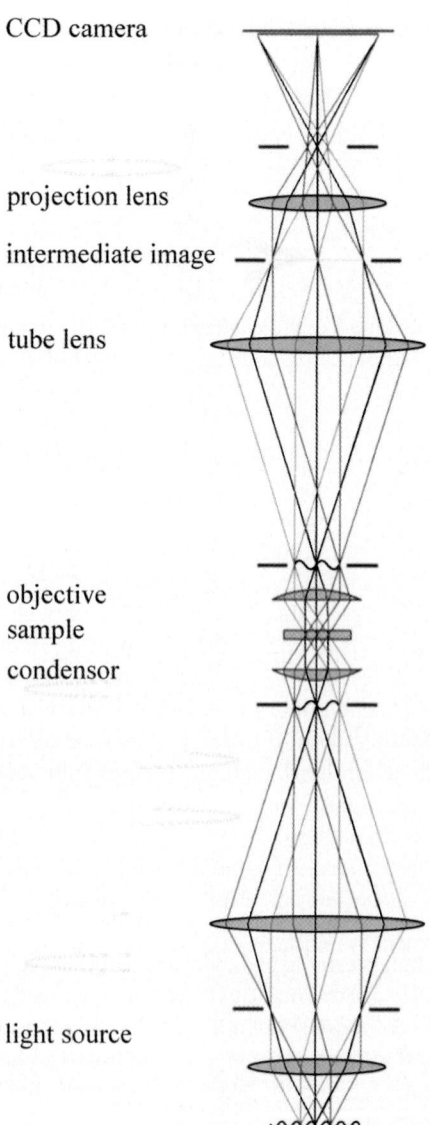

Fig. 8.1 Ray diagram of a regular transmitted light microscope in Koehler illumination (adapted from Hammond and Heath 2006). The hot filament of the light source is denoted by *a sinusoidal line*. It radiates light in all directions. Three representative rays from three different locations on the filament are shown. Whenever rays from a common origin on the filament (indicated by the same shade of *gray*) cross, the filament is in focus. Rays emanating from a common spot in the sample recombine at the plane of the intermediate image and (further enlarged) at the camera plane

microscope. Disks that are close to each other in the image, representing objects in close proximity in the specimen, will overlap and eventually become indiscernible. The better the microscope, the smaller the Airy disks and the finer the detail of the specimen discernible. Objectives that yield better correction result in smaller Airy disks than objectives that produce less correction, and so they have higher resolving power. Numerical aperture (NA) is another factor that determines the diameter of the Airy disk. The NA is a measure of the opening angle of the objective. It quantifies the ability of an objective to include rays of light from all directions from the specimen. Objectives with lower opening angles include a narrower cone of light compared to objectives with higher angular apertures. Specifically, the numerical aperture (NA) is given by $NA = n\sin(\alpha)$, where n is the index of refraction of the material between the specimen and the front lens of the objective ($n = 1$ for air with a "dry" lens, $n = 1.5$ for the oil in an oil immersion lens, and $n = 1.3$ for a water immersion lens). $\sin(\alpha)$ is the sine of half the opening angle of the objective. An objective with a higher NA will produce narrower Airy disks than an otherwise similar counterpart with a lower NA. The resolution power of the best dry lens is by definition lower than that of a water or oil immersion lens because of its inherently lower NA. Another factor influencing the resolving power of a microscope is the wavelength of the light. The shorter the wavelength, the higher the achievable resolution. In summary, the traditional resolution criterion for a light microscope is the Rayleigh criterion r, where $r = 0.61\lambda/N.A.$ r is the smallest distance between objects that still are discernible in the image; λ is the wavelength of the light (Fig. 8.2).

Cardioid annular condenser illumination: Despite the success of fluorescence light microscopy, transmission light microscopy (TLM) is also still important in the life sciences. For example, the evaluation of stained tissue sections is ubiquitous in animal and human pathology. When used in combination with fluorescence light microscopy, TLM allows some of the elements of the ultrastructure of the cell with which the fluorescently labeled elements associate to be identified. In this case, the highest possible resolution is of particular interest, as many elements of the ultrastructure of the cell are only just resolvable by regular TLM or cannot be visualized this way. In TLM, the light enters the sample from a source on the opposite side of the objective, and the quality and numerical aperture of the objective are not the only concerns in high-resolution imaging; the illumination light path needs to be factored in too. In the common approach known as "Koehler illumination," a lens projects an enlarged image of a light source (e.g., a hot filament) into focus at the front focal plane of a condenser. The condenser is a lens system with distortion corrections that should match the quality of the objectives. The light emerging from the condenser travels through the specimen in parallel rays from every angle within the numerical aperture of the condenser such that the sample is evenly illuminated from all directions. To fully realize the numerical aperture of the objective, the condenser must have a matched NA. This may require that the condenser also needs to be used with immersion oil. In this configuration, the diffraction-limited resolution (Rayleigh criterion) is $r \sim 240$ nm at best. Note that this is insufficient for resolving most cellular ultrastructure.

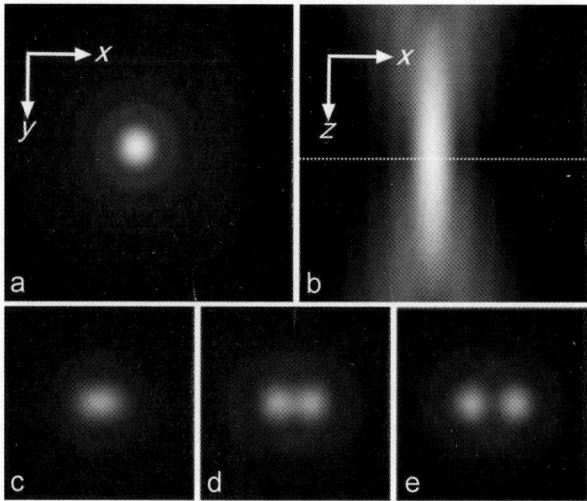

Fig. 8.2 (**a**) Even an infinitesimally small structure in a sample is spread out into an Airy disk in the image plane (x, y plane), according to the point spread function (PSF) of the optical system. (**b**) The spread is even more pronounced along the optical axis (z direction; the image plane is indicated by a *dashed line*). No feature of the sample is imaged beyond the precision permitted by the PSF and the resolution power is thus limited. The lower row shows the Airy disks of two structures too close to each other to be resolved (**c**); far enough apart to be barely resolved (**d**), and well resolved (**e**)

Higher resolution and better contrast can now be obtained by modifying the light path used in Koehler illumination. CytoViva markets an optical illumination system that replaces the standard condenser with a cardioid annular condenser (Fig. 8.3). In this configuration, the sample is illuminated by a hollow cone of parallel rays. This improves optical resolution; indeed, a resolution of 90 nm has been demonstrated. This increased resolution results from the narrower Airy disk that arises under coherent illumination for an annular aperture compared to that arising from the circular aperture under regular Koehler illumination (Vainrub et al. 2006).

Cardioid annular condenser illumination also appears to inherently provide contrast for unstained "transparent" cell or tissue samples (through a mechanism that is yet to be described). Note that regular Koehler illumination requires one to either stain the samples or employ methods to enhance the contrast of the unstained samples, such as differential interference contrast (DIC).

Optical sectioning microscopy; confocal and deconvolution light microscopy: Fluorescence light microscopy is arguably the most important microscopy used in the life sciences. In the following sections, we first describe technical advancements that have been instrumental to its success and then point out new directions that can take this microscopy into the realm of the ultrastructure of the cell (Conchello and Lichtman 2005).

In fluorescence light microscopy, the light used to excite fluorescent dyes is focused into the sample by the same objective that also collects the fluorescence

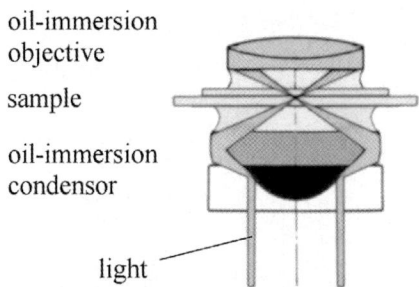

oil-immersion
objective

sample

oil-immersion
condensor

light

Fig. 8.3 Cardioid annular condenser. A hollow cylinder of light is deflected by a sperical mirror (*black*) outwards and is reflected back in again by another mirror onto the sample in a hollow cone. To ensure a large angle of illumination, the condensor is optically coupled to the glass slide of the sample with immersion oil. The light penetrating the sample and the coverslip enters the oil immersion objective far from the optical axis. This set-up improves resolution and offers a strong contrast for unstained samples that are barely visble in regular transmitted light micrsocopy

produced by the dyes. This configuration is termed epi-illumination. As with Koehler illumination, a cone of (excitation) light enters the sample. The fluorescent light is separated from the illuminating light by a filter set (Fig. 8.4).

Unfortunately, the illuminating light causes the excitation of not only the fluorescent dyes within the focal plane but also the fluorophores well above and below this plane. This is because the intensity of the illumination light does not drop sharply below and above the focal plane, according to the PSF of the optical system in the axial direction (see Fig. 8.2b). The haze caused by this out-of-focus light strongly obscures the in-focus image and reduces the contrast.

Optical sectioning microscopy, achieved using either specific hardware or through computational methods, eliminates the out-of-focus light and thus strongly improves the contrast in the images. It constitutes the single most important improvement in this type of microscopy. Moreover, for all methods, rather than taking a single image of the sample region, a stack of images ("optical sections" or frames) are acquired over the depth of the sample. Three-dimensional reconstructions of the sample volume are then obtained by combining the individual images from the stack in the computer.

In laser scanning confocal microscopy (LSCM, also termed LSM), optical sections of a sample are obtained by raster scanning an individual image frame at a given focal height, then moving incrementally to a new focal height and acquiring the next frame, etc., until a full stack of images has been acquired. The excitation light is brought into focus within the sample, and this focal spot is scanned. The fluorescent light produced travels back through the objective. It is separated from the excitation light by a filter and projected into a light detector. The image is then produced in the computer from the fluorescence intensity for each location within the image frame. To prevent out-of-focus light from entering the detector (i.e., fluorescence produced in the illumination cones above and below the focal spot), the fluorescence light stemming from the focal spot is brought into focus in front of the detector. At this location, the light must pass through a pinhole aperture

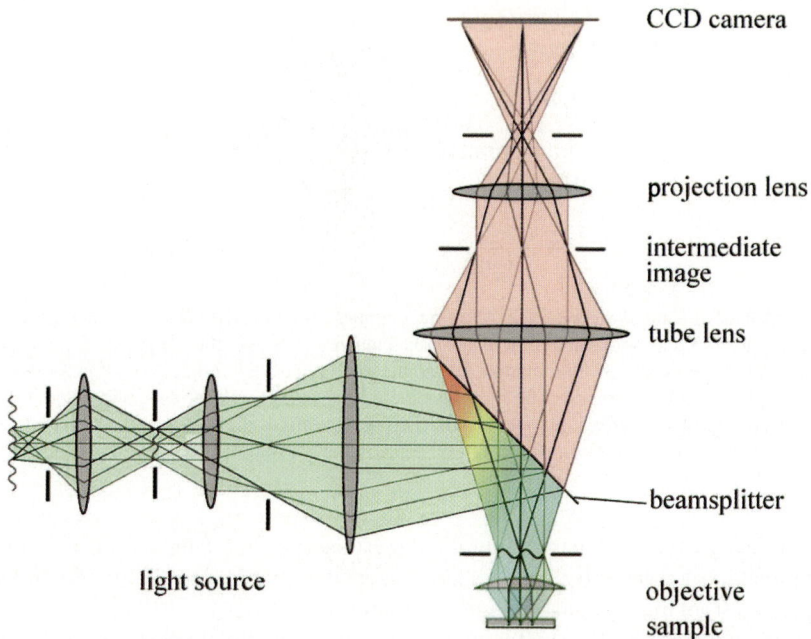

CCD camera

projection lens

intermediate image

tube lens

beamsplitter

light source

objective

sample

Fig. 8.4 Ray diagram of a fluorescence light microscope (adapted from Hammond and Heath 2006). Note that the illumination (excitation) light enters the sample from the same side as it is collected (epi-illumination). The excitation light (*denoted green*) and the emitted fluorescence (*denoted red*) are inherently of different wavelengths. This allows one to stop any excitation light from entering the camera by using a dichroic mirror (beam splitter) as well as additional filters (not shown)

that is just large enough to get through. Fluorescence produced above or below the illumination spot inside the sample will reach the pinhole somewhat out of focus and will thus spread out. Most of it will therefore be blocked out by the pinhole aperture. Because the illumination spot inside the sample and the fluorescence emanating from this spot at the pinhole in front of the detector are always in focus together, this type of microscopy is called "confocal." The pinhole not only eliminates the out-of-focus light and so strongly improves contrast; it also leads to a noticeable, albeit small, improvement in resolution.

The first commercial LSCM was designed at the MRC Laboratory of Molecular Biology in Cambridge, UK (Biorad MRC 600; BioRad Laboratories, Hercules, CA, USA) and scanned a single illumination spot using mirrors. One of the few drawbacks of this type of LSCM, particularly for live cell imaging, is slow image acquisition due to the time needed to scan an image frame with a single laser spot. In comparison, in a "wide-field" (i.e., regular) microscope, the entire image is acquired simultaneously, in which case the frame rate of the camera determines the speed. However, the problem of long acquisition times has now been overcome in another implementation of the confocal microscope, the "spinning-disk confocal microscope" (SDCM) (Fig. 8.5). Here, the image acquisition rate may be up to

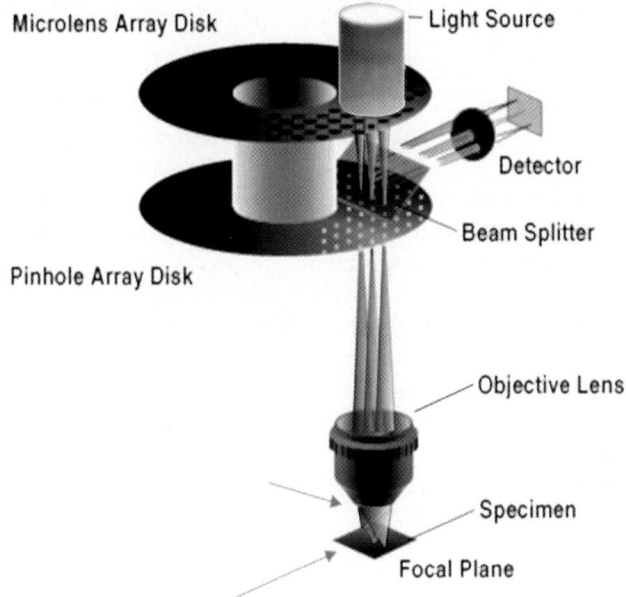

Fig. 8.5 Principle of the microlens-enhanced SDCM. An expanded, collimated laser beam illuminates the upper disk, which contains ~20,000 microlenses. Each lens focuses an individual beam onto a corresponding pinhole. The beams passing through each of the pinholes fill the aperture of the objective lens and are focused on the focal plane. The fluorescence generated from the specimen for each beam is focused back onto its pinhole, deflected by a dichroic mirror and focused into the image plane in the camera. As compared to conventional LSCM, multibeam scanning requires a significantly lower level of light intensity per unit area, which results in significantly reduced photobleaching and phototoxicity in live cells (adapted from http://www.yokogawa.com)

2,000 frames per second (CSU-X1, Yokogawa Co., Tokyo, Japan). In SDCM, rather than having a single laser spot scanning the sample, up to a thousand spots within a single image frame at any given time move simultaneously through the sample. This is accomplished by using a rapidly spinning disk (the Nipkow disk) that contains up to about 20,000 pinholes which cut perpendicularly through the illumination beam and break it up into individual, fast-moving beams. Newer set-ups use two disks on a common axis that spin together. One disk contains as many microlenses as there are pinholes in the other disk. Each of these lenses focuses part of the illuminating light into the corresponding pinhole on the other disk, where it is focused by the objective into the sample. The fluorescence from each illumination spot travels back into the objective, is focused into its corresponding pinhole, and progresses into a CCD (charge-coupled device) camera or an eyepiece. The pinholes reject the out-of-focus light and only allow the in-focus light into the camera. Image formation is so fast that it appears to be instantaneous, as in wide-field microscopy. Another advantage of the spinning-disk system over single-spot LSCM, particularly for live cell imaging, is the lower light intensity required per unit area of the sample. This results in reduced photobleaching and lower phototoxicity, both of which are problems that arise with a single spot scanner.

An alternative way of removing out-of-focus light from the LSCM is "digital deconvolution microscopy." A stack of regular wide-field fluorescence images are acquired, and the out-of-focus light is removed from each optical section afterwards by computational methods rather than by using pinholes. This method was derived based on information about the process of image formation. The earliest and computationally least demanding implementation is termed "nearest neighbors deconvolution." It subtracts the blur stemming from the sections immediately adjacent to (above and below) the current section. The method works well if a specimen consists of small fluorescent spots or thin filaments, but fails for larger fluorescent areas such as cell membranes or larger volumes. In these latter cases, it breaks the contiguous structures up into many spots. A more sophisticated method, termed "frequency-based deconvolution," first decomposes images computationally using a Fourier transform into series of periodic structures of different frequencies, amplitudes and phases (where smaller details give rise to higher frequencies and large structures to low frequencies). The frequency components of a microscopic image represent the "true frequency components of the specimen" multiplied by the "optical transfer function" (OTF) of the microscope. The OTF is the Fourier-transformed point spread function of the microscope. Hence, dividing the frequency components of the image by the OTF should undo the blur produced by the microscope within the resolution limit imposed by the impracticality of dissecting the overlapping Airy disks from objects too close to each other. When they are then superimposed (using a reverse Fourier transform), these corrected periodic structures should give rise to a corrected image. Unfortunately, this operation is not directly applicable because the OTF is zero for high frequencies (small details), beyond the resolution limit of the microscope. Division by zero is undefined. This makes more sophisticated methods necessary that circumvent division by zero, but unfortunately all of these introduce various image artefacts. These artefacts are minimized by also including information obtained a priori about the image in "constrained deconvolution" algorithms. These algorithms work iteratively and are typically time-consuming and computationally demanding. Different implementations are distributed commercially by Applied Precision, Scanalitics, AutoQuant Imaging, or are freely available (http://www.omrfcosm.omrf.org).

Super-resolution light microscopy: Currently, fluorescence microscopy is undergoing another revolution and a number of new methods have improved resolution by a degree that only recently was considered impossible. As a result, true ultrastructure imaging has now become possible. However, this super-resolution optical microscopy has so far provided only a few novel biological results. However, this situation is bound to change as the technologies become more readily available. Some of the new methods, with one prominent example being STED (stimulated emission depletion, see below), are technically demanding and require specialized equipment that is only now becoming commercially available. Another method of achieving super-resolution imaging only requires a regular microscope and depends on specialized but easily obtainable fluorescent labels, as well as special microscope control and data processing (this method has been published by three independent groups as STORM, PALM and FPALM, respectively).

STED uses a set-up similar to the laser scanning confocal microscope described above, but employs a sequence of two laser pulses for each point in the scanned region (Schmidt et al. 2008: Vainrub et al. 2006; Willig et al. 2006) (Fig. 8.6). First there is a pulse to excite any fluorophores in the focal spot, just as in regular LSCM (excitation pulse). Immediately following this, a second pulse with a slightly longer wavelength returns the excited fluorophores to their previous nonfluorescent ground states (depletion pulse). Unlike the excitation pulse, the depletion pulse is not focused into a spot. Instead, it is shaped into a cylinder that tightly encloses and cuts into the excitation spot, thus narrowing the region from which the fluorescence emanates down to a few nanometers in diameter. Thus, the lateral resolution has been improved by close to an order of a magnitude! The 20–70 nm lateral resolution realized with STED has been used to map proteins inside cells and on the plasma membrane.

More recently, the inventors of STED have come up with a STED scheme where an excitation spot is surrounded not just in the sample plane (x, y), but also in the axial direction (z), so that an isotropic fluorescent spot about 50 nm in diameter is achieved (Schmidt et al. 2008) (Fig. 8.6). For this, the sample was mounted in-between two oil immersion objectives positioned opposite to each other in a set-up termed 4-pi (due to the theoretical numerical aperture of 4-pi, or 360°, offered by such system). STED is now being offered by Leica Microsystems (TCS STED). It is integrated into the confocal platform TCS SP5. Note that the current commercial system is not a 4-pi microscope, so the improvement achieved by STED in this case will occur in the image plane only. However, Leica also offers a 4-pi system, so one can reasonably expect the full three-dimensional system to become commercially available too.

STORM (stochastic optical reconstruction microscopy) (Rust et al. 2006; Perfect Storm 2008; Vainrub et al. 2006; Huang et al. 2008), photo-activated localization microscopy (PALM) (Betzig et al. 2006; Manley et al. 2008; Shroff et al. 2008), and fluorescence photo-activated localization microscopy (FPALM) (Hess et al. 2006, 2007) are acronyms for essentially the same simple and yet ingenious method for accomplishing super-resolution in light microscopy (Fig. 8.7). The three implementations were published by three different groups at about the same time. As discussed above, the point spread function of a light microscope describes how light from a point source in the sample spreads into an Airy disk in the image plane, thus making it impossible to separate two light spots very close to each other. However, for any single light spot (from a single fluorophore) that is well separated from any other spot, the absolute position can be determined with much higher accuracy than the Raleigh criterion suggests by simply fitting a Gaussian to the Airy disk to locate the center and thus the location of the fluorophore. PALM, FPALM and STORM all make use of this strategy in combination with special dyes that first need to be photoactivated by a pulse of light before becoming fluorescent. Photoactivation is stochastic, and with a sufficiently weak activation pulse only a small subset of all of the dye molecules in the field of view are activated. Thus, most fluorescing

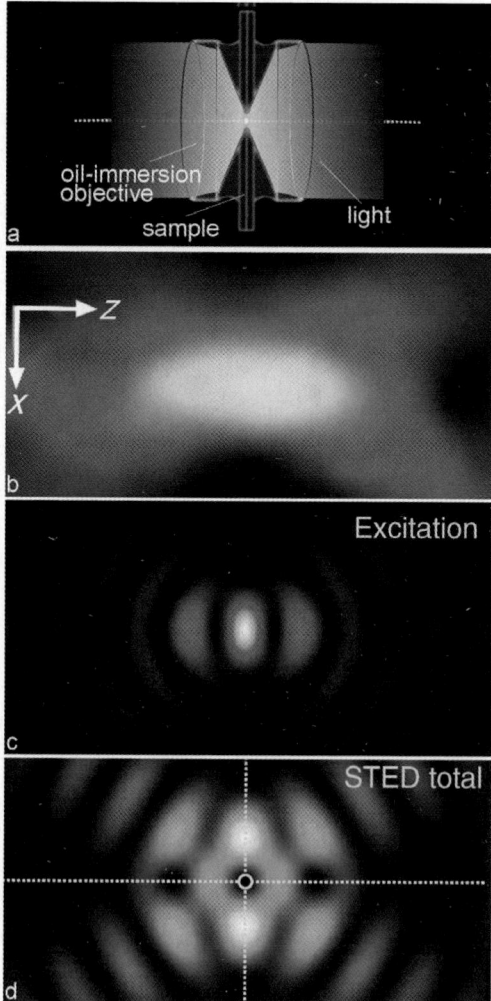

Fig. 8.6 (**a–d**) Focal intensity distributions and formation of the STED point spread function. (**a**) 4-pi microscopy. Two oil immersion objective lenses are mounted opposite to each other with the sample positioned in-between them. This results in twice the numerical aperture of a single lens, and thus improved resolution. Moreover, the focal spot can be further narrowed through constructive interference between the beams coming from opposite directions, similar to interferometry. (**b–d**) Close-ups of the light intensity distribution in the focal region. (**b**) A regular cigar-shaped focal spot from a single lens (shown for comparison). (**c**) Constructive interference pattern of excitation light in the 4-pi set-up. This light is used to excite the fluorescence in the sample. Note the favorable intensity distribution as compared to the regular PSF shown in (**b**). The situation is further improved through STED. (**d**) STED intensity distribution. This light is of a longer wavelength than the excitation light (shown in **c**) and serves to extinguish fluorescence. The STED light cuts into the excitation spot almost isotropically from all directions, sparing only a small central sphere. This sphere, with a diameter of only about 40 nm (denoted by a ring in **d**), is the only region from which fluorescence emanates, thus leading to the dramatic increase resolution of this microscopy over traditional confocal fluorescence microscopes (adapted with the permission of Macmillan Publishers Ltd. from Schmidt et al. 2008)

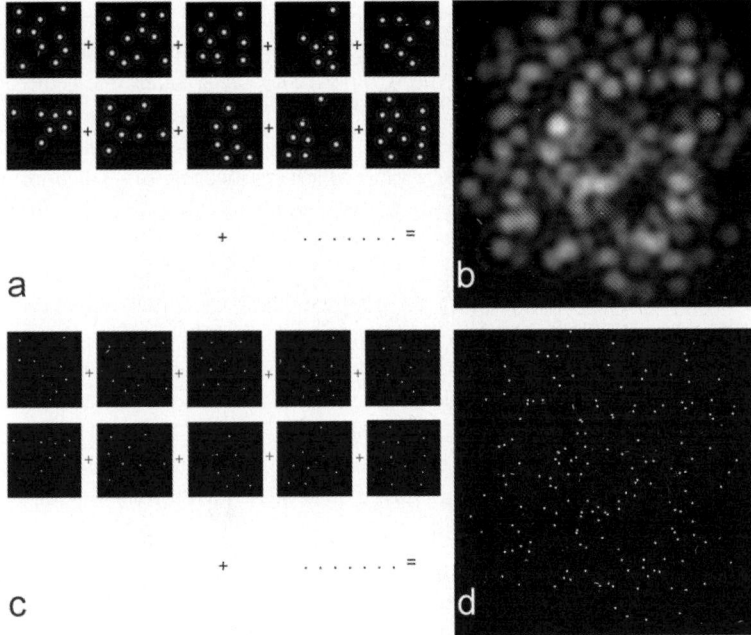

Fig. 8.7 (**a–d**) Principle of STORM/PALM/FPALM. (**a**) An activation laser pulse stochastically excites a small subset of photoactivatable fluorescent dyes to their fluorescent states. The dyes are imaged for one or a few frames until they become bleached (PALM, FPALM) or are returned to the dark state (STORM). This process is repeated, with a different subset of the dyes observed each time. Conditions are chosen so that the dye molecules are well separated from each other in the individual frames. Airy disks in individual frames are reduced to dots (**c**). The aggregate image of all frames shows that the locations of the dyes are well resolved (**d**). (**b**) A regular image of the same dye molecules included for comparison

molecules are far enough from each other to determine their locations with high accuracy.

In STORM, after acquiring a frame, all fluorescing molecules are returned to their dark states and another subset of the dyes is activated in order to acquire another frame from the same region. This process is then repeated many times (e.g., 10,000 times). Each individual frame is reduced computationally to a map of locations of individual dye molecules. Adding the maps yields the super-resolution image. The inventors of STORM report the use of a photoswitchable cyanine dye. Cy5, a common cyan dye, can be switched between a fluorescent and a stable dark state ("triplet state") by red laser light, which also excites the fluorescence. Exposure to green laser light converts Cy5 back to the fluorescent state, but only when Cy3 is close-by. Hence, the Cy3–Cy5 pair comprises an optical switch. In PALM and FPALM, photoactivatable fluorescent protein molecules are turned on and localized (to 2–25 nm). After a few consecutive frames, the dyes become bleached. The method can be used to achieve high resolution not just in the lateral dimension but also the axial direction. This is accomplished by making the objective slightly astigmatic.

An astigmatic objective focuses light in one lateral direction more strongly than in another. Any point source imaged with such lens will produce a round Airy disk only when it is in focus between the strongest and the weakest focal directions. If it is in focus in the stronger focal direction, it will be spread out in the weaker focal direction, thus producing an ellipsoid, and vice versa. Using this information, the distorted shape of the Airy disk enables the true location of each fluorophore within the image frame to be gauged not just laterally (with a precision of 20 nm) but also axially (to an accuracy of 60 nm), thus achieving resolutions well below the diffraction limit (Fig. 8.8).

All of the super-resolution light microscopy methods described above are inherently limited in terms of speed by the low rate of photon emission from a small spot as compared to the larger spots associated with regular microscopy. However, live cell imaging has still been achieved in STED at 28 frames s^{-1} for the example of the movement of synaptic vesicles in cultured neurons. This has been accomplished by reducing the spot size to a level where just enough photons from the features of interest are emitted within the given short acquisition time to allow them

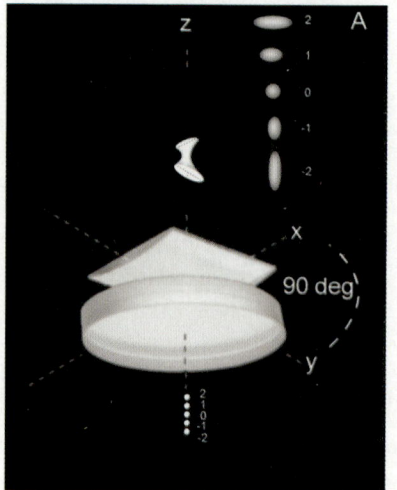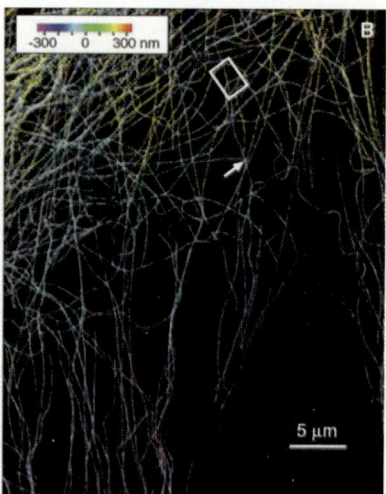

Fig. 8.8 (**a–b**) Three-dimensional (3D) STORM. (**a**) Principle. A weak cylindrical lens is added to the objective to make it astigmatic. In an astigmatic objective, the focal length in one direction (here the x-direction) is shorter than in another direction (y-direction). A fluorophore is thus imaged into an elliptical Airy disk with its main axis along the x-direction when it is in focus in the x-direction (close to the objective, denoted by 2), a circular disk when at an intermediate focal length (0), and an elliptical disk with its main axis in the y-direction when it is even further away from the objective (Vainrub et al. 2006) and in focus in the y-direction. The shapes of the Airy disks above the objective and the corresponding fluorophores below the objective are denoted by corresponding numbers. Therefore, a single image frame in 3D STORM contains not only information about the locations of all the fluorescent dyes in the image plane (x, y), but also information on the distance of each fluorophore from the objective (z-direction), as derived from the shape of its Airy disk. (**b**) Image of microtubules in a cell. The z-position information is color-coded according to the *color scale bar* (from Huang et al. 2008 and used with permission)

to be discerned from the background. PALM and FPALM have also been used in live cell imaging. The time resolution of PALM has been relatively slow for live cell imaging, and a value of 500 s/image (with an image being the aggregate information from 10,000 individual frames taken at 20 frames s^{-1}) has been reported. However, in their live cell application of PALM, Manley et al. did not use only the aggregate information. Rather, reoccurring fluorescence from individual dyes over a number of consecutive frames (before they became bleached) was mapped and the trajectories of the tagged proteins were plotted as in conventional "single-particle tracking" (spt). However, rather than tracking single ensembles of molecules, the photoactivatable fluorophores allowed multiple ensembles of molecules to be activated, imaged and bleached. Thus, several orders of magnitude more trajectories per cell were obtained as compared to conventional single-particle tracking. Hess et al. (FPALM) also made use of the reoccurrence of fluorescence from single molecules over two to four frames to study molecular dynamics. The PALM/FPALM/STORM principle has now been licensed by Zeiss and will be offered on their laser scanning confocal microscope platform LSM 710.

In summary, recent improvements in light microscopy have extended the resolution well into the realm of the ultrastructure of the cell. Lateral resolution on the order of 20 nm and axial resolution on the order of 60 nm have been convincingly demonstrated, even in living cells. At the same time, molecular motion has been observed at up to 20 frames s^{-1}. Future improvements in all super-resolution microscopy methods are entirely possible, with respect to both spatial and time resolution. Such improvements may be brought about or at least depend on brighter fluorophores.

Image processing: Image processing in advanced light microscopy is pursued in order to enhance 3D light microscopy datasets by deconvolution (discussed above), to provide a comprehensive display, and to perform a statistical data analysis, including the colocalization of cellular components in three dimensions and in time series of 3D images (termed 4D). Unlike two-dimensional images, we cannot directly display three-dimensional images on a computer screen. Therefore digital image processing is an essential component of any 3D microscope system. Moreover, quantitative data assessment is often indispensable when answering a research question.

Principally, all of the abovementioned traditional optical sectioning microscopy and super-resolution techniques produce datasets that require image processing. The following section briefly describes what image processing involves, starting with the raw data and achieving quantitative, statistically meaningful numbers. Examples are taken from one of the common image processing software packages (*Imaris* from bitplane, Zurich, Switzerland). In image processing, the first task usually consists of visualizing the data. Imaris as well as other packages offer a palette of fast operations to project the data onto a 2D screen, a process often referred to as "3D reconstruction" or "rendering" (Fig. 8.9). The projection methods include *maximum intensity projection* (MIP), which prioritizes the bright structures, and *blending* or *ray-tracing*, which prioritizes the structures close to the point of view. Software enables users to adapt the transparency, colors, and lighting to the needs of the study

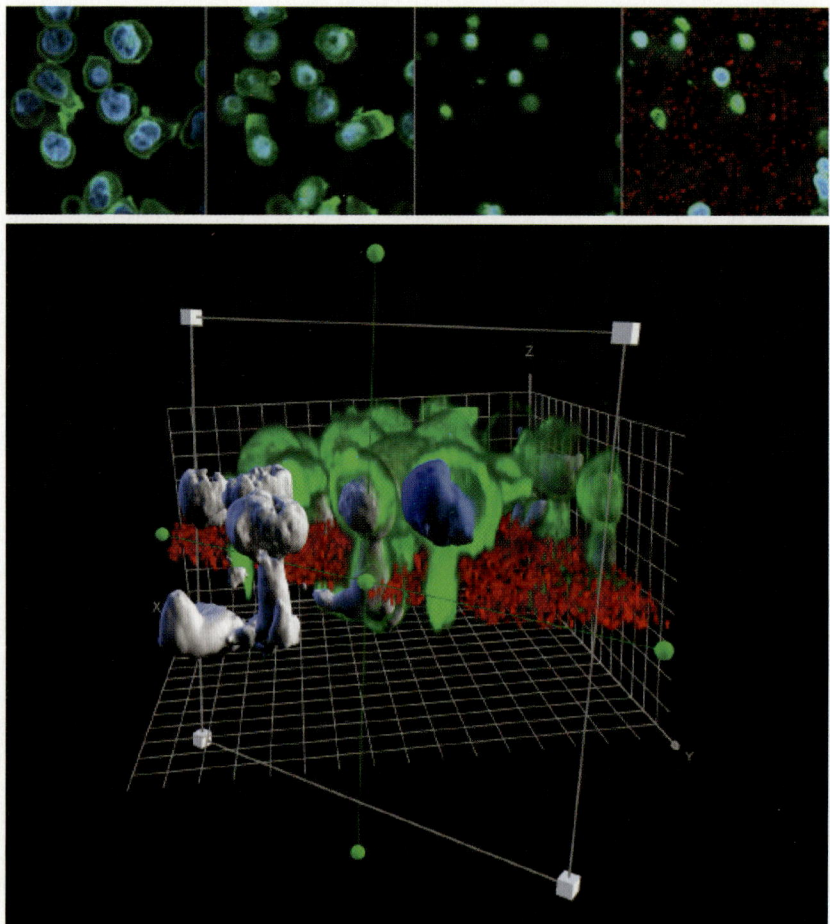

Fig. 8.9 (*Top row*) Confocal optical slices taken at four different z-planes 4.7 µm apart from each other. Actin is shown in *green*, DAPI stain (a stain of the nucleus) is shown in *blue* and an artificial membrane is shown in *red* (*bottom row*). A 3D reconstruction of 64 slices, of which four are shown above, is depicted. Only in 3D does it become apparent that the cells have migrated across the membrane. A clipping plane is positioned vertically to the plane of the membrane to expose a cross-section through the cells (image display by Imaris, data courtesy of Dr. Tomoko Shibutani, DAIICHI Pharmaceutical, Tokyo)

and the dataset. If the purpose was visualization, we may end up with a set of rendered pictures and animations showing important aspects of the 3D image. Quite often, however, visualization is used to assess the image before it is subjected to further quantitative analysis.

For many studies involving quantitative analysis, the first step is to identify objects in the image. This operation is usually referred to as "segmentation," and is one of the major steps in the analysis. The user can choose the object "detector" that best matches the image. For instance, tiny structures such as vesicles or nuclear FISH signals would typically be detected in Imaris by the "Spots" detector component, whereas dendrite trees of neurons are detected using the Imaris "Filament" detector. The segmentation step is also where fluorescent microscopy has a large advantage over other methods such as transmission or reflection microscopy.

The ability to selectively label structures of interest combined with its suitability for 3D imaging has enormously facilitated the segmentation step. Rather than analyzing the complex patterns and edges of transmission micrographs, software can simply localize high-intensity elements in the image; these are inherently the areas of interest. Segmentation is one of the reasons why interactive software packages such as Imaris are generally favored over noninteractive software—they offer manual editing of automatically obtained results. Once segmentation has occurred, quantitative analysis of the objects can be pursued, volumes can be calculated, distances can be measured and the branching of processes can be evaluated, to name just a few examples for quantitative data analysis (Fig. 8.10).

Electron Tomography

Only recently, ultrastructure imaging was the domain of transmission electron microscopy (TEM) alone. Electron microscopy utilizes a beam of electrons that are accelerated by a high voltage and directed into vacuum column with electromagnetic lenses, rather than visible light and glass lenses. TEM thus exploits the very

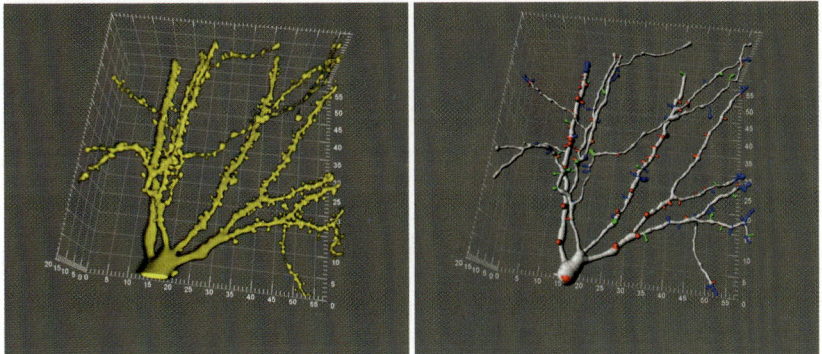

Fig. 8.10 (*Left*) A blending reconstruction of a lucifer yellow-injected single nerve cell displaying dendrites and spines. While visually informative, the software is unable to obtain quantitative information from data in this form. The figure on the (*right*) was created from the same data after automatic segmentation. The dendrites and spines were automatically segmented using FilamentTracer from Imaris, a software module designed for the analysis of longitudinal, branched structures. The data is now available as an object graph and the software is able to count branch points and to measure the lengths and diameters of each of the dendrite segments; hence the information is much more readily available for quantification (dimensions are in microns; courtesy of Anne McKinney, McGill University, Montreal)

short wavelength of the electron beam to boost the resolution of the optical system way beyond that offered by visible light. While TEM is now receiving competition from super-resolution light microscopy, it will continue to play an important role in revealing the spatial relationships of the macromolecules in cells and tissues. This is because, in light microscopy, macromolecular identity is only revealed after label-ing. The large number of different proteins and other macromolecules at work at any time in a cell makes it inconceivable to extend this approach to deduce even a substantial subset of all of its macromolecular relationships by methods that depend on labeling. Moreover, even the cellular architecture of compartments separated by membranes, the cytoskeleton and the nucleus with its substructure all need labeling to be well resolved, even when using the most advanced light microscope. TEM on the other hand reveals all of the compartments in the cell directly, and much of the cytoskeleton is identified without labeling as well. Moreover, TEM is now making a strong resurgence as electron tomography (ET), a novel data collection strategy, enables explorations of cellular ultrastructure in three dimensions at very high resolution (Kurner et al. 2005; Richter et al. 2008; Marsh et al. 2001; McIntosh et al. 2005; Lucic et al. 2005).

Electron tomography has been developed over more than 20 years. However, the technology has only just now matured to the point where it can be routinely applied to specimens in a fully automated fashion. This not only makes the recording of tomographic datasets much less cumbersome, but it also makes it possible to mini-mize the exposure of the specimen to the electron beam, a major concern in the high-resolution TEM of inherently beam-sensitive biological specimens. In tradi-tional TEM, a single projection image (micrograph) of the region of interest in a thin section is acquired with the specimen positioned perpendicular to the electron beam. In the process of electron tomography, on the other hand, several hundred digital micrographs are acquired, each from a different viewing direction, as the specimen is rotated around the region of interest from about 70° tilt angle from its normal to the beam, through zero, to −70°. The resulting set of micrographs is com-putationally back-projected into a single 3D object, according to the same concepts as used in medical computer tomography or other tomographic methods (Fig. 8.11).

The tomogram allows for a much more comprehensive structure-based under-standing of the role of the proteins of interest in their cellular context than traditional electron micrographs. In addition to revealing ultra- and molecular structure in three dimensions (which is far more meaningful), it greatly (or significantly) increases the information obtained in the viewing direction over the traditional single projection. In a single micrograph, all of the structures in the viewing direction overlap and the resolution is basically reduced to the thickness of the complete specimen, e.g., the thickness of a plastic section of some cells or tissue, whereas the resolution is almost isotropic in all directions after tomography. The current resolution of cell organelles, subcellular assemblies and, in some cases, whole cells permitted by ET is in the range 5–20 nm: suitable for identifying supramolecular multiprotein structures in their cellular context (Fig. 8.12). However, in order to identify it in a tomogram, the shape of such supramolecular structure must be known beforehand; from X-ray crystallography, NMR spectroscopy, EM crystallography or single-particle analysis

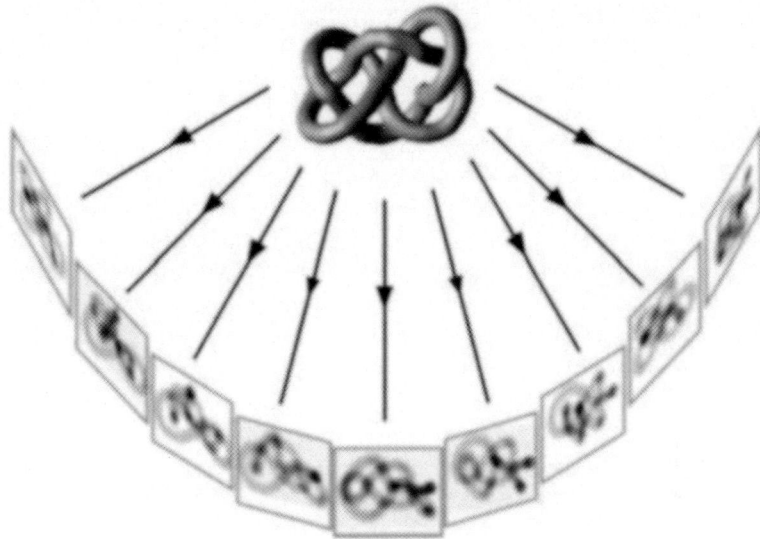

Fig. 8.11 In electron tomography (ET), many projected images of a sample are acquired. In order to achieve this, the sample is tilted with respect to the beam in-between images. The 3D structure of the object is then reconstructed from the individual micrographs through computational "back projection" (from Lucic et al. 2005 and used with permission)

(Medalia et al. 2002). The known shape of the complex is then used as a "mask" in an algorithm that is designed to locate copies of it within a tomogram via cross-correlation. ET also reveals the context of protein complexes with other cellular components, such as DNA and membranes.

The high resolution offered by ET naturally depends on adequate sample preservation for the improved resolution to be meaningful. During processing, samples are rendered compatible with the high vacuum of the microscope, and they are often cut into sections that are sufficiently thin for the electron to beam to penetrate. Tissues or cells may be fixed chemically, embedded in plastic and cut into thin sections by an ultra microtome. Some heavy metal staining is usually introduced to surround the molecules and thus improve contrast. Modern TEM performed with an "intermediate" acceleration voltage for the beam electrons of 200–300 kV allows samples up to about 200–300 nm thick to be penetrated. Similar, although thinner, sections are also used widely in traditional TEM to study the morphology, ultrastructure, and contents of cells and their subcellular organelles. However, chemical fixation may not sufficiently preserve the structures for ET. To greatly improve structure preservation, samples can be rapidly frozen employing special procedures to prevent ice crystals from forming: this is an artefact-free fixation method. Frozen specimens can then be processed by "freeze substitution." In this process, the water in the frozen specimen is gradually replaced with a plastic resin at low temperature so that all structures are retained in place throughout. After polymerizing the resin, the samples are sectioned in a similar way to chemically fixed samples. In a variant of this

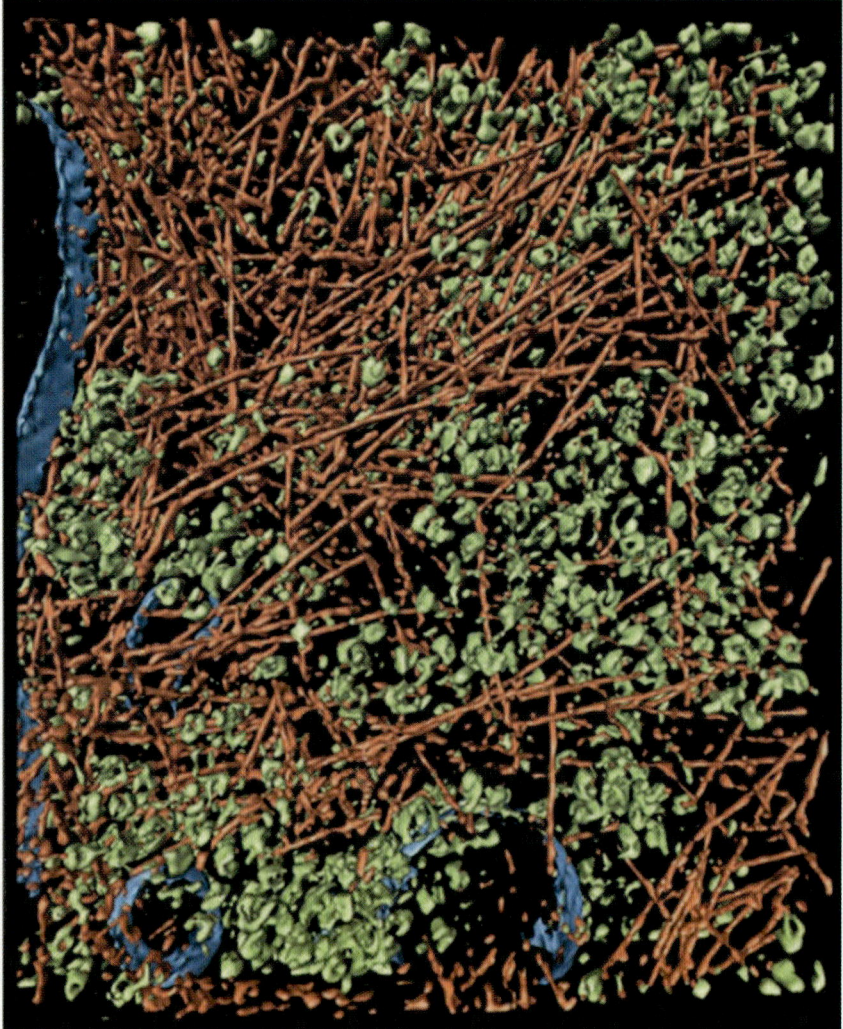

Fig. 8.12 Surface rendering of an actin network, membranes, and cytoplasmic macromolecular complexes (815 × 870 × 97 nm). The actin filaments are *reddish*, the ribosomes *green* and the membranes *blue* (from Medalia et al. 2002 and used with permission)

approach, the frozen sample can also be cut directly into thin sections for cryo-TEM without plastic embedding or staining. However, cryosections are difficult to obtain as they tend to crack because of the compression caused by commonly used diamond knifes. Better results have recently been achieved by carving sections out of a frozen sample by focused ion beam thinning (i.e., exposing the frozen sample to an ion beam in order to sputter material). Notwithstanding the great technical challenges posed by obtaining and imaging cryosections, excellent results have been obtained from bacteria samples, particularly in terms of the preservation of membrane structure and extracellular glycoprotein architecture (Marko et al. 2007). In other cases, the specimens are within the required 200–300 nm thickness or are not much greater than this, so sectioning is not required. This may be true for extended processes of cells or for large regions of epithelial or endothelial cells. The cells can be cultured directly on a thin carbon or plastic film used as a support during the microscopy; this is then frozen and introduced into the microscope (Medalia et al. 2002).

ET has matured at a time when electron microscopes have also improved greatly, meaning that the information about the object carried by the electron beam is almost fully exploited. However, the ultimate resolution of electron microscopy and hence ET is inherently limited due to the damage that occurs when electrons interact with the cellular structure. Current resolutions may indeed be close to a "theoretical" limit, and this would mean that the secondary and tertiary structure of an individual protein or polypeptide would remain out of bounds. The identification of proteins that are not in large complexes would therefore continue to require (immunogold) labeling.

Atomic Force Microscopy

AFM does not depend on light or optics. In AFM, the topology of a sample is traced by a sharp stylus that is scanned line by line over the sample. The stylus sits at the free end of a cantilever spring and lightly touches the sample. Increases in elevation of the sample cause the stylus to move up and bend the cantilever upward while depressions allow the stylus to move down. The stylus and cantilever are usually microfabricated from silicon or silicon nitrite. The cantilever is typically a fraction of a millimeter long and a few micrometers thick. AFM lends itself well to imaging molecular and cellular specimens under physiological conditions in buffer. The AFM cannot look inside an intact cell, but it can be used to either investigate elements of the ultrastructure that have been isolated or to probe the outer surfaces of cells.

When imaging isolated molecular assemblies, AFM offers very high resolution at a high signal-to-noise ratio. Biological membranes for example can be imaged in their native state at a lateral resolution of 0.5–1 nm and a vertical resolution of 0.1–0.2 nm. Conformational changes that are related to functions can be resolved to a similar level. This is well beyond the capabilities of any other microscope (Fig. 8.13).

In addition to imaging the structure, AFM can map properties of the sample such as local stiffness, electrical conductivity or electric surface potential (Fig. 8.14). In the example given in Fig. 8.14, the map of the electric surface potential helped to reveal the local molecular arrangement of pulmonary surfactant (Leonenko et al. 2007a, b).

Fig. 8.13 ATP synthase is a proton-driven molecular motor with a stator (seen here) and a rotor. It has not yet even been possible to count the number of individual proteins inside the ring structure based on electron micrographs (adapted with the permission of Macmillan Publishers Ltd. from Seelert et al. 2000)

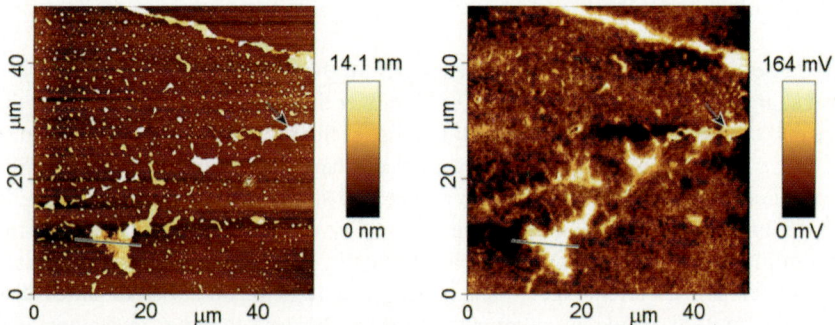

Fig. 8.14 A film of pulmonary surfactant. The topographical image (*left*) shows a pattern of lipid monolayer and scattered multilayer regions (stacks of lipid bilayers). In the potential map (*right*), large stacks of bilayer patches are at a potential of up to 200 mV above the monolayer. We associated the positive surface potential of the stacks with the presence of the surfactant-associated protein C (SP-C). The helical structure of SP-C constitutes a strong molecular dipole (taken from Leonenko et al. 2001 with permission; the image was taken with a NanoWizardII AFM from JPK Instruments, Berlin)

AFM started out as an imaging tool, and early applications demonstrated the resolution power and high signal-to-noise ratio that it offers at the macromolecular level. Although unintentionally at first, it soon became clear that the probe could be used to manipulate and probe macromolecular structures individually, thus opening up a wide range of applications for this technology. The most notable of these is termed "force spectroscopy" (Fig. 8.15). It is based on the fact that the force between

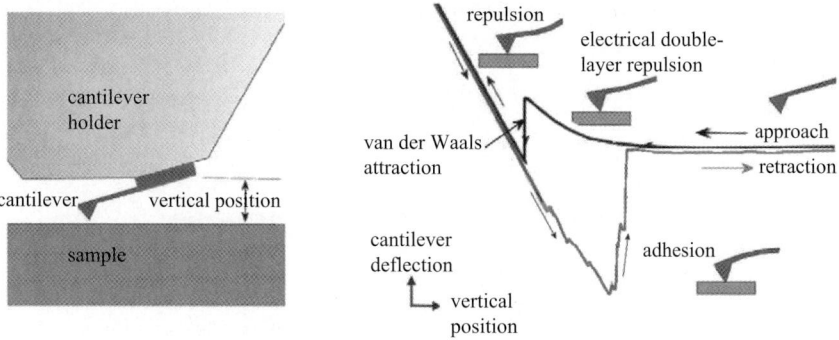

Fig. 8.15 Force vs. distance curve. For the example shown here, the tip first experiences a long-range repulsive force upon approaching the sample, even before the tip and sample are in physical contact. Close to the sample, the tip becomes strongly attracted by the van der Waals force. In this instance, the attractive force gradient becomes greater than the force associated with the cantilever spring. This causes the tip to snap into physical contact with the sample (the perpendicular part of the approach curve). Once physical contact has been made, the cantilever is deflected linearly by the approaching scanner until the scanner stops. On the way back, the tip often sticks to the sample by adhesion until the pull from the cantilever forces it out of contact. The adhesion curve is often the most interesting part. It can reveal the folding path of a protein, be used to measure immune cell activation, or show the time dependence and progression of nanoparticle uptake by lung epithelial cells (Fig. 8.16), to name but a few examples

the stylus and sample derives from not only the active loading of the stylus onto the sample, but also the interactions that occur between the sample and the tip, including van der Walls, ionic and hydrophobic interactions. These interactions are the same as those that rule molecular interactions in cells and tissues, including nonspecific interactions and highly specific receptor–ligand interactions (Israelachvili 1997). The forces they produce typically range from a few piconewtons to a few nanonewtons and are thus well within the measurement accuracy of the AFM.

For a number of examples, molecular images have been obtained with sufficient resolution to individually recognize single macromolecules and then address the molecules individually with the stylus of the AFM. Single-molecule force spectroscopy combined with single-molecule imaging has provided an unprecedented means to analyze intramolecular and intermolecular forces, including those involved in the folding pathways of proteins. Probing the self-assembly of macromolecular complexes and measuring the mechanical properties of macromolecular "springs" are other examples of where AFM has made substantial contributions to the life sciences.

The imaging of living cells with AFM is performed in conjunction with local probing of the sample (Figs. 8.16 and 8.17). The images by themselves are usually only used to obtain proper experimental control. The extreme flexibility of the cell membrane means that the images often show a combination of the cell topography and the mechanical stiffness of cytoskeletal fibers and vesicles below the cell surface. The in situ probing of the responses of cells to electrical, chemical or mechanical stimulation is also possible.

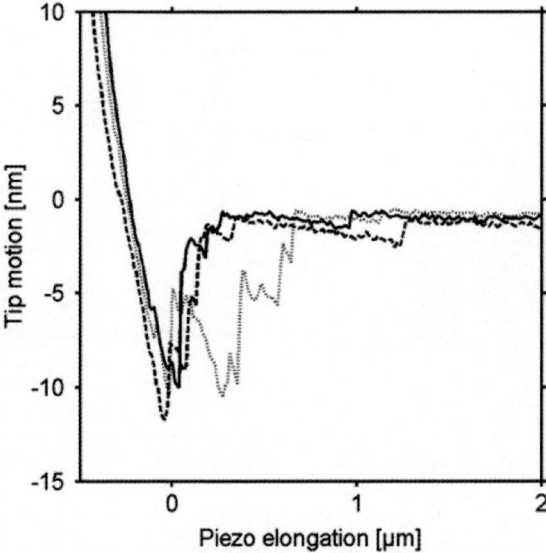

Fig. 8.16 Force vs. distance curve of a chemically modified AFM stylus on a lung epithelial cell (adapted from Leonenko et al. 2007a, b and used with permission from Elsevier). Here, the AFM tip serves as a model of a nanoparticle that enters the lung via the ambient air and is taken up by the body. The study revealed the important aspects of the particle uptake by measuring the time-dependent adhesion of the particle to the plasma membrane

Future Directions in Ultrastructure Imaging

Knowing at any given time where every atom of every molecule resides in an organism holds the key to answering any research question in the life sciences. Our current understanding is that the combination of crystallography or NMR (which reveal the atomic structures of macromolecules), electron tomography (which allows the determination of the precise subcellular positions and interactions of macromolecular machines), and fluorescence light microscopy (which describes the dynamic nature of cellular processes) should lead to a profound understanding of the cellular processes associated with the physiological functions of tissues and organs and the pathogenesis of disease. However, the sheer number of macromolecules in an organism or even in a single cell that are interacting and being modified and turned at any given time, not to mention the small molecules of the "metabolome," makes the notion of obtaining the entirety or even a substantial fraction of the mechanics of a cell, tissue or organism with current techniques quite inconceivable. Gene array technology that is designed to reveal cellular responses to even small interferences with normal cell function typically returns hundreds of changes in gene activity, some of which are anticipated but many of which are entirely unexpected and in no way understood. This is testament to the level of integration and interdependence of the molecular construction of cells. Hence, by observing hundreds or thousands

Fig. 8.17 A neuron cultured on an electronic chip. The chip is designed to pick up an action potential from the cell. The image demonstrates the proper tracing of the cell surface. In a future application, an appropriately designed stylus could be used as an additional electrode to excite or record an action potential at any location on the cell body or a neuronal cell process

of different molecular species at a time by microscopy, one ought to be able to do justice to the complexity of life, or, in more practical terms, efficiently unravel the molecular pathways associated with a function of interest. Such an approach would be conceptually similar to the systematic, unbiased approaches of genomics and proteomics. No single existing microscopy technique that is currently in existence or even any combination of all of them has the potential to achieve this. On the other hand, current analytical tools for identifying large numbers of molecular species at a time—with mass spectrometry being one important example—are not designed to determine the part of a cell in which a particular molecule is located. An ideal microscope would combine the high level of localization of today's ultrastructure imaging tools and the capabilities of mass spectrometry (for example, the ability to identify

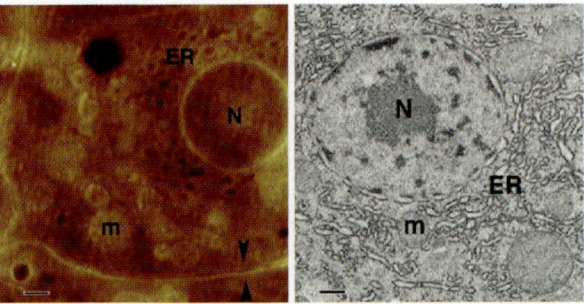

Fig. 8.18 AFM of cell sections (*left*); for comparison a TEM micrograph of the same specimen is also shown (*right*). AFM and TEM both give a highly detailed image of the cell's ultrastructure. *N*, nucleus; *ER*, endoplasmatic reticulum; *m*, mitochondria (adapted from Matsko and Mueller 2004 with permission)

molecular species based on their own characteristics rather than via labeling). Such a microscope is yet to be invented.

Among all current microscopic techniques, AFM may offer one avenue to such a microscope. It has recently been demonstrated that thin sections of fixed and embedded tissue and cell samples make the cell's interior accessible to AFM. High-resolution cell images give a clear image of the cell's ultrastructure, sometimes reaching down to the macromolecular level (Fig. 8.18). While the AFM is unlikely to allow macromolecules to be identified by shape, it could be combined with powerful local spectroscopy. The stylus could then act as both an imaging tool and a local spectroscopic probe. One can speculate that spectroscopy might lend the microscopy the ability to identify macromolecular specimens under the tip directly.

In summary, the past few years have seen dramatic improvements in life science microscopy. Super-resolution fluorescence light microscopy has permitted the ultra-structure imaging of living cells. In electron tomograms, macromolecular complexes can be identified directly by shape, without the need for labeling. Atomic force microscopes image isolated elements of the ultrastructure, including membranes and macromolecules, at submolecular resolution. This technology also makes molecular interactions directly accessible through the measurement of binding forces. Finally, there is still room for further invention—a microscope that does not require the use of labeling to identify molecules, from metabolites to macromolecules within the ultrastructure of the cell, is not yet on the horizon.

Acknowledgments The author is grateful to Dr. Mary Brindle and Dr. Michael Schoel for their critical reading. We thank Dr. Marius Messerli for his assistance in preparing the section on image processing and for providing the figures associated with that section. The "neuron on a chip" shown in Fig. 8.17 was prepared by the laboratory of Dr. Naweed Syed. We acknowledge the permission to reproduce the diagrams and images shown in Figs. 1, 4–6, 8, 11–14, 16 and 18.

References

Betzig E, Patterson GH, Sougrat R, Lindwasser OW, Olenych S, Bonifacino JS, Davidson MW, Lippincott-Schwartz J, Hess HF (2006) Imaging intracellular fluorescent proteins at nanometer resolution. Science 313:1642–1645

Conchello JA, Lichtman JW (2005) Optical sectioning microscopy. Nat Methods 2:920–931

Hammond C, Heath J (2006) Symmetrical ray diagrams of the optical pathways in light microscopes. Microsc Anal 115:5

Hess ST, Girirajan TPK, Mason MD (2006) Ultra-high resolution imaging by fluorescence photoactivation localization microscopy. Biophys J 91:4258

Hess ST, Gould TJ, Gudheti MV, Maas SA, Mills KD, Zimmerberg J (2007) Dynamic clustered distribution of hemagglutinin resolved at 40 nm in living cell membranes discriminates between raft theories. Proc Natl Acad Sci USA 104:17370–17375

Huang B, Wang W, Bates M, Zhuang X (2008) Three-dimensional super-resolution imaging by stochastic optical reconstruction microscopy. Science 319:810–813

Israelachvili JN (1997) Intermolecular and surface forces. Academic Press, London

Kurner J, Frangakis AS, Baumeister W (2005) Cryo-electron tomography reveals the cytoskeletal structure of spiroplasma melliferum. Science 307:436–438

Leonenko Z, Gill S, Baoukina S, Monticelli L, Doehner J, Gunasekara L, Felderer F, Rodenstein M, Eng LM, Amrein M (2007a) An elevated level of cholesterol impairs self-assembly of pulmonary surfactant into a functional film. Biophys J 93:674–683

Leonenko Z, Finot E, Amrein M (2007b) Adhesive interaction measured between AFM probe and lung epithelial type II cells. Ultramicroscopy 107:948–953

Leonenko Z, Finot E, Leonenko Y, Amrein M (2008) An experimental and theoretical study of the electrostatic interactions between pulmonary surfactant and airborne, charged particles. Submitted for publication

Lucic V, Forster F, Baumeister W (2005) Structural studies by electron tomography: From cells to molecules. Annu Rev Biochem 74:833–865

Manley S, Gillette JM, Patterson GH, Shroff H, Hess HF, Betzig E, Lippincott-Schwartz J (2008) High-density mapping of single-molecule trajectories with photoactivated localization microscopy. Nat Methods 5:155–157

Marko M, Hsieh C, Schalek R, Frank J, Mannella C (2007) Focused-ion-beam thinning of frozen-hydrated biological specimens for cryo-electron microscopy. Nat Methods 4:215–217

Marsh BJ, Mastronarde DN, Buttle KF, Howell KE, McIntosh JR (2001) Organellar relationships in the golgi region of the pancreatic beta cell line, HIT-T15, visualized by high resolution electron tomography. Proc Natl Acad Sci USA 98:2399–2406

Matsko N, Mueller M (2004) AFM of biological material embedded in epoxy resin. J Struct Biol 146(3):334–343

McIntosh R, Nicastro D, Mastronarde D (2005) New views of cells in 3D: an introduction to electron tomography. Trends Cell Biol 15:43–51

Medalia O, Weber I, Frangakis AS, Nicastro D, Gerisch G, Baumeister W (2002) Macromolecular architecture in eukaryotic cells visualized by cryoelectron tomography. Science 298:1209

Perfect Storm A (2008) Colour-controllable LEDs, three-dimensional fluorescence nanoscopy, medical nanotags, and more. Science 319:810–813

Richter T, Floetenmeyer M, Ferguson C, Galea J, Goh J, Lindsay MR, Morgan GP, Marsh BJ, Parton RG (2008) High-resolution 3D quantitative analysis of caveolar ultrastructure and caveola-cytoskeleton interactions. Traffic 9(6):893–909

Rust MJ, Bates M, Zhuang X (2006) Sub-diffraction-limit imaging by stochastic optical reconstruction microscopy (STORM). Nat Methods 3:793–795

Schmidt R, Wurm CA, Jakobs S, Engelhardt J, Egner A, Hell SW (2008) Spherical nanosized focal spot unravels the interior of cells. Nat Methods 5:539–544

Seelert H, Poetsch A, Dencher NA, Engel A, Stahlberg H, Muller DJ (2000) Structural biology. Proton-powered turbine of a plant motor. Nature 405:418–419

Shroff H, Galbraith CG, Galbraith JA, Betzig E (2008) Live-cell photoactivated localization microscopy of nanoscale adhesion dynamics. Nat Methods 5:417–423

Vainrub A, Pustovyy O, Vodyanoy V (2006) Resolution of 90 nm (lambda/5) in an optical transmission microscope with an annular condenser. Opt Lett 31:2855–2857

Willig KI, Rizzoli SO, Westphal V, Jahn R, Hell SW (2006) STED microscopy reveals that synaptotagmin remains clustered after synaptic vesicle exocytosis. Nature 440:935–939

Chapter 9
Optical Projection Tomography

James Sharpe

Abstract Optical projection tomography (OPT) is a relatively new technology that is especially well suited to the 3D imaging of "mesoscopic" specimens (those from about 1 to 10 mm across). It is fundamentally different from optical sectioning techniques such as confocal microscopy, since it does not attempt to limit data acquisition to a narrow focused 2D plane. Instead, it is an optical equivalent of computed tomography (CT), in which projection images are captured for many angles around the specimen and the 3D results are calculated using a back-projection algorithm. Volumetric data sets can be generated from both bright-field and fluorescent images. OPT has seen the development of a wide range of applications over the last five years, especially in the field of developmental biology, and increasingly for the analysis of whole mouse organs (such as the pancreas, brain and lungs). Within these contexts, it is particularly useful for mapping gene expression patterns at both RNA and protein levels. In this chapter, both the principles of the technology and the range of applications will be introduced. A few potential directions for the future will be summarized at the end.

Introduction

Optical projection tomography (OPT) is still a relatively young technique (Sharpe et al. 2002), but has seen steady growth over the last five years in terms of both technical improvements and the development of applications. It can be caricatured as an optical version of CT scanning, but the use of visible photons instead of X-rays results in numerous differences between the two technologies—not only in terms of applications, which are quite dramatically different, but also with respect to the technology itself, and both of these issues will be addressed in this chapter.

J. Sharpe
ICREA (Catalan Institute for Advanced Research and Education), EMBL-CRG Systems Biology Unit, Centre for Genomic Regulation, UPF, Barcelona, Spain, e-mail: james.sharpre@crg.es

C.W. Sensen and B. Hallgrímsson (eds.), *Advanced Imaging in Biology and Medicine.*   199
© Springer-Verlag Berlin Heidelberg 2009

OPT was originally developed due to the recognition of an "imaging gap" in the spectrum of existing 3D imaging techniques (Sharpe 2003). At one end of the spectrum are *optical sectioning* approaches, such as confocal laser-scanning microscopy (Kulesa and Fraser 1998) and multiphoton microscopy (Potter et al. 1996), which have generally been geared towards providing very sharp, clean, submicron-resolution images of microscopic specimens, rather than large volumetric 3D data sets. They are excellent for imaging small biological samples such as cells and tissues, but display a couple of limitations when considering whole organs or whole embryos: (1) size—high-quality images are usually only obtained up to a depth of a few hundred microns; (2) dependence on a fluorescent signal—the very sharply focused confocal scan point which allows the generation of high-quality 3D voxel stacks is dependent on the use of fluorescent excitation. Although a "transmission mode" is sometimes available on a confocal microscope, this is not able to create genuine 3D voxel stacks: the lack of localized fluorescent excitation in this mode reduces the system back to a standard bright-field microscope. An alternative optical sectioning approach, SPIM (single-plane illumination microscopy), has recently been developed (Huisken et al. 2004) and is able to image much larger specimens than confocal, but it is also limited to fluorescent signals.

Because visible photons do not traverse large specimens easily, is inevitable that the other end of the size spectrum contains non-optical techniques such as MRI and CT. These have traditionally been aimed at imaging whole organisms, but have been increasingly adapted towards smaller specimens such as embryos (Dhenain et al. 2001; Schneider et al. 2004; Johnson et al. 2006). Their main limitations are almost the opposite of those of confocal microscopy: MRI cannot achieve a spatial resolution anywhere close to that of confocal, and both the techniques lack a range of flexible, efficient, targetable contrast agents—in particular, for the goal of monitoring gene expression levels within the context of the whole organism. CT images depend on spatial differences in the absorption of X-rays, so compounds with heavy metals (such as osmium tetroxide) are sometimes used to increase the contrast of anatomical structures (Johnson et al. 2006). However, a targetable version of heavy metal labeling suitable for recording the distributions of gene expression patterns has not been developed. In principle, MRI has more available options; for example, it has been shown that dynamic changes in the expression of a transgenic reporter construct can be observed over space and time within a *Xenopus* embryo (Louie et al. 2000). The reporter gene encoded for the β-galactosidase enzyme and the specimen was injected with a substrate with a caged high-contrast agent. However, this technology has not become widespread, presumably due to both the cost and technical difficulties involved. Instead, both MRI and CT are increasingly used in combination with molecular imaging technologies such as PET and SPECT, because although they have lower resolution than optical techniques like confocal, they nevertheless display significantly higher resolution than PET/SPECT (which instead have the ability to image labeled molecular distributions; Massoud and Gambhir 2003).

Apart from the common 3D imaging tools mentioned above, it should be noted that not all optical techniques are limited to fluorescent signals. Optical coherence

tomography (OCT) uses a very different principle to generate 3D images of nonfluorescent samples: interferometry and a coherent light source (Huang et al. 1991).
Nevertheless, both OCT and confocal are each restricted to one mode (either fluorescent or nonfluorescent imaging). By contrast, OPT has the significant advantage
that it can produce 3D images in both transmission and fluorescent modes.

OPT Technology

This section will provide an overview of the standard OPT setup, the consequences
of using light for computed tomography (rather than for example X-rays or electron
beams), and the basics of the reconstruction and visualization approaches.

Data Capture

The apparatus used for OPT data capture (Fig. 9.1) shares many similarities with a
micro-CT scanner: a stepper motor is used to rotate the specimen to precisely controlled angles, a 2D array detector (CCD) is used to capture the signal, and photon
sources are included for illumination. In transmission mode, photons are projected
through the specimen from a light source on one side and captured by the array
detector on the other—very much like a microCT scanner—and the data captured
essentially represents linear projections through the sample.

 However, despite these similarities with microCT, OPT displays a few very
important differences (Sharpe 2004). In the CT approach, the use of X-rays allows a
sharp quantitative shadow to be formed simply by placing the specimen in-between
a point source and the CCD. This is because, although X-rays can be absorbed by
dense materials such as bone, those rays which do manage to pass through the sample continue to travel in straight lines, i.e., although absorption occurs, scattering or
diffraction is negligible. By contrast, it is the use of photons in the visible part of the
spectrum for OPT which leads to the biggest difference between the two technologies: the majority of photons passing through the sample undergo at least a small
degree of scattering, refraction or diffraction, such that they follow slightly diverged
paths and the resulting shadow is a blur. OPT therefore uses an optical system of
lenses (essentially a macroscope) to create a focused image of the specimen, rather
than capturing a shadow.

 This explains one of the important advantages that OPT displays over many other
forms of optical imaging: it is capable of being used in two different modalities:
transmission OPT (tOPT), and emission OPT (eOPT)—a modality which is not
analogous to CT, but is in fact closer to SPECT or PET (Massoud and Gambhir
2003). In the visible part of the spectrum, a wide range of fluorescent dyes are
available; upon excitation, these will emit photons with characteristic wavelengths.
Since the photons are emitted in all directions from within the specimen, without
a focusing technology (i.e., lenses for visible photons) it would not be possible to

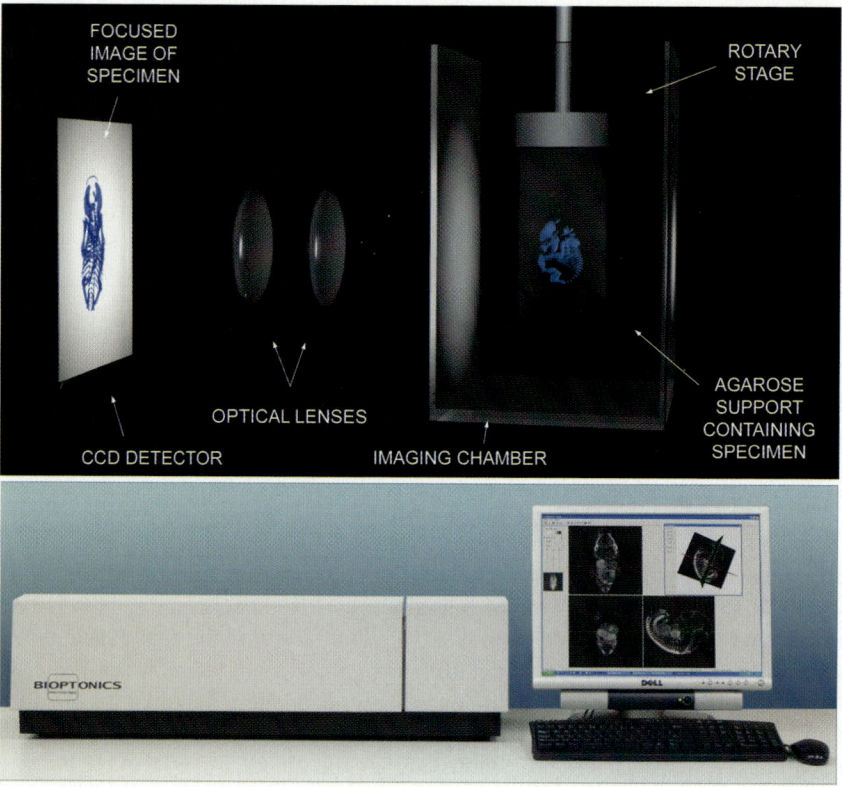

Fig. 9.1 Main features of an OPT scanner. The specimen (*blue*) is usually supported in a cylinder of agarose, although this is not essential and certain specimens can be glued directly to the mount of the rotary stage. The specimen is rotated about a vertical axis and suspended in the imaging chamber, which maintains it in suitable imaging fluid (for example BABB for most ex vivo imaging—not shown in figure). In the example shown, transmission illumination is coming from the right hand side of the figure, and data capture consists of a focused image of the specimen on the CCD (left hand side of the diagram). All information from one horizontal section through the specimen is captured by one horizontal row of pixels on the CCD. 3D graphic created and kindly provided by Seth Ruffins (Caltech: sruffins@caltech). The *photo below* shows the current commercial OPT scanner (http://www.bioptonics.com)

capture projection data. However, the use of diffraction-based optics in OPT allows a sharp image to be focused onto the CCD, allowing projection data to be captured. These projections are not exactly equivalent to the shadow projections of CT—the beam attenuation, rather than being solely attributable to absorption, is the combined result of absorption, scattering and refraction. Nevertheless, experience with OPT images has proven that this hybrid form of attenuation results in biologically useful images.

Another consequence of using lenses and capturing a focused image instead of a shadow is that a point source (as typically used in CT) is not a suitable illumination for OPT. A photon point source in a tOPT system would appear as an out-of-focus but localized disc of light behind the specimen that interferes with the image and results in nonuniform illumination. Instead, the whole field of view (FOV) must have uniform back-lighting, and this is best performed using a white-light diffuser. For fluorescence imaging, an arrangement including mercury source, filter wheels and a focusing lens is also included—fluorescence illumination is usually provided on the same side as the objective lens (as with a standard fluorescent microscope), so that maximum excitation occurs within the focused region, and that beam attenuation is kept to a minimum.

For the most common applications of OPT, i.e., analysis of ex vivo samples, the specimen (and agarose support) is suspended in an index-matching liquid to (1) reduce photon scattering and (2) reduce heterogeneities of refractive index throughout the specimen. This allows photons to pass through the specimen in approximately straight lines, and a standard filtered back-projection algorithm can be employed to calculate relatively high-resolution images. The clearing agent most often used is BABB or Murray's Clear (a 1:2 mixture of benzyl alcohol and benzyl benzoate; see Sharpe et al. 2002). Once in the OPT scanner, the sample is suspended within the liquid and rotated through a series of precise angular positions (usually less than 1° apart), and a focused image is captured on the CCD at each orientation. The rotary stage must allow careful alignment of the axis of rotation to ensure it is perpendicular to the optical axis. This allows the simplest reconstruction scheme in which the projection data coming from each virtual plane through the sample is collected by a linear row of pixels on the CCD of the camera.

The fact that image-forming lenses are required for OPT (rather than capturing simple shadow data) leads to another significant difference with microCT: there is only a limited depth of focus, and this usually cannot encompass the entire specimen. Reducing the numerical aperture of the system to increase the depth of focus has limited success; the raw images soon lose their sharpness due to the reduced resolving power of the system. The most convenient scheme for OPT imaging therefore involves a depth of focus that is deep enough to encompass half the specimen (from the edge of the specimen closest to the objective lens to the axis of rotation). This means that every image contains both focused data from the "front half" of the specimen (the half closest to the lens) and out-of-focus data from the "back half" of the specimen. Most OPT results to date have been achieved using this raw data in a standard filtered back-projection algorithm. However, an improved algorithm has recently been developed which takes this issue into account (see the section "CT Reconstruction"). The current OPT scanner design usually takes a series of 400 projection images for a full 360° rotation. Although the images taken 180° apart are similar to inverted copies of each other, issues such as the noncentralized focal plane mean that they are not identical. A good-quality CCD camera is used, and the current standard for the images is 12-bit with $1,024^2$ pixels.

The choice of wavelength used for OPT can be important in obtaining a good image. As specimens become more opaque, or more heavily stained (more absorbing), shorter visible wavelengths penetrate less efficiently. However, shifting up towards the infrared end of the spectrum has proven to significantly improve the image quality (Sharpe 2005). This is particularly useful for specimens stained with a colored precipitate (which are imaged using transmission mode) and for larger adult tissues, which tend to be more opaque due to the more differentiated state of the cells.

CT Reconstruction

As described above, the collection of projection data for OPT exhibits significant differences compared to X-ray CT. However, the algorithm for reconstructing this data into a 3D voxel stack—filtered back-projection—is very similar to the CT case (Kak and Slaney 1988). Minor modifications to the approach include correcting for the gradual fading of a fluorescent signal during the scanning procedure, and correcting for the small random intensity fluctuations caused by the mercury light source (Walls et al. 2005). The only significant modification explored so far relates to the limited depth of field mentioned above. In conventional CT there is no focal plane, so the only difference between two images captured 180° apart is due to the fan-beam or cone-beam arrangement of the projections (Sharpe 2004). In OPT, the fact that there is a focal plane, and that this is not centered on the axis of rotation (see the previous section) means that a second difference exists between the 180° rotated images: in each case a different region of the specimen will be in focus. In principle therefore, considering the opposing views to be identical mirror-image versions of each other means that out-of-focus information can slightly degrade the final image quality. Until recently this potential problem has not been taken into account in the reconstruction process, and in fact this simplification still results in high-quality images (most images referred to in this chapter have been reconstructed in this way). However, recently Walls et al. (2007) explored an improved approach that employs a frequency–distance relationship (FDR) to perform the frequency filtering of raw projection data before it is back-projected into the reconstruction. This minimizes the influence of out-of-focus data and can improve the spatial resolution of the resulting images. Examples of the high-detail images resulting from this algorithm can be seen in Fig. 9.2, which shows a developmental sequence of vasculature in mouse embryo heads (Walls et al. 2008).

Reconstruction of a 512^3 data set currently takes about 5 min on a single PC (for a single channel), which is approximately equal to the time it takes to perform the scan. Due to this efficiency, although higher-resolution OPT scans ($1,024^3$ voxels) are typically four times slower to scan, and take about 45 min to reconstruct, they are now routinely performed due to the improved detail visible in the final image.

Visualization

The 3D voxel data sets generated by OPT are similar to those produced by other techniques such as MRI, and therefore in principle the same kinds of software can often be used for visualization. Virtual sections through the data can be sliced at any

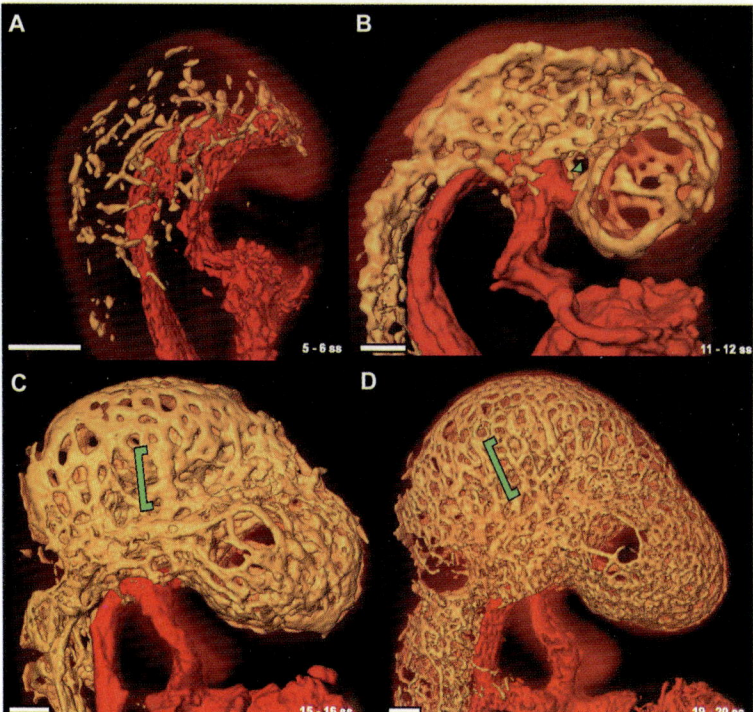

Fig. 9.2 (a–d) Formation of the cephalic vasculature during mouse development is revealed by fluorescent immunohistochemistry and the FDR-deconvolution algorithm to increase the spatial resolution of this complex structure. Embryo stages are: **(a)** 5 somites; **(b)** 11 somites; **(c)** 15 somites; **(d)** 19 somites. The dorsal aorta and heart are surface-rendered *red*; other vasculature revealed by PECAM expression throughout the cephalic mesenchyme is surface-rendered *orange*; and the autofluorescence of the mouse embryo is volume-rendered with a HoTMetaL color map. The development of the cephalic veins can be appreciated (*green brackets*). All scale bars represent 100 μm. Reproduced with permission from Walls et al. (2008)

orientation through the specimen (e.g., Fig. 9.3), or the results can be rendered as a 3D object—either as isosurfaces (e.g., Fig. 9.2), volume renderings, or a combination of both (e.g., Fig. 9.4). However it is important to note that OPT data tends to be different from both MRI/CT data on the one hand, and also from confocal data on the other. In the first case, although both MRI and CT are 3D, they data tend to contain only one "channel;" i.e., each voxel has a single 16-bit value associated with it, whereas OPT data tends to have multiple fluorescent channels, and therefore multiple 16-bit data points for each voxel. In the second case, although confocal data is also multichannel, it tends not to be so three-dimensional—it usually consists of high pixel dimensions along the X and Y axes (e.g., one or two thousand), but a much smaller dimension along the Z-axis (e.g., 100). So far, most visualization software has been aimed at one or the other of these two classes (either very 3D, or multiple channels), and therefore generating good-quality renderings of OPT results has

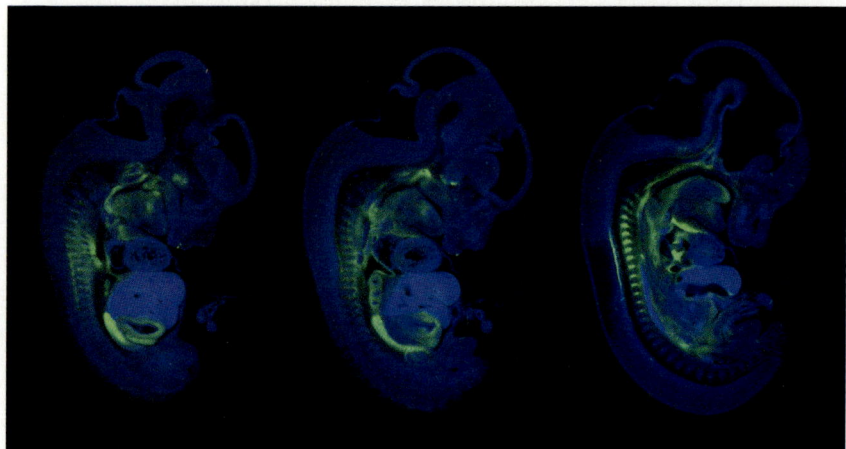

Fig. 9.3 Virtual histology and nonfluorescent gene expression analysis. Three virtual sagittal sections are shown from a transgenic E12.5 mouse embryo carrying a "knocked-in" LacZ reporter showing the expression pattern of the *Pkd1* gene. Because this data is from a 3D voxel stack, virtual sections like these can be displayed at any angle through the same specimen. The embryo was whole-mount-stained with the X-Gal substrate, which produces a blue precipitate wherever the reporter gene is active. This light-absorbing stain was captured and reconstructed through transmission OPT (or bright-field OPT), and is false-colored here in green. The anatomy of the specimen was captured using fluorescence OPT, by choosing a wavelength that picks up autofluorescence from the fixed tissue. Virtual sagittal sections can be viewed through the entire specimen, and the quality of signal localization deep within the embryo can be appreciated despite the fact that this signal is not fluorescent (see Alanentalo et al. 2007 for more information about this specimen)

not always been straightforward. One option for high-quality rendering has been to employ "offline" precalculation of images (we currently use the package VTK). In this case, an interactive phase allows the user to define rendering parameters (such as colors, transparency values and angles), and the software then renders the images or movies overnight. However, the most recent software developments mean that a first wave of genuinely interactive algorithms are appearing which are tailored to take advantage of the latest generation of affordable computer graphics cards. Software such as Bioptonics Viewer is able to interactively display 3 different color channels, each at $1,024^3$ resolution, and with channel-specific interactive clipping (http://www.bioptonics.com).

Ex Vivo Applications

By definition, OPT is limited to samples which allow photons to traverse them along substantially straight paths. (This is inherent in the use of the word *projection* within the name of the technique.) If too much photon scattering occurs within the sample, then predictable, deterministic paths are no longer possible and photon transport becomes probabilistic. Tomographic imaging for these scattering

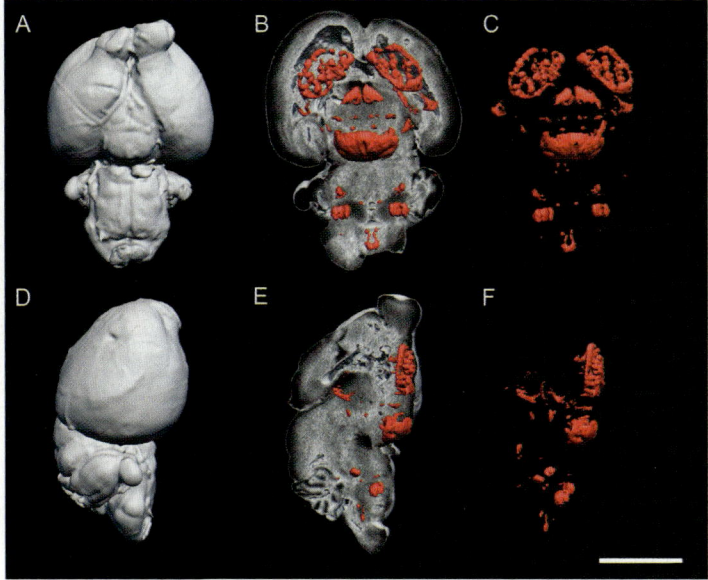

Fig. 9.4 (a–f) Whole-brain imaging. 3D visualizations of a wild-type mouse brain showing clean, specific molecular labeling of discrete regions of the organ (ventral view **a–c** and lateral view **d–f**). The first column (**a, d**) shows the outline of the organ as imaged from the autofluorescent channel. The middle column (**b, e**) illustrates virtual clipping of the organ to display the second fluorescent channel: Isl1-labeled regions (*red*). The last column (**c, f**) shows isosurface reconstructions of the Isl1-labeled structures only. Scale bar is ∼3 mm. First published in Alanentalo et al. (2007)

specimens is known as diffuse optical tomography (DOT; see Gibson et al. 2005) or fluorescence mediated tomography (FMT; see Ntziachristos et al. 2005) and necessarily results in reconstructions with lower resolution.

There are essentially two factors that determine how much scattering will occur. First are the intrinsic optical properties of the tissue (certain tissues are more optically dense than others) and second is the sample size. As criteria for successful imaging, these two features are not independent. Given a collection of samples that are the same size but different tissue types, the imaging resolution will decrease as the optical density increases. Depending on the scientific question, a density threshold will therefore exist above which useful imaging is not possible. Importantly though, because OPT uses lenses to collect a focused image (rather than the quantitative shadows of CT) a small degree of scattering within the sample actually contributes to useful biological contrast (because slightly diverging paths from a single point within the sample will be collected by the same acceptance cone and therefore directed onto the same pixel of the CCD). The importance of scattering has another practical consequence for imaging—if the sample can be imaged ex vivo, then efficient clearing agents such as BABB can be employed to dramatically improve imaging results. Regarding sample size, a given tissue type will display a constant probability of photon scattering per unit pathlength, and the overall degree of scattering will therefore increase exponentially with tissue depth. Again, depending on the goals of imaging, an upper size limit will therefore exist for each specimen type.

These two factors explain why developmental biology has benefited particularly strongly from OPT: embryonic tissues generally contain very undifferentiated cells that tend to be optically more transparent than adult differentiated tissue (presumably due to the abundance of more specialized structural proteins, extracellular matrix, collagens, etc.). An embryo of a given diameter will allow higher-resolution imaging than an adult tissue sample of the same size. Additionally, an entire midgestation vertebrate embryo, despite being generally much more complex than a single organ, is nevertheless smaller than 1 cm across, which, when combined with efficient clearing agents, results in highly detailed 3D images. Ex vivo OPT is therefore routinely performed on mouse embryos at E9.5 up to E12.5, and subregions are scanned for older specimens, for example the developing limb or brain at E14.5 (Baldock et al. 2003; DeLaurier et al. 2006; Wilkie et al. 2004). The Edinburgh Mouse Atlas Project (EMAP) has also performed purely anatomical scanning (of autofluorescence) for a few whole embryos up to an age of ~E15. Younger and smaller embryos can also be imaged by OPT, but due to the existence of previous techniques (such as confocal) which were already able to capture these smaller specimens, it has not been considered as important an area for OPT itself. A variety of other standard laboratory models have also been successfully analyzed by OPT, including chick (Fisher et al. 2008; Tickle 2004), zebrafish (Bryson-Richardson et al. 2007), *Drosophila* (McGurk et al. 2007) and plants (Lee et al. 2006). In the case of *Drosophila*, successful imaging to pinpoint neurological lesions in the brain of the adult fly was possible without any special bleaching of the chitinous exoskeleton.

Despite the potential difficulties of imaging adult differentiated tissues, nevertheless the last couple of years have seen imaging success with a number of whole organs taken from the adult mouse. Results have now been published for brain (Alanentalo et al. 2007; Hajihosseini et al. 2008a), pancreas (Alanentalo et al. 2007), kidney (Davies and Armstrong 2006), and lungs (Hajihosseini et al. 2008b). An important technical point here is that autofluorescence can be a potential problem for imaging more differentiated adult tissues—not only are such tissues less transparent, but they also tend to display much stronger nonspecific background fluorescence. For this reason, the development of protocols for bleaching autofluorescence, or for minimizing its appearance in the first place, have become important for improving the imaging of these types of samples (Alanentalo et al. 2007; Sakhalkar et al. 2007). One logical direction for whole-organ imaging is preclinical disease research, and accordingly the whole pancreas imaging was performed to quantitatively compare the mass of β-cell tissue in normal vs. diabetic specimens, using the NOD mouse model (see the section "Preclinical Disease Research").

Before starting the list of specific applications, it is important to mention that many of them take advantage of dual modalities of OPT, i.e., there are a number of cases in which both transmission and fluorescent channels are taken of the same specimen (tOPT and eOPT). This can be either because the molecular signal may not be fluorescently labeled (for example X-Gal staining of the LacZ transgenic reporter gene), or because the optical density of the specimen means that a transmission scan achieves a better representation of the global anatomy of the specimen than autofluorescence. It should also be mentioned that although in general the range of labels that can be imaged by OPT is broad (essentially any light-absorbing or light-emitting dye), practical issues such as the stability of the dyes upon dehydration are of course very important. Notable examples which are not ideal for OPT imaging are: (1) lipophilic dyes such as DiI which integrate into the lipid membranes of cells and therefore wash out of the tissue due to disruption of the membranes when the specimen is dehydrated, and (2) fluorescent proteins such as GFP, whose fluorescence tends to decrease over a few hours in alcohol. However, the latter does not tend to be a serious problem, as the distribution of GFP protein can easily be imaged by whole-mount antibody labeling (DeLaurier et al. 2006).

Virtual 2D Histology

The original meaning of the word "tomography" was to illustrate a *slice* or *cut* through an object, and although one of the most striking features of OPT imaging is its ability to produce complete 3D models of complex mesoscopic samples, in some cases the ability to rapidly create 2D "virtual histology" is a useful feature in itself. An embryo can be imaged and reconstructed in minutes, providing a complete set of virtual sections all the way through the specimen, whereas cutting the equivalent series of real paraffin sections, mounting them on glass slides and digitally photographing each one takes days. For a given specimen, OPT cannot achieve the same spatial resolution as cutting microtome sections (real sections can be imaged at submicron resolution, while OPT of a mouse embryo will provide resolution of approximately 5–10 μm; Sharpe et al. 2002). However, as illustrated in Fig. 9.3, the achievable resolution is generally considered ideal for histology-level or anatomy-level studies (Yoder et al. 2006) as opposed to cellular-level studies. Also, in addition to being much faster than conventional histology, another advantage of OPT is that virtual sections can subsequently be viewed at any angle through the specimen without having to go back to the real specimen or rescan it. Although the standard reconstruction algorithm generates the full 3D data set by calculating a series of independent 2D sections (perpendicular to the axis of rotation), the perfect alignment of the sections results in a genuinely 3D voxel set which can be digitally resliced in any orientation. This is a significant benefit compared to real sections; here, once an angle for cutting has been chosen, it cannot be changed later. This advantage can be very important to histologists and anatomists, because correctly identifying tissues can be very dependent on the angle of the section through the specimen (e.g., whether the section is a genuine sagittal or transverse plane, rather than a slightly oblique angle).

Gene Expression Analysis

Possibly the most important role of OPT imaging is the discovery, analysis and mapping of gene expression patterns. Such information is a vital clue in our attempts to assign functions to the thousands of genes in the genome. Molecular labeling techniques have been developed and refined over the last 20 years that allow whole organs or embryos to be stained to reveal the activity of endogenous genes, either at the RNA level (by whole-mount in situ hybridization—WMISH) or at the protein level (using specific antibodies). The former has the advantage that an RNA probe can be easily generated for any gene in the genome, whereas raising specific antibodies is expensive, time-consuming and may not be possible for many proteins of interest. On the other hand, immunohistochemistry has the advantage that it can easily be performed fluorescently, which allows double- or triple-labeling experiments, whereas RNA in situ is still mostly limited to producing a light-absorbing precipitate at the site of expression, which limits accurate 3D imaging to one gene per specimen. In addition to exploring endogenous gene expression, a related goal is to analyze the expression patterns of reporter constructs from transgenic specimens, for example the popular LacZ gene (Fig. 9.3; see also Yoder et al. 2006).

As mentioned previously, one major advantage of OPT is that it can create 3D voxel data from nonfluorescent signals. The biggest practical advantage of this is seen for the labeling techniques just mentioned: (1) gene expression analysis at the RNA level, because the most common protocols for whole-mount in situ hybridization (WMISH) result in the production of a purple precipitate (using the substrates BCIP and NBT) rather than a fluorescent dye, and (2) the blue precipitate generated by the β-Gal reporter protein (from the LacZ gene). Although fluorescent versions are being developed, these two nonfluorescent staining protocols remain in widespread use due partly to their high reliability. The largest OPT-based gene expression study to date performed whole-mount in situ hybridization on developing mouse limb buds for 29 different genes, namely all the 20 members of the Wnt signaling pathway, and all the nine members of the frizzled family of Wnt receptors (Summerhurst et al. 2008). The speed of OPT made it feasible to generate comprehensive maps for all these genes in a reasonable time.

The advantages of OPT as a tool for gene expression analysis are very similar to its use in generating virtual 2D histology (described above): the combination of speed and comprehensiveness. Indeed, OPT data on gene expression is most often examined as a series of 2D sections (for example Fig. 9.3), rather than exploring the 3D geometry of these domains. This is largely because we do not yet have the experience of understanding this information in 3D—we are far more familiar with identifying anatomical structures from their 2D histological appearance in a normal section. Therefore, the technical advantages mentioned in the previous section for histology—such as being able to adjust the section orientation to arbitrary angles— apply equally to this kind of molecular analysis. Also, the ability to scroll through every single section makes the approach very powerful for documenting complete expression patterns. However, it is hoped that the ability to explore this data in 3D will gradually become more feasible, and Fig. 9.4 shows one example of the 3D appearance of this data.

In addition to exploring and documenting the tissue distributions of new genes, 3D imaging can use specific molecular labels for another purpose: to highlight tissues, organs or other anatomical features within the context of a larger, unlabeled structure. This is described in the next section.

3D Morphology

Imaging the 3D shapes of complete undistorted samples is one of the keys roles of OPT technology. As will be elaborated in subsequent sections, seeing the true 3D shapes of anatomical structures—especially the ability to interactively explore these structures on a computer screen—is an invaluable tool for many types of research. In developmental biology they are used in two ways: first to accurately describe the complex sequence of 3D morphogenic movements during normal development, which has led to the creation of digital atlases of development (see "3D Localization of Labeled Cells"), and second to document the phenotypic changes seen when certain genes are mutated (see "Atlases of Development"). 3D phenotyping can also be an essential part of preclinical disease research (see "Phenotyping"). Another example which has not yet been fully developed is the use of OPT for neurobiology. The intrinsically complex spatial organization of the brain means that 3D imaging is likely to be essential for a complete understanding, and some preliminary studies show that OPT could be a suitable tool for this field (Fig. 9.4). In particular, OPT imaging has already managed to obtain full volumetric data sets for one-month-old mouse brains, and protocols to allow penetration of antibodies into the core of the organ have also been developed Alanentalo et al. 2007. The red staining in Fig. 9.4 shows the spatial distribution of the expression of Islet1, detected by fluorescent whole-mount immunohistochemistry. Although this pattern does not highlight nerve tracts, it illustrates the degree of 3D detail obtainable for a whole brain, suggesting that the resolution will be sufficient for more complex neurobiological studies.

The fact that OPT is an optical modality and is therefore able to utilize the existing wealth of fluorescent and/or absorbing molecular labels is not only useful for gene expression analysis. It is equally common to use these molecular labels (especially at the protein level) as a "molecular marker" to highlight a particular subset of the 3D anatomy or morphology of a specimen. For example, in Fig. 9.2 an antibody against the protein PECAM has been used, but not for the purpose of exploring an unknown gene expression pattern. On the contrary, it is precisely because we understand exactly where this gene is expressed (i.e., in all blood vessels) that the antibody can be used for a purely anatomical reason—to highlight the complex 3D sequence of vascular development (Walls et al. 2008; Coultas et al. 2005). Other examples include the diabetic pancreas imaging seen in Fig. 9.5, in which the anti-insulin antibody was used to highlight the Islets of Langerhans.

212

J. Sharpe

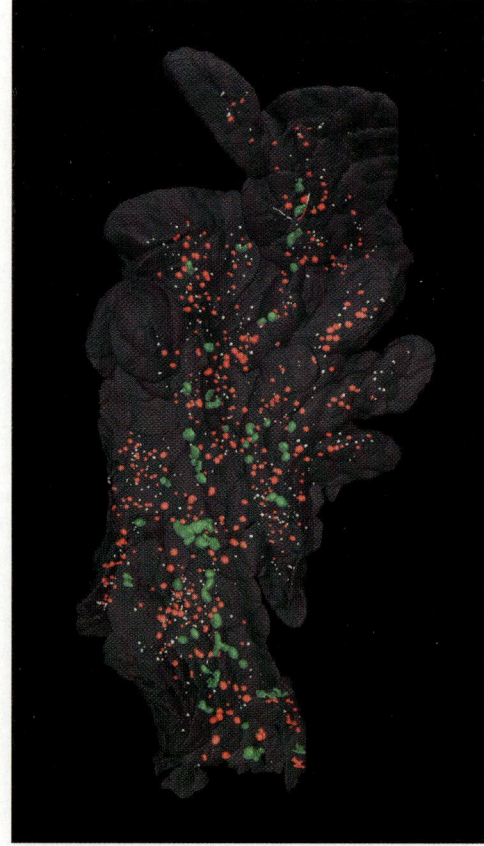

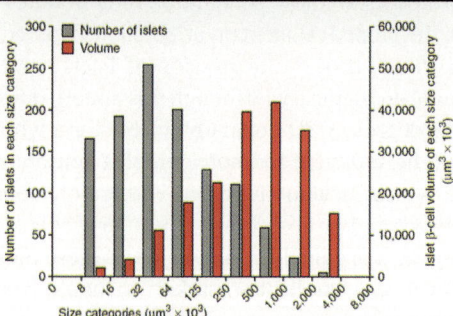

Fig. 9.5 OPT imaging of an entire adult mouse pancreas. By developing whole-mount immuno-histochemistry for the pancreas using antibodies against the insulin protein, it is now possible to label all the β-cells within the Islets of Langerhans, and to quantify this information from a single OPT scan. *Upper panel* shows the 3D image captured by OPT with Islets color-coded according to three size categories. The *lower panel* shows a more detailed quantification (from a different pancreas), in which the Islets are quantified for each size category (from small Islets of 8–16,000 μm^3 up to large Islets of 2–4,000,000 μm^3). *Gray bars* represent the number of Islets in each category, and *red bars* represent the total β-cell volume composed of Islets from each size category. Adapted from Alanentalo et al. (2007)

Manual inspection of 3D models using interactive computer graphics has so far been the main tool for most analysis of morphology (for discovering new mutant phenotypes, e.g., Alanentalo et al. 2007; Hecksher-Sørensen et al. 2004; Asayesh et al. 2006; Lickert et al. 2004, or even new aspects of wild-type morphology, e.g., Walls et al. 2008). However this data is also the perfect substrate for morphometrics—quantifying aspects of morphology, such as the natural variation of an organ shape, or measuring the degree of abnormality in mutant or diseased organisms. The only example developed so far of this more automated analysis was the quantification of the distributions of Islets of Langerhans mentioned above (see "Phenotyping").

3D Localization of Labeled Cells

OPT does not usually achieve single-cell spatial resolution (apart from the rare cases where individual cells are especially large, such as the zebrafish notochord and certain plant cells; Bryson-Richardson et al. 2007, Lee et al. 2006). However, if one or a few labeled cells are surrounded by unlabeled cells they can be detected, and their position within the context of the sample can be localized. This is a rather specific application compared to the others in this section, but it could become important in preclinical cell-therapy experiments where modified cells are injected into a diseased organ with the goal of repairing it. In these cases, the cells usually have a unique genetic marker that could be labeled quite easily, and OPT could then allow an accurate description of the final 3D distribution of cells throughout the organ. This particular idea has not yet been achieved, but two published examples of clonal analysis illustrate this concept very well.

In Wilkie et al. (2004) a clonal analysis was performed within the developing mouse brain. A transgenic construct was employed which contained a ubiquitously expressed but nonfunctional version of the popular reporter gene LacZ. The loss-of-function mutation in the gene (known as LaacZ) reverts to normal activity at a known low frequency during embryo development. This produces rare clonal labeling events in which all subsequent LacZ-expressing cells are descended from the single cell in which the spontaneous reversion event occurred. Although neither the timing nor the location of the clonal labeling can be controlled, after screening enough embryos, specimens can be found which display clones in the organ of interest. In this example, OPT scans of the resulting embryos displayed clones in the developing neocortex. Analysis of the 3D data revealed clear evidence for widespread tangential movements of the neural progenitors. This highlighted two of OPT's strengths: the ability to capture the whole organ, and the fact that the labeling was not fluorescent. Staining the specimen with X-Gal produces a blue precipitate wherever the reporter gene is expressed. Comparison of results from 2D serial sections proved that it was very difficult to arrive at the same conclusions without a genuinely 3D visualization.

The second example concerns mouse limb development. In this case, a more controllable version of clonal LacZ labeling was employed in which the timing and frequency of clonal events could be adjusted by exploring injections of different concentrations of tamoxifen (Arques et al. 2007). The clonal analysis proved for the first time the existence of a dorsoventral lineage-restricted compartmental boundary within the mesenchyme of the limb bud. OPT again allowed 3D data sets to be created from nonfluorescent samples, which helped prove that the new boundaries did indeed form a 2D curved surface running between the dorsal and ventral surfaces of the limb bud.

Atlases of Development

One of the foundations of developmental biology is descriptive embryology—it is impossible to study a complex 4D process such as morphogenesis, or even to know which questions to ask, without a good understanding of where interacting tissues are in relation to each other. Traditional embryological atlases, still essential to the modern biologist, are large books filled with hundreds of detailed photographs of microtome sections, and labeled with thousands of annotations which identify the array of tissues and organs therein. Essential as these paper-based atlases are, they are not without significant practical problems. Probably the two most important of these are a lack of comprehensiveness and the fixed orientation of each section. Both of these problems lead to the same consequence for the researcher: when trying to identify an unknown tissue from your own specimen, it will often be very difficult to match up what you see in your own 2D section with what is displayed in the atlas. It will either be in a section which doesn't exist in the atlas (because only a subset of the sections can be included), or the section will be at an angle that makes it unrecognizable.

To deal with these problems, the Edinburgh Mouse Atlas Project (EMAP) has developed the most detailed and comprehensive digital atlas for mouse development (Baldock et al. 2003). Until the advent of OPT, the digital data were gathered only from physically cut paraffin sections of mouse embryos. Digital photographs were taken of every single section, and software was developed to allow the virtual sections to be digitally stacked on top of each other to reconstruct the full 3D anatomy of each embryo. This is both very slow and also inaccurate—it is impossible to recreate the correct 3D geometry from a series of 2D sections alone (Hecksher-Sorensen and Sharpe 2001). By including OPT into the model-building process, EMAP was able to solve both of these problems. Real paraffin sections are still cut, because of the higher spatial resolution compared to OPT, but before cutting the embryo a 3D OPT scan is made which can subsequently be used as an accurate 3D geometric template. The 3D stack containing hundreds of 2D sections can then be "morphed" back into the 3D template, to recreate the correct shape of the whole sample. The atlas currently contains 3D models of 17 mouse embryos starting with Theiler Stage 7 ($\sim$E5) up to Theiler Stage 23 ($\sim$E15).

A human atlas of embryo development has also adopted OPT—the Electronic Atlas of the Developing Human Brain (EADHB), based in Newcastle (UK), which contains 3D models from Carnegie Stage 12 up to Stage 23 (Kerwin et al. 2004). In this case, the project took advantage of both the accurate geometry (as described above) and also the nondestructive nature of OPT. The specimens for the human atlas comprise a very precious and carefully controlled resource, and the project wished to create 3D representations of the embryos while still keeping the specimens for further analysis. Indeed, they proved that after performing an OPT scan, these embryos could subsequently be sectioned as normal and used for gene expression analysis at the levels of both RNA (using in situ hybridization) and protein (using immunohistochemistry). By using alternate sections for different genes, they were able to maximize the information provided by each embryo (Sarma et al. 2005).

The third vertebrate atlas based on OPT imaging is FishNet (Bryson-Richardson et al. 2007), which documents the development of the zebrafish—another standard model for developmental biology. FishNet has been created by the Victor Chang Cardiac Research Institute in Sydney, Australia, and contains 18 grayscale data sets covering the period from a 24-h larva up to the adult fish (17 mm in length). As with the previous atlases, it can be accessed by an interactive website which allows the user to view the virtual sections in any of the three orthogonal orientations. A subset of sections in each orientation have been manually annotated to aid with the identification of anatomical features.

Websites for OPT-based atlases of development include http://genex.hgu.mrc. ac.uk, http://www.ncl.ac.uk/ihg/EADHB/ and http://www.fishnet.org.au/FishNet/ sc.cfm.

Phenotyping

There is increasing realization that an accurate, efficient and economical platform for phenotyping mutant mouse embryos would provide a significant benefit to the mouse genetics community. This application has clear links with the previous section on atlases: to know when a organ looks abnormal you first need a good record of what the normal morphology looks like. Consequently, phenotyping is a field that is also looking for suitable 3D imaging approaches. Technologies that have recently been explored include microCT, microMRI and OCT (micro-computed tomography, micromagnetic resonance imaging and optical coherence tomography). MicroCT has been proposed as an economical high-throughput 3D screen for large-scale screens (Johnson et al. 2006). To obtain sufficient contrast (i.e., sufficient absorption of X-rays), the samples must be prepared with heavy-metal compounds (osmium tetroxide). It has been demonstrated specifically for imaging brain defects (in a transgenic mouse expressing the Pax3:Fkhr oncogene; Johnson et al. 2006), but could in principle be used for many developmental abnormalities. MicroMRI has been proposed as a similar platform, and one specifically for use in heart phenotyping (Schneider et al. 2004). It tends to generate images with lower spatial resolution but has been successfully used to identify ventricular septal defects, double-outlet right ventricle, and hypoplasia of the pulmonary artery and thymus in E15.5 embryos which were homozygous null for the *Ptdsr* gene. Finally, OCT has more recently also been proposed for heart phenotyping, but this has only been performed on hearts that are physically dissected out of the embryo (Jenkins et al. 2007).

These three techniques all display one particular limitation in common for the purpose of phenotyping: their inability to image molecular labels such as antibody staining, in situ hybridization or reporter gene expression. Even the last of these, OCT—an optical technique, is unable to take advantage of fluorescent labels. The most useful information for a morphometric phenotyping system to extract from a specimen is accurate 3D data on the morphology of your organ of interest. The lack of molecular imaging for microCT, microMRI and OCT means that the shape of the organ can only be extracted by either (1) physically dissecting the organ away from the rest of the embryo (as was done for OCT; Jenkins et al. 2007), which is time-consuming and can introduce artefactual distortions into the shape, or (2) scanning the whole embryo, but then manually defining which pixels/voxels correspond to the heart (as done for microMRI; Schneider et al. 2004), which is also time-consuming and can introduce manual errors. By contrast, it is more efficient, accurate and objective to use a molecular label (for example organ-specific antibody) to automatically and consistently define the voxels that belong to the studied organ. Any technique that can perform fluorescence imaging will therefore allow a "molecular dissection" of the structure out of the 3D embryo.

For this reason OPT, is an excellent tool for phenotyping mouse embryos (Ruijter et al. 2004; Dickinson 2006). Although it has not yet been employed in a high/medium-throughput screen, it has nevertheless been used as a phenotyping tool in a number of studies, and these point to its possible use for screens in the future. In 2002, OPT was able to discover a new phenotype in the development of the stomach of *Bapx1* mutants (Sharpe et al. 2002). As suggested above, it was the ability to use a fluorescently labeled molecular marker for the gut epithelium that made the abnormal anatomy of the mutant stomach obvious once it was visible as a 3D model. OPT also helped characterize the heart-defect phenotype of the Baf60c knock-out (Lickert et al. 2004), in which a molecular label was also used as an anatomical marker to avoid the problems mentioned above. In this case, the nonfluorescent imaging capabilities of OPT were highlighted again, because the molecular marker used was visualized by WMISH, producing the typical purple precipitate of BCIP/NBT. Indeed, heart defects have proven a popular phenotype for OPT analysis (Hart et al. 2006; Risebro et al. 2006), complementing other studies which have characterized defects in limbs (DeLaurier et al. 2006), lungs (Hajihosseini et al. 2008a, b), and pancreas (Hecksher-Sørensen et al. 2004).

Preclinical Disease Research

Preclinical disease research shows promise as an important future area for OPT imaging. At a general level, some of the previously described applications (for example gene expression analysis and phenotyping) are clearly very relevant tools for a range of preclinical research (for characterizing mouse models, etc.). However, in some cases OPT may develop a more specific imaging role, as illustrated by the following example of pancreas imaging.

OPT has recently been used to study whole adult mouse pancreases in a common mouse model of diabetes—the NOD strain (Alanentalo et al. 2007). This application is closely related to the phenotyping mentioned in the previous section, because it involves the use of a molecular marker to extract information about a particular tissue subset of the organ studied. Antibodies against insulin were used to label all of the β-cell tissue within the Islets of Langerhans (Fig. 9.5). A single OPT scan provides enough information to produce a detailed quantification of the size distribution of Islets within the entire pancreas. Only the very smallest Islets are missed, and currently the only alternative is the painstaking sectioning and stereology of many physically cut sections from each specimen. In a preliminary study, the speed and convenience of OPT allowed quantitative comparisons between ten normal and diabetic pancreata in a reasonable time. Because the results are digital voxel data sets, it allowed automated analysis on both the size distribution of the Islets and their spatial distribution with respect to the pancreas as a whole. This represents an important new source of information for understanding the nature of insulitis during the onset of diabetes. In particular, it was possible to suggest that the absolute volume of β-cell mass may be a more important predictor of diabetic onset than the number of Islets per se (Alanentalo et al. 2007).

In Vivo/In Vitro Applications

In the ex vivo applications listed above, OPT is able to generate high-resolution images because efficient optical clearing (by organic agents such as BABB) means that photons can pass through the sample in fairly straight lines. This explains why ex vivo applications have so far been more actively explored than in vivo projects. However, despite the lower resolution achievable from a diffuse specimen, tracking the dynamics of living tissue often yields important information not obtainable from a series of static fixed samples. A couple of projects have therefore explored the potential of OPT for 4D time-lapse imaging of growing specimens in vitro.

Dynamic Imaging of Plant Growth

Although many parts of a growing plant are opaque due to chlorophyll and other compounds, the roots of many plants are semitransparent and are 1–2 mm across.

This structure therefore provided an ideal opportunity to try a proof-of-concept experiment for in vivo time-lapse OPT, and results from the growing root of an *Arabidopsis* plant were the first successful example (Lee et al. 2006). A germinating seed was embedded in agarose and introduced into a normal OPT scanner. The agarose concentration was chosen to be strong enough to hold the specimen in the correct positions for tomographic imaging, but also soft enough for the root to push its way through during extension. So far, this approach has been used to perform 4D time lapse of nonfluorescent wild-type plants, highlighting again one of the strengths of OPT—that it can perform genuine 3D reconstructions using the transmission ("bright-field") mode. At each time point of interest, the specimen was rotated through a full 360°, capturing the usual 400 raw projection images. The transparency of the root allowed very high quality reconstructions, in which many of the individual cells can be clearly seen (Lee et al. 2006).

This proof of concept was achieved using only transmitted light, but nevertheless allows detailed tracking of tissue movements over time for an individual specimen (and even some cell movements). It also suggests that monitoring dynamic gene expression patterns through the use of transgenic plant strains carrying a GFP reporter gene should be possible in future studies.

Tracking Global Tissue Movements of Mouse Limb Bud Development

Although in vitro organ culture is more complicated for mammalian embryos than other model systems, rodents are nevertheless the primary research models for biomedical research. Research has therefore been pursued to merge OPT technology with in vitro organ culture techniques with the goal of monitoring a developing organ in 4D (Boot et al. 2008). The organ chosen for this project was the developing limb bud of the mouse embryo, since it is visually accessible (unlike the heart for example, which is concealed by other tissues) and is about 1 mm in size, therefore allowing reasonable transmission of photons.

Imaging is performed within a chamber heated to 37°C. Previous standard protocols for culturing limb buds involved dissecting a portion of the trunk and positioning this at the air–liquid interface (Zuniga et al. 1999). For time-lapse OPT, the specimen must be fully submerged to avoid refractive distortions from the surface of the limb, and so a new method was introduced for supplying extra oxygen to the sample—a layer of perfluorodecalin (or "artificial blood;" Rinaldi 2005) saturated with oxygen which is gradually released during the experiment. As with previous OPT experiments, the axis of rotation was vertical; however, a number of instrumentation changes were necessary to allow the attachment and alignment of this rather delicate living specimen. In particular, embedding the sample in agarose is unlikely to yield good results as it would: (1) increase the handling time between dissection and the start of time-lapse imaging; (2) act as a physical restraint against normal growth; (3) restrict access to oxygen and nutrients within the medium. As a result, living tissue samples were pinned directly to the apparatus though a region of the tissue not required for the imaging. As this process makes it very difficult to correctly control the angle of the sample with respect to the axis of rotation, a new micromanipulator was designed allowing careful adjustments to be made after the specimen has already been transferred into the culture chamber (Boot et al. 2008).

The goal of the experiment was to build up a dynamic picture of normal limb growth by tracking the tissue-level movements of ectoderm; however, the limb bud contains no natural landmarks to act as reference points. Therefore, prior to placing the sample inside the apparatus, fluorescent microspheres were distributed around the ectoderm. The specimen was imaged from 200 angles (every 1.8°) at 15-min intervals, in both fluorescent and transmission modes—the former to track the artificial landmarks and the latter to measure the overall shape of the organ. From this data, both the shape changes and the tissue direction could be extracted (Fig. 9.6), providing novel information on this process and discovering for the first time that a twisting motion is involved in normal limb development.

Dynamic Gene Expression Imaging in 4D

Although surface movement tracking is very useful, the most useful role for 4D OPT would be to monitor changing gene expression activities throughout the 3D volume of a living tissue while an interesting biological process occurs. So far, a single example of this has been achieved, using the same model system as above: the developing mouse limb bud. Organogenesis is a highly dynamic process involving spatial pattern formation. A fairly simple, homologous ball of cells organizes itself over a matter of hours into a complex shape with a specific spatial arrangement of gene expression. It is therefore an ideal subject for experiments of 4D OPT because the pattern changes significantly during the time window available for organ culture protocols and because the biological process is complicated enough that it benefits from the advantages of a comprehensive 4D description. A transgenic mouse line was created which expresses the GFP protein under the control of the Scleraxis promoter (Pryce et al. 2007), and limb bud culture from these mice was performed in the same time-lapse OPT device as described above (Boot et al. 2008).

The resulting OPT analysis describes how the GFP expression dynamically changes its 3D spatial pattern over time (Fig. 9.6). At the beginning of the culture, the GFP signal is seen on the dorsal and ventral sides of the autopod, restricted to the center of the limb bud (specifically the medial regions where tendon specification is starting). During 19 h of culture the GFP domain increases in size and changes shape—it extends three new zones of expression into the more distal mesenchyme. These new regions of expression correspond to the first three digits to be patterned, and the increased fluorescence levels represent active upregulation of the fluorescent reporter gene in the cells in these domains. To illustrate the changing 3D shape of this dynamic domain, a single-threshold isosurface was chosen (the green surface in Fig. 9.6) which captures the spatiotemporal dynamics of the process, in particular

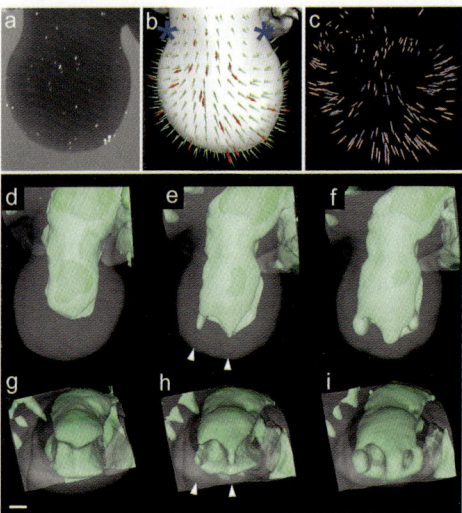

Fig. 9.6 (a–i) 4D time-lapse optical projection tomography. *Top row* shows experiments to track tissue movement during limb growth. **(a)** A raw projection image at the beginning of culture shows the distribution of fluorescent microspheres over the limb bud ectoderm. During the in vitro culture experiment, the limb bud grows and the positions of the microspheres move accordingly. These positions can be tracked over time, producing the velocity vectors shown in **(b)**. *Red arrows* show the movement of the original microspheres, and *green arrows* show an interpolation of this movement over the whole field (calculated using 3D radial-basis functions Boot et al. 2008, which revealed a double-inverted vortex flow in which tissue rotates around two almost fixed points (*blue asterisks*). Similar data were obtained for younger limb buds **(c)**, and results from different experiments can be aligned in 3D (*pink and blue arrows*), confirming the repeatability of the experiments. The two lower rows **(d–i)** show OPT imaging of a dynamic gene expression pattern. A limb bud from a transgenic mouse embryo carrying the GFP reporter gene under the control of the *Scleraxis* promoter (*Scx-GFP*) was cultured in the adapted OPT scanner. The three columns **(d, g)**, **(e, h)** and **(f, i)** show a time course of the experiment (at 0, 13 and 19 h, respectively). The two rows show different 3D perspectives on a graphic rendering of this time course: a view from the dorsal side **(d–f)** and from the distal end **(g–i)**. The emergence of new domains of GFP expression can be seen, in particular for digits 2 and 3 (*white arrowheads*) in the *middle column*. All images were first published in Boot et al. (2008)

showing how the emergence of digits 4 and 3 clearly precedes that of digit 2. These experiments represent the first direct 4D observation of dynamic spatial patterning of the mesenchymal tissue in a vertebrate limb. Assessment of the image quality was performed by comparing the virtual sections with real physically cut sections (Boot et al. 2008), and this emphasized that the observed spatial changes in the expression domain appear to involve gene activation in the digital domains, rather than large-scale tissue movements. The ability to monitor growth and gene expression in 3D over time is a technical step forward which should become invaluable to a full understanding of organogenesis and serve as a quantitative basis for computational modeling of organ development.

Future Prospects

The next five years of OPT are likely to see significant developments, both in terms of applications and the technology itself. Of the existing applications, probably developmental biology is the only field that has already explored most of the advantages OPT has to offer, especially for gene expression analysis, 3D morphology, atlases, and phenotyping. It is likely therefore that within this field the use of OPT will become more widespread due to the recent availability of a commercial OPT scanner. In other fields, such as preclinical disease research, the full possibilities have not yet been explored. Based on preliminary scans of rodent lungs (http://www.bioptonics.com), it appears that OPT could be valuable for studying lung diseases such as emphysema, and in fact any disease which predominantly affects soft tissues or internal organs might also benefit. Similarly, although the concept of pinpointing the 3D distribution of labeled cells within an organ has been proven, exploring whether this could benefit cell therapy research has not yet been attempted.

In addition to elaborating on applications that have already been tested, there is another interesting potential use of OPT which is still to be explored: analysis of biopsies from human patients. There are two main advantages this technique could have over conventional approaches, both of which reflect OPT's normal strengths. (1) *Comprehensiveness*. Standard protocols for assessing extracted lymph nodes from cancer patients involve examining sections cut roughly every 1–2 mm through the tissue. The goal is to determine whether any secondary metastases have developed in each node, and the patient's prognosis will be heavily influenced by the result. The potential advantage of OPT is that a certain category of micrometastases (those which are roughly 50–200 μm across and consist of a few hundred cells) may be missed during conventional examination, but could be more reliably found by rapidly browsing through the complete set of virtual OPT sections (which may be 10 μm thick). (2) *3D geometry*. A less certain but nevertheless potential advantage for certain types of biopsy could be the ability to examine the true shape of the tissue. It is possible that morphological changes for certain tissues provide extra information for the clinician regarding progression of the disease, although it must be emphasized that this idea is still speculative, not least because biopsies have simply not been investigated in this way before.

Regarding 4D time-lapse OPT, the technical challenges illustrated by the published studies on mouse limb buds suggest that for mammalian systems this application may remain a rather specialized technique for some time. However, zebrafish, which are the ideal model system for time-lapse imaging, could indeed benefit from OPT imaging, and projects are underway to explore these possibilities. Also, the results from growing plant roots are very encouraging for further development, and this may become a standard approach in the future.

The likely technical improvements to OPT over the next few years can be divided into two categories: incremental improvements, and the adoption of new concepts from other optical fields. Regarding the first category, despite some improvements in algorithms so far, it is likely that there is still scope for further optimization. The FDR deconvolution approach mentioned in the section "CT Reconstruction" has suggested that tailoring the general-purpose CT algorithm towards the specific technical arrangement of OPT (i.e., the consequences of using lenses) can be productive, and future adjustments of this kind are likely to continue.

Regarding the second category, although the OPT approach to extracting 3D information is very different from other techniques (computed tomography rather than optical sectioning; e.g., Kulesa and Fraser 1998; Huisken et al. 2004), in fluorescence mode it relies on the same principles regarding wavelengths as many other forms of microscopy. It may therefore see a number of improvements taken from these other fields, including multispectral linear unmixing (Dickinson et al. 2001), and time-resolved fluorescence-lifetime imaging (Requejo-Isidro et al. 2004). Both of these approaches could help to extract more intrinsic contrast from samples; for example, allowing the identification of different tissue types without the need for labeling. This could be especially useful in the potential future application of biopsy analysis. As has been seen in the field of confocal microscopy, multispectral imaging could alternatively allow the separation of more molecular dyes within a single specimen—allowing more genes to be simultaneously monitored, or more fluorescent proteins from transgenic model species, Thus, although OPT has shown major developments over the last five years and is no longer a "new" approach, it is likely that the technique has scope for many more improvements in the foreseeable future.

References

Alanentalo T, Asayesh A, Morrison H, Loren CE, Holmberg D, Sharpe J, Ahlgren U (2007) Tomographic molecular imaging and 3D quantification within adult mouse organs. Nat Methods 4:31–33

Arques CG, Doohan R, Sharpe J, Torres M (2007) Cell tracing reveals a dorsoventral lineage restriction plane in the mouse limb bud mesenchyme. Development 134:3173–3722

Asayesh A, Sharpe J, Watson RP, Hecksher-Sørensen J, Hastie ND, Hill RE, Ahlgren U (2006) Spleen versus pancreas: strict control of organ interrelationship revealed by analyses of Bapx1–/– mice. Genes Dev 20:2208–2213

Baldock RA, Bard JBL, Burger A, Burton N, Christiansen J, Feng G, Hill R, Houghton D, Kaufman M, Rao J, Sharpe J, Ross A, Stevenson P, Venkataraman S, Waterhouse A, Yang Y, Davidson DR (2003) EMAP and EMAGE: a framework for understanding spatially organized data. Neuroinformatics 1:309–325

Boot M, Westerberg H, Sanz-Esquerro J, Schweitzer R, Cotterell J, Torres M, Sharpe J (2008) In vitro whole-organ imaging: Quantitative 4D analysis of growth and dynamic gene expression in mouse limb buds. Nature Methods 5:609–612

Bryson-Richardson RJ, Berger S, Schilling TF, Hall TE, Cole NJ, Gibson AJ, Sharpe J, Currie PD (2007) FishNet: an online database of zebrafish anatomy. BMC Biol 5:34

Coultas L, Chawengsaksophak K, Rossant J (2005) Endothelial cells and VEGF in vascular development. Nature 438:937–945

Davies JA, Armstrong J (2006) The anatomy of organogenesis: novel solutions to old problems. Progr Histochem Cytochem 40:165–176

DeLaurier A, Schweitzer R, Logan M (2006) Pitx1 determines the morphology of muscle, tendon, and bones of the hindlimb. Dev Biol 299:22–34

Dhenain D, Ruffins S, Jacobs RE (2001) Three-dimensional digital mouse atlas using high-resolution MRI. Dev Biol 232:458–470

Dickinson ME (2006) Multimodal imaging of mouse development: tools for the postgenomic era. Dev Dyn 235:2386–2400

Dickinson ME, Bearman G, Tilie S, Lansford R, Fraser SE (2001) Multi-spectral imaging and linear unmixing add a whole new dimension to laser scanning flourescence microscopy. Biotechniques 31:1274–1278

Fisher ME, Clelland AK, Bain A, Baldock RA, Murphy P, Downie H, Tickle C, Davidson DR, Buckland RA (2008) Integrating technologies for comparing 3D gene expression domains in the developing chick limb. Dev Biol 317:13–23

Gibson AP, Hebden JC, Arridge SR (2005) Recent advances in diffuse optical imaging. Phys Med Biol 50:R1–R43

Hajihosseini MK, Langhe S, Lana-Elola E, Morrison H, Sparshott N, Kelly R, Sharpe J, Rice D, Bellusci S (2008a) Localization and fate of Fgf10-expressing cells in the adult mouse brain implicate Fgf10 in control of neurogenesis. Mol Cell Neurosci 37:857–868

Hajihosseini MK, Duarte R, Pegrum J, Donjacour A, Lana-Elola X, Rice D, Sharpe J, Dickson C (2008b) Evidence that Fgf10 contributes to the skeletal and visceral defects of an Apert syndrome mouse model. Dev Dyn (online publication: 4 Sep 2008)

Hart AW, Morgan JE, Schneider J, West K, McKie L, Bhattacharya S, Jackson IJ, Cross SH (2006) Cardiac malformations and midline skeletal defects in mice lacking filamin A. Human Mol Genet 15:2457–2467

Hecksher-Sørensen J, Sharpe J (2001) 3D confocal reconstruction of gene expression in mouse. Mech Dev 100:59–63

Hecksher-Sørensen J, Watson RP, Lettice LA, Serup P, Eley L, De Angelis C, Ahlgren U, Hill RE (2004) The splanchnic mesodermal plate directs spleen and pancreatic laterality, and is regulated by Bapx1/Nkx3.2 Development 131:4665–4675

Huang D, Swanson E, Lin C, Schuman J, Stinson W, Chang W, Hee M, Flotte T, Gregory K, Puliafito C, Fujimoto J (1991) Optical coherence tomography. Science 254:1178–1181

Huisken J, Swoger J, Del Bene F, Wittbrodt J, Steltzer E (2004) Optical sectioning deep inside live embryos by selective plane illumination microscopy. Science 305:1007–1009

Jenkins M, Pankti P, Huayun D, Monica MM, Michiko W, Andrew MR (2007) Phenotyping transgenic embryonic murine hearts using optical coherence tomography. Appl Opt 46:1776–1781

Johnson J, Mark SH, Isabel W, Lindsey JH, Christopher RJ, Greg MJ, Mario RC, Charles K (2006) Virtual histology of transgenic mouse embryos for high-throughput phenotyping. PLoS Genet 2(4):e61

Kak AC, Slaney M (1988) Principles of computerized tomographic imaging. IEEE, New York

Kerwin J, Scott M, Sharpe J, Puelles L, Robson S, Martínez-de-la-Torre M, Feran JL, Feng G, Baldock R, Strachan T, Davidson D, Lindsay S (2004) 3-Dimensional modelling of early human brain development using optical projection tomography. BioMedCentral Neurobiology 5:27

Kulesa PM, Fraser SE (1998) Confocal imaging of living cells in intact embryos. In: Paddock SW (ed) Confocal microscopy: methods and protocols (Methods in Molocular Biology 122). Humana, Totowa, NJ

Lee K, Avondo J, Morrison H, Blot L, Stark M, Sharpe J, Bangham A, Coen E (2006) Visualizing plant development and gene expression in three dimensions using optical projection tomography. Plant Cell 18:2145–2156

Lickert H, Takeuchi JK, Von Both I, Walls JR, McAuliffe F, Adamson SL, Henkelman RM, Wrana JL, Rossant J, Bruneau BG (2004) Baf60c is essential for function of BAF chromatin remodelling complexes in heart development. Nature 432:107–112

Louie AY, Huber MM, Ahrens ET, Rothbacher U, Moats R, Jacobs RE, Fraser SE, Meade TJ (2000) In vivo visualization of gene expression using magnetic resonance imaging. Nat Biotechnol 18:321–325

Massoud TF, Gambhir SS (2003) Molecular imaging in living subjects: seeing fundamental biological processes in a new light. Genes Dev 17:545–580

McGurk L, Morrison H, Keegan LP, Sharpe J, O'Connell MA (2007) Three-dimensional imaging of Drosophila melanogaster. PLoS ONE 2(9):e834

Ntziachristos V, Ripoll J, Wang LV, Weissleder R (2005) Looking and listening to light: the evolution of whole-body photonic imaging. Nat Biotechnol 23:313–320

Potter SM, Fraser SE, Pine J (1996) The greatly reduced photodamage of 2-photon microscopy enables extended 3-dimensional time-lapse imaging of living neurons. Scanning 18:147

Pryce BA, Brent AE, Murchison ND, Tabin CJ, Schweitzer R (2007) Generation of transgenic tendon reporters, ScxGFP and ScxAP, utilizing regulatory elements of the Scleraxis gene. Dev Dyn 236:1677–1682

Requejo-Isidro J, McGinty J, Munro I, Elson DS, Galletly NP, Lever MJ, Neil MA, Stamp GW, French PM, Kellett PA, Hares JD, Dymoke-Bradshaw AK (2004) High-speed wide-field time-gated endoscopic fluorescence-lifetime imaging. Opt Lett 29:2249

Rinaldi A (2005) A bloodless revolution—a growing interest in artificial blood substitutes has resulted in new products that could soon improve transfusion medicine. EMBO Rep 6:705–708

Risebro CA, Smart N, Dupays L, Breckenridge R, Mohun TJ, Riley PR (2006) Hand1 regulates cardiomyocyte proliferation versus differentiation in the developing heart. Development 133:4595–4606

Ruijter JM, Soufan AT, Hagoort J, Moorman AF (2004) Molecular imaging of the embryonic heart: fables and facts on 3D imaging of gene expression patterns. Birth Defects Res C Embryo Today 72:224–240

Sakhalkar HS, Dewhirst M, Oliver T, Cao Y, Oldham M (2007) Functional imaging in bulk tissue specimens using optical emission tomography: fluorescence preservation during optical clearing. Phys Med Biol 52:2035–2054

Sarma S, Kerwin J, Puelles L, Scott M, Strachan T, Feng G, Sharpe J, Davidson D, Baldock R, Lindsay S (2005) 3D modelling, gene expression mapping and post-mapping image analysis in the developing human brain. Brain Res Bull 66:449–453

Schneider X, Böse J, Bamforth S, Gruber AD, Broadbent C, Clarke K, Neubauer S, Lengeling A, Bhattacharya S (2004) Identification of cardiac malformations in mice lacking Ptdsr using a novel high-throughput magnetic resonance imaging technique. BMC Dev Biol 4:16

Sharpe J (2003) Optical projection tomography as a new tool for studying embryo anatomy. J Anat 202:175–181

Sharpe J (2004) Optical projection tomography. Annu Rev Biomed Eng 6:209–228

Sharpe J (2005) Optical projection tomography: imaging 3D organ shapes and gene expression patterns in whole vertebrate embryos. In: Yuste X, Konnerth X (eds) Imaging in neuroscience and development. Cold Spring Harbor Laboratory Press, Cold Spring Harbor, NY

Sharpe J, Ahlgren U, Perry P, Hill B, Ross A, Hecksher-Sørensen J, Baldock R, Davidson D (2002) Optical projection tomography as a tool for 3D microscopy and gene expression studies. Science 296:541–545

Summerhurst K, Stark M, Sharpe J, Davidson D, Murphy P (2008) 3D representation of Wnt and Frizzled gene expression patterns in the mouse embryo at embryonic day 11.5 (Ts19). Gene Expression Patterns 8:331–348

Tickle C (2004) The contribution of chicken embryology to the understanding of vertebrate limb development. Mechanisms Dev 121:1019–1029

Walls JR, Sled JG, Sharpe J, Henkelman RM (2005) Correction of artefacts in optical projection tomography. Phys Med Biol 50:1–21

Walls JR, Sled JG, Sharpe J, Henkelman RM (2007) Resolution improvement in emission optical projection tomography. Phys Med Biol 52:2775–2790

Walls J, Coultas L, Rossant J, Henkelman M (2008) Three-dimensional analysis of early embryonic mouse vascular development. PLoS ONE 3(8):e2853

Wilkie A, Jordan SA, Sharpe J, Price DJ, Jackson IJ (2004) Widespread tangential dispersion and extensive cell death during early neurogenesis in the mouse neocortex. Dev Biol 267:109–118

Yoder BK, Mulroy S, Eustace H, Boucher C, Sandford R (2006) Molecular pathogenesis of autosomal dominant polycystic kidney disease Expert Rev Mol Med 8:1–22

Zuniga A, Haramis AP, McMahon AP, Zeller R (1999) Signal relay by BMP antagonism controls the SHH/FGF4 feedback loop in vertebrate limb buds. Nature 401:598–602

Chapter 10
Medical Imaging Modalities – An Introduction

Jörg Peter

Abstract This chapter provides an overview of the fundamental physical principles and technical concepts of in vivo imaging instrumentation as applied in cancer
diagnostics. Included are sections on X-ray tomography (X-ray CT), magnetic resonance imaging (MRI), single photon emission computed tomography (SPECT),
positron emission tomography (PET), ultrasound, and optical imaging (bioluminescence imaging, fluorescence-mediated imaging). The aim is to provide the reader
with an understanding of these modalities; in particular, how the signal is created
and how it is detected. It is not intended to be a comprehensive study of any of these
imaging modalities. The technical description is provided in combination with the
current state of the art in imaging instrumentation.

> "Time does not flow on emptily;
> it brings, and takes, and leaves behind ..."
> *Wilhelm von Humboldt*

In 1895, Wilhelm Konrad Röntgen was experimenting with electric current flow in
order to generate high voltages in a partially evacuated glass tube—a type of tube we
now call a cathode ray tube—when he noticed that crystals of barium platinocyanide
scattered on a nearby table began to gleam (Röntgen 1896). Further experiments
revealed that these strange invisible rays—which he called X-rays because of his
uncertainty over the observation—penetrate objects such as paper, aluminum, or a
coil in a wooden box, as well as lead glass to a much lesser degree, and can be
recorded if phosphorescent crystals are placed upon a photographic plate. Röngten
was so fascinated by his observations that he soon asked his wife to place her hand
in the beam from that cathode ray tube; the resulting recorded projection was the
first medical image.

J. Peter
Department of Medical Physics in Radiology, German Cancer Research Center, Im Neuenheimer
Feld 280, D-69120, Heidelberg, Germany, e-mail: j.peter@dkfz.de

C.W. Sensen and B. Hallgrímsson (eds.), *Advanced Imaging in Biology and Medicine.*
© Springer-Verlag Berlin Heidelberg 2009

The discovery caused much excitement in the scientific and medical communities, and many scientists started to experiment with these new rays and with ways to detect them. One of them was Antoine Henri Becquerel. While investigating phosphorescence in uranium salts in 1896, he rather unexpectedly observed that the photographic plate he was using was exposed even when there was no light that could trigger phosphorescence. This finding was the first *accidental* detection of the spontaneous emission of nuclear radiation. X-rays soon found many applications in science and, particularly, in medicine. The *purposeful* use of radioisotopes in medicine was not considered until George de Hevesy's experiments with ^{32}P uptake in rats in 1935 (Chievitz and Hevesy 1935), which laid the ground for what is known today as nuclear medicine. Fundamental physical breakthroughs such as the invention of the cyclotron by Ernest Lawrence (1930) and the discovery of the positron by Carl David Anderson (1931) eventually sparked the idea of using nuclear decay for the diagnosis of disease, long before the technology for their detection was available.

In 1948, Robert Hofstadter discovered that thallium-activated sodium iodide (NaI(Tl)) made an excellent scintillation material. Four years later, Hal Anger invented the γ-ray scintillation camera by attaching a NaI(Tl) crystal to a photographic plate (Anger 1953), and later, because of its better sensitivity, to a grid of photomultiplier tubes—this detection principle is still found in today's nuclear medicine imaging modalities. Images acquired with a γ-ray camera represent a two-dimensional projectional mapping of a three-dimensional photon-emitting source distribution, as detected from a specific viewing angle. To resolve the distribution three-dimensionally, a series of projections need to be collected, and a mathematical image reconstruction algorithm is then applied. This is called computed tomography (CT), and was first achieved by monitoring radioisotopes through single photon emission computed tomography (SPECT) in 1960 by David Kuhl and Roy Edwards (Kuhl and Edwards 1963). However, it took another ten years for the first prototype X-ray CT scanner to be built by Sir Godfrey Newbold Hounsfield.

Stimulated by the work of Gordon Brownell and William Sweet, who built the first prototype positron scanner designed specifically for brain tumor imaging, in 1975 Ter-Pogossian and his colleagues built the first prototype of a fully three-dimensional cylindrical positron emission tomograph (PET) (Brownell and Sweet 1953; Hoffman et al. 1975), the second imaging modality used in nuclear medicine after SPECT. As the name indicates, PET uses positron emitters rather than photon emitters as the signal source. When emitted in tissue, positrons will undergo complete annihilation within a range of a few millimeters. This process yields two coincidence photons, each with an energy of 511 keV, which are omitted in antiparallel directions. Because there are two photons, as opposed to only one in SPECT, different methods of collimation—electronic coincidence selection—can be applied, resulting in PET having a much higher signal sensitivity than SPECT.

The unforeseen also played a role in the development of another imaging modality: magnetic resonance imaging (MRI). Isidor Isaac Rabi, an American physicist, was studying the nature of the force binding protons to atomic nuclei using the molecular beam magnetic resonance detection method he invented in the early 1930s. Occasionally he observed atomic spectra he thought were an indication of a malfunction of his instrument, and so did not consider them meaningful. However, his observations became the object of intensive research few years later. In 1946, Felix Bloch and Edward Purcell independently found that when nuclei of certain atoms were placed in a magnetic field they absorbed energy and re-emitted this energy when the nuclei were transferred to their original state (Purcell et al. 1946; Bloch et al. 1946). They both also found a direct proportionality between the strength of the magnetic field and the absorbed radiofrequency spectrum, i.e., the angular frequency of precession of the nuclear spins. This phenomenon was termed nuclear magnetic resonance (NMR), and NMR spectroscopy soon became an important analytical nondestructive method for studying the compositions of chemical compounds. In the late 1960s and early 1970s, Raymond Damadian, an American medical doctor, demonstrated that the NMR relaxation time of tumor was significantly higher than that of normal tissue.

Having heard about Hounsfield's CT machine, Paul Lauterbur, a professor of chemistry, began investigating methods for the spatial localization of magnetic resonance, and in 1973 he published an article in which he envisioned the basic concept of magnetic resonance imaging (Lauterbur 1973). He experimentally accomplished the detection of a spatially resolved NMR signal from water tubes by superimposing a weak magnetic field gradient on a strong main magnetic field. Ten years after Lauterbur's experiments the first commercial MR scanner was installed. Since then, medical doctors have been able to choose between tomographic anatomical imaging modalities, X-ray CT and MRI (note that MRI is now also used for functional / molecular imaging), and nuclear medicine modalities, PET and SPECT, that provide functional and molecular information. Technological advancements have improved steadily in terms of sensitivity and spatial and temporal resolution for all modalities. A common trend today is to combine modalities in order to detect complementary information in a single imaging procedure; mainly by overlaying low-resolution functional with high-resolution anatomical information.

Another diagnostic imaging modality that is useful in cancer research is ultrasonography (US). Ultrasonography is effective at imaging soft tissues of the body. Ultrasound was originally used in the field of medicine as a therapeutic tool (ultrasonics), which involved utilizing the heating and disruptive effects of directing high-intensity acoustic energy into tissues. Karl Dussik, an Austrian medical doctor, was the first to apply ultrasound as a diagnostic tool in 1942 (Dussik 1942), even though his experiments were later shown to be unsuccessful. The first systematic investigation into the use of ultrasound as a diagnostic tool that used a so-called A-mode (amplitude pulse-echo) scanner was performed by George Ludwig in the mid-1940s (Ludwig and Struthers 1949). In the early 1950s, Joseph Holmes and Douglass Howry experimented with a two-dimensional immersion tank ultrasound system, a so-called B-mode (brightness mode) scanner, and were able to demonstrate an ultrasonic echo interface between tissues, such as that between breast and other soft tissue (Howry et al. 1954). Still, their technique required the patient to be partially immersed in water, though new inventions soon emerged to overcome this limitation. Enlightened by the work of John Julian Wild who, in 1955, showed that malignant tissue was more echogenic than benign tissue (Wild and Reid 1952), Ian

Donald and Tom Brown propelled the development of practical, clinically usable ultrasonic technology, and were the first to build a hand-held two-dimensional scanner in 1957. Further improvements were necessary before ultrasonography finally became a widely accepted diagnostic tool in the early 1970s. Today, fast ultrasonic imaging systems with array transducers and digital beamforming as well as Doppler capabilities serve as inexpensive imaging tools for the diagnosis and management of many diseases, including cancer.

Any overview of imaging modalities would not be complete without including the emerging field of optical imaging (OI), an inherently molecular modality. Whereas nuclear medicine involves photon- or positron-emitting radionuclides, light is the signal carrier in optical imaging. Light is made of photons too, although while radionuclides release photons with an energy of several tens to hundreds of kiloelectron volts (keV), the energy of light photons is only on the order of two to three electron volts (eV). Also, photons of light are created very differently to those from radionuclides: either through bioluminescence (bioluminescence imaging, BLI) or fluorescence (fluorescence-mediated imaging, FMI). This impacts not only on the design and application of chemical probes, but also on the way that these photons behave in tissue, i.e., the interaction processes that these particles experience. High-energy PET and SPECT photons (just like X-ray photons) have a much lower likelihood of being scattered than optical photons, and so they can be detected from sources deep within large volumes of tissue. Once a particle undergoes scattering, it becomes much more difficult to estimate its original point of emission. While scattering is a problem in both PET and SPECT, causing statistically impaired images, the scattering of light is so severe in optical imaging that the applicability of this modality is limited to small animals (nude mice in particular) and spatially confined applications with larger subjects. Even in small animals, and using a rather sophisticated data acquisition setup, optical imaging in vivo is still primarily planar, because the mathematical complexity involved with solving the inverse problem—a prerequisite for image reconstruction and, hence, tomography—is very high. On the other hand, detector technology is rather straightforward. Nonetheless, in vivo optical imaging is an intensely studied field and exhibits high potential for molecular imaging applications because of advances in molecular and cell biology techniques, the availability of activatable probes, its lack of ionizing radiation, and the fact that it is comparatively inexpensive.

Performance characteristics (spatial resolution, sensitivity, and temporal resolution) for all of the modalities included in this overview are listed in Table 10.1. In the rest of this chapter, the basic principles of image formation and basic technical components are summarized for each modality. As space is limited in this introductory chapter, the physical and instrumental aspects of imaging for cancer diagnosis are only touched upon here. For much more comprehensive information on this subject, the curious reader should refer to textbooks and other more detailed works (see Oppelt 2005; Knoll 1989; Cherry et al. 2003; Christian and Waterstram-Rich 2007; Kalender 2006; Smith and Lange 1997; Liang and Lauterbur 1999; Kremkau 2005; Vo-Dinh 2003).

Table 10.1 In vivo imaging modalities and approximate values for their performance characteristics

Modality	Spatial resolution	Sensitivity	Temporal resolution
CT	<100 µm (<500 µm)	mM	<1 s
MRI	<100 µm (<1 mm)	µM	s
PET (^{18}F)	≈1 mm (<5 mm)	pM	s min
SPECT (^{99m}Tc)	<500 µm (<7 mm)	<1 nM	min
US	µm (mm)	1 bubble	ms
BLI	mm	≈100 cells/voxel	min
FMI	mm	pM	ms

Spatial resolution is provided for the highest achievable values in small animals; values in parentheses reflect the performances of state-of-the-art clinical systems. Sensitivity values are provided for molecular imaging applications (contrast media: *CT*, iodinated contrast; *MRI*, Gd-based contrast; *US*, microbubbles)

X-Ray Computed Tomography

Principles of X-Ray Imaging

A controlled beam of electromagnetic radiation (X-ray) that is differentially attenuated by an exposed object is the carrier of signal content. When the object of interest is positioned between an X-ray-sensitive detector and the X-ray source, the transmitted X-ray fluence (which is most likely modified by the X-ray attenuation characteristics of the object) can be recorded. If the X-ray beam is scanned or broadened to form a two-dimensional transmission projection, it depicts a superimposed shadow (X-ray image) of the internal anatomy of the imaged object. By acquiring multiple, angular-dependent projections of the same object, a three-dimensional representation of the anatomy can be generated using a computer (X-ray CT).

Technical Realization

The X-Ray Tube

X-rays for medical diagnostics are generated by an X-ray tube. The X-ray tube houses a pair of electrodes, the cathode and anode, within an evacuated environment. The cathode, which acts as an electron beam emitter, consists of a filament, while the anode consists of a metal that acts as a target. Typically, such targets are made of tungsten because of its high atomic number ($Z = 74$) and its high melting point. An electrical current through the cathode filament causes the filament to anneal due to its electrical resistance. This causes electrons to be released by thermionic emission; the number of emitted electrons correlates with the filament temperature, which, in turn, depends on the filament current. When a high potential difference (tube voltage) is applied between the cathode and anode, typically between 50 and 150 kV, the released electrons are immediately accelerated towards the anode, attaining a kinetic energy (in keV) that corresponds to the applied tube voltage.

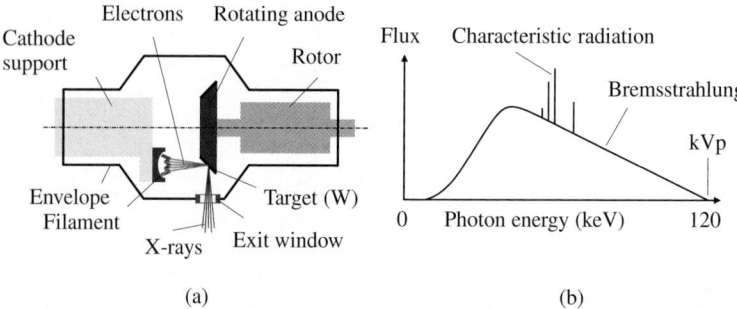

Fig. 10.1 Schematic showing a cross-section of a rotating-anode diagnostic X-ray tube (**a**). Idealized typical X-ray spectrum from a tube with a tungsten target (**b**)

The electron fluence establishes a tube current which is expressed in milliampere (mA) units, where 1 mA equals 6.24×10^{15} electrons $^{-1}$. For clinical X-ray CT systems, the tube current is typically in the range of 50–300 mA. The current in tubes used in dedicated small-animal imagers is typically lower due to the reduced amount of absorption. In addition to tube voltage and tube current, exposure duration is the remaining X-ray tube parameter. The latter two are combined into the product tube current–exposure time (in mAs, mAs/slice for CT).

Upon collision with the target, the electrons are suddenly decelerated, and their kinetic energy is converted into polyenergetic bremsstrahlung or, if the electrons have sufficient energy, it causes an electron from an inner shell of the target metal atom to be knocked out. If the latter happens, electrons from higher orbital states fill the vacancy and, during this process, characteristic X-ray photons with discrete, monoenergetic energies are emitted. Bremsstrahlung is characterized by a continuous distribution of radiation which becomes more intense and shifts towards higher energies when the energy of the bombarding electrons (i.e., the tube voltage) is increased. A continuous bremsstrahlung energy spectrum is formed because the deceleration of the electrons, and hence the X-ray photon energy generated, depends on distance at which they interact with the target nuclei. The closer the interaction occurs to the nucleus, the higher the X-ray energy. However, because the (tungsten) atom is spherical in shape, the likelihood of an interaction occurring decreases as the nucleus is approached. Therefore, bremsstrahlung forms a continuous spectrum of X-ray energies with a maximum X-ray energy (in keV) that is determined by the peak potential difference (in kVp), after which the spectrum linearly decreases towards the peak energy. Because lower-energy X-ray photons are attenuated to a greater degree upon passing through the tube window, the radiation spectrum decreases towards low energy. For clinical CT scanners, the average X-ray energy in a typical X-ray spectrum is usually 50–70 keV. If the incident electrons have sufficient energies to kick inner-shell electrons out of the target atom, characteristic X-ray photons will be created during such interactions (because each element has a unique set of electron binding energies). The interaction will leave a vacancy in the electron shell, which causes the atom to become energetically unstable and so an electron from a higher shell will move to fill the K-shell vacancy. As a result, a monoenergetic X-ray photon is created with an energy equal to the difference in binding energies. In the case of tungsten, the binding energies of the K, L_{I-III}, and M_{I-III} atomic orbitals are: 69.5, 12.1, 11.5, 10.2, 2.8, 2.6, and 2.3 keV, respectively. Hence, an incident electron can interact with and remove a K-shell electron if it has a kinetic energy of at least 69.5 keV, and the subsequent L_I–K electron transition (which is dominant in the energy spectrum) produces a characteristic X-ray of $69.5 - 10.2 = 59.3$ keV.

One important characteristic of an X-ray tube is the size of the electron beam when it reaches the target surface. A smaller focal spot for the beam produces less geometric blurring on the detector (but also a higher temperature on the target, so the anode is often spun during tube operation). Focal spot sizes of clinical CT systems are on the order of about 0.2 to > 1mm; in low-power tubes like those used for small-animal imaging, the focal spot can be smaller than 50 μm. The X-ray beam is oriented through a tube window towards the imaged object. An (adjustable) lead collimator assembly is typically placed in front of the window to define the geometry of the X-ray beam with respect to the object.

The X-Ray Detector

The X-ray detector records the spatial distribution of the X-ray photons transmitted through the object. To convert the energy of the X-ray beam into visible information, the detector usually utilizes a luminescent screen. This screen is a thin layer of tiny phosphor crystals mixed with a suitable binder. It absorbs the X-ray photons and releases light. The most common phosphor materials are based on gadolinium ($Z = 64$), e.g., terbium-activated gadolinium oxysulfide ($Gd_2O_2S(Tb)$). Modern digital detectors convert the light into an electrical signal, either by combining a thin-film transistor array with the luminescent screen, or, more recently, by employing an X-ray photoconductor. In either case, the X-ray detector consists of a pixel array (single row, multirow, or flat panel) with a pixel size ranging from about 50 μm to >1 mm, depending on the imaging applications intended for the system. Each pixel consists of a photon sensor, a capacitor to store the charge, and active matrix addressing organizing the readout of the signal before it is directed to external electronics which amplify, digitize and display the image. X-ray detectors are energy integrators; in other words, the energy of the incident photon is not discriminated (in order to reject scattered photons, for instance). This is due in part to the polyenergetic spectrum involved but it is also a consequence of the high X-ray photon fluence (the γ-ray fluences detected in nuclear medicine modalities are several orders of magnitude lower).

The CT Scanner

In most clinical systems, X-ray detectors do not directly produce a two-dimensional projection image. Rather, the beam field generated by the X-ray tube is restricted to form a narrow (typically 0.3–10 mm axially) fan-beam geometry axially, which is still wide enough to completely cover the patient transaxially. Two major scanner designs can be found in clinical systems. In one of them, a curved detector array consisting of several hundreds or thousands of independent detector elements is mechanically coupled and placed opposite to the X-ray source on a common gantry, so they both rotate mutually around the fan beam isocenter. Thin tungsten septa focused on the X-ray source can be placed between each detector in order to reject scattered photons. In the other design, only the X-ray source with its fan beam rotates about the isocenter, while the detector array, which is normally greater in diameter than the circular trajectory drawn by the X-ray tube, consists of a full ring of statically assigned independent detector elements. CT scanners of this design, also known as fourth-generation scanners, require more detector elements and were initially introduced to correct for ring artifacts due to electronic drift in the detectors, which was a problem in previous, third-generation scanner designs (note that this problem was eventually resolved by using improved detector electronics and preprocessing, so most commercial CT systems use a third-generation scanner design). Also, because of the relatively non-stationary positions of the X-ray source and the detectors during imaging, septa cannot be used in this design.

X-ray CT scanners are typically equipped with a symmetrically shaped absorption filter called a bow-tie filter, which is placed in front of the X-ray tube. Absorption filtration modifies a number of X-ray beam properties, such as the effective energy, the flux and statistics, in order to reduce radiation dose. In its simplest form, an absorption filter is made of a thin aluminum foil that sieves out low-energy photons that do not contribute to image contrast from the energy distribution. A bow-tie filter performs these modifications nonuniformly across the fan- or cone-beam field of view, thus reducing intensity variations across detector elements in the presence of the patient's anatomy.

To primarily increase the speed of volume coverage, multiple detector slices are placed together (this is known as multislice CT). To increase the temporal resolution, some modern CT systems contain several, circumferentially offset, X-ray source–detector units on the same gantry. More importantly within the context of cancer diagnostics, if the X-ray tubes create photons of different energies, the average atomic number in a voxel can be estimated, enabling improved differentiation of tissues (e.g., normal tissues from cancerous ones).

Two modes of CT image acquisition are possible: step-and-shoot and helical. In the step-and-shoot mode, the X-ray source–detector system rotates to acquire a single *slice*. Once the slice has been obtained, the scanner table moves in order to position the patient appropriately for the next slice. When a faster scan speed is needed, in particular for fast three-dimensional imaging, helical scanning is the preferred mode (at the expensive of a slightly lower axial spatial resolution). Helical CT scanners represent sixth-generation systems and are characterized by the implementation of slip-ring interconnects, which allow the stationary elements of the CT to be decoupled from the moving parts, thus permitting continuous rotation of the gantry. In the helical mode, multiple images are acquired while the patient is moved through the gantry in a smooth, continuous motion.

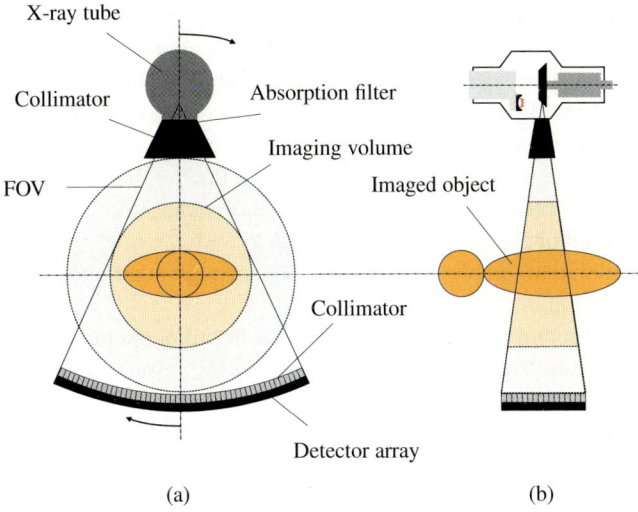

X-ray tube

Collimator

Absorption filter

Imaging volume

FOV

Imaged object

Collimator

Detector array

(a)   (b)

Fig. 10.2 Schematic of a cross-section of a third-generation X-ray clinical CT scanner design: transaxial view (**a**); axial view (**b**)

The next step in CT design may be the introduction of flat-panel X-ray detectors into clinical systems. Flat-panel CTs are mainly currently available for small-animal systems. Such detectors facilitate an increased axial field-of-view and allow very high isotropic spatial resolution, on the order of 40–150 μm. However, current major challenges include the relatively poor temporal resolution and the computational performance required to transmit and store the large number of images.

Magnetic Resonance Imaging

Principles of Nuclear Magnetic Resonance

A prerequisite for MR imaging is the presence of nuclear spin, which is nonzero in about $\frac{2}{3}$ of all nuclei. The spin of hydrogen (^{1}H) is typically used in MRI, which is given (in quantum mechanical units) as $I = \frac{1}{2}$. In a classical description, the spin corresponds to an axis-aligned circumferential charge distribution in the nucleus, yielding an associated dipole magnetic moment that is collinear with the spin. If we consider a macroscopic tissue volume, such as that corresponding to a single voxel element resolvable by an MRI scanner, there are a large number of protons, all of which have their own associated dipole magnetic moment. In total, they produce a collective (net) magnetization $\boldsymbol{M} = M_x \hat{\imath} + M_y \hat{\jmath} + M_z \hat{\kappa}$, the strength of which corresponds to the vector sum of the individual dipole moments within the considered volume. When there is no magnetic field, the spatial orientations of the proton spins, and hence the associated dipole moments, are uncorrelated and randomly distributed, resulting in zero net magnetization.

In the presence of a static magnetic field, $\boldsymbol{B_0}$, the proton spins tend to align with $\boldsymbol{B_0}$ because this corresponds to a lower total energy. At room temperature and $\boldsymbol{B_0} = 1.5\,\mathrm{T}$, however, the energy contained in the Brownian motion of the spins is much higher than the magnetic energy, so that on average only about 5×10^{-5} spins contribute to the net magnetization. However, the number of protons within a macroscopic tissue volume is so huge—e.g., 1 µl of water holds about 6.7×10^{19} protons—that the number of excess protons aligned with the field is about 3×10^{15} at $\boldsymbol{B_0} = 1.5\,\mathrm{T}$ within that sample. This ratio is large enough to produce a measurable net magnetization, $\boldsymbol{M_0}$.

By inducing a second magnetic field, $\boldsymbol{M_1}$, that is orthogonal to $\boldsymbol{B_0}$ and rotates about $\boldsymbol{B_0}$ at an angular frequency $\omega_0 = \gamma \boldsymbol{B_0}$, called the resonance or Larmor frequency (γ is the gyromagnetic ratio, $\gamma = 42.58\,\mathrm{MHz\,T^{-1}}$ for hydrogen), a precessing magnetization that is orthogonal to $\boldsymbol{B_0}$ is generated. The shape and duration of $\boldsymbol{M_1}$ can be varied, causing it to act as an electromagnetic radiofrequency (RF) pulse, and since $\boldsymbol{M_1}$ is in resonance with the protons, they can absorb that energy, causing the net magnetization vector $\boldsymbol{M}$ to rotate in the direction of the transverse plane. The degree of rotation, called the flip angle, between the longitudinal z-axis and $\boldsymbol{M}$, depends on the shape and duration of the $\boldsymbol{M_1}$-induced RF pulse. Once the excitation pulse is turned off, $\boldsymbol{M}$ begins to align with $\boldsymbol{M_0}$.

Following a $90°$ excitation RF pulse, the following dynamics can be recorded. First, the longitudinal magnetization, M_z, approaches the equilibrium $M_z = \boldsymbol{M_0}(1 - \exp(-t/T1))$. This recovery rate is characterized by the time constant $T1$, the longitudinal or spin lattice relaxation time. Since the value of $T1$ depends on the tissue type, it enables differentiation between tissues. Second, immediately after the RF pulse has been turned off, the phase coherence of all magnetic moments gradually diminishes as the spins precess at slightly different rates due to random spin–spin interactions. As the individual magnetic moments become more and more incoherent, M_{xy} starts to decrease in magnitude according to $|M_{xy}| = \boldsymbol{M_0}\exp(-t/T2)$. This decay is characterized by the time constant $T2$, called the transverse relaxation time. Like $T1$, the value of $T2$ also depends on the tissue, and in particular its chemical environment.

As the $\boldsymbol{B_0}$-aligned spins rotate, their dipole magnetic moments induce an electromagnetic field that is circularly polarized about the axis of precession. This electromagnetic RF signal is recorded using a receiver coil that resonates at ω_0 (at $\boldsymbol{B_0} = 1.5\,\mathrm{T}$ the resonance frequency of hydrogen is $63.86\,\mathrm{MHz}$). This is the essence of nuclear magnetic resonance signal detection. In an MR image, contrast is produced by differences in the relaxation times, $T1$, $T2$, respectively, of protons in different tissues. For example, at $\boldsymbol{B_0} = 1.5\,\mathrm{T}$, gray/white matter in the brain have the following approximate relaxation times: $T1 = 1{,}000\,\mathrm{ms}/650\,\mathrm{ms}$, $T2 = 100\,\mathrm{ms}/70\,\mathrm{ms}$. As $\boldsymbol{B_0}$ increases, the excess number of protons aligned with the magnetic field increases too, which enhances the signal-to-noise ratio.

Technical Realization

The Magnet

The magnet is the core part of any MRI system. The B_0 field has some key physical characteristics: high intensity, spatial homogeneity, and temporal stability. In most clinical systems used today, the field strength is in the range 0.2–3 T, or even higher for dedicated small-animal imaging systems. As some clinical applications and environments impose further constraints on the scanner, such as patient accessibility or maintenance costs, various types of magnets are used in today's MRI instruments.

Permanent magnets are used for field strengths of below 0.3 T. They have the advantage of requiring almost no maintenance costs, since the magnetic field is present without the need to supply electrical energy. The material most often used in this case is neodymium–boron–iron (NdBFe) because of its relatively high magnetic energy product and remanence field strength. Permanent magnets are designed and manufactured mostly in a two-pole geometry whereby the poles are shaped so as to create a field between the poles that is as homogeneous as possible. The imaged subject is positioned between the poles, with its long axis orthogonal to the poles. Disadvantages of permanent magnets include their weight (which increases with field strength) and the dependency of the field strength on the ambient temperature (NdBFe: -0.15 %/K).

Another type of magnet used for low field strengths (typically <1 T) is the resistive magnet. Such an electromagnet generates its magnetic field with a coil that is typically made of copper wire. Due to resistivity, most of the electrical energy is transformed not into the desired magnetic field but into unwanted heat. This degrades efficiency and requires cooling. Also, a constant current supply is essential in order to maintain the magnetic field. Most electromagnets contain an iron core to establish a vertically oriented magnetic field with an advantageously confined fringe field.

To generate higher field strengths, which are generally needed for most cancer-related imaging applications, superconducting magnets are used. Superconducting magnets are similar in design to resistive magnets. The wire consists of a super-conducting material which completely loses its electrical resistance at temperatures below a material-specific transition temperature. The majority of superconducting magnets are made of niobium–titanium (NbTi) alloys which have a transition temperature of approximately 9.3 K. The transition temperature of the material decreases further as the electrical current passing through it increases. Therefore, extensive cooling to below the transition temperature is necessary which, in most cases, is achieved by enclosing the magnet within a liquid-helium cryogenic system. In most clinical MRI instruments, the superconducting magnet is designed as a solenoid with a free inner bore diameter of about (or which is larger than) 60 cm.

The magnetic field needs to be as homogeneous as possible inside the solenoid, because this is the space in which the imaged object is placed, with its long axis aligned with the longitudinal axis of the solenoid. To optimize the homogeneity of the static B_0 field, compensation techniques known as shimming are applied.

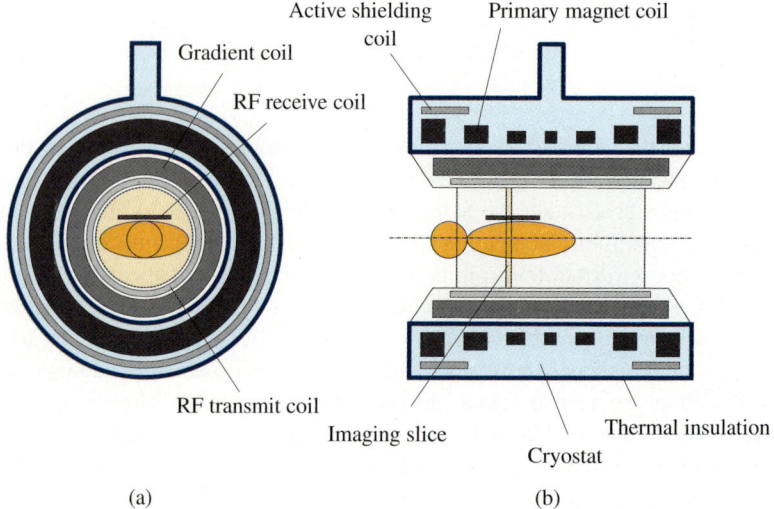

Active shielding   Primary magnet coil
coil
Gradient coil

RF receive coil

RF transmit coil

Imaging slice   Thermal insulation
Cryostat

(a)   (b)

Fig. 10.3 Schematic showing a cross-section of a clinical MRI system containing a superconducting magnet: transaxial view (**a**); axial view (**b**)

Passive shimming by means of additional magnet plate tuning is used to compensate for field distortions caused by manufacturing tolerances within the MRI system. Patient-specific, active shimming is performed as a "shim" measurement right after the patient is placed in the MR scanner in order to correct for field inhomogeneities caused by susceptibility differences at tissue interfaces.

The Gradients

Gradient coils are used to generate magnetic field gradients that are superimposed onto the static B_0 field in order to spatially encode the positions of protons. These coils are located between the magnet and the imaging volume and are switchable along each spatial direction. Two coils with opposite polarities are placed along each direction; these decrease and increase the static magnetic field, respectively. This yields a linear change in B_0 in the direction of the turned-on gradient, resulting in a linear change in precessional frequency that is directly proportional to the distance from the center of the magnet. Therefore, an MRI signal is only generated in the slice section within the imaging volume where the Larmor frequency matches the frequency of the oscillating magnetic field.

The most important properties of a magnetic gradient are its maximum strength and slew rate. Typically, clinical MRI systems have maximum gradient strengths of up to $40\,\mathrm{mT\,m^{-1}}$ at bore diameters of $60\,\mathrm{cm}$. The higher the gradient field strength, the higher the current flow within the gradient coils (up to $500\,\mathrm{A}$ at $2{,}000\,\mathrm{V}$ for the example given) and thus the higher the heat generation within the wires, which requires dedicated water-cooling.

Switching the current in a coil induces an electromagnetic field in the circuit, which itself creates a current that induces a secondary magnetic field. This secondary field also changes with time and so creates a secondary magnetic flux. As this flux is also dynamic, it opposes the change in magnetic flux that creates the electromagnetic field (Lenz's law). Therefore, the rise time for switching a gradient coil on (or off) cannot be made infinitely short, even though this would be desirable for fast signal localization and ultimately higher image resolution. The ratio of maximum gradient strength to coil rise time is called the slew rate. Clinical MRI systems have slew rates of between 10 and 200 $mT\,m^{-1}\,(ms)^{-1}$.

Gradient-induced currents can also occur in the human body; they cause peripheral nerve stimulation, which is another limitation on clinically applicable maximum slew rates. Also, switching gradient systems causes strong time-varying mechanical forces that can be heard as acoustic noise, with sound pressure levels in excess of 100 dB. These forces are proportional to the gradient current and, because they are generated by the interaction of the gradient field with the static magnetic field, to the field strength of B_0.

The Radiofrequency System

The radiofrequency field M_1 is created by the transmission chain of the MRI system. This consists of a synthesizer to create an electrical signal that oscillates with the Larmor frequency, a pulse shaper to modulate the RF pulse onto the signal, and an RF power amplifier to amplify the signal to a peak power of about 10 kW for a clinical system. This signal is transferred to a radiofrequency coil via a specifically designed and shielded cable. This transmit coil is an antenna that transmits the RF signal to the imaged subject. The same coil can also be used to receive the echo signal created by the spins. A typical example of a transmit/receive coil is the body coil integrated into most superconducting MRI systems. Often, better signal reception is achieved with dedicated receive coils (in which case the integrated body coil is used to apply the M_1 field) or transmit/receive coils, such as RF coils to image the head, or surface coils to image anatomical regions close to the body surface. To improve the signal-to-noise ratio, or to allow parallel imaging and thus reduce the number of phase-encoding steps, so-called phased-array coils have also been introduced.

In order to detect NMR signals, the entire MRI system must be placed in a radiofrequency (RF) cabin. This cabin is a Faraday cage that suppresses any external electromagnetic waves. The NMR signal received by the coil is a very weak high-frequency electric signal which needs to be amplified, demodulated, and finally digitized for further processing. Demodulation is necessary to remove the high-frequency Larmor signal component, thus yielding an analog signal that only contains the low-frequency information imposed by the gradients. When tomographic imaging is performed, the last step is to reconstruct the images.

Single Photon Emission Computed Tomography

Principles of SPECT Imaging

Gamma-ray imaging refers to the detection of an injected substance labeled with a photon-emitting radionuclide. A γ-camera intercepts the emitted photons that are traveling in its direction, measures their energies, and forms an image of the photon distribution—a two-dimensional projection of the radioactivity indicating the presence or absence of a specific physiologic function. This type of (planar) recording is called scintigraphy. In order to elucidate the three-dimensional biodistribution of the injected radiopharmaceutical, the γ-camera rotates around the object and acquires projections from (equally spaced) angular intervals. The distribution of the radioactivity within the imaged volume is then estimated from the projection data, typically using an iterative reconstruction technique. This type of tomographic recording is called SPECT.

A great deal of progress is currently being made in the field of developing pixellated solid-state detectors for γ-ray detection. These novel detectors consist of advanced materials such as semiconductors that potentially facilitate improved intrinsic spatial detector resolution and much better energy resolution. However, the transition to such detectors has not been finalized, mainly due to their cost, reliability, and ease of operation. Still, almost all commercially available γ-cameras for nuclear medicine applications are scintillation-based detectors that employ photomultiplier tubes (PMT) as photon sensors.

Technical Realization of a Scintillation-Based Detector

The Scintillator

A scintillation detector consists of an optically transparent scintillator crystal which absorbs γ-ray photons and converts the photon's high kinetic energy into proportional amounts of detectable light. Because the energy conversion is based primarily on photoelectric absorption, a dense crystal material with a high atomic number (Z) is desired. The scintillator most often used to detect photons emitted by radionuclides in nuclear medicine is sodium iodide doped with thallium (NaI(Tl)). Since NaI(Tl) is a relatively dense material ($\rho = 3.67\,\mathrm{g\,cm^{-3}}$) with a relatively high effective Z of 50, its ability to absorb keV photons is high, which makes it a very efficient converter. The amount of light produced depends on the incoming photon's energy. NaI(Tl) yields about one photon of light per 30 eV of absorbed photon energy. Aside from their high detection efficiency and high light yield, another important characteristic of scintillators is their signal decay, which relates to the time period during which light is present in the crystal after photon conversion. The smaller this constant, the "faster" and more efficient the scintillator. NaI(Tl) has a decay time of 230 ns. Depending on the camera design, the crystal can either be fabricated to form a single, large-area crystal (as used in most clinical systems) or to form a detector that is composed of optically separated small voxelized crystal elements (often seen in high-resolution small-animal imaging systems). The thickness of the crystal is being selected based on the expected photon energy and the desired intrinsic spatial resolution. The thicker the crystal, the higher its detection efficiency. However, as the crystal becomes thicker, its spatial resolution decreases, mainly because of increased light spread (large-area crystals) and parallax effects. For most γ-cameras employed in nuclear medicine, the NaI(Tl) crystal thickness is on the order of 6–12 mm.

A significant fraction of the emitted photons will be scattered within the body. Most scattered photons are caused by Compton interactions of the photons with weakly bound electrons. Compton scattering not only deflects the photons, causing considerable degradation in image contrast, it also causes an (scattering-angle-dependent) diminution in photon energy. The light produced in the scintillator scales with photon energy. However, the amount of light generated by photons of identical energy varies because of variations in the conversion process. This quality is characterized by the energy resolution. The typical energy resolution of a clinical NaI(Tl)-based γ-ray detector is about 8–10% FWHM at 140 keV. Therefore, if the energy diminution is large enough, the scattered photons can be sorted out by the detector. Energy resolution is the main limitation in SPECT. Therefore, new scintillator materials are being investigated primarily to optimize the energy resolution.

The Photomultiplier Tube

The scintillator is optically coupled to a grid of photomultiplier tubes (PMTs). A PMT converts the energy from photons of light into an electrical signal; the amplitude of the signal is proportional to the number of photons. It consists of a glass vacuum tube containing a photocathode and a cascade of dynodes, usually 8–12 dynode stages. Once the light created in the scintillator strikes the photocathode, a number of photoelectrons proportional to the light intensity will be ejected from the photocathode due to the photoelectric effect. Attracted by the large potential differences between the photocathode and the first dynode, as well as those between all subsequent dynodes, the ejected electrons will be accelerated towards each dynode within the cascade. Secondary electrons are produced when the electrons impact on each dynode, yielding an overall signal gain at the final dynode (the tube's anode) of typically 10^7 electrons per photoelectron. Every photocathode material has a specific response spectrum which ought to match the scintillator light output spectrum. This material, usually an alloy of an alkali metal and antimony, also defines the fraction of photoelectrons released to the number of incident light photons, called quantum efficiency. Typical photocathode quantum efficiency is in the range of 10–30%. The signal at the anode is flown through a load resistor to form a voltage pulse which constitutes the output signal of the PMT.

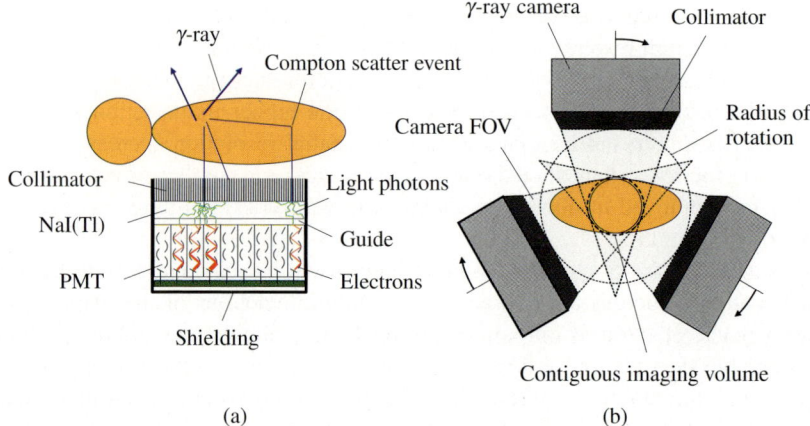

Fig. 10.4 Simplistic cutaway view of a conventional γ-ray camera (**a**). Transaxial schematic view of a typical clinical SPECT system involving three γ-ray cameras mounted on a common rotating gantry (**b**). Multiple camera heads allow for faster data acquisition. The cameras are equipped with fan-beam collimators and they image the head region

The Electronic Processing Unit

A preamplifier is connected to each PMT in order to amplify the rather weak analog output signal. In most cases, the signal is then transmitted to pulse-shaping circuits and to a pulse-height analyzer. The purpose of the latter is primarily to discriminate Compton-scattered and background photons from unscattered (primary) photons. Because the amplitude of the voltage pulse is proportional to the amount of energy deposited in the scintillator crystal, it is possible to determine the energy of the detected photon by the height of the electrical signal it produces. In most modern γ-ray camera systems, pulse-height analyzers are implemented in the form of multichannel analyzers (mainly analog-to-digital converters, ADCs) that divide the output signals into channels according to their amplitudes. Each analyzer channel is connected to a corresponding storage location, and every time the signal amplitude falls within a specific channel, the counter for that channel is incremented.

The Collimator

The γ-ray detector described thus far would detect any γ-ray photons that deposit energy in the scintillator. However, the photon itself does not carry information about its point of origin or possible site of interaction. In order to ascertain the photon's angle of incidence, a collimator is placed in front of the detector. A collimator is a unit of dense, high-Z material, most commonly lead or tungsten containing a large number of holes (producing a septum pattern); only photons traveling within the acceptance cones of the collimator holes are detected. All other photons impinging on the collimator septum will be absorbed in the collimator (some do penetrate through it, however, causing a reduction in image contrast) and so are prevented from being detected. Thus, the collimator defines the camera's field-of-view (FOV) and forms a projected image of the three-dimensional photon distribution on the detector surface. While the principle of collimation is crucial to scintigraphy and SPECT imaging, the collimator unfortunately absorbs most of the γ-ray photons, resulting in a relatively low detection efficiency and poor image quality.

Various types of collimators are available for different imaging applications, and they directly determine the spatial resolution and detection efficiency of the γ-ray camera. The parallel beam (PB) collimator is most commonly used for whole-body clinical applications. It consists of a large number of straight holes with parallel axes. For example, when used for high-resolution, low to medium energy imaging (e.g., with ^{99m}Tc), the thickness (hole length) of a PB collimator made of lead might be 30 mm with holes 1.2 mm in diameter and a septum thickness of 0.16 mm. The higher the energy rating of the collimator, the greater all these geometric values will be, and so the lower its spatial resolution. As all of the holes are oriented parallel to each other, the projected detector image is the same size as the source distribution. If the holes are cast in a way that the extended hole axes merge to a focal line (fan beam, FB) or to a focal point (cone beam, CB), the source distribution is mapped onto the detector in a smaller (diverging) or a larger (converging) projection, respectively. For converging-hole collimators, the focal line/point lies in front of the collimator at a focal length that is normally greater than the diameter of the imaged object. For fan-beam collimators, the minimization/maximization occurs along one axis (generally the object's transverse axis), while for cone-beam collimators it occurs along both axes. Converging collimators are used to image small objects or portions thereof with a larger camera, permitting full utilization of the detector area and thus yielding improved spatial resolution and geometric efficiency. If the detector is smaller than the imaged object, a diverging-hole collimator can be used, although this yields lower spatial resolution and efficiency than either a parallel-hole or a converging-hole collimator. Therefore, diverging-hole collimators are seldom used. If the imaged object is very small compared to the detector size (e.g., in small-animal imaging or imaging of the thyroid), a pinhole (PH) collimator yields the highest magnification and hence the highest spatial resolution, particularly for objects very close to the pinhole. However, the sensitivity is comparatively low, and as the distance between the object and the camera becomes larger, the geometric efficiency drops significantly. The pinhole aperture, which is typically 0.5–2 mm in diameter, is placed at the end of a lead cone in front of the detector, with the distance chosen according to the desired magnification.

Given the various collimator types available, a SPECT camera is a very versatile instrument and can be—when used in concert with application-specific camera orbits and various camera tilting modifications—adapted to a broad range of imaging applications.

Positron Emission Tomography

Principles of PET Imaging

PET imaging is the detection and spatial reconstruction of an injected tracer that is labeled with a positron-emitting radionuclide. The positron is the antimatter equivalent of the electron. Positron (or β^+) decay is a radioactive decay process in which a proton decays into a neutron, positron, and neutrino ($p \rightarrow n + \beta^+ + \upsilon$). For ^{18}F, the positron emitter most often used in PET imaging, the β^+ decay can be expressed as $^{18}F \rightarrow {}^{18}O + \beta^+ + \upsilon$.

As a positron travels through tissue or other materials it undergoes scattering due to Coloumb interactions with electrons, causing the positron to give up some of its kinetic energy during each scatter event. When the positron finally reaches thermal energy, it (basically) interacts with another electron by annihilation, producing two antiparallel 511 keV photons ($\beta^+ + e^- \rightarrow \gamma + \gamma$).

There are two factors involved in this process that inherently limit the spatial resolution of positron detection. First, the positron travels for some time before it annihilates. The annihilation interaction marks the point at which it is detectable, but not the site of its emission. In tissue, the mean free path length until annihilation can be several millimeters, depending on the energy of the positron. For instance, the maximum kinetic energy of ^{18}F is 635 keV, which would enable the positron to travel about 2.6 mm in water (which is about the density of tissue). On average, positrons from ^{18}F travel about 0.22 mm before annihilation. Second, the emerging 511 keV photons are only exactly antiparallel in the rather unlikely event that both of the annihilating particles are at rest (in the center of momentum frame). Most 511 keV photon pairs do not travel at exactly 180° to each other, although the angular deviation is very small, mostly less than 1°. Both effects can be neglected for clinical PET systems because of the limited spatial resolution and minimal acceptance angle of most detectors. However, the degradation in spatial resolution due to the range of the positron is a limiting factor in small-animal imaging applications when some higher-energy positron emitters are involved.

Technical Realization

Scanner Geometry and the Block Detector

Two γ-ray detectors (including coincidence electronics, as explained below), positioned opposite to each other, are required to detect a pair of simultaneously emitted annihilation photons from within an imaging volume located between both detectors. These detectors, as a pair, can be rotated in order to acquire annihilation photons from all angles, hence enabling tomography. Still more efficient is the positioning of stationary detectors around the imaged object, which enables the acquisition of annihilation photons from all angles simultaneously. Most of today's dedicated PET systems utilize a cylindrical allocation of small block detectors.

A block detector is similar to a γ-ray camera (as described in the previous section) in that a scintillator crystal is used to convert the incident photon's energy into visible light, which then is transformed into an electrical signal by a PMT. However, in its specific configuration, it uses a different design. For example, as the photon energy is generally higher than in scintigraphy and SPECT, the scintillator crystal is made of a different material. Dense materials with a high Z number are preferred. This characteristic is best satisfied by bismuth germanate (BGO, $Z = 74$), a scintillator that can be found in many PET scanners. Unfortunately, BGO has a rather long decay time of 300 ns. Therefore, manufacturers have developed alternative scintillators, such as cerium-doped lutetium oxyorthosilicate (LSO), which has a decay time of 40 ns. This material also has a higher light yield than BGO (75% of the light yield of NaI(Tl); BGO yields 15% of the light yield of NaI(Tl)).

Ideally, an array of crystal elements of some size should be coupled to an array of photon sensors of equal area. In typical clinical PET systems, however, relatively large PMTs (with tube diameters of about 1.5 cm) are still the sensor of choice, although specifically designed PET block detectors have been developed that yield an intrinsic spatial detector resolution that is smaller than the sensor size (PMT diameter). The crystal of a block detector, as large as or smaller than (in front area size) the array of 2×2 PMTs to which it is optically coupled, has partial cuts of different lengths that act as light guides and segment the crystal into (typically) 8×8 elements. These cuts are designed to produced a controlled spread of the light created in one crystal element onto all adjoining elements. Hence, signals are generated in all of the PMTs, which can then be electronically analyzed to resolve the crystal element that houses the interaction.

To give an example, a whole-body PET scanner, the Siemens ECAT EXACT HR$^+$, has the following parameters. The block detector has 64 crystal elements (8×8) with dimensions of 4.39 mm $\times$ 4.05 mm $\times$ 30 mm (axial $\times$ transaxial $\times$ depth). There are four detector rings, defining an axial FOV of 15.5 cm, with 72 block detectors per ring, mounted to give an inner ring diameter of 82.7 cm. This yields 18,432 distinct crystal elements in total (576×32).

Electronic Coincidence Selection

Because there are always two annihilation photons traveling in opposite directions, a PET system has extensive electronics that monitors the output signals of all of its detectors for the occurrence of coincident signal pulses. Every time two detectors create a photon detection pulse near-simultaneously (more precise, within a limited time window of about 10 to 20 ns for typical clinical PET systems) this information is saved and used for the localization of the causing annihilation process. Given only one such coincidence trigger, the information is sufficient to provide a line of response (LOR) in the space between the two pulsing detectors along which positron annihilation took place. By collecting a large number of coincidence events, and by applying image reconstruction after data acquisition, a statistical solution can be found for the spatial distribution of the annihilation sites, and thus the positron

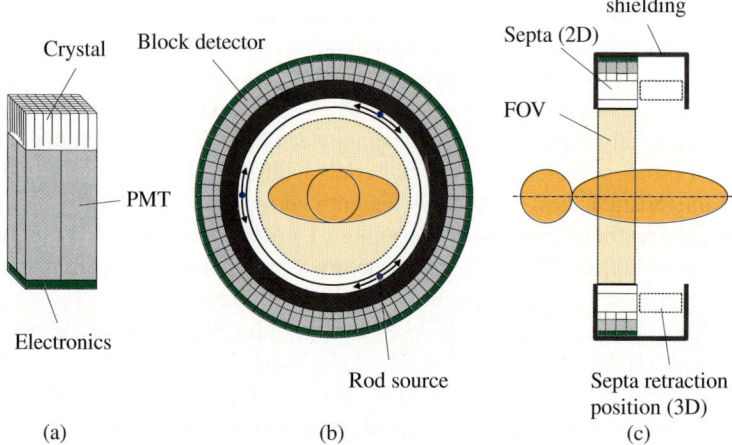

Fig. 10.5 PET block detector design (**a**) consisting of an 8×8 array of crystal elements cut into the block. The crystal is coupled to four (2×2) PMTs. Schematic of the cross-section of a clinical PET system involving three rotating line sources for transmission scanning: transaxial view (**b**); axial view (**c**)

density within the imaged volume, because regions with greater radioactivity produce more LORs. Electronic coincidence selection fulfills the same purpose as the collimator in a γ-ray camera, although it enables many more photons to reach the sensor. Therefore, PET has much higher sensitivity than SPECT (by $\approx$2–3 orders of magnitude).

Pulses generated by the PMTs are passed through a constant fraction discriminator (CFD). This generates a digital timing pulse when the output signal reaches a constant fraction of the peak pulse height. Energy signals from the PMTs are also sent to energy discrimination circuits (pulse-height analyzers) in order to reject events of significantly lower energy than 511 keV resulting from Compton scattering interactions. Timing pulses associated with photons accepted by the energy discrimination circuits are used for coincidence detection. When a coincidence is detected, the circuitry of the scanner or a computer analyzing the signal determines the LOR that connects the two sites of detection.

2D and 3D Acquisition Modes

Unfortunately, only about 15–45 % of the detected LORs are *true* coincidences. The largest fraction, 40–55 %, of the coincident events are induced by an annihilation photon pair in which at least one photon has undergone at least one scattering interaction prior to detection, but both are detected. The remaining fraction of assigned coincidences, 15–30 %, are created by *random* coincidences that occur when two photons from different β^+ decays are detected simultaneously within the coincidence time window. Scattering and random coincidence events cause wrong LOR assignments which subsequently add noise to the true coincidence rate and degrade image contrast. The rates of both scattering and random coincidences depend on the volume and attenuation characteristics of the imaged object, and on the scanner geometry. The number of random coincidences increases with the activity of the substance injected into the imaged object. Both of these types of unwanted coincidence events can be corrected for, though the efficiency of the camera is affected by increasing dead-time losses.

Even though positional location is performed by electronic coincidence selection, most clinical PET systems include thin annular collimating septa made mostly of tungsten that can be moved in front of the block detectors and coaligned with the crystal rings. When these are in place, coincidences are only recorded between detectors within the same ring or within closely neighboring rings. This is called the 2D mode of operation. It is used in order to suppress oblique, out-of-transverse-plane LORs and thus reject scattered photons, and also to reduce the counting rate, thereby reducing the number of random coincidence photons. In the 3D mode, the annular septa are retracted and coincidences are recorded between detectors in any ring combination. This results in a significant increase in sensitivity to true coincidences, making it possible to administer a less-active substance to the imaged object. At the same time, there are increased scattering and random fractions and there is also a higher sensitivity to random coincidences from outside the FOV, which can be severe when imaging regions near organs with a high accumulation of activity, such as the bladder. Therefore, 3D acquisition is more useful in low-scatter studies, such as pediatric, brain, and small-animal studies.

Data Correction

A number of data correction strategies are applied to PET (and SPECT) imaging—most importantly attenuation correction. As the photons travel through tissue (or any material), a fraction of them will be absorbed according to the density of the material. Therefore, knowledge of the attenuation coefficients of the object being imaged is a prerequisite for quantitative imaging. Attenuation is a bigger problem in PET than in SPECT because both annihilation photons must escape from the object to cause a coincidence event to be registered. To correct for photon attenuation, two additional scans are necessary as well as the emission scan. For these measurements, most PET systems incorporate up to three retractable rod sources that are typically filled with a long-lived radioisotope, such as ^{68}Ge. First, a blank scan is performed without the object being present in the scanner (this blank scan is typically performed once a day, and is also used for a daily assessment of scanner quality). Second, the object is placed in the scanner and a transmission scan is performed. In both of these measurements, the rod source(s) revolve around the imaging volume, so the attenuation is measured along all possible lines of response. Finally, object-specific attenuation correction factors are derived by dividing the count map from the blank scan by the count map from the transmission scan.

Ultrasonic Imaging

Principles of Ultrasonic Imaging

Ultrasonics is concerned with the propagation of acoustic waves (sound pulses) with frequencies above the range of human hearing (>20 kHz) in various media. The lowest sound frequency commonly used in the context of diagnostic imaging is about 1 MHz. When pulses of high-frequency sound are transmitted into the human or animal body, these pulses are absorbed, reflected, refracted, and scattered by tissue, all of which lead to the attenuation of the ultrasound beam, which increases with penetration depth. While there is almost complete reflection at the boundaries between soft tissue and air and soft tissue and bone, rather less reflection occurs at the boundary between different soft tissues. The scattered energy that returns to the transmitter as the ultrasound travels through tissues is called the backscatter. Since different kinds of tissues and interfaces each have their own individual scattering properties, backscatter is the most relevant information used for diagnostic imaging applications. It can be used to locate the tissue interface (by monitoring the time taken for the scattered energy to return), to detect the size and the shape of the interface (by measuring the scattered energy)—for instance that of tumor tissue—and for the detection of movement (by measuring the Doppler shift of the scattered energy).

The axial spatial resolution achievable depends on the wavelength: the higher the ultrasound frequency, the higher the spatial resolution. In most soft tissues, the wavelength is about 1.5 mm at 1 MHz. Hence, in order to distinguish between two interfaces parallel to the beam path spaced at 1 mm, an ultrasound beam with a frequency of at least 1.5 MHz needs to be transmitted into the object. However, as the frequency increases the attenuation of the ultrasonic signal also increases, which has an inverse effect on the depth of ultrasound penetration. Therefore, applications that require deep penetration (such as tumor imaging in humans) typically use frequencies that are generally not larger than 5 MHz, while ultrasound frequencies of greater than 40 MHz have been applied to enable very high spatial resolution (up to 30 μm) at small penetration depths for the imaging of small animals.

Technical Realization

The Transducer

Almost all ultrasonic instruments used in medical diagnostics are based on the pulse-echo technique. In this, an ultrasound transducer converts electrical energy into acoustic waves (actuator) which are transmitted into the body being imaged, and it then receives (sensor) and reconverts the acoustic wave echoes into an electric signal using a piezoelectric element. Lead zirconate titanate (PZT), a ceramic with a perovskite structure, is commonly used for this purpose because of its strong piezoelectricity and high dielectric constant.

While a single transducer could be used to scan a certain region within tissue by mechanically steering the beam, the use of multielement (array) transducers is common in clinical ultrasonic systems. Transducer arrays allow the ultrasonic beam to be focused and steered very rapidly electronically. However, in multielement alignment a transducer not only needs to possess high sensitivity, spatial resolution, and an electrical impedance that is matched to the transmitter, but also a wide angular response in the steering dimensions as well as low cross-coupling. The one-dimensional transducer arrays currently used for two-dimensional imaging typically contain at least 128 (in some instruments up to 512) coresonating transducer elements that are evenly spaced over an area of about 7–12 cm. A single array element is typically about 500 µm in lateral size, small enough to behave like a point source that radiates a spherically shaped wavefront into the medium. As each transducer element has its own pulsing circuit, transmit–receive switch and signal amplifier–detector, variable timing schemes can be applied across the transducer aperture in order to create a variety of specific steering and focusing patterns for the ultrasonic beam.

To focus an ultrasound beam, a group of adjacent elements (typically 8–16) are pulsed in conjunction. When the pulsing of the inner elements is delayed with respect to that of the outer elements, a focused beam can be generated. By further adjusting the pulse delay, the focused beam can be established at various depths. As the same delay parameters are also applied to the elements during the echo receiving phase, a single scan line in an ultrasound image is formed. By applying a linear time gradient along the aperture, the ultrasonic beam can be steered. To steer a focused beam, a linear gradient superimposed with a cylindrical timing profile is added to the pulse delay pattern. By shifting the group of associated adjacent elements along the transducer array and repeating the focusing procedure, another line can be scanned. Depending on the multiarray transducer design, the set of all lines scanned yields an ultrasonic image which is either a 2D sector or rectangle.

By arranging transducer elements in a two-dimensional layout, it is possible in principle to acquire three-dimensional image data simply by steering the ultrasonic beam in terms of both azimuth and elevation, yielding three-dimensional ultrasonic images. However, manufacturing and operating 2D transducer arrays is a complex task because of the high number of channels (e.g., a 128×128 element array requires 16,384 channels). Therefore, three-dimensional ultrasonic imaging is mostly performed by oscillating a convex shaped one-dimensional transducer array mounted on a mechanical wobble.

Ultrasound Imaging Modes

A-mode (amplitude).

Here, an ultrasound beam pulse is send into an object and the reflections that come back over time are recorded. As the beam encounters different tissues the amplitude of the echo varies. Major signal spikes are generated at tissue boundaries.

(c)

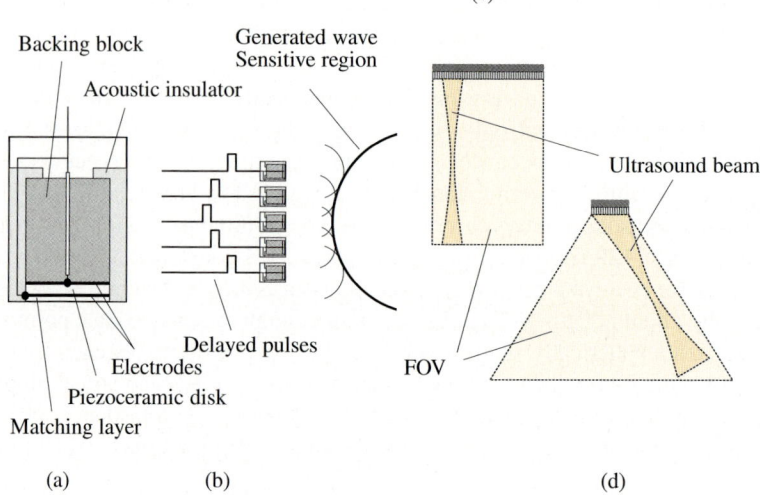

Fig. 10.6 Ultrasound piezoelectric crystal transducer (**a**). Phased-array concept for generating a focused ultrasound beam (**b**). Linear transducer array (**c**). Phased-array transducer (**d**)

The distance between these spikes can be measured by dividing the speed of sound in tissue ($\approx$1,540 m s^{-1}) by half of the sound's travel time.

B-mode (brightness).

This gives a two-dimensional (spatial) representation of A-mode data whereby the signal amplitudes are converted into brightness-modulated entities.

M-mode (motion, real time).

Sequential B-mode pulses are applied across the transducer array in order to generate up to 100 images per second. The ability to localize structures within a two-dimensional ultrasonic image is greatly enhanced by visualizing the changes that occur to the structures within that image over time (e.g., cardiac motion). Most modern ultrasound devices are 2D real-time imaging systems.

Doppler ultrasound.

The Doppler effect (named after the Austrian physicist Johann Christian Doppler) is the change in the received sound frequency due to the relative motion of the sound source with respect to the sound receiver (or reflector). In medicine, Doppler ultrasound is used to detect and measure blood flow, in which case the major reflector is the red blood cell. The Doppler shift is dependent on the frequency of the incident sound beam, the velocity of the blood, and the angle between the sound beam and the direction of movement of the blood. There are various types of blood flow measurement: color Doppler, pulsed Doppler, and power Doppler.

Optical Imaging

Light Distribution in Tissue

The fundamental limits on optical imaging are determined by the optical properties of tissue, which is a turbid medium that both scatters and absorbs photons. Absorption depends on the type of tissue, but is increased significantly by the presence of hemoglobin, which absorbs light strongly at wavelengths of less than about 600 nm, and by water, which attenuates light strongly at wavelengths of greater than about 900 nm. In the intermediate wavelength band of about 600–900 nm, optical photons have a higher likelihood of penetrating (and escaping from) tissue, even tissue several centimeters thick. However, photon trajectories are very erratic in tissue because scattering is very pronounced. Therefore, the fluency of emission created by an internal or external source becomes quickly highly diffuse. The absorption coefficient, μ_a, is defined as $dI = -\mu_a I \, dx$, where dI is the differential change in the intensity I of a collimated light beam traversing an infinitesimal path dx through a homogeneous medium with absorption coefficient μ_a. The reciprocal, $1/\mu_a$, is called the absorption path length and equals the mean free path a photon travels between consecutive absorption events. The scattering of light in biological tissue is caused largely by refractive index mismatches at microscopic boundaries, such as cell membranes, organelles, etc. The scattering coefficient, μ_s, is defined as $dI = -\mu_s I \, dx$. Similarly, the reciprocal, $1/\mu_s$, is called the scattering path length and equals the mean free path that a photon travels between consecutive scattering events. The scattering anisotropy, g, can be characterized in terms of the mean cosine of the scattering angle, which is defined as $g = 0$ for perfectly isotropic scattering and $g = 1$ for complete forward scattering. The scattering cosine depends on the size, shape and refractive index mismatches at the microscopic boundary. Scattering anisotropy is typically in the range of 0.7–1 in biological tissues, i.e., photons of light are predominantly forward-scattered. The g-factor is often used to simplify the directional effects of scattering by introducing the reduced scattering coefficient, defined as $\mu_{s'} = \mu_s(1 - g)$.

Measuring the optical properties in vivo is a very difficult task, if possible at all. However, a knowledge of them (or at least estimates of tissue-specific absorption and reduced scattering coefficients) is crucial to any type of optical imaging, particularly quantitative imaging and tomography, and is a prerequisite for the derivation and application of models that describe the diffusion of optical photons in tissue. The optical properties of tissue are often only available from ex vivo tissue samples and may not represent in vivo conditions. For example, the following optical properties have been measured for muscle tissue at 515 nm: $\mu_a = 1.1 \, \text{mm}^{-1}$, $\mu_s = 51.3 \, \text{mm}^{-1}$ (Vo-Dinh 2003). The corresponding mean free scattering path length yields 0.02 mm. Note that scattering is the dominant interaction mechanism for optical photons in tissue ($\mu_s \gg \mu_a$); in the example given above, the scattering coefficient is close to 50 times greater than the absorption coefficient. In comparison, the mean free path length of keV photons (CT, SPECT, PET) is several orders of magnitude greater (several centimeters in tissue, depending on the energy).

Moreover, absorption is the dominant interaction effect for detected keV photons in tissue. Therefore, standard emission data models or methods for reducing scattering, such as those very successfully applied in nuclear imaging, cannot be used.

In order to spatially resolve a three-dimensional distribution of optical signaling probes in vivo quantitatively, a parametric model that takes into account the optical properties of tissue needs to be applied on order to estimate photon propagation. Such a model, commonly derived using a diffusion approach, is part of any tomographic image reconstruction strategy. Because of the severity of light scattering in turbid media, and because of the uncertainty over (if not unavailability of) estimates for optical media properties, particularly when heterogeneous media such as small animals are imaged, the modeling of photon propagation is an extremely difficult mathematical problem. Further, projection images are surface-weighted, i.e., light sources closer to the surface of the imaged object appear brighter than sources deeper. Therefore, the diagnostic potential of optical imaging does not seem to be limited by physical detector issues, but it would benefit from a deeper insight into the optical features of biological tissues, and their accurate measurement. Further, quantitative analysis must be approached with caution, and validation for each specific application is necessary. Nevertheless, optical imaging offers unique possibilities for in vivo imaging applications, especially in the context of molecular imaging in small animals (nude mice in particular).

Fluorescence Imaging

The success of fluorescent methods in biological research has provided a great incentive to apply fluorescent probes to the in vivo molecular imaging of cellular molecules as well. Fluorophores, in a conventional sense, are organic substances composed of either chemically synthesized fluorescent dyes or genetically encoded fluorescent proteins. Fluorescence results from a process that occurs when fluorophores absorb photons of light. A fluorescent molecule typically consists of a chain of carbon atoms between two aromatic structures; such a molecule acts as an optical resonator. When a photon of a certain energy is absorbed by a fluorophore, the molecule attains an excited state as a result of electron transfer from a ground state to a higher energy level. As the excited electron quickly drops back to its ground state, it emits a quantum of fluorescent light at a characteristic wavelength. This physical process is different from chemiluminescence, in which the excited state is created by a chemical reaction. A fluorescent molecule has two characteristic spectra, the excitation spectrum and the emission spectrum. The difference in wavelength between the maxima of both spectra is called the Stokes shift, and for most organic molecules this typically ranges from under 20 to several hundred nanometers. The probability of light absorption by a fluorescent molecule is called the molar extinction coefficient. The emission intensity directly relates to the quantum yield, which is the ratio of the light emitted to the light absorbed (fluorochrome efficiency). The product of the extinction coefficient and the quantum yield is the fluorescence intensity. For example, green fluorescent proteins (GFPs) are commonly used as reporters of gene expression. In fact, GFPs were the first generally available molecular imaging probes, which create a signal solely by molecular biological processes. Wild-type GFP emits bright green fluorescence (emission peak at 508 nm) when illuminated by blue light (excitation peak at 395 nm) with a fluorescence intensity of about $22,000 \, cm^{-1} M^{-1}$. As discussed previously, red-shifted GFP variants are generally preferred for in vivo labeling because of the lower absorption in tissue. Nevertheless, when choosing an optical probe, its optical properties should be weighed against its fluorescence intensity, labeling efficiency, probe toxicity, and stability.

Recently, fluorescent semiconductor nanoparticles—so-called quantum dots (inorganic semiconductor crystals consisting of only tens of thousands of atoms)—have been found to exhibit very attractive optical properties compared to fluorophores, such as larger absorption coefficients across a wide spectral range, higher absorbance of incident light per unit concentration of dye, and larger extinction coefficients.

Fluorescence Optical Diffusion Tomography

If a diffusion equation that accurately models the propagation of light through tissue can be derived, then it should also be possible to calculate volumetric images using surface measurements of the scattered light from various sources and detectors positioned around the imaged object. An unknown optical probe distribution can be estimated by solving the three-dimensional distribution of absorption and scattering coefficients (diffusion model) that could have produced the measurements of scattered light. This is referred to as fluorescence optical diffusion tomography (FODT), which is an emerging imaging modality. However, the image reconstruction framework for FODT is very complex mathematically: it represents a so-called ill-posed inverse problem, which has only been solved for a few experimental settings so far.

Bioluminescence Imaging

Bioluminescence is a special form of chemoluminescence. Bioluminescence imaging (BLI) is based on the detection of visible light photons that are emitted when a bioluminescent enzyme, e.g., firefly luciferase, metabolizes its specific substrate, e.g., luciferin. Light emission from the firefly luciferase-catalyzed luciferin reaction has a relatively broad spectrum (530–640 nm) and peaks at 562 nm. Since the signal to be detected is created by a chemical process, no excitation light is necessary for BLI. This means that there are no unwanted background signals (from either autofluorescence or from filter leakage) either, so the images achieved are of a rather high quality, even when acquiring bioluminescence signals from whole animals. The efficiency with which internal light sources can be detected is, however, dependent on several factors, including the level of luciferase expression, the depth of the labeled cells within the body, and the sensitivity of the detection system.

Technical Realization

The Light Source

To excite the fluorescent optical probe a light source is required. Characteristic properties of light sources are the output light spectrum and power, as well as the temporal characteristics of the light. Two major light source types are commonly used: broad-wavelength sources (such as arc lamps), and line sources with discrete wavelengths (such as lasers). Most instruments employed for generic usage utilize high-pressure arc lamps which emit light over a broad spectrum. An excitation filter, typically with a short- or band-pass characteristic, or a monochromator is necessary to restrict the output spectrum to the range of wavelengths used to excite the dye. Lasers, on the other hand, emit coherent light in a narrow, low-divergence beam with a narrow spectrum. Various laser types are available, among them gas lasers, solid-state lasers, and semiconductor lasers (laser diodes).

Fluorescence imaging can be classified into three groups depending on the temporal characteristics of the light. Sources generating a continuous light wave are commonly employed when the sole purpose of imaging is to measure the light intensity emitted or transmitted from or through the object. While continuous-wave imaging is the (inexpensive) standard mode of operation for two-dimensional applications, investigations of the optical properties of highly scattering heterogeneous media or fully three-dimensional tomography of such media may require additional constraints. Therefore, imaging systems are also operated in the time or frequency domains. For time-domain measurements, an ultrashort laser pulse is injected into tissue and its temporal distribution is measured. As the light pulse travels through tissue, its peak amplitude decreases and its profile widens. Both effects can be associated with specific μ_a and μ_s' values. Time-domain measurements require very fast light sources as well as highly sensitive sensors, necessitating expensive instrumentation. Instead of using a short light pulse, systems working in the frequency domain use intensity-modulated light sources, typically working in the frequency range of 100–1,000 MHz. During measurement, the phase shift and demodulation of the light transmitted through the object relative to the incident light are measured as a function of frequency.

The Photon Detector

The light emitted from the object is measured by a photon sensor and converted into an electronic signal. Characteristic properties of photon sensors are spatial resolution, quantum efficiency, temporal resolution, and dynamic range. Most imaging systems tailored for continuous-wave imaging employ position-sensitive integrating sensors, such as those based on CCD or CMOS arrays, which are not in contact with the imaged object. The intrinsic spatial resolution capabilities of these sensors are excellent, as most sensors have light-sensitive pixels that have edge lengths of less than 50 μm. Various methods of improving upon the remaining characteristics are applied. The quantum efficiency, which is typically on the order of about 40% at peak sensitivity, can be increased to close to 85% if the sensor is reduced in thickness by etching so that the light can pass through the back layers of the chip (back-illuminated sensor). Dark current, a major cause of noise, can be reduced by cooling the sensor chip. Temporal resolution (the minimum light integration time), which is typically on the order of milliseconds, can be reduced to picoseconds for gated imaging applications by incorporating a time-modulated image intensifier. An image intensifier is an evacuated tube which consists of a photocathode, a microchannel plate (MCP), and a phosphor screen. When an incident photon hits the photocathode, a photoelectron is emitted and is accelerated towards the MCP by a high potential difference. If the photoelectron has sufficient energy, it discharges secondary electrons from the MCP. As there is another electrical field between the MCP and the phosphor screen, the dislodged electrons undergo acceleration towards the screen. Upon impact, light is generated in the phosphor substrate by phosphorescence. The overall gain with an MCP-based intensifier is about 10^5; the degree of electron multiplication depends on, and can be gated by, the gain voltage applied across the MCP. The output of the image intensifier is optically coupled to the photon sensor.

When time-resolved single-photon counting or the highest possible temporal resolution is desired, which is particularly true of time-domain measurements, photomultiplier tubes (as described in the section "Positron Emission Tomography") are preferably used. Because of the sizes of PMTs, these very sensitive and fast sensors are normally used in connection with fiber-based imaging systems. In a fiber-based imaging setup, an appropriate spatial arrangement of optical fibers (typically in direct contact with the imaged object) is used to deliver excitation light to the object and also to collect the transmitted light. Fiber-based systems are also commonly used for frequency-domain imaging applications and for fluorescence optical diffusion tomography.

To separate the excitation light from the emission fluence, a band- or long-pass filter is mounted in front of the detector when fluorescence imaging is performed. Because optical filters often exhibit sigmoidal rather than ideal bandwidth separation, filter leakage superimposes a background signal onto the desired emission light.

References

Anger HO (1953) A new instrument for mapping the distribution of radioactive material in vivo. Acta Radiol 39:317–322

Bloch F, Hanson WW, Packard M (1946) Nuclear induction. Phys Rev 69:127

Brownell GL, Sweet WH (1953) Localization of brain tumors with positron emitters. Nucleonics 11:40–45

Cherry SC, Sorenson JA, Phelps ME (2003) Physics in nuclear medicine. Saunders, Philadelphia, PA

Chievitz O, Hevesy G (1935) Radioactive indicators in the study of phosphorus metabolism in rats. Nature 136:754–755

Christian PE, Waterstram-Rich K (2007) Nuclear medicine and PET/CT technology and techniques. Mosby, St. Louis, MO

Dussik KD (1942) über die möglichkeit, hochfrequente mechanische schwingungen als diagnostisches hilfsmittel zu verwerten. Z Neurol Psychiatr 174:153–168

Hoffman EJ, Ter-Pogossian MM, Phelps ME, Mullani NA (1975) A positron emission transaxial tomograph for nuclear medicine imaging (PETT). Radiology 114:89–98

Howry DH, Scott DA, Bliss WR (1954) The ultrasonic visualization of carcinoma of the breast and other soft-tissue structures. Cancer 7:354–358

Kalender WA (2006) Computed tomography: fundamentals, system technology, image quality, applications. Wiley, New York

Knoll GF (1989) Radiation detection and measurement. Wiley, New York

Kremkau FW (2005) Diagnostic ultrasound: principles and instruments. Saunders, London

Kuhl DE, Edwards RQ (1963) Image separation radioisotope scanning. Radiology 80:653–662

Lauterbur PC (1973) Image formation by induced local interaction; examples employing magnetic resonance. Nature 242:190–191

Liang ZP, Lauterbur PC (1999) Principles of magnetic resonance imaging: A signal processing perspective. Wiley, New York

Ludwig GD, Struthers FW (1949) Considerations underlying the use of ultrasound to detect gallstones and foregn bodies in tissue (Tech Rep 4). Naval Medical Research Institute, Bethesda, MD

Oppelt A (2005) Imaging systems for medical diagnostics. Publicis, Erlangen

Purcell EM, Torrey HC, Pound RV (1946) Resonance absorption by nuclear magnetic moments in a solid. Phys Rev 69:37–38

Röntgen WC (1896) On a new kind of rays. Nature 53:274–276

Smith RC, Lange RC (1997) Understanding magnetic resonance imaging. CRC, Boca Raton, FL

Vo-Dinh T (2003) Biomedical photonics handbook. CRC, Boca Raton, FL

Wild JJ, Reid JM (1952) Application of echo-ranging techniques to the determination of structure of biological tissues. Science 115:226–230

Part II
Software

Chapter 11
Volume Visualization Using Virtual Reality
Medical Applications of Volume Rendering in Immersive Virtual Environments

Anton H.J. Koning

Abstract In this chapter we take a look at the possible applications of volume visualization in immersive virtual environments, and how they can be implemented. Nowadays, the availability of many kinds of 3D imaging modalities, like CT, MRI and 3D ultrasound, gives clinicians an unprecedented ability to look inside a patient without the need to operate. However, while these datasets are three dimensional, they are still presented on 2D (flat) screens. While most systems and workstations offer a volume rendering option to present the data, when projected onto a normal screen (or printed on paper or film), the images are not truly 3D and are often termed "2.5D." By using a virtual reality system that immerses the viewer(s) in a truly three-dimensional world, we hope to improve the understanding of these images.

Introduction

With the invention of the CT scanner by Hounsfield and Cormack in the early 1970s, doctors were no longer limited to the flat, 2D world of X-ray imaging when they wanted to obtain information on the "insides" of their patients without operating on them. CT scanners allowed them to obtain a number of "slices" through the patient's anatomy, essentially creating a rectangular volume of data describing the density of the tissues at a large number of points (also known as voxels) in three dimensions. This was a huge improvement over the X-ray photograph, which compresses all of the information on the third dimension into a flat image, making it impossible to tell what is in front of what.

In addition to the CT scanner, MRI and 3D ultrasound imaging modalities have now been introduced, which also generate volumes of data that image anatomy in three dimensions. However, 3D imaging technology is no longer limited to anatomy;

A.H.J. Koning
Erasmus MC University, Medical Centre, Rotterdam, The Netherlands,
e-mail: a.koning@erasmusmc.nl

C.W. Sensen and B. Hallgrímsson (eds.), *Advanced Imaging in Biology and Medicine.*   257
© Springer-Verlag Berlin Heidelberg 2009

it also allows us to look at the actual biochemical processes taking place in the patient's body, using, for instance, SPECT, PET and functional MRI (fMRI) scans.

When these imaging modalities were introduced, radiology departments were equipped to handle the X-ray films and the graphical capabilities of computers were almost nonexistent. Therefore, the only viable option was to examine the data volumes generated by the scanners by printing them slice by slice on film, and then examine them on light-boxes like ordinary X-rays. While light-boxes have since been replaced (initially) by dedicated workstations with CRT monitors and (more recently) with PCs and LCD panels, decades of radiological experience mean that even today the majority of radiological exams are still based on a slice-by-slice evaluation of the datasets. Only in a limited number of cases are volume visualization techniques used to render 3D views of the data, and even these are presented on what are essentially 2D media: either computer monitors or prints on paper. This means that in almost all cases, fully three-dimensional data is being reduced to two dimensions. To make matters worse, the manipulation of these images, normally with a computer mouse, is also restricted to 2D methods.

In this chapter we will try to show the added value of using so-called immersive virtual reality (VR) systems, which offer true 3D (stereoscopic) images and 3D interaction for the visualization and investigation of medical volumetric datasets. The best-known immersive VR systems are those based on a head-mounted display (HMD), also known as the VR helmet, that immerse the wearer in a 3D virtual world by placing small screens (typically LCD panels a few inches in size) right in front of the eyes. While current display technology can provide images of high enough quality for most applications, HMDs still have two major drawbacks: first and foremost, by shutting out the "real world" they prevent meaningful collaboration with other people, as interaction between avatars in the virtual world is still severely limited with current technology; second, despite increases in image quality and responsiveness, the users of HMDs frequently experience a condition known as "simulator sickness," an affliction not unlike motion sickness, where the user is affected by nausea and dizziness. To overcome these drawbacks, the Electronic Visualization Laboratory of the University of Chicago developed the concept of the CAVE^{TM} (Cruz-Neira et al. 1993) in the early 1990s.

Systems like the CAVE^{TM} immerse viewers in a virtual world by surrounding them with three-dimensional (i.e., stereoscopic) computer-generated images. The images are projected onto the walls, floor and sometimes ceiling of a small "room" in which the observers stand. Current LCD and DLP projection technology allow high-resolution images with a considerable dynamic range to be projected onto screens ranging from 2.5 to 4 m wide. The stereoscopic effect can be attained by using a passive system with filters (based on either polarization or color shifts) that simultaneously projects the left and right images, or an active system that alternately projects the left and right images and uses LCD shutter glasses to separate the images. The head position of one of the users is usually tracked by an electromagnetic or optical tracking system, which allows the system to provide that particular user with the correct perspective. As the images are generated in real time, this also generates so-called motion parallax, which further enhances depth perception.

Interaction with the virtual world is usually by means of a "joystick" or "wand," a hand-held device equipped with a number of buttons; the position and orientation of this device are also tracked.

Most of the methods we will describe in this chapter are also applicable to smaller VR systems, for instance those based on a single projection screen or on a 3D monitor. These may also be used for clinical visualizations, and are indeed more suited for integration into daily clinical practice.

Volume Rendering Techniques

In order to be of use for the visualization of medical volumes, like those produced by CT and MRI scanners, the VR system needs software that is capable of generating stereoscopic images. In order to be of real value, the software must also be able to interact with the tracking system and I/O devices (typically a wireless joystick or a "wand"), otherwise the system will be nothing more than just a stereo display add-on. Standard offline medical image analysis packages that scanner vendors offer cannot be used with VR systems. Unfortunately, some smaller turnkey systems aside, most VR systems do not come with any application software, let alone an application that is suitable for medical volume visualization. As the use of VR systems for this purpose is still very limited, there are no out-of-the box commercial solutions as yet.

Some scientific visualization applications like Amira (http://www.amiravis.com) and VTK (http://www.vtk.org) can be configured to run on a virtual reality system, but they generally lack an immersive interface that will allow clinicians (as opposed to scientists or developers) to make efficient use of the application. Therefore, someone interested in using an immersive VR system for medical volume visualization can currently either try to find an open source application on the Internet, for instance VOX, the Volume Explorer developed at Brown University (http://www.calit2.net/~jschulze/projects/vox), or they will have to develop their own software.

Two approaches to visualizing volumetric data exist. Originally, all 3D computer graphics algorithms, and therefore all specialized computer graphics hardware, were based on the rendering of polygons, i.e., triangles. Objects were defined as collections of triangles, described by their vertices. Transformations defined the positions and orientations of objects in 3D space. Therefore, a traditional rendering pipeline used to generate a 2D image of a 3D object consists of transforming the 3D vertices into the 2D "screen space," followed by a rastering process that maps the transformed triangles to the pixels of the image.

Therefore, the first algorithms developed to render volumetric data extracted surfaces, i.e., meshes, created from large numbers of polygons, and used the standard rendering pipeline to generate images of these surfaces. The marching cubes algorithm (Lorensen and Cline 1987) is the best-known example of this type of algorithm, and is implemented in many visualization applications. While surface extraction is frequently used even today, it does have some severe limitations. While earlier problems like the extremely large number of triangles that can be generated from a high-resolution volumetric dataset have more or less been solved by the increase in computational performance (with respect to both CPU and graphics), the fact that the "correct" surface cannot be defined in many cases has led to the development of so-called "direct volume rendering" algorithms.

Direct volume rendering algorithms do not generate a model created from polygons, but instead directly map the voxel data to image (or screen) pixels, using a "transfer function" to change the color and transparency of the original voxels. Direct volume rendering algorithms can be divided into several categories: ray casting, splatting, shear-warp and texture mapping. Many scientific and medical visualization applications use the texture mapping approach (Wilson et al. 1994). In this technique, the volume is loaded onto the graphics card and used as a texture on polygons generated by intersecting the volume's rectangle with a large number of planes. By drawing these polygons back-to-front and compositing the colors and transparencies derived from the texture data, one can obtain a high-quality rendering of most types of volumetric data. The current state-of-the-art graphics accelerator cards with 256 Mb or more of memory allow the rendering of all but the largest medical datasets at truly interactive rates (i.e., 20–60 frames per second). This high frame rate is especially important when targeting a VR system, as a low frame rate will make precise interaction impossible, and will also lead to fatigue and simulation sickness.

Visualization toolkits like Amira and VTK usually implement direct volume rendering algorithms with and without hardware acceleration, and can be used as the basis for an immersive application, but extensive development work will be required to add an immersive user interface. An alternative may be to use a third party library, either open source or commercial, which implements direct volume rendering. However, one must be careful when combining APIs, as many libraries set up their own graphical environment, which may lead to conflicts. Most commercial VR systems come with a library that sets up both the graphical environment and handles the I/O with the tracking systems and input devices. These libraries are generally incompatible with scene graph libraries that try to do the same.

When choosing a third party API to do the volume rendering, continued support by the developers or vendor is very important. Graphics hardware technology is still making rapid advances in performance and capabilities, which means that when upgrading to a new graphics card older applications may fail, and also that newer versions of driver software for older cards may have features removed, causing incompatibilities with existing software. This may lead to a situation where one cannot upgrade the hardware or update the drivers or operating system, because the reliance of the application on third party software that is no longer being actively developed or supported means that it will no longer function. In this case, the only solution will be to rewrite the application using a different API that does support the new hardware or driver. Needless to say, this can be very costly.

User Interfaces

Possibly even more important and costly in terms of effort required is the implementation of an immersive user interface. While in theory it would be possible to have the application being controlled by an operator sitting behind a normal 2D screen, this would severely limit the benefits that a VR system has to offer. Unfortunately, there are no accepted standards for the implementation of 3D user interfaces. While everybody is used to the WIMP (Windows, Icons, Mouse, Pointer) user interfaces we have on our computers to manipulate the 2D world of our desktop, there are many different approaches when it comes to 3D user interfaces for VR systems.

A good user interface can make or break an application, and this is especially true for virtual reality applications. People are generally apprehensive about new technology, and this is especially true for clinicians, who on average have very little computer experience. If you want these people to use your application independently, it has to have a simple, easy-to-understand user interface. One of the pitfalls of developing a user interface for a virtual reality application is that of trying to implement it such that it is completely analogous to the real world. This type of user interface generally uses virtual 3D objects, like scissors and knives, and may include complex gesture recognition as well. Ideally, this would lead to a very intuitive user interface, but unfortunately the current state of the technology prevents this: 3D objects can be hard to recognize and identify due to the limits in screen resolution, the tracking is too slow and inaccurate for intricate manipulation and gesture recognition, and there is generally no force feedback with these type of systems. The result is often a "larger than life" interface, and operating the application becomes the equivalent of a workout at the gym.

Therefore most successful VR applications implement a user interface similar to that of a desktop application, as this is what users will be familiar with. A simple menu system with one or two word items will be easier to understand and operate than a collection of 3D objects. It is a matter of preference as to whether this menu is placed at a fixed position (usually to the side in a CAVE^{TM}-like environment), or whether it is free-floating near the user. However, care must be taken to ensure that the menu does not interfere with the object(s) being visualized.

While implementing a menu system for a volume visualization VR application is relatively straightforward, there will also be a need to implement to implement interactions with the volume data itself. For instance, users will want to rotate, scale and translate the volume to obtain the views they need. Here a choice has to be made between direct and indirect manipulation. With direct manipulation, the volume is directly linked to the users input device, and follows the movements of the device. This is comparable to dragging things in a desktop environment, or using the "trackball" mode that is available in visualization packages for rotating 3D objects. Indirect manipulation uses the input device to set parameters like rotation angles along predefined axes. The input device may have a hat-switch (a small secondary joystick), one or more dials, or use button presses to change a parameter. Using a virtual pointer to operate virtual dials or sliders also falls into this category.

Both approaches have their merits. Direct manipulation is very intuitive, while indirect manipulation is more precise and generally less tiring when used for longer periods. Some users may prefer one method over the other, so it can be a good idea to implement both options. A combination is also possible: the volume's position and orientation may be set indirectly, while tools such as clipping planes and measuring devices are manipulated directly. For actions like setting a transfer function, it will be necessary to implement some sort of "widget," a graphical user interface object that can be manipulated with the input device and shows the current settings. When implementing multiple volume manipulation tools, it is common to run out of input device buttons to assign functionality to. This means that a "modal" interface, where the result of pressing a button is dependent on the mode that is currently selected, cannot be avoided. Therefore, giving the user clear feedback on which mode(s) is (are) currently being used is an important route to preventing mistakes and unnecessary mode switching.

Data Handling

Given that you have succeeded in obtaining or implementing a volume visualization application for your VR system, the next step is to get the data you want to visualize into it. Unfortunately, this is not as easy as it sounds. Even if your application is capable of loading DICOM 3.0 formatted files directly, and the chances are that it is not, there are still some pitfalls you may encounter. First of all, if you are new to medical data, DICOM 3.0 is the medical standard for image data, which covers both a network protocol and a file format. CT and MRI scanners and ultrasound equipment all use this standard to communicate with the PACS (Picture Archiving and Communication System) system, which is the central repository for digital image data in a hospital. When datasets need to be sent to a GP or to another hospital, they are generally written to CD in DICOM format.

DICOM uses a hierarchy of patients, studies, series and images. For standard CT or MRI datasets, this means that a series is the equivalent of a stack of slices, also known as a volume. When the application has a DICOM load option, you should be able to load all of the files in the appropriate series directory, and the application should be able to figure out things like slice distance, bit depth and the correct order of the slices. If not, you will have to convert the data to a format that your application can read. There are several utilities that will let you extract the raw pixel data from DICOM files, or convert them to a more common image format like GIF, TIFF or PNG. Avoid converting the data to JPEGs, as you will introduce artifacts and you may lose small details.

If you need to use the raw pixel data, you will have to look at the header information to determine the image's size and bit depth. DICOM is a so-called tagged image format, not unlike TIFF, that stores information in fields that have a unique identifier, like (0018,0088). The list of tags and their meaning is called the DICOM dictionary. You can get a list of the tags in a file and their values using an application like dcmdump from the DICOM toolkit (http://dicom.offis.de/dcmtk.php.en). Apart from a long list of predefined tags, it is also possible for manufacturers to use their own private tags to store additional information in the images. While some of these may be of interest to people trying to do something different with the data than just look at the slices on a PACS workstation, manufacturers are usually very secretive about their private tags, and it may be impossible to obtain the necessary information.

CT data normally consists of 12-bit gray values (ranging from 0 to 4,096), stored as 16-bit integers. Normally these values will be in Hounsfield units, but sometimes you will encounter data that has been modified by a radiologist, and the values will no longer cover the standard range. MRI data may be 8 bit or more, depending on the scanner. Unfortunately, most MRI scanners will report 16 significant bits, while the highest value in the data is actually only around 1,000–1,500. If data with a smaller than standard range is converted to an 8-bit format without using the actual highest value, the dynamic range is dramatically reduced, and it may be impossible to use the transfer function to obtain the required images.

Another pitfall when converting individual slices lies in the fact that when you scale each slice down using its own maximum, you will end up with a stack in which a certain tissue will have a different gray value in each slice. This will prevent you from using gray-level windowing to segment the data. It is therefore advisable to use the overall maximum value of the image stack.

Most MRI and CT scans are anisotropic. This means that the pixels in the images are smaller than the distance between the slices. If you don't correct for this anisotropy and just use cubic voxels, you will end up with an image that appears compressed in one dimension. You will need to extract this information from the DICOM tags in the files. Unfortunately, there are several tags that contain information on pixel size and slice distance, and not all scanners use the same method to store this information. Most will provide you with the pixel dimensions in tag (0028,0030)—PixelSpacing. The inter-slice distance should be stored in (0018,0088)—SliceSpacing, but this is often not the case. Sometimes you will find that tag (0018,0050)—SliceThickness is being used instead. Other manufacturers just store the slice's location relative to some origin in tag (0020,1041)—SliceLocation, and you will have to calculate the inter-slice distance yourself from two consecutive slices.

You can also use the SliceLocation to stack the images in the correct order. Alternately you can look at tag (0020,0013)—InstanceNumber to determine the order, as there is no guarantee the filenames will correspond to the physical order. Most datasets you encounter will have "transversal" slices, which means that they are made perpendicular to the long (head-to-toe) axis of the body. Something to remember when converting CT and MRI data is that radiologists are used to looking at these slices "from below," which means that your volume rendering will be reversed in the left-right direction if you do not correct for it.

Another imaging modality that we have found to benefit greatly from visualization in a virtual environment is 3D ultrasound, also known as "echo" data. Ultrasound is frequently used in prenatal medicine and in cardiology. Most cardiological

3D ultrasound data will actually consist of time series, and is therefore also called 4D ultrasound. Unfortunately, there is no DICOM standard for ultrasound data. While there are some systems that can store uncompressed Cartesian (i.e., rectangular) volumes of ultrasound data, most will just store a 2D screen dump as pixel data in the DICOM file and use private tags to store the original data in a proprietary format. To get access to this data you will need to obtain from your vendor either a separate application, which is capable of exporting the data in a nonproprietary format, or information on their proprietary format.

Depending on your volume rendering application, rendering a 4D ultrasound dataset with the correct frame rate may be a challenge. Fortunately, most cardiologists will not be bothered by the fact that the data is being displayed at a somewhat slower than real time frame rate. Reducing the resolution of the dataset can help you to achieve an acceptable frame rate, especially when you are using a 3D texture mapping approach, where you will be limited by the speed at which you can upload your data to the graphics card. If you have enough texture memory, you may be able to store the entire time series on the graphics card, circumventing the upload bottleneck.

Functional imaging modalities like SPECT, PET and fMRI can also be volume rendered, but their limited resolution and lack of anatomical information means that you will need to combine them with a normal CT or MRI dataset to be of any real use. Finally, in a research environment you may find confocal or 3D electron microscopy data that are volume rendered. Usually these consist of separate slices that are either already in an easily accessible format like TIFF or that can be converted to such a format with the application that is normally used to view the images.

If you plan to do more than some occasional visualization, and are not part of the radiology or other clinical department with direct access to the datasets, it may be a good idea to install a small PACS system, for example Conquest (http://www.xs4all.nl/~ingenium/dicom.html) or K-PACS (http://www.k-pacs.net), or at least a so-called Storage-Service Class Provider application, as this will allow people to send images to you via the network from the scanners or the central PACS. Depending on your status within the hospital and the intended use of the data, you may be required to anonymize the data you store locally. A good place to start looking for DICOM software is the "I Do Imaging" website (http://www.idoimaging.com).

Applications

The Department of Bioinformatics at the Erasmus MC operates a Barco I-Space, which is a four-walled CAVE-like virtual reality system. Images are projected onto three walls and the floor using eight Barco SIM4 projectors with a resolution of $1,280 \times 1,024$ pixels, a light output of 1500 ANSI lumen and a contrast ratio of more than 1,300:1. The 3D effect is obtained using the so-called passive stereo method.

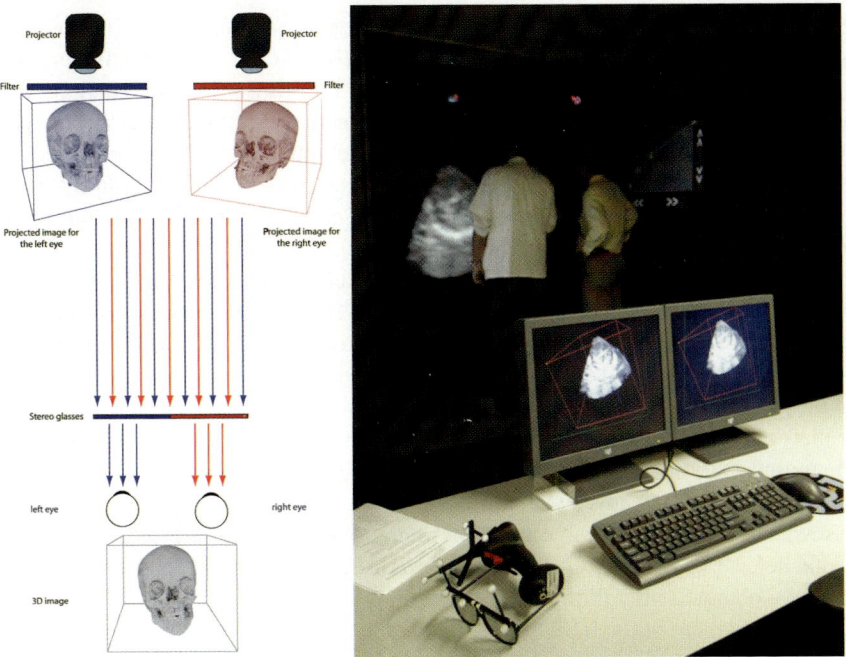

Fig. 11.1 (**a–b**) Stereoscopic images (**a**) surround clinicians, while a wireless tracking system provides the correct perspective, motion parallax and interaction in the Barco I-Space virtual reality system (**b**)

Here, two projectors per projection surface are used, with circular polarizing filters. The observers wear lightweight "sunglasses" with polarizing lenses, which ensure that each eye only sees the images from one set of projectors, resulting in depth perception (Fig. 11.1a).

The images are generated by an SGI Prism visualization system. This is a shared memory multiprocessor computer system, with eight Itanium2 processors, 12 GB of memory and eight ATI FireGL X3 graphics cards. The operating system is based on Suse Linux, with some SGI-specific extensions. We use the CAVORE volume visualization application (Koning 1999) which has been implemented using the C++ programming language and the OpenGL Volumizer, CAVElib and OpenGL APIs. CAVORE was originally developed by the author at SARA Computing and Networking Services as a general-purpose volume-rendering application.

CAVORE can be used to generate volume-rendered images of most medical imaging modalities: CT, MRI, 3D ultrasound, SPECT, PET and also confocal microscopy data. In addition to static volumes, time series (e.g., echocardiography datasets) can also be rendered and animated, providing insights into both anatomy and functionality.

In order to evaluate the potential of this system for improving the interpretation of 3D echocardiography images, we undertook two studies. The first study was aimed at evaluating mitral valve function and involved six datasets (two normal subjects and four patients with mitral valve pathology). Ten independent observers were asked to assess the datasets (acquired on a Philips Sonos 7500 echo-system) after having received ten minutes of instruction on the operation of the virtual reality application with regard to manipulating the volume and the clipping plane. The Sonos 7500 is able to save the ultrasound data directly as a Cartesian volume, and it was relatively easy to determine which private tags described the dimensions of the volume.

In this study, all ten observers could correctly assess the normal and pathological mitral valves, with an average diagnosis time of 10 min. The various substructures of the mitral valve apparatus (chordae, papillary muscles and valve leaflets) could be visualized and identified by all observers. Different pathologies, such as a prolapse of the posterior leaflet or the commissure between the superior and inferior bridging leaflets in the case of a patient with an atrioventricular septal defect, were correctly diagnosed.

The second study investigated the use of the VR system in the postoperative assessment of tricuspid valve function after the surgical closure of a ventricular septal defect (VSD) (Fig. 11.2). Twelve intraoperative epicardial echocardiographic datasets, obtained on a Philips iE33 ultrasound system during five different operations after closure of the VSD, were included in this study. The data were analyzed using both the two-dimensional (2D) images on the screen of the ultrasound system and the virtual reality system, with the latter being used to specifically assess the tricuspid valve leaflet mobility. Conversion of the data from the iE33 required the use of a research version of the Q-Lab application offered by Philips, which allowed the data to be exported in the older format used by the Sonos 7500.

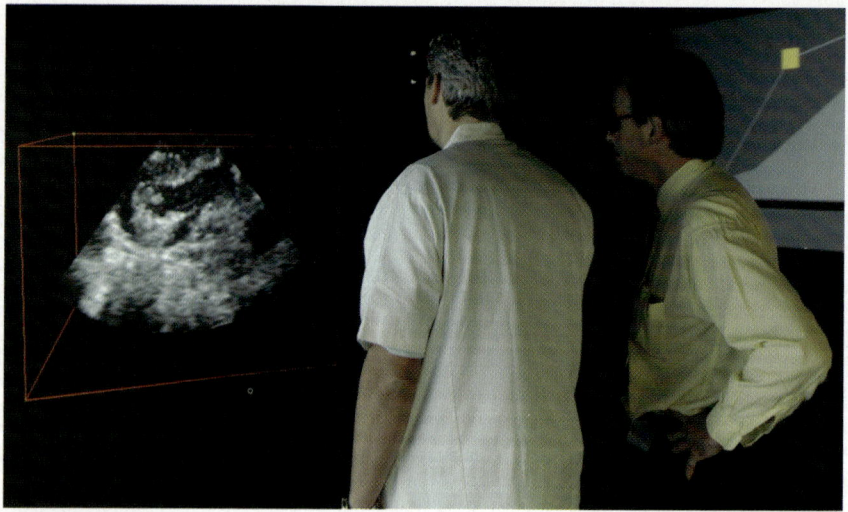

Fig. 11.2 Looking into your heart: two cardiothoracic surgeons evaluating heart valve functionality in the Erasmus MC I-Space

All of the datasets in the second study could be used for 2D as well as for 3D analysis, and the tricuspid valve could be identified in all datasets. The 2D analysis showed no triscuspid valve stenosis of regurgitation, and leaflet mobility was considered normal. However, when analyzed using the I-Space VR system, three datasets showed a restricted mobility of the septal leaflet, which was not appreciated on the 2D echocardiogram. In four datasets, the posterior leaflet and tricuspid papillary apparatus were not completely included in the recorded volume.

Another important field for the application of our virtual reality system is research into the growth and development of early pregnancies, meaning approximately the first three months of pregnancy. We have evaluated its value in obtaining insight into both normal and abnormal embryonic development using two different approaches. First, we analyzed the visibility of external morphological features of the embryos and used these to stage the embryos according to a well-known staging system for human embryos, the Carnegie Staging System (Fig. 11.3). Secondly, we used the well-documented approach of measuring different parts of the embryo, like the greatest length and the head circumference, to evaluate prenatal growth.

Stages are based on the apparent morphological state of development, and hence are not directly dependent on either chronological age or on size. In humans, the Carnegie Staging System describes approximately the first nine weeks of pregnancy. Stages are numbered from 1 to 23 and are based on internal and external physical characteristics of the embryo. At stage 23, all essential internal organ systems are present. This stage therefore represents the end of the embryonic period.

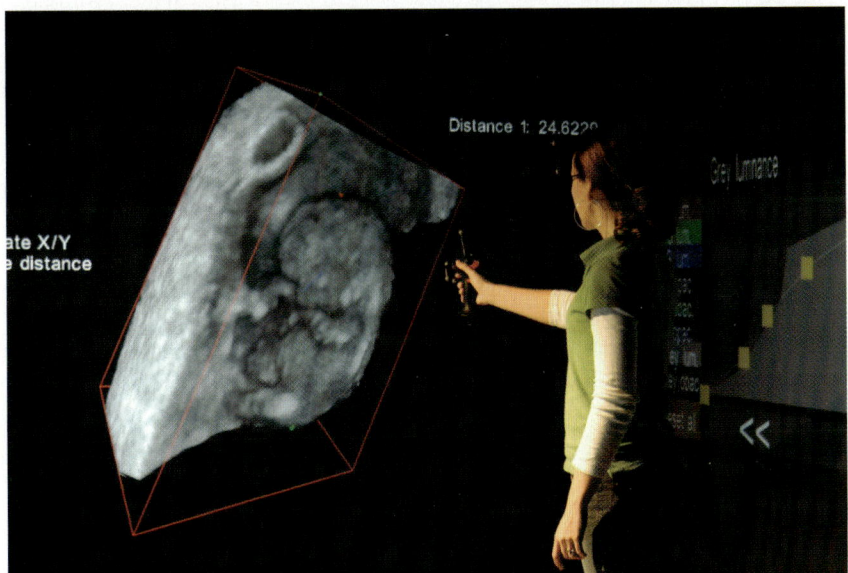

Fig. 11.3 Determining an embryo's developmental stage using a VR-based volume rendering system

To evaluate the possibility of assigning Carnegie Stages to embryos visualized with 3D ultrasound, we used datasets from 19 different IVF/ICSI pregnancies, as the exact gestational ages of these cases are known. In total we obtained 48 ultrasound scans between the sixth and tenth weeks of gestational age on a GE Voluson 730 Expert system. The 4DView application was used to convert the data to a Cartesian format. The embryos were staged according to the description of external morphological features, mainly limb development, from the Carnegie Stages described by O'Rahilly and Müller (1987). After staging, the greatest length of the embryo was measured as well. Stage, age and length were compared with the classical data on embryology described by O'Rahilly and Müller. Staging and measuring only took a couple of minutes per embryo and proved to be highly accurate.

To determine the reliability of biometry measurements in virtual reality, we performed a number of measurements that are frequently used in the embryonic and fetal period in 28 patients with gestational ages varying from 6 to 14 weeks. A comparison was made between measurements made using specialized 3D measuring software and 3D measurements using CAVORE in our BARCO I-Space. We then established the intra- and inter-observer reproducibility of these standard measurements by calculating intraclass correlation coefficients from repeated measurements by two observers. While both methods proved to be highly reliable, with ICCs of 0.96 or higher, the virtual reality approach consistently scored higher than the workstation.

In addition to these large studies, we have also evaluated several cases of congenital abnormalities in order to determine whether visualization using a VR system would benefit prenatal diagnosis and counseling in these cases. While sample size and conditions preclude statistically significant results, we can conclude from our observations that small details that stay hidden during normal examination on a 2D display are easily seen in 3D, and that the depth perception helps to distinguish between normal and abnormal anatomy.

Apart from these studies involving ultrasound data, we have also successfully visualized numerous CT and MRI datasets. These datasets range from MRI brain scans in which lesions are identified to CT scans of living donor liver transplantations in which the correct liver resection is determined. Three-dimensional blood vessel structures in general lend themselves very well to evaluations using virtual reality. Other possible applications include virtual autopsies using total body scans, virtual endoscopies, radiotherapy and surgical planning, and in general any clinical or research question involving complex 3D anatomical or functional information.

Conclusions

The successful application of VR techniques in order to answer medical imaging questions depends on a number of factors. Some of those are not related to computer science at all. For instance, clinicians tend to have very busy agendas and are not willing to adapt their daily routine to use a VR system. This means that having the system within walking distance of their office is a requirement.

This busy agenda, and probably a healthy dose of skepticism with regard to this "Star Trek-like" technology, will also mean that they are not willing spend hours or days figuring out how to use the system. Keeping the user interface simple and intuitive is one of the most important aspects of designing the system. This also applies to the data handing: having to burn a CD for every single patient quickly becomes tiresome. Interfacing with the existing PACS infrastructure will be essential.

Our first study with cardio data showed that observers with echocardiographic experience have no trouble adapting to the new system, and were able to make correct diagnoses with minimal training, i.e., in less than 10 min. Learning to effectively use a workstation-based visualization application typically requires a week-long course, and several months of learning on the job. The time required for the diagnosis was comparable to that needed when using a normal workstation, around 10 min.

The prenatal biometry study shows that measurements made with a VR system are highly accurate, and this opens the door to measurements of 3D structures that cannot be effectively measured in 2D, such as blood vessels and the umbilical cord. The second cardio and the smaller prenatal studies actually showed that the use of VR can have an advantage over using a traditional display, as the 3D images provided greater insight into anatomy and functionality than the 2D images.

References

Cruz-Neira C, Sandin DJ, Defanti T (1993) Surround-screen projection-based virtual reality: the design and implementation of the CAVE. In: Proceedings of the 20th Annual Conference on Computer Graphics and Interactive Techniques. ACM Press, New York, pp 135–142

Koning AHJ (1999) Applications of volume rendering in the CAVE. In: Engquist B, Johnsson L, Hammill M, Short F (eds) Simulation and visualization on the grid. Springer, Berlin, pp 112–121

Lorensen WE, Cline HE (1987) Marching cubes: a high resolution 3D surface construction algorithm. Comput Graphics 21(4):163–169

O'Rahilly R, Müller F (1987) Developmental stages in human embryos (Publication 637). Carnegie Institution of Washington, Washington, DC

Wilson O, Van Gelder A, Wilhelms J (1994) Direct volume rendering via 3D textures (technical report UCSC-CRL-94-19). University of California, Santa Cruz, CA

Chapter 12
Surface Modeling

Andrei L. Turinsky

Abstract Surface modeling plays a crucial role in image analysis, surgical simulations, and a myriad of biomedical applications. Surface models represent three-dimensional structures, pathological deviations from the norm and statistical variability across populations. Equipped with simulation features, they are essential tools that help researchers to visualize the progress of a disease and the likely scenarios resulting from available treatment options. We introduce the concept of surface modeling in life sciences and trace the data analysis steps that lead from volumetric imaging to models. We survey a number of methods frequently used in three-dimensional image segmentation, skeletonization, registration and warping. We describe the atlas-based approach to template fitting and surface modeling. Finally, we review some of the prominent applications of surface modeling in biomedical research, teaching and practice.

Abbreviations

2D	two-dimensional
3D	three-dimensional
CT	computed tomography
ECG	echocardiography
fMRI	functional magnetic resonance imaging
MRI	magnetic resonance imaging
μCT	micro-computed tomography

A.L. Turinsky
Hospital for Sick Children, 555 University Avenue, Toronto, Ontario, Canada M5G 1X8
e-mail: turinsky@sickkids.ca

C.W. Sensen and B. Hallgrímsson (eds.), *Advanced Imaging in Biology and Medicine*.   271
© Springer-Verlag Berlin Heidelberg 2009

Imaging: From Data to Models

Experimental Imaging

The proliferation of new technologies has created a number of ways to capture biomedical imaging data. Older techniques such as those based on X-rays are now commonplace in medicine and dentistry, even in small rural hospital units, thanks to their simplicity and affordability. Newer methods, such as magnetic resonance imaging (MRI), computed tomography (CT) and ultrasound, remain relatively expensive but are rapidly gaining acceptance as standard medical diagnostic procedures, at least in developed countries. They are also routinely used for medical research with both human patients and animal organisms. Imaging procedures are becoming more elaborate and often lead to novel breakthrough versions. For example, advances in MRI technology have been instrumental in the development of functional MRI (fMRI), which captures the real-time activity of the brain rather than its static geometry. Similarly, advances in CT have led to micro-computed tomography (μCT), which has been used extensively to visualize microscopic structures.

As a result, the amount of diverse biomedical imaging data collected by hospitals, research laboratories and commercial companies is growing rapidly and these data take a variety of forms and formats. This trend is common to virtually all of the life sciences, but is especially pronounced in imaging. For example, even a "simple" three-dimensional CT scan with a resolution of $1{,}024 \times 1{,}024 \times 1{,}024$ contains over a billion volume elements (voxels, or three-dimensional pixels), each with its own grayscale intensity value between 0 and 255. Normally, this massive amount of information would far exceed the human brain's ability to process it with any degree of efficiency. But when it is presented in a familiar visual form, a trained expert can immediately recognize salient features and patterns captured within a 3D image and use them as a basis for medical conclusions and diagnostics. Therefore, the development of efficient imaging and visualization techniques deserves special consideration due to their inherent user-friendliness and intuitiveness.

Modeling Requirements

One of the key goals of biomedical imaging efforts is to create a unified visual model of the pathology in question. As an illustration, let us consider the process of heart-disease development. Accurate three-dimensional images of the heart may be captured using echocardiography (ECG) but they do not yet constitute a model. Modeling the disease process will likely involve a number of additional aspects, as follows.

The first step in the construction of a heart model requires the creation of an anatomically correct representation of the heart, which defines the visual context for all further analysis. Second, the model should be equipped with morphing features in order to be able to represent morphological variability between patients or across specific population groups. Third, based on the observed variability, the model should assist the expert user in determining whether a specific individual's heart falls outside the norm and hence should be qualified as pathological. Fourth, the model should help the researcher to simulate and visualize the progress of a disease and the likely scenarios resulting from each of the available treatment options.

Finally, the model should be able to represent and quantify the functional aspects of the organ. In the case of the heart, this amounts to modeling the biomechanical movement of the heart walls and internal structures during the process of pumping the blood. The goal is to capture the differences between the mechanical properties of a normal heart and those of a pathological heart. The complexity of functional modeling depends substantially on the organ in question. For organs such as the heart, the lungs, the bladder or the muscles, it focuses predominantly on mechanical properties and can in principle be accomplished using the experimental imaging data alone. In contrast, accurate modeling of the functionality of the brain, the liver, the kidneys, the prostate and many other organs that are routinely investigated by clinicians is far more complex and is therefore beyond the scope of purely imaging-based methods.

Models in 2D and 3D

This chapter focuses on surface modeling derived from three-dimensional imaging data. The potential benefits of three-dimensional modeling are apparent. Experimental imaging merely captures separate snapshots of the affected anatomical tissues, whereas a true model presents a conceptual understanding of the biomedical phenomenon under study, such as the functionality of normal and abnormal hearts and their likely response to possible treatments. Individual imaging data sets can then be viewed as instances of the general model. The availability of a modeling framework greatly facilitates the formulation of new hypotheses, the simulation of predicted outcomes, and the validation of the hypotheses through new experiments.

Successful modeling of biological systems does not necessarily have to be based on three-dimensional techniques. Two-dimensional cross-sections have historically been highly effective in education as well as in clinical applications. To date, medical books and atlases with selected anatomical cross-sections have been the dominant teaching tools used in medical curricula. Some studies have shown that under certain conditions, three-dimensional models provide no discernible benefits over the traditional teaching tools, despite the substantial investment required (Garg et al. 1999). Depending on the nature of the studied phenomena, two-dimensional imaging may be able to capture much of the information about the biomedical process at only a fraction of the cost of a full 3D model.

Furthermore, 2D imaging techniques can be performed faster and with smaller latency. This property is extremely useful whenever the medical event under investigation is highly dynamic and lasts no more than a few seconds—as in the case of the heart beat, muscle contractions, coughs, soft-tissue movement under direct surgical manipulation, etc. In this case, a sequence of snapshots must be rapidly taken in order to capture the full temporal dynamics of the event, and some of the existing 3D imaging techniques may be too slow to achieve this.

Nevertheless, the ability to model biomedical processes in three dimensions adds tremendous power for researchers and educators alike. Interestingly, the authors of the previously mentioned study arrived as similar conclusions in one of their later projects (Garg et al. 2001). Many anatomical organs, including the heart, have highly complex inner structures that are best examined in 3D (Fig. 12.1).

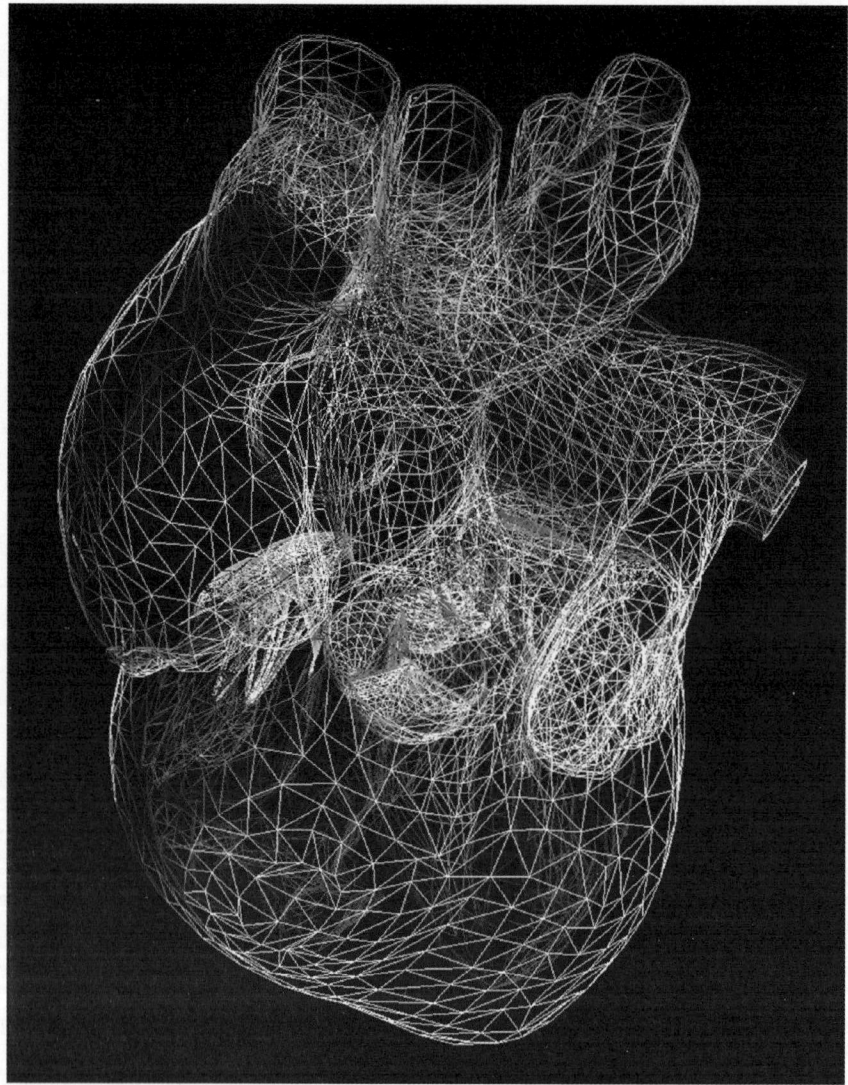

Fig. 12.1 Surface model of a heart represented by a triangulated polygonal mesh

Furthermore, some biomedical phenomena defy simplistic computational analysis based on cross-sections because of the presence of highly complex and nontrivial spatial arrangements (Cooper et al. 2003). Adding a temporal component leads to even more elaborate models that are truly four-dimensional (3D plus time), and arguably provides the closest match to the intuitive understanding of medical phenomena by human experts.

Segmentation

Data Acquisition

Accurate modeling of medical phenomena requires the original data to be "real"; that is, acquired from clinical data sets. In the case of 3D imaging, such data sets are typically accumulated in the form of *volumetric* images, which represent the entire three-dimensional volume of interest. Not surprisingly, much of the imaging research has been focused on the brain. Accurate automated analysis of brain images is extremely important to cancer research and other clinical efforts. Every year, approximately 20,000 people are newly diagnosed with malignant brain tumors in United States alone (Louis et al. 2002). Currently, most tumor diagnoses are derived through MRI imaging. Patients may first experience headaches, seizures, or focal neurological deficits (DeAngelis 2001). Despite numerous research studies to date, modeling the brain remains a very challenging undertaking, not least because a human brain has a highly complex geometry that defies simplistic image-processing algorithms. However, its importance compels the research community to continue investing considerable efforts in this area.

Volumetric data are typically fuzzy, often contain an appreciable amount of image noise, and require relatively large memory resources to maintain and process hundreds of millions or even billions of voxels. The clarity of the image, and hence the amount of data processing needed, depends on numerous environmental and technical conditions during the experiment. It also depends on the imaging modality: for example, a high-quality CT scan may be sufficient to capture even minor geometric details of skeletal structures, but is inadequate for representing soft tissues of internal organs. The use of MRI imaging is substantially more appropriate in the latter case.

Segmentation Methods

After a three-dimensional image of the volume of interest is captured and postprocessed for noise, the first task is to perform the segmentation of the image. Segmentation is a general process of partitioning an image into its constituent parts. In brain imaging, segmentation is used to identify the internal suborgans: e.g., gray matter, white matter, caudate, putamen, thalamus, and pathological development such as tumors (Grimson et al. 1998).

Automated image segmentation is an active area of research and a large number of techniques have been developed. The most common way to segment a 3D image is by grayscale thresholding, where the histogram of the grayscale is built and used to split the grayscale range into intervals (Johannsen and Bille 1982; Kittler and Illingworth 1986). The partition points may be chosen simply as valleys in the histogram, in which case each of the intervals will roughly correspond to modes in a multimodal distribution. However, simple thresholding has been found to be inadequate in a number of situations, primarily due to the inability of the grayscale alone to capture spatial neighborhood information. In brain imaging, image noise may place isolated voxels labeled as white matter into the area that clearly represents gray matter (Jain and Dubuisson 1992).

Figure 12.2 illustrates the difficulty of applying a single global threshold in a simple two-dimensional scenario. The main difficulty is the uneven lighting, which requires additional preprocessing of the image background. Similar technical difficulties occur routinely in volumetric imaging, but are greatly complicated by the

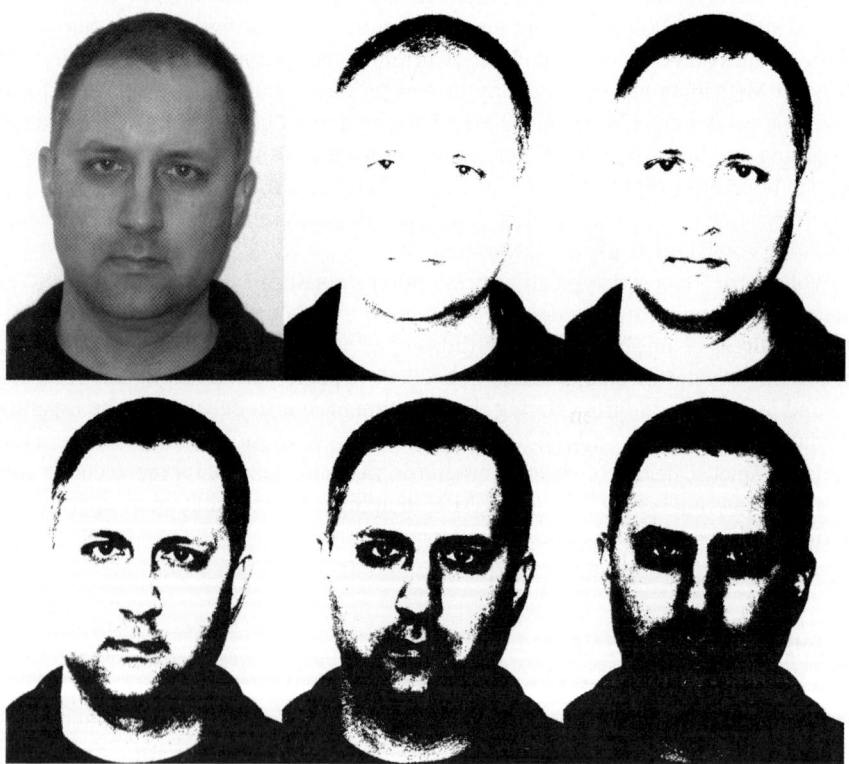

Fig. 12.2 A series of global thresholds applied to a grayscaled facial image

fact that images are three-dimensional and often require an expert to recognize flaws in the image processing output.

An effective way to achieve segmentation using global thresholds is the so-called double thresholding, whereby the grayscale range is partitioned into class intervals separated by intervals of uncertainty that require additional analysis (Oh and Lindquist 1999). Unlike single-threshold methods, double thresholding is robust to image noise. For example, in a bimodal voxel distribution in brain imaging, two classes and therefore three grayscale intervals are created: darker voxels are labeled as gray matter, lighter voxels as white matter, and medium-gray voxels are then assigned to one of the two classes based on the analysis of spatial proximity.

Another option is to use a smoothing technique such as the wavelet method of Simoncelli and Adelson (1996) to remove noisy elements before segmentation. Unlike voxel-based smoothing, which is largely local, wavelet smoothing is able to rank the contributions of image features on a range of scales, from local to image-wide. After smoothing, segmentation may be accomplished by optimal threshold selection, as described in Otsu (1979). Other possible approaches to segmentation include statistical modeling of a mixture of distributions (Delignon et al. 1997) and gradient methods of boundary detection (Cumani 1991; Lee and Cok 1991).

Data Curation

A notable factor in the area of 3D image analysis is the lag between the user's ability to accumulate large data sets and the ability to process and analyze these data. This lag is not only due to the inadequate computational resources per se; in fact, the rate of technological innovation in 3D graphics is quite substantial and new graphics resources, such as graphics processing units and other tools, are being routinely developed. Rather, the lag is mostly due to the fact that volumetric data are notoriously difficult to process in a fully automatic fashion: at present, 3D imagery is still unable to represent complex anatomy with adequate clarity, making automated segmentation rather challenging. As we have already seen, even fundamental image-processing tasks can be extremely noise-sensitive, requiring manual curation.

Manual segmentation, although a very tedious task, is considered the most reliable for MRI images. It may take several hours, and possibly days, to process a volumetric image manually. Accurate analysis of volumetric images requires substantial expertise and frequent manual intervention by trained users. Despite the availability of sophisticated algorithms, many of the low-level decisions about image denoising, thresholding, segmentation and post-processing are routinely taken with a degree of subjectivity, based on the previous experiences of particular users or on other ad hoc considerations.

This tedious process involves identifying areas of interest in each 2D image slice of the volumetric stack, segmenting them, labeling the results, and removing image noise from each slice; for example, using a free 3D Slicer software package (Navabi et al. 2003). Another possibility is a hybrid semiautomated segmentation procedure, where the initial segmentation is created automatically in 3D and then presented to the human expert for curation. If the quality of the automated segmentation is reasonably high, the time-consuming task of manual segmentation is reduced to minor corrections. At the same time, the hybrid process provides a better guarantee of the validity of the final segmentation compared to fully automated procedures.

Structural Models

Shape Modeling

Regardless of the segmentation procedure employed, the result of the segmentation effort is typically an image structure model, which constitutes a higher-level description of the image and shows how image elements relate to each other spatially. Segmentation methods may be classified into low-level and high-level techniques. Low-level techniques convert an original imaging data set into another imaging data set of essentially the same kind. A typical example is thresholding, in which a three-dimensional stack of grayscale images is converted into a three-dimensional stack of binary (black-and-white) images. The binary stack is then used for extracting image parts, surface models and other structural features.

In contrast, high-level segmentation methods start by developing a conceptual model of the data, mapping it onto the experimental imaging data set, and guiding the segmentation by examining the matching subparts. The high-level segmentation approach is illustrated by the broad family of algorithms based on direct shape modeling. These methods include snake contours (Cootes et al. 1994), dynamic front propagation (Malladi et al. 1995), probabilistic deformable models (McInerney and Terzopoulos 1996), template fitting (Lipson et al. 1990), and spectral transforms (Székely et al. 1995). Although computationally expensive, direct shape modeling methods incorporate spatial information into the analysis and are able to produce structurally consistent segments.

Topology Modeling

A popular method of creating a conceptual structural model in the form of an abstract graph is *skeletonization*. This technique extracts a representative inner layer from each of the image segments, thereby creating an image skeleton graph. The resulting skeleton may be used as an abstraction of the original image, from which further analysis can extract a number of useful measurements. In 3D, the skeleton consists of surface elements and curve elements, or may be reduced further to contain only curve elements. Skeletonization is used frequently in medical imaging, e.g., in Golland et al. (1999) and Selle et al. (2000).

Three main approaches are currently used for skeletonization: voxel peeling methods that thin the 3D image by progressively removing outer layers of voxels (Lam et al. 1992); medial axis transform methods that identify the central voxels in the image directly by approximating their distance from the image boundary (Choi et al. 1997); and Voronoi tessellation methods that partition the space into Voronoi cells and produce the skeleton by examining the boundary elements between the cells (Ogniewicz and Ilg 1992). We point out that Voronoi tessellation requires extensive skeleton pruning and should be avoided in 3D due to excessive computational cost. Alternative ways to extract image structure include run length graphs (Kropatsch and Burge 1998), line adjacency graphs (Iliescu et al. 1996), and versions of clustering for brain imaging (Zhou and Toga 1999).

We have used skeletonization in the development of a new method of medical image analysis (Cooper et al. 2003). It has been observed that the traditional structure analysis methods used for trabecular bones could not be applied to cortical bones. In our approach, 3D cortical bone images were recreated by high-resolution μCT scans, after which skeletonization was used to extract the structure of the pore canals. The resulting skeleton allowed us to perform previously unavailable measurements of the pore canal connectivity, which is one of the key factors used in the medical diagnostics of osteoporosis. To implement the skeletonization, we used a robust voxel peeling schema and also utilized some elements of the medial axis transform approach for skeleton post-processing.

Surface Models

From Volumes to Surfaces

The desire to capture all possible details of the volume of interest in a 3D image leads to massive volumetric data sets that are difficult to process computationally. On the other hand, light graph-based models obtained through skeletonization and similar methods provide a high-level overview of the data but eliminate most of the details that may be especially important to a clinical expert. Surface-based models usually achieve a good balance between the two approaches: they are typically much lighter than the original volumetric data, yet they provide a comprehensive description of the 3D images. Surfaces are usually represented using polygonal meshes that have special structural properties to reduce memory and improve performance. For example, triangulated meshes are arranged into so-called triangle strips and triangle fans (Fig. 12.3), in which neighboring triangles share vertices. As a result, adding a new triangle polygon to a triangle strip requires only one (rather than three) additional vertex to be defined, allowing for faster rendering performance by the graphics card.

3D surfaces represent the boundaries between homogeneous regions within the original volumetric image, determined by the segmentation. Surface models are built

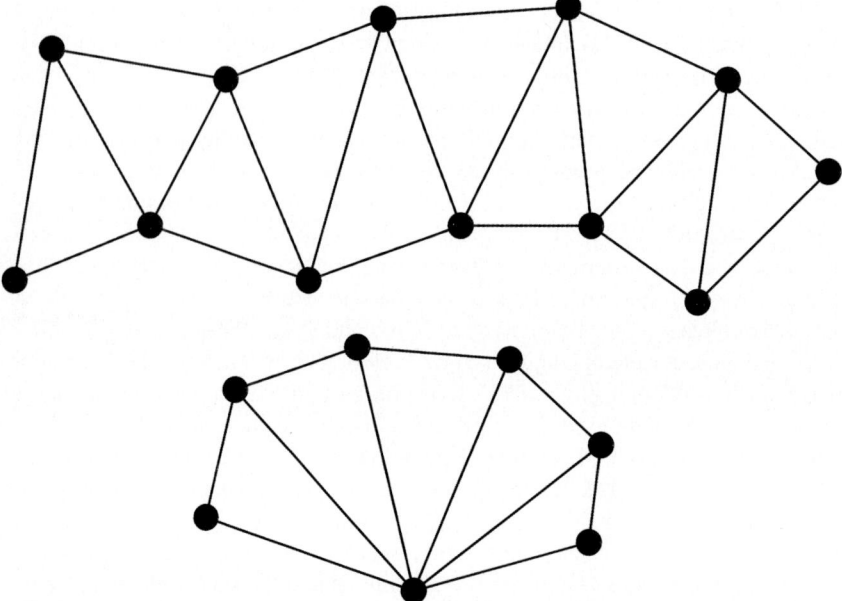

Fig. 12.3 Examples of a triangle strip (*top*) and a triangle fan (*bottom*)

using edge-detection algorithms that examine the gradient of the image intensity and locate its local extrema. Perhaps the most widely used is the marching cubes algorithm (Lorensen and Cline 1987). The resulting surfaces in effect serve as the summaries of distinct shapes and patterns observed in the original 3D image. They are often sufficient for the conceptual understanding of the image structure. Another advantage of surface-based modeling is the ability to organize the surfaces into a logical hierarchy, and use only the parts that are necessary for a particular application. For example, the high-level surface models of a skull and the brain hemispheres may suffice for a basic car-crash simulation. However, a much larger collection of surface models for the internal substructures of the brain may be available, and may be loaded interactively by the user.

Surface Quality

Manual correction may be applied after three-dimensional surface models have already been extracted automatically from the volumetric data using image-processing algorithms. In this case, the methodological limitations of the imaging technologies, as well as the impact of image noise, may be fully evident even to a non-expert observer. The obvious concern is the relative roughness of automatically extracted surfaces, which might be aesthetically unpleasant: indeed, such

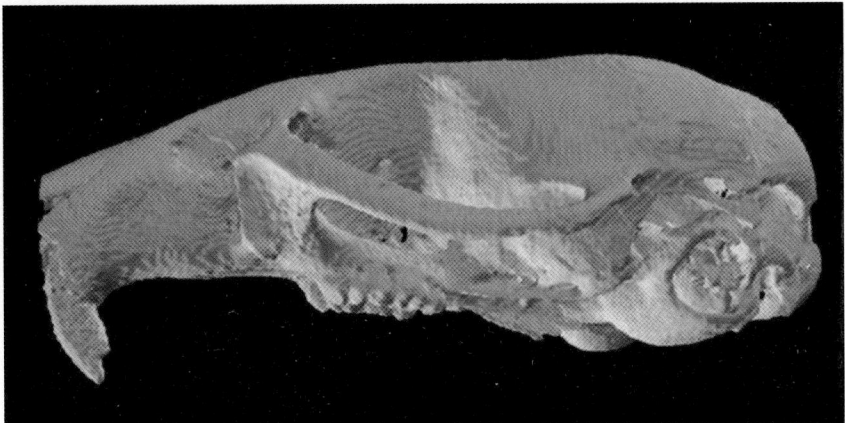

Fig. 12.4 Surface model of a mouse skull extracted from a three-dimensional CT dataset (CT data provided courtesy of Benedikt Hallgrímsson)

surfaces are rarely as smooth as the medical objects that they depict. Figure 12.4 shows an example of a mouse-skull surface model, which was extracted automatically from a CT data set using a popular Visualization ToolKit software package (http://www.vtk.org). The surface produced by the segmentation procedure is relatively rough and it is possible to distinguish individual voxels that defined the boundary of the skull. Subsequent smoothing and manipulation of surface normals may alleviate this deficiency.

A much more serious concern is the structural defects of such surfaces, which are common. A typical example is the inability of automated techniques to differentiate between anatomical structures that are logically distinct but appear in close proximity to each other in the body. In an automatically extracted surface model, adjacent structures such as neighboring vertebrae, the carpal bones of the wrist or the joints of the foot, and other bone joints, are typically merged into a single 3D object. The same effect is common with full body scans, in which automated image-segmentation techniques often join together the upper thighs of a person or merge the upper arms to the torso, unless the limbs are deliberately spread wider apart (Xi et al. 2007). Further difficulties arise if the image quality is poor due to low resolution or high noise level: in this case automated thresholding may be unable to separate even nonadjacent structures, such as neighboring ribs or the bones of hands and feet. The opposite problem arises when a 3D surface contains gaps and holes due to the inability of the surface-extraction method to detect the boundary of the 3D object. In the case of creating 3D surfaces, manual editing would require breaking, merging, smoothing, and otherwise manipulating the 3D polygonal mesh of the surface model.

Atlas Models

Structural Templates

The main drawback of naïve automated image-processing techniques is the absence of a structural template a priori. An examination of color intensity alone is unable to distinguish carpal bones, but an expert knowledge of their general shape and configuration would facilitate their extraction from the image. In a simple case, a user may identify these subparts manually to assist the segmentation. A more sophisticated procedure may involve seeding the image with the likely locations of individual wrist bones and them performing a dynamic front propagation (Malladi et al. 1995) until the contour of each bone is identified. This procedure is not only computationally expensive but also requires expert supervision to minimize the effects of image noise and to ensure that the contours are anatomically feasible.

The processing of original data into models is facilitated by using atlases. The term *atlas* may be interpreted in a number of ways and may assume drastically different implementations: there are volumetric atlases, surface-based atlases, or even atlases in the form of printed media such as medical books. In this chapter, we understand an atlas to be a collection of three-dimensional surface models that are properly labeled and have an internal structural organization. A typical example is a 3D atlas of a human anatomy in which individual surfaces are labeled with standard anatomical terms, contain predefined 3D landmark points, and are organized hierarchically into organ systems.

The atlas-based approach to image modeling is a generalization of template fitting, which is a well-established technique in image analysis (McInerney and Terzopoulos 1996; Lipson et al. 1990). A template encapsulates the knowledge known a priori about the phenomenon under study, such as an organ of interest. It therefore allows automated algorithms to model individual shapes without compromising their structural complexity. Viewed as a set of deformable surface templates, the atlas provides both higher-quality models and the parametric representation of specific organs, organ systems, or an entire body. The main technical task of atlas-based modeling is image registration between a volumetric data set and a surface-based atlas. Different sets of parameters define different surface shapes; the goal is therefore to find a set of parameters that matches the template to the shape defined by a specific volumetric data set.

Let us consider the example of creating a 3D surface model of a brain from a volumetric data set. Without prior knowledge, this task requires accurate edge detection and noise removal, which (as we saw) may be quite challenging. However, the availability of a template replaces this error-prone effort with a relatively more precise task of 3D registration and warping: the newly acquired 3D imaging data are compared to the existing template in order to identify the matching parts and patterns within the data. This requires translating, rotating and deforming the template until it fits within the volumetric image.

Registration and Morphing

Nonlinear image registration, also known as image warping, refers to finding a transformation that morphs one image into another. For example, a meaningful comparison of spatial and biochemical patterns in two different brains is difficult due to the underlying anatomical differences. It is greatly facilitated when both brains are mapped to a common reference image, within which spatiotemporal patterns of cell proliferation or tumor growth may be meaningfully compared.

A number of nonlinear registration methods exist, of which several implement a general three-step procedure. The first linear transformation aligns the two images, the next nonlinear short-range elastic body transformation matches coarse image features, followed by a nonlinear long-range transformation to match fine details. A thorough warping algorithm of this nature is presented in Christensen et al. (1997), and a more simplistic, although computationally efficient, alternative approach is described in Chen et al. (1998). Another well-established method for image registration is mutual information alignment, which only provides a rigid body alignment and is therefore much less computationally expensive than elastic warping. Despite its deficiencies, it may be the method of choice for aligning highly similar images, such as two MRI images of the same patient (Viola and Wells 1995). A commonly used technique is also the so-called thin-plate spline, which achieves nonlinear registration based on matching pairs of landmarks in two images (Fig. 12.5) while minimizing the bending energy of the transformation (Bookstein 1989).

Nonlinear image warping is a computationally intensive task and may take considerable time to perfect. It often helps to pre-compute the parameters of warping between each new 3D brain image and the brain atlas: this allows the querying system to display warping results in real time. Warping also allows users to create smooth spatiotemporal visual sequences of brain images. The users are then able to

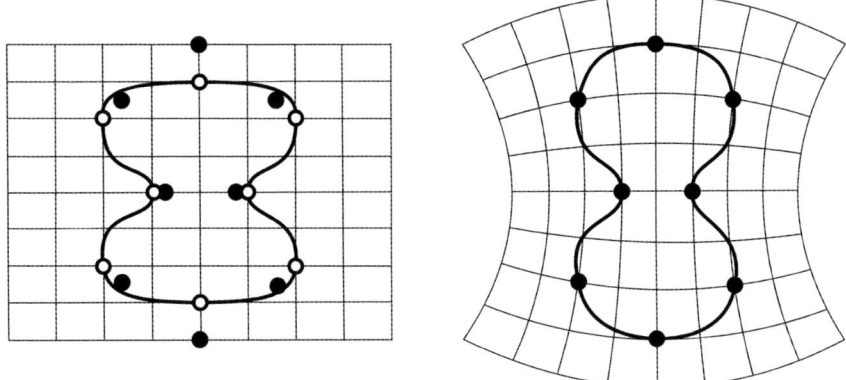

Fig. 12.5 Image registration based on pairwise matching of template landmarks (*white circles*) and the landmarks of the desired shape (*black circles*), showing the resulting nonlinear warping of the space

inspect the progress of a particular condition over time. Some useful insights regarding the time-varying visualization of lesion growth in brain imagery are provided in Tory et al. (2001).

Application of Surface Modeling

Parametric Knowledge Base

An important benefit of the template-based modeling of 3D images is the ability to provide a parametric description of a family of shapes, rather than a single shape. This contrasts favorably with storing clinically acquired volumetric data sets, which typically represent individual volumes but lack generalization features. Atlas templates allow generalization seamlessly, both by design and through repeated application. Indeed, the morphing of a surface template model into an individual shape is often encoded by pairwise matching of the predefined landmarks. In the process of registering the template to multiple volumetric data sets, the available range of movement of these landmarks in numerous individuals is elucidated. The resulting sets of morphing parameters and landmarks provide a quantitative description of a whole population.

The true power of the surface modeling comes from the fact that it need not be limited only to the experimental data. Instead, by varying the combinations of parameters, one may simulate shapes and patterns that have not been observed in experiments. This feature is often used for the modeling of population averages: for example, a popular method used to create an average brain model is to construct transformations of a brain template into several individual brain data sets, then compute the average of these transformations and apply it to the template. The resulting 3D shape retains the minor details from the model and yet presents the average of all individual brains (without necessarily matching any of them). In contrast, techniques based on averaging of the grayscale intensities of the original volumetric images, rather than averaging the transformation parameters, would typically smooth out all of the small but important features of the images, such as the locations of sulci, etc.

Medical Applications

Surface modeling based on parametric descriptors allows researchers and educators to use the model for simulations. This is a crucial feature, which gives the ability to simulate not only average behaviors but also various abnormalities, outlier shapes, and pathological developments. The modeling of an abnormality should typically involve matching the template model to individual scans obtained from either the human patients or animal subjects with known pathologies. From a technical standpoint, the focus of such modeling is on the elucidation of the parameter sets that correspond to disease, and their deviation from the parameter sets that represent normality. A useful application of this approach is the creation of probabilistic disease atlases, which show the variation in spatiotemporal patterns across the population (Toga and Thompson 2001; Toga et al. 2001).

Surface modeling is used extensively for surgical simulation, education and training, but is only slowly expanding into the area of real-time clinical procedures (Liu et al. 2003; Satava 2008). The more technologically advanced types of simulations are based on the use of a standardized human body, in the form of computerized human-body atlases, virtual patient models and similar mechanisms. The limitations of these new technologies are still so profound that animal models or human cadavers currently remain indispensable in practice. An important aspect that should be captured by a surface model is the organ's response to a direct physical contact or to an invasive procedure, such as incision, insertion of a medical scope, and a host of other possible clinical manipulations. Figure 12.6 demonstrates the need to accurately model deformation, breakage and force propagation, not only in the model directly affected by the manipulation but also in surrounding surfaces. Accurate modeling of force propagation across the surface mesh, surface tension and resistance remains a challenging computational problem, especially for real-time applications.

This strategy is required, in particular, when developing haptic capabilities for surgical simulations. Haptic feedback is essential for advanced surgical simulators: it prevents the trainee from mistakenly damaging surrounding tissues while performing surgical tasks. Simulations in a haptic-enabled environment involve such surface-modeling tasks as the interaction between different tissue types and surgical instruments, the movement of surrounding soft tissues, the manipulation of endoscopes, and a number of other complex tasks. Recent advances and requirements in haptic technology are surveyed in van der Putten et al. (2008).

Despite a large body of research in surface modeling, the application of surface modeling techniques in the clinical setting has generally been rather limited. Ideally, imaging data collected from a patient should be processed into an anatomically correct, interactive and quantifiable model of the patient's body in real time. Such a model would be a major step forward in medical diagnostics as well as in surgical procedures. A crucial goal is to be able to guide a robotic surgical tool along the optimal path in a patient's body, using the imaging data collected from a patient to model the body. Doing so in an automated fashion would make it possible in the future to deliver sophisticated medical procedures in places where highly trained surgeons are unavailable: rural areas, battlefields, developing countries, etc. However, real-time surgical navigation is still an emerging field that needs to overcome numerous technical challenges. Its application remains limited, and in most cases requires direct supervision by a trained human expert.

Nevertheless, computerized surface simulations are able to provide many features that cannot be obtained using animal models, cadavers or live patients. Simulations are interactive, reproducible and quantifiable. Using a simulator, a trainee

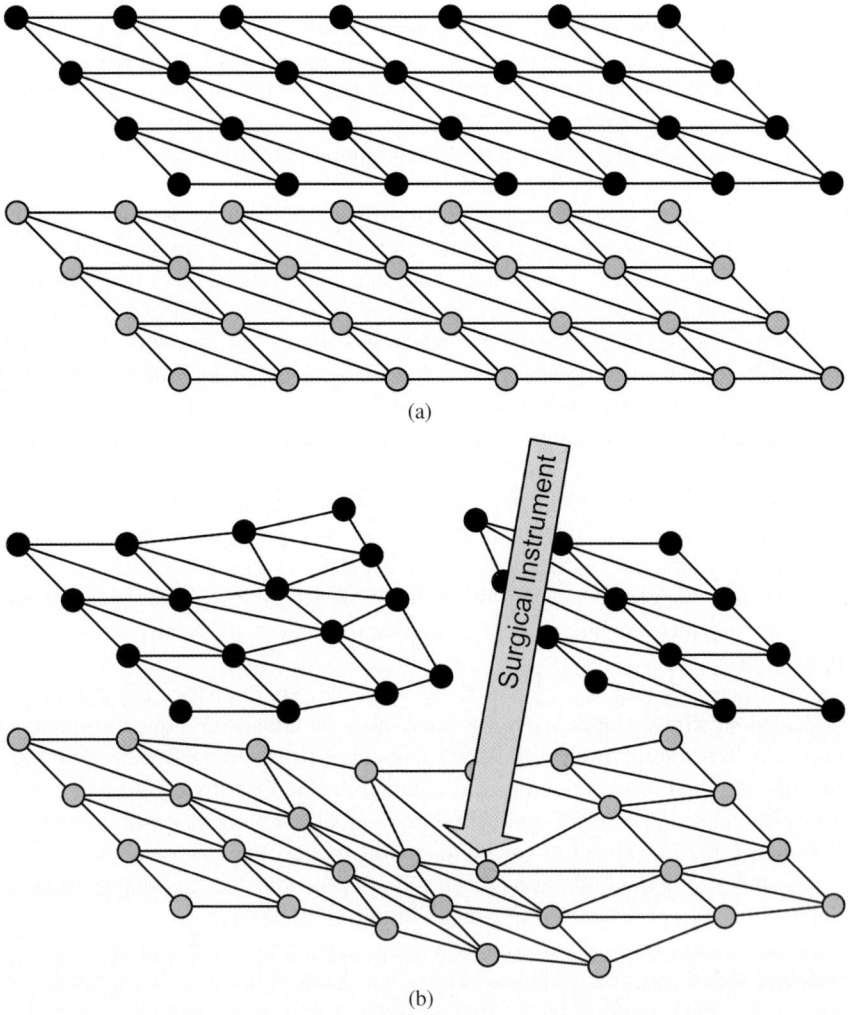

Fig. 12.6 (**a–b**) Simulated force-propagation behavior in two neighboring surface models (**a**) before manipulation and (**b**) after a surgical tool has cut through the first surface and applied force to the second surface

is able to perfect the same procedure after many repetitions and learn from previously made mistakes. Learning scenarios may be built into the simulation with ease. One may reasonably hope that in the future, most (if not all) cadavers will be replaced by their virtual equivalents. Perhaps the biggest advantage of using virtual simulators is the ability to make predictive models. This is especially evident in light of the fact that it is increasingly difficult to get new medical treatment through the approval process defined by governmental and regulatory bodies. With the rising cost of medical research and the difficulty involved in validating new medical procedures, the ability to simulate and predict the effects of a medical intervention in advance has become increasingly important to practitioners. It remains one of the key enabling factors for the development of novel applications of virtual-reality surface modeling in medicine.

Acknowledgements The author would like to thank Prof. Christoph W. Sensen for his helpful suggestions and guidance, and Prof. Benedikt Hallgrímsson for the mouse-skull CT data set used in one of the figures, as well as for many fruitful discussions.

References

Bookstein FL (1989) Principal warps: thin-plate splines and the decomposition of deformations. IEEE Trans Pattern Anal Machine Intell 11:567–585

Chen M, Kanade T, Rowley HA et al. (1998) Anomaly detection through registration. In: Proc IEEE Conf Computer Vision and Pattern Recognition, 23–25 June 1998, pp 304–310

Choi HI, Choi SW, Moon HP (1997) Mathematical theory of medial axis transform. Pacific J Math 181:57–88

Christensen G, Joshi S, Miller M (1997) Volumetric transformation of brain anatomy. IEEE Trans Med Imag 16:864–877

Cooper DML, Turinsky AL, Sensen CW et al. (2003) Quantitative 3D analysis of the canal network in cortical bone by micro-computed tomography. Anat Rec 274:169–179

Cootes T, Hill A, Taylor C et al. (1994) Use of active shape models for locating structures in medical images. Image Vision Comput 12:355–366

Cumani A (1991) Edge detection in multispectral images. CVGIP: Graph Models Image Process 53:40–51

DeAngelis LM (2001) Brain tumors. N Engl J Med 344:114–123

Delignon Y, Marzouki A, Pieczynski W (1997) Estimation of generalized mixture and its application in image segmentation. IEEE Trans Image Process 6:1364–1375

Garg A, Norman G, Spero L et al. (1999) Learning anatomy: do new computer models improve spatial understanding? Med Teacher 21:519–522

Garg A, Norman G, Sperotable L (2001) How medical students learn spatial anatomy. Lancet 357:363–364

Golland P, Grimson W, Kikinis R (1999) Statistical shape analysis using fixed topology skeletons: Corpus callosum study. Proc 16th Int. Conf Inf Process Med Imaging (LNCS 1613), Visegrád, Hungary, 28 June–2 July 1999, pp 382–387

Grimson W, Ettinger G, Kapur T et al. (1998) Utilizing segmented MRI data in image-guided surgery. Intl J Pattern Recogn Artif Intell 11:1367–1397

Iliescu S, Shinghal R, Teo RYM (1996) Proposed heuristic procedures to preprocess character patterns using line adjacency graphs. Pattern Recogn 29:951–976

Jain AJ, Dubuisson M (1992) Segmentation of X-ray and C-scan images of fiber reinforced composite materials. Pattern Recogn J 25:257–270

Johannsen G, Bille J (1982) A threshold selection method using information measures. In: Proc 6th Int Conf Pattern Recognition (ICPR'82), Munich, Germany, 19–22 Oct, pp. 140–143

Kittler J, Illingworth J (1986) Minimum error thresholding. Pattern Recogn 19:41–47

Kropatsch WG, Burge M (1998) Minimizing the topological structure of line images. Proc Joint IAPR Intl Workshops on Advances in Pattern Recognition (LNCS 1451), Sydney, Australia, 11–13 Aug. 1998, pp 149–158

Lam L, Lee S-W, Suen CY (1992) Thinning methodologies—a comprehensive survey. IEEE Trans Pattern Anal Machine Intell 14:869–885

Lee HC, Cok DR (1991) Detecting boundaries in vector field. IEEE Trans Acoustic Speech Signal Process 39:1181–1194

Lipson P, Yuille AL, O'Keefe D et al. (1990) Deformable templates for feature extraction from medical images. Proc 1st Eur Conf Computer Vision (LNCS 427), Antibes, France, 23–27 April 1990, pp 413–417

Liu A, Tendick F, Cleary K, Kaufmann C (2003) A survey of surgical simulation: applications, technology, and education. Presence: Teleoperators Virtual Env 12:599–614

Lorensen WE, Cline HE (1987) Marching cubes: A high resolution 3D surface construction algorithm. Proc 14th Intl Conf Computer Graphics and Interactive Techniques, Anaheim, CA, 27–31 July 1987, pp 163–169

Louis DN, Pomeroy SL, Cairncross G (2002) Focus on central nervous system neoplasia. Cancer Cell 1:125–128

Malladi R, Sethian JA, Vemuri BC (1995) Shape modeling with front propagation: A level set approach. IEEE Trans PAMI 17:158–175

McInerney T, Terzopoulos D (1996) Deformable models in medical images analysis: a survey. Med Image Anal 1:91–108

Nabavi A, Gering DT, Kacher DF et al. (2003) Surgical navigation in the open MRI. Acta Neurochir Suppl 85:121–125

Ogniewicz R, Ilg M (1992) Voronoi skeletons: theory and applications. In: Proc IEEE Conf Computer Vision and Pattern Recognition, Champaign, IL, 15–18 June 1992, pp 63–69

Oh W, Lindquist WB (1999) Image thresholding by indicator kriging. IEEE Trans Pattern Anal Mach Intell 21:590–602

Otsu N (1979) A threshold selection method from gray level histograms. IEEE Trans Syst Man Cybernet 9:62–66

Satava RM (2008) Historical review of surgical simulation—a personal perspective. World J Surg 32:141–148

Selle D, Spindler W, Preim B et al. (2000) Mathematical methods in medical imaging: analysis of vascular structures for liver surgery planning. In: Engquist B, Schmid W (eds.) Mathematics unlimited—2001 and beyond. Springer, Berlin

Simoncelli E, Adelson E (1996) Noise removal via Bayesian wavelet coring. In: Proc IEEE Int Conf Image Processing, Lausanne, Switzerland, 16–19 Sept. 1996, 1:379–382

Székely G, Kelemen A, Brechbühler C et al. (1995) Segmentation of 3D objects from MRI volume data using constrained elastic deformations of flexible Fourier surface models. In: Proc 1st Int Conf Computer Vision, Virtual Reality and Robots in Medicine (LNCS 905), Nice, France, 3–6 April 1995, pp 493–505

Toga AW, Thompson PM (2001) Maps of the brain. Anat Rec 265:37–53

Toga AW, Thompson PM, Mega MS et al. (2001) Probabilistic approaches for atlasing normal and disease-specific brain variability. Anat Embryol 204:267–282

Tory M, Möller T, Atkins MS (2001) Visualization of time-varying MRI data for MS lesion analysis. In: Proc SPIE Int Symp Medical Imaging (LNCS 4319), San Diego, CA, USA, 18–20 February 2001, pp 590–598

van der Putten EP, Goossens RH, Jakimowicz JJ, Dankelman J (2008) Haptics in minimally invasive surgery—a review. Minim Invasive Ther Allied Technol 17:3–16

Viola P, Wells WM (1995) Alignment by maximization of mutual information. In: Proc 5th Int Conf Computer Vision, Boston, MA, 23–23 June 1995, 1:16–23

Xi P, Lee WS, Shu C (2007) Analysis of segmented human body scans. In: Proc Graphics Interface (ACM Intl Conf Proc Series 234), Montreal, Canada, 28–30 May 2007, pp 19–26

Zhou Y, Toga AW (1999) Efficient skeletonization of volumetric objects. IEEE Trans Visual Comput Graphics 5:196–209

Chapter 13
CAVEman, An Object-Oriented Model of the Human Body

Christoph W. Sensen(✉) and Jung Soh

Abstract We have created the world's first object-oriented atlas of the male adult human anatomy. The system, which is called CAVEman, is a surface model based on the ontology contained in *Terminologia Anatomica*. CAVEman, which was built using Java 3D^{TM}, contains more than 3,000 independent objects, including all bones and muscles, as well as internal organs and skin. We have begun to equip the CAVEman model with functionality that allows the mapping of volumetric information onto the atlas and subsequent morphing, the combination of gene expression data with the anatomy, as well as the overlaying of pharmacological information onto the model. The surface model of the human anatomy opens up many new avenues for research and development. Soon, the system will be used for patient consultation, the mining of gene expression and metabolic data sets and probably surgical planning.

Introduction

The last 20 years have seen the introduction of many multidimensional diagnostic imaging technologies, such as magnetic resonance imaging (MRI), computed tomography (CT) and four-dimensional (4D) ultrasound imaging. Today, these techniques are routinely used for the imaging of patients. Typically, the information generated in the imaging procedure is of a volumetric nature and so is not immediately accessible to an automated and computerized interpretation. Therefore specialists have to inspect the information and derive the patient's diagnosis, making it impossible to automatically mine large imaging data collections.

C.W. Sensen
University of Calgary, Faculty of Medicine, Department of Biochemistry and Molecular Biology, Sun Center of Excellence for Visual Genomics, 3330 Hospital Drive NW, Calgary, Alberta, Canada T2N 4N1, e-mail: csensen@ucalgary.ca

C.W. Sensen and B. Hallgrímsson (eds.), *Advanced Imaging in Biology and Medicine*.
© Springer-Verlag Berlin Heidelberg 2009

With the completion of the human genome sequence and the progress made in studying metabolic disorders, complex genetic diseases and developmental patterns, large data sets containing spatiotemporal information have recently become available. Aside from specialists, who usually only study a few aspects of these datasets, the data produced have been largely unutilized, as the general scientific community does not have access to tools that provide access to such data in a meaningful way to non-experts.

To overcome these challenges, new tools need to be developed which can be used to automatically index volumetric information as well as provide a template onto which spatiotemporal information can be mapped. We have developed the world's first object-oriented computer model of the human anatomy. This surface model complements, rather than replaces, current approaches to the understanding of volumetric data.

Creation of the Human Body Atlas

Our initial plan for the construction of the human body atlas focused on utilizing the Visible Human dataset distributed by the U.S. National Library of Medicine (http://www.nlm.nih.gov/research/visible/visible_human.html). It became evident early on that automated feature extraction using only 8-bit information, which contains only 256 shades of gray, would lead to very low resolution images (Turinsky and Sensen 2006). We therefore decided to team up with Kasterstener Inc. of Red Deer, Alberta, to get a team of artists to construct the atlas de novo using information derived from the Visible Human Project, anatomy books, expert input and the dissection of human cadavers.

The ontology (i.e., controlled vocabulary) underlying the atlas was derived from *Terminologia Anatomica* (TA), an anatomical taxonomy that is the de facto international standard for the ontology of the human body (FCAT 1998). Over a time period of six years, this effort resulted in the creation of a complete atlas of the male adult anatomy. Initially, the anatomical objects were created using the 3D painting software Maya (Lanier 2006) and 3ds Max (Ross and Bousquet 2006), which are the software most commonly used to create graphic objects intended for use in virtual reality environments. Subsequently, all objects generated from the software tools were converted into the Java 3D format (.j3f files). This format allows flexible indexing of the objects and enables partial loading/unloading of organs when an organ (such as the heart or the kidney) is modeled as multiple geometry groups organized in a hierarchical fashion. Table 13.1 provides an overview of the objects comprising the human body atlas of CAVEman and their level of detail.

The Java 3D format provides a hierarchical structure for the body model, which essentially recreates the hierarchical structure that is represented in TA, where organs are hierarchically organized into separate systems. This allowed us to follow a top-down approach for the generation of the models. The resulting set of

Table 13.1 Human body model

Systemic anatomy as defined in TA	Number of modeled organs	Total number of polygons
Skeletal system and articular system	214	385,107
Muscles; muscular system	1,365	853,199
Alimentary system	26	712,628
Respiratory system	68	272,830
Urinary system	32	262,920
Genital systems: male genital system	24	81,182
Endocrine glands	7	218,243
Cardiovascular system	154	1,490,485
Lymphoid system	29	127,563
Nervous system	338	1,673,322
Sense organs	46	265,681
The integument	32	84,408
All systems	2,335	6,427,568

The model consists of 2,335 modeled organs at full resolution (many organs are modeled at lower resolutions as well), which are organized into a hierarchy of 12 systems, based on TA

objects can be complemented by an additional set of objects at any time. Figure 13.1 show the hierarchical structure of the objects in the human body atlas underlying the CAVEman model.

Once in their final Java 3D format, the objects can be displayed on any Java 3D-capable computational platform. This includes personal computers and workstations as well as computers equipped with stereo displays, all the way up to virtual reality rooms. No changes to the software itself are necessary to move from one display system to another. The only change which has to occur is a revision of Java 3D's configuration file for the *ConfiguredUniverse* package, which is necessary to transmit the display parameters, including the parameters for the tracking device, to the system. The display/viewing environment and the software are therefore effectively separated. This essentially allows CAVEman to be used independent of the computing platform, as long as Java 3D is supported. As Java is often the first programming language learned by computer science students, even beginners are able to add to the software environment. This approach is therefore in stark contrast to a proprietary system, which often requires highly trained specialists for any modification.

JABIRU: A Middleware Layer for the Manipulation of the Human Body Atlas

Due to the hierarchical nature of the system, we were able to build a middleware layer, called JABIRU (Stromer et al. 2005), which can be used to manipulate the individual objects (e.g., a tooth or a muscle) independently, as well as the entire model simultaneously. Our goal was to build a system in which objects could be

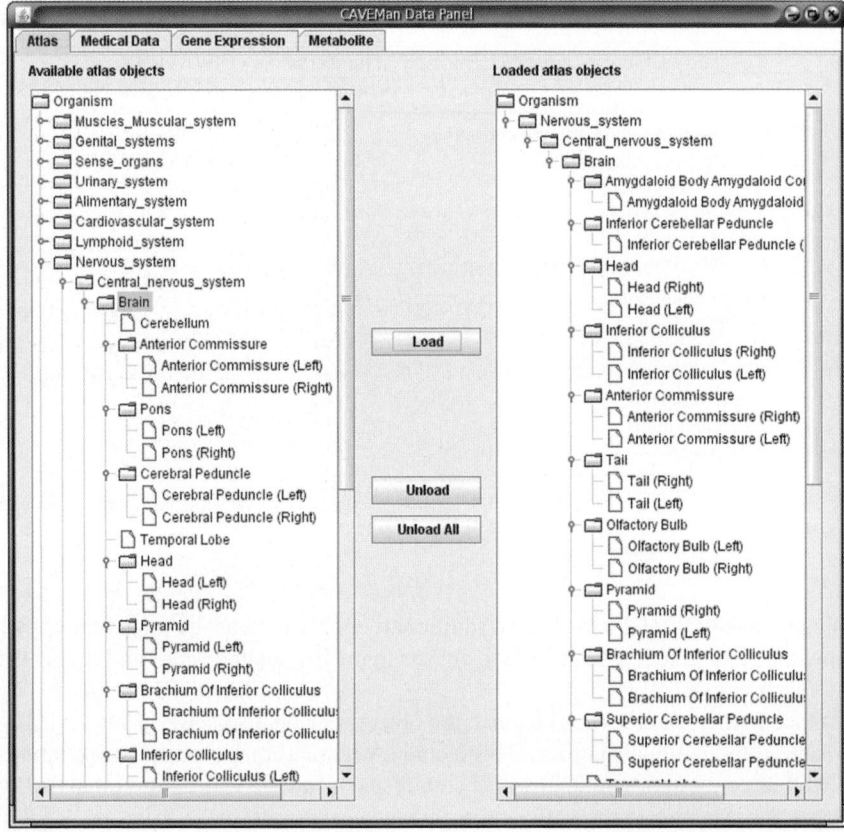

Fig. 13.1 CAVEman hierarchy in Java 3D. Upon selecting an atlas, a tree of systems and organs becomes available. Individual organ objects as well as body systems from the atlas can be loaded and unloaded. In the snapshot, the "Brain" object (consisting of several objects) from "Central nervous system" has been loaded into CAVEman

manipulated in natural ways. Using the CAVE automated virtual reality environment at the University of Calgary, users can perform all manipulations using a magic (i.e., virtual) wand with six degrees of freedom. Such manipulations on objects include picking, dragging, dropping, zooming, translating, rotating in an orbit, rotating around an axis of object, and adjusting surface transparency. On single-screen 2D visualization platforms, much of the wand experience can be recreated by replacing the wand operations with the combination of a mouse and a keyboard.

Simple operations include the rotation of objects, a zoom function and load/save functionality. More complex operations allow objects to be dissected and objects to be made transparent in order to be able to explore the underlying structures.

The JABIRU middleware is now being expanded to include the ability to move objects within their natural constraints (such as the movement of the lower jaw or an arm), the provision of a user interface for CAVEman's morphing system, and

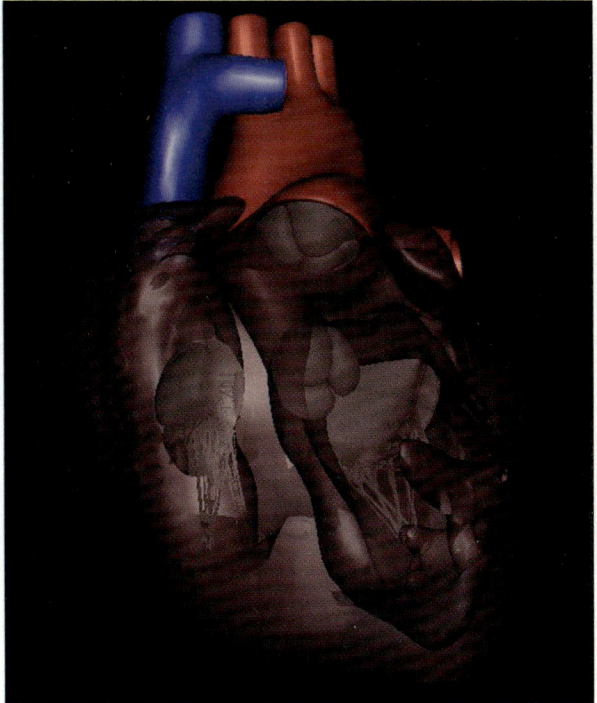

Fig. 13.2 Dissected CAVEman object. A heart object is shown with an outermost wall taken apart to reveal the structure inside the heart wall

assistance in the automated segmentation and indexing of volumetric information. Figure 13.2 shows a dissected subcomponent of the CAVEman system.

Integration of Volumetric Information with the Human Body Atlas

One of the major goals of the CAVEman project is the merging of patient data with standardized anatomy. This includes data derived from MRI, CT or 4D ultrasound. In all instances, the raw data is comprised of a stack of image files that contain images of slices of the body. The images are usually in TIFF format, and are often embedded in a DICOM (Bidgood et al. 1997) format file.

To map this information successfully onto the surface model, the information contained in the scan images needs to be indexed. The first step in an automated indexing process is the identification of landmarks within the volumetric information. The most obvious landmarks in volumetric scans are usually the bones and the outermost shape of the scanned object. Without the assistance of an atlas, a

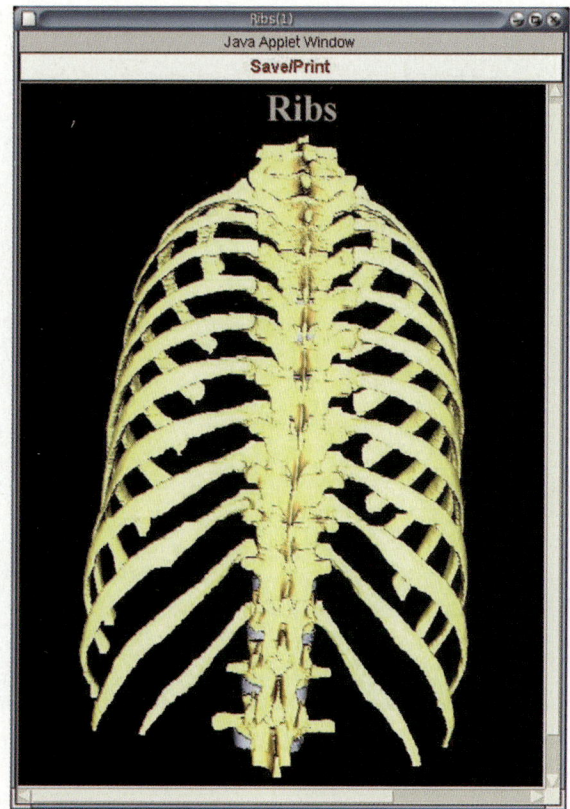

Fig. 13.3 Direct extraction of volumetric information. A result of the automatic extraction of bone information from the Visible Human dataset, viewed from AnatQuest Viewer (http://anatquest.nlm.nih.gov/AnatQuest/ViewerApplet/aqrendered.html) provided by the US National Library of Medicine

maximalist approach needs to be taken to outline the bone material within the object, which usually leads to an overprediction (see Fig. 13.3).

Using the information contained in the human body atlas, we are now able to recreate the shape of the bone material from a minimalist information set, which just uses a few shades of gray to define bone matter within the volumetric information. In addition, contaminations of the volumetric information outside the general bone area (i.e., regions of the image that share the same gray shades as the bones but are not part of the bones) can be automatically excluded during the reconstruction, as the overall shape of all bones is known. Figure 13.4 shows volumetric information derived from an MRI scan fitted with the anatomical atlas.

It is our goal to develop the indexing system to the point where regular cases of individual patient data can be measured automatically. This would of course exclude special cases, such as the location of the heart on the wrong side or the existence of only one or three kidneys in a patient. These unusual cases would be presented to a

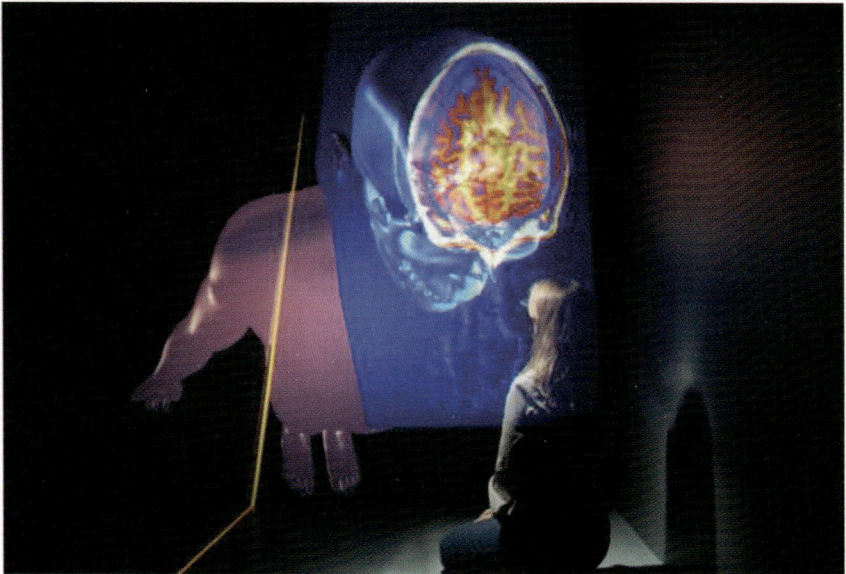

Fig. 13.4 Combination of MRI scan and atlas. A 3D MRI image of an individual's brain (*shown in blue and orange*) is embedded in the context of the 3D skull object of the atlas

specialist, who can subsequently store the new knowledge as part of the CAVEman system.

If successful, this kind of a system might be able to circumvent the need for the assessment of volumetric information by experts to a large extent, and it could allow the mining of large existing collections of volumetric information.

Merging Gene Expression Information and the Human Body Atlas

The ontology underlying the anatomical atlas can be used to combine spatiotemporal information such as gene expression patterns or metabolomic information with the human atlas. Ideally, the spatial information of data collected in gene expression experiments would be organized according to the information contained in *Terminologia Anatomica*. This is unfortunately almost never the case, as molecular biologists are usually not trained anatomists. Therefore, the inclusion of data from public repositories for gene expression patterns such as Gene Expression Omnibus (Edgar et al. 2002) requires a conversion system that transforms the anatomical information contained in the data sets into the TA format. Once this is achieved, the data could theoretically be mapped directly onto the anatomical structures, but the complexity of the information requires additional filtering of the highlights in order to assist the user in the discovery process.

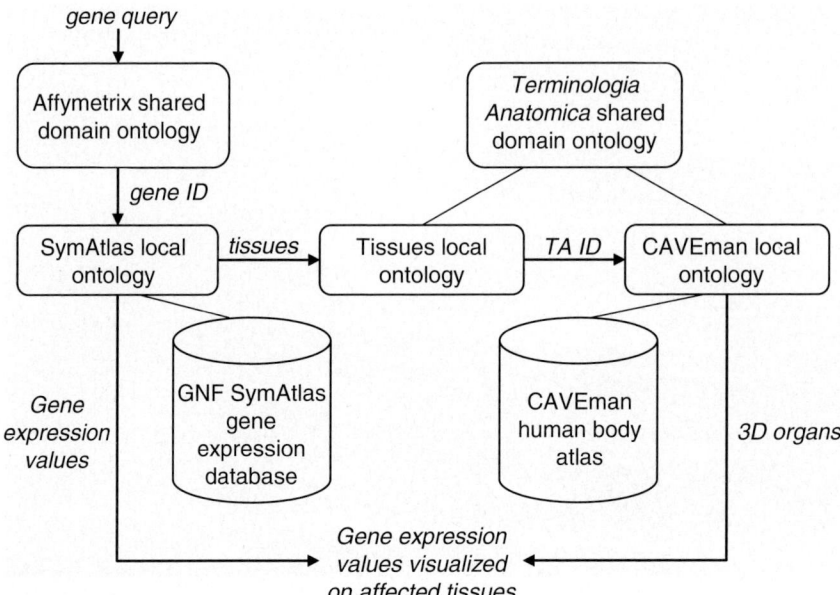

Fig. 13.5 CAVEman gene expression workflow. Ontologies are used as the basis for mapping gene expression information onto the relevant parts of the human body atlas

We have integrated the Java-based TIGR Multi-Experiment Viewer (TIGR MeV) (Saeed et al. 2003) package with the CAVEman system in order to facilitate the processing of the gene expression data and the filtering of the relevant information, which can subsequently be displayed in context with the anatomical model. Figure 13.5 shows the workflow through the system, using gene expression patters from public repositories. When the user loads the GNF SymAtlas (Su et al. 2004) data set and queries a gene using the Affymetrix ID from the CAVEman system, the symatlas.owl local ontology is searched for matching Affymetrix terms. Matching terms are retrieved along with the associated information, including the tissue in which the expression was measured and the spot intensities. From this information, the mapping of the tissue names to TA identifiers is accomplished through tissues.owl. Finally, the caveman.owl ontology provides the mapping of TA identifiers to specific 3D objects of the CAVEman model.

Initially, the system allows us to display the knowledge in four dimensions (space and time). Figure 13.6 shows an example of the mapping of gene expression data onto the CAVEman atlas. Gene expression datasets are initially analyzed using TIGR MeV integrated into CAVEman. Gene expression patterns of interest are visualized as animated color maps directly on the chosen organs, representing the expression levels of a gene or a cluster of genes in those organs.

We are now implementing a system that automatically provides the users with the experimental "highlights." Once completed, this system will provide the user with an interface for the rapid discovery of new knowledge within large gene expression experiments.

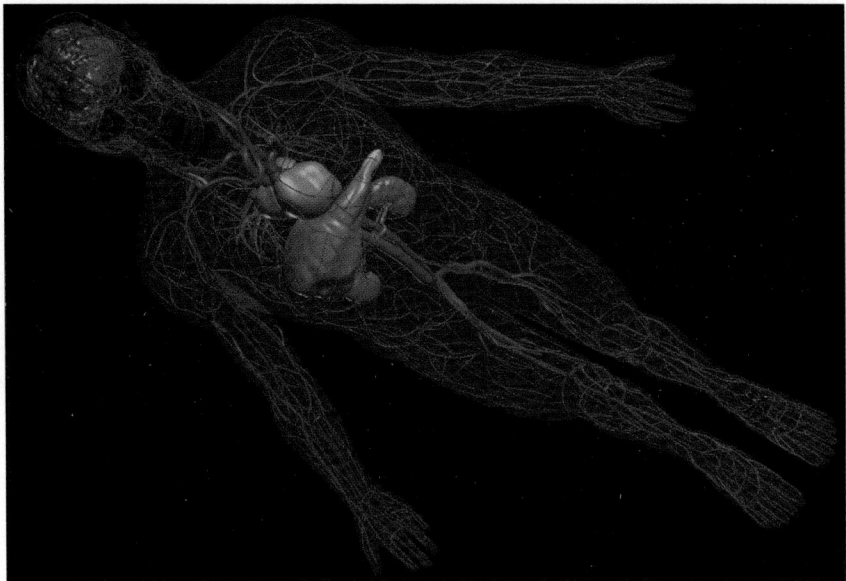

Fig. 13.6 CAVEman and gene expression pattern. For a set of genes clustered by TIGR MeV, gene expression patterns in brain, heart, liver, and kidneys are mapped onto the partially loaded human body atlas, with colors of organs visually representing levels of gene expression in those organs

Merging Metabolomic Information and the Human Body Atlas

Similar to gene expression experiments, large datasets are generated when new drugs are being developed. We have begun to map information derived from metabolomic studies, especially information regarding pharmacokinetic events, onto the anatomical atlas. Spatiotemporal mapping of such information in their corresponding anatomical contexts will bring many benefits, such as effective teaching of pharmacokinetic principles and real-time exploration of hypotheses on drug absorption, distribution, metabolism and excretion (ADME).

Unlike gene expression data, which are often obtained from public repositories, much of the metabolomic information is not yet deposited in machine-readable form and has to be extracted manually from the literature. We used the data generated by the Human Metabolome Project (Wishart et al. 2007) through an ADME modeling Java program, called ADME Modeler. Figure 13.7 shows the workflow of mapping an ADME model of a drug onto the CAVEman human body atlas. The user first provides ADME Modeler with requested parameters, such as the chemical compounds involved, the route of administration (e.g., oral or intravenous), various rate constants of the ADME process, time intervals, and the dosage. ADME Modeler then calculates the temporal distribution, concentration, and location of each drug or metabolite. For example, ADME modeling of acetylsalicylic acid (i.e., Aspirin) results in pharmacokinetic data that essentially tells us that the

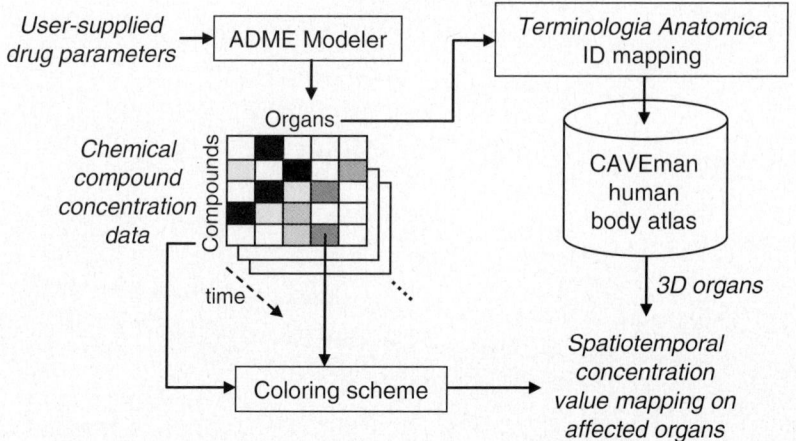

Fig. 13.7 Metabolomic information mapping onto the human body atlas. ADME modeling results are mapped both spatially and temporally onto the affected parts of the human body atlas, with colors representing different drug metabolites and their concentrations

acetylsalicylic acid is first converted to salicylic acid within the alimentary and cardiovascular systems, which is then either directly excreted or further converted to one of the four known derivatives before eventual excretion through the urinary system. This data contains numeric concentration values of chemical compounds (i.e., drugs or drug metabolites) across both spatial (distribution across different body systems at a specific moment in time) and temporal (conversion into different derivatives over time) domains.

As for the visual mapping results, the visualization of metabolomic data onto the CAVEman model is quite similar to the gene expression case, except that it changes over time to express the time-dependent change in compound concentrations across different parts of the human body. Figure 13.8 shows an example of the mapping of pharmacokinetic information for acetylsalicylic acid onto the CAVEman model objects of the stomach, small intestine, arterial structures, brain and kidneys. Different metabolites are assigned different colors and their concentrations are represented by different shades of the corresponding color. Similar to the gene expression pattern studies, we are planning to move from the visualization of known facts to the mining of large and complex data sets.

Outlook

We have finally reached the point in time where the computer "knows" about human anatomy. Using the CAVEman model, we are able to load for instance a kidney and display it as such. This is unprecedented and opens up the possibility of developing a large number of applications.

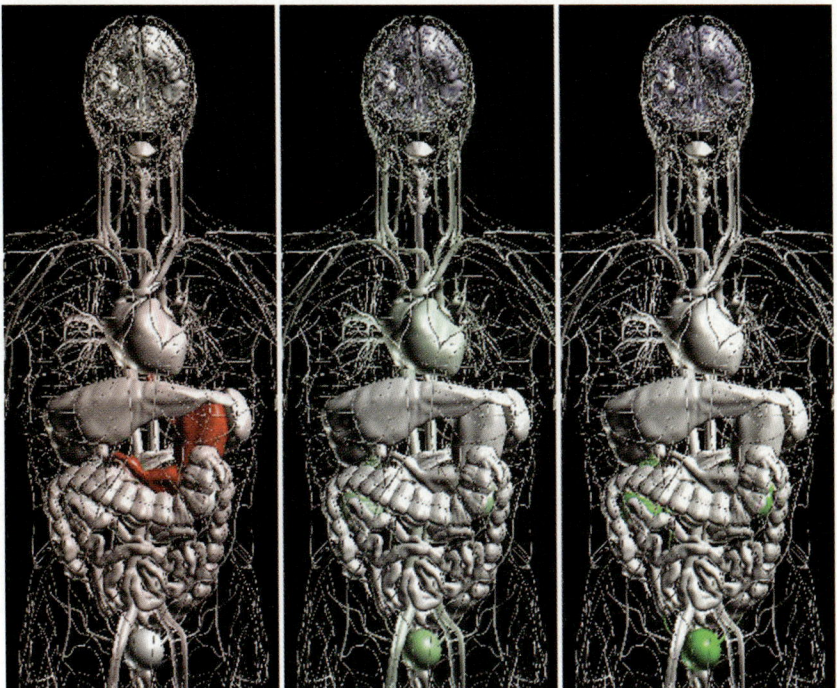

Fig. 13.8 CAVEman and metabolomic information. Visualization of parts of the human body atlas at three different time instants of the ADME process of acetylsalicylic acid: initial absorption into the alimentary system (*left*); distribution and metabolism through the cardiovascular system (*middle*); and excretion of by-products through the urinary system (*right*)

Certainly, the ability to merge volumetric data and a standardized anatomical atlas will soon lead to applications that allow patient data to be displayed using low-cost stereo displays. This will lead to the development of patient consultation tools, which will finally help patients understand the extent of their disease using imaging rather than verbal descriptions, which many laypeople are unable to understand.

In the same vein, the model can be used to teach medical students about anatomy, biochemistry and other phenomena around the human body. Once the modeling component is fully developed, the model can be used to complement the usual cadaver work during medical studies. One example of how this could be useful is the modeling of rare diseases for which no specimen exists in the local collection at the medical school.

In the not-too-distant future, we also expect to expand the model to include tactile feedback (via a haptics system) and audio feedback. Such a system could be used for surgical planning and for training surgeons in new procedures, as well as for certification.

The gene expression and metabolomics toolkits will be useful for mining existing data as well as providing assistance to new studies. For example, many experiments have been conducted that were only analyzed to characterize the behavior of a few genes, but gene chips representing the entire human genome were employed to do this. Re-mining such data using automated tools will identify the functions of genes for which no function is currently known, and unravel metabolomic networks automatically.

Many other applications for CAVEman might exist, such as ergonomic studies, machine development, or even the study of one's golfing performance, but these would need to be discussed in another context.

Acknowledgments This work has been supported by the Genome Canada through Genome Alberta; the National Research Council of Canada's Industrial Research Assistance Program; the Alberta Science and Research Authority; Western Economic Diversification; the Government of Canada, and the Government of Alberta through the Western Economic Partnership Agreement; the iCORE/Sun Microsystems Industrial Research Chair program; the Alberta Network for Proteomics Innovation; and the Canada Foundation for Innovation.

References

Bidgood WD, Horii SC, Prior FD, Van Syckle DE (1997) Understanding and using DICOM, the data interchange standard for biomedical imaging. J Am Med Inform Assoc 4:199–212

Edgar R, Domrachev M, Lash AE (2002) Gene expression omnibus: NCBI gene expression and hybridization array data repository. Nucl Acids Res 30(1):207–210

FCAT (1998) Terminologia anatomica: international anatomical terminology. Thieme, Stuttgart

Lanier L (2006) Advanced Maya texturing and lighting. Sybex Publications, Hoboken, NJ

Ross AF, Bousquet M (2006) Harnessing 3ds Max 8. Autodesk, San Rafael

Saeed AI, Sharov V, White J, Li J, Liang W, Bhagabati N, Braisted J, Klapa M, Currier T, Thiagarajan M, Sturn A, Snuffin M, Rezantsev A, Popov D, Ryltsov A, Kostukovich E, Borisovsky I, Liu Z, Vinsavich A, Trush V, Quackenbush J (2003) TM4: a free, open-source system for microarray data management and analysis. Biotechniques 34(2):374–378

Stromer JN, Quon GT, Gordon PMK, Turinsky AL, Sensen CW (2005) Jabiru: harnessing Java 3D behaviors for device and display portability. IEEE Comput Graphics Appl 25:70–80

Su AI, Wiltshire T, Batalov S, Lapp H, Ching KA, Block D, Zhang J, Soden R, Hayakawa M, Kreiman G, Cooke MP, Walker JR, Hogenesch JB (2004) A gene atlas of the mouse and human protein-encoding transcriptomes. Proc Natl Acad Sci USA 101:6062–6067

Turinsky AL, Sensen CW (2006) On the way to building an integrated computational environment for the study of developmental diseases. Int J Nanomed 1(1):89–96

Wishart DS, Tzur D, Knox C, Eisner R, Guo AC, Young N, Cheng D, Jewell K, Arndt D, Sawhney S, Fung C, Nikolai L, Lewis M, Coutouly MA, Forsythe I, Tang P, Shrivastava S, Jeroncic K, Stothard P, Amegbey G, Block D, Hau DD, Wagner J, Miniaci J, Clements M, Gebremedhin M, Guo N, Zhang Y, Duggan GE, Macinnis GD, Weljie AM, Dowlatabadi R, Bamforth F, Clive D, Greiner R, Li L, Marrie T, Sykes BD, Vogel HJ, Querengesser L (2007) HMDB: the human metabolome database. Nucl Acids Res 35:D521–D526

Chapter 14
Image-Based Finite Element Analysis

Steven K. Boyd

Abstract Finite element (FE) analysis is a nondestructive simulation tool that can estimate mechanical properties of biomaterials when combined with 3D imaging modalities such as micro-computed tomography. This chapter will review state-of-the-art FE methods that use micro-CT to generate subject-specific models for application to large cohorts of experimental animal studies, and most recently for patient studies. Methods used to automatically generate FE meshes, and recent developments that improve the accuracy of these meshes, as well as advances in the acquisition of material properties for FE modeling and the incorporation of constitutive material properties into models will be discussed. The application of this technology for in vivo micro-CT is particularly exciting because it provides a method to noninvasively estimate strength, and this can provide valuable information for monitoring disease progress and treatment efficacy. The field is still in the early stages, and there are significant opportunities to advance this unique combination of imaging and modeling technologies to provide new insights into strength-related issues.

Introduction

Micro-computed tomography (micro-CT) is an ideal tool to provide detailed 3D architecture data, and these data can be assessed quantitatively using standard morphological approaches. It provides a unique opportunity to closely monitor disease processes and treatment effects, particularly when applied in vivo. The effect of an anti-osteoporosis drug, for example, may be to preferentially target the cortical bone vs. the trabecular compartments, and this can be directly observed through

S.K. Boyd
Department of Mechanical and Manufacturing Engineering, Schulich School of Engineering, University of Calgary, 2500 University Drive, NW, Calgary, Alberta, Canada T2N 1N4, e-mail: skboyd@ucalgary.ca

C.W. Sensen and B. Hallgrímsson (eds.), *Advanced Imaging in Biology and Medicine.*   301
© Springer-Verlag Berlin Heidelberg 2009

the 3D measurements provided by micro-CT. In disease processes, monitoring the reduction in cancellous connectivity may cause irreversible damage to the architecture, and this knowledge is critical for understanding when to treat. The standard quantitative analysis methods available include a wide spectrum of possibilities to describe the architecture; however, an important aspect that cannot be measured using morphological tools alone are the implications for mechanical behavior. Understanding the morphological changes that occur to bone, for example, in the context of osteoporosis is valuable, but it is important to remember that bone *strength* that is the "bottom line" from the patient's perspective. A fragility fracture can have an enormous impact on quality of life, so the critical clinical concern is the assessment of fracture risk, and the measurement of bone strength is an important contributor affecting that risk.

In traditional engineering fields, mechanical properties are often determined by performing experimental tests on representative samples. For example, the design of a bridge would include the assessment and characterization of the mechanical properties of the construction material (e.g., steel). In the clinical field, however, this is not practical. Although bone biopsies can be extracted and mechanical tests can yield useful information from that sample, it is an invasive and uncomfortable procedure and the results from the biopsy may not represent the rest of the body (Eckstein et al. 2007). Furthermore, follow-up measurements are extremely difficult, if not impossible, to perform. The essential problem is that mechanical testing is an invasive and destructive approach, and it is for this reason that computational methods of estimating mechanical properties have received considerable interest.

The finite element (FE) method is a computational approach where the mechanical properties of a structure can be determined and/or simulated. When combined with micro-CT, where detailed geometric information can be provided noninvasively and nondestructively, it offers a unique approach to characterizing the mechanics without tissue damage. The basic inputs for a finite element model include geometry, material properties, and boundary conditions describing the loading configuration. Because the approach is nondestructive, it is possible to perform many tests on the same bone under different loading conditions and with different material characteristics. Furthermore, through the use of micro-CT, it is possible to automatically generate finite element models so that patient-specific estimations of bone strength are possible.

In the following sections of this chapter, aspects related to the principles and application of image-based finite element modeling will be described in detail, including descriptions of both established and leading-edge applications. Before discussing issues related specifically to finite element modeling, it is important that some basic terminology and definitions related to the mechanics of materials are defined.

Mechanics of Materials Primer

There are many scales at which biological materials can be considered, ranging from the whole bone structure to the fine details of the cellular matrix. In the context of micro-CT measurements, it is possible to measure whole bones as well as detailed trabecular architecture, and even to approach the cellular level at extremely high resolutions (via synchrotron radiation micro-CT). At the whole bone level, we define the mechanical properties as the *structural* properties. Sometimes these are referred to as *apparent* or *continuum* properties when considering a representative tissue sample; both the structural and apparent properties refer to how the corresponding structure as a whole behaves. At the trabecular scale, we refer to the *tissue* properties. The tissue properties describe the bone matrix that comprises the bone. The tissue properties may vary within an individual trabecula or cortical wall, and may vary throughout the whole bone. It is the distribution of those tissue properties *and* the organization of the trabeculae that affect the structural properties of bone. An analogy that is often used is the Eiffel Tower: the *tissue* properties are akin to the mechanical characteristics of the steel in the beams, and the *structural* properties depend on how the beams are interconnected to provide strength.

The determination of *tissue* and *structural* properties is commonly performed by experimental mechanical testing. One commonly performed test is called the axial test. Here, a specimen is fixed at two ends and incrementally stretched while the load is recorded, often until failure. The test can be performed in tension or compression, and the relation between the displacement of the specimen and the resulting force provides important information about its mechanical behavior. When such a test is applied to a whole bone, the *structural* properties are determined. Similarly, but at a different scale, *tissue* properties can be determined by preparing very small specimens representing either trabeculae or small pieces of cortical bone; however, these tests are challenging and less frequently performed (Fig. 14.1). As will be discussed later, there are other methods that can be used to indirectly determine tissue properties. Although not discussed here, there are many important considerations when performing mechanical testing that need to be addressed to provide high-quality experimental data (Keaveny and Hayes 1993; Turner and Burr 1993).

Characterization of Mechanical Properties

Generally, once force–displacement data have been collected from a mechanical test, the force is normalized by the cross-sectional area of the specimen to provide stress data (σ, N mm^{-2} or Pa), and the displacement is normalized by the change from the specimens original length to give strain data (ε, dimensionless, or percent). The so-called stress–strain plot for a given test result provides important mechanical information (Fig. 14.2). When interpreting the stress–strain results, the test indicates a linear region where the stress and strain are linearly proportional; the slope of this region defines the stiffness of the bone. The modulus of elasticity (E) represents

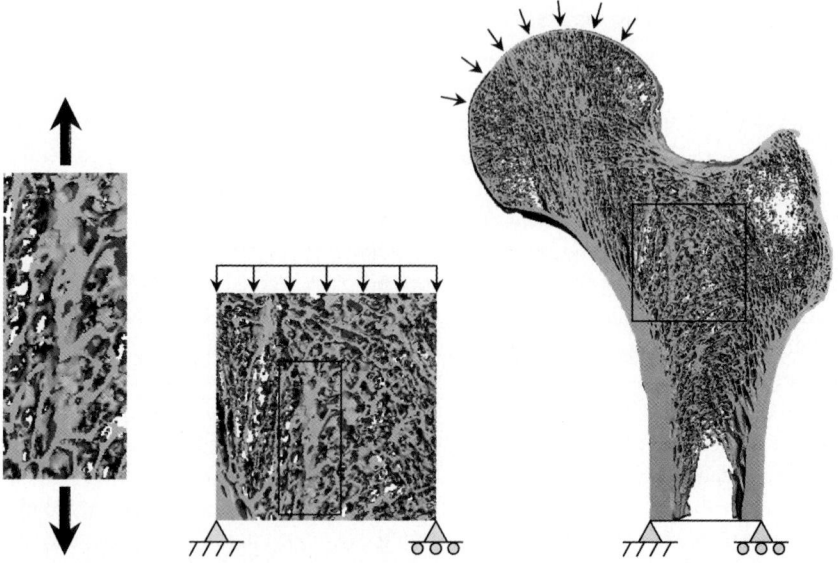

Fig. 14.1 A human femur (4.1 mm slice) scanned by micro-CT (82 μm nominal isotropic resolution). The loading configuration on the whole femur (*right*) illustrates a test to characterize the structural properties; a subvolume (*middle*) illustrates a test to determine the apparent properties of the cancellous bone; and a subvolume illustrating the concept of a single trabecular test (*right*) indicates the characterization of tissue properties. The subvolumes (*left*; *middle*) were taken from the regions indicated by the *rectangles*

the linear relation between the stress and strain, and for a test of an intact specimen it represents the apparent (structural) modulus (E_a). The stiffness of the whole bone depends on the trabecular and cortical organization, and the properties of those tissues. A structure loaded within the linear range will return to its original length upon loading. The linearity of the relation between stress and strain falters at the point defined as the yield stress (σ_Y), and at this point sufficient damage has been caused to the structure that upon removing the load it will not return back to its original length. If loading continues beyond the yield stress, eventually the structure will begin to fail until it cannot sustain any more load, and this point is called the ultimate stress (σ_u). The ultimate stress reflects the failure of a structure—the point at which it is no longer useful for providing support. Finally, with continued loading, catastrophic failure will eventually occur. Generally, the day-to-day physiological loading of bone is such that it is within the bottom-most region of the stress–strain curve (i.e., 0.1%) (Whiting and Zernicke 1998).

A mechanical test provides important information about the structural stiffness, the yield strength and the ultimate strength. In the context of finite element analysis, mechanical tests can perform two valuable functions. The results of the tests can be used to determine fundamental constitutive properties that are important inputs for FE analyses, and experimental tests are often used to validate finite element models. Often, the validation of finite element models is performed with cadaver studies

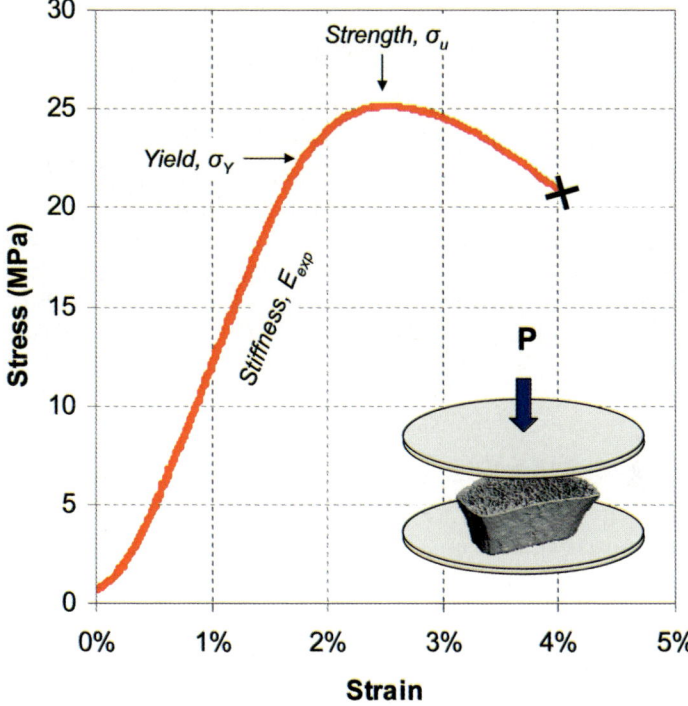

Fig. 14.2 A typical stress–strain curve is shown for an axial compression test (*inset*). In a compression testing device, displacements are applied and reaction loads are measured. The displacements and loads are normalized to strain and stress, and are plotted here to illustrate the linear region of the curve, the yield and ultimate stress points, as well as the failure of the specimen (marked by an "X")

where, after predicting the mechanical properties from a finite element model, it is validated through the destructive testing of the same bones experimentally. The important mechanical characteristics to consider are the apparent modulus, the yield stress and the ultimate stress, and these terms will be used throughout this chapter. It should be noted that although we defined these terms here using an experimental test as an example, a finite element model (that is well validated) could simulate the same procedure used to estimate mechanical properties.

Tissue Properties

The measurement of tissue-level properties is a challenging task and has necessitated the development of specialized techniques. Traditional testing methods have been employed on individual trabeculae, including tensile testing (Ryan and Williams 1989) and three- or four-point bending experiments (e.g., Choi and

Goldstein 1992). For assessments of cortical bone, small test specimens can be carefully machined for tensile tests. The challenge of tissue-level tests is that trabeculae are difficult to define and isolate (and only possible if the structure is rod-like), and both trabecular and cortical tissue test preparations are extremely time-consuming. Alternatives to traditional tension/compression/bending test configurations include methods such as nano-indentation using a micro-indenter probe (Zysset et al. 1999), acoustic microscopy (Turner et at. 1999), or a combination of tensile testing and acoustic microscopy (Rho et al. 1993). The moduli of elasticity for trabecular and cortical bone span a wide range of values. Its value for trabecular bone has been found to range from 1.3 to 18 GPa, and cortical bone from 15 to 26 GPa (Rho et al. 1993; Van Rietbergen et al. 1995; Zysset et al. 1999; Cowin 2001; Bayraktar et al. 2004; Hoffler et al. 2005).

Finite Element Modeling

Analytical approaches can be used to determine the relation between loads applied to a structure and their deformations and associated stress, but these are only practical for simple geometries and loading conditions (e.g., three-point bending of a cylindrical beam, axial compression of a cube). The finite element method breaks down a complex problem into small pieces (i.e., the elements) that can be easily solved individually, and compiles them into a large system of equations to solve problems with complex geometries, material characteristics and loading conditions. The generation of a finite element model requires three basic components: geometric information, boundary conditions defining the loads and constraints, and a constitutive model (i.e., material properties).

Meshing and Boundary Conditions

Geometric data is represented in the FE model as a mesh comprising individual elements (e.g., hexahedrons, tetrahedrons) and the nodes that define them. The geometric information obtained from 3D imaging data is useful for providing the basis for an FE mesh. Imaging can be performed by magnetic resonance imaging (MRI) (Van Rietbergen et al. 2002), but here we focus on CT because it can provide good spatial data, particularly for bone, as well as density information that can augment FE models. The current standard for generating FE meshes from 3D image data is to use the voxel-conversion technique (Müller and Rüegsegger 1995), because this approach has the advantage of being fast, efficient and fully automated. Thus, patient- or specimen-specific FE meshes can be generated with ease, and large investigations can include the determination of mechanical properties in entire cohorts.

In contrast to the voxel-conversion method, the classical approach to meshing is to use the extracted surface data to generate a custom mesh with elements that are optimally configured to represent shape with the minimum number of elements necessary. It is an advantage from a computational perspective to reduce the number of elements, because quite often FE models become too large to solve with available computer hardware—this is particularly a problem with image-based meshes. The main limitation of the classic approach is that it requires user input and is therefore impractical for creating patient-specific models for large cohorts. Furthermore, when based on micro-CT data, it is often not possible to reduce the mesh size without discarding valuable geometric information.

Thus, the motivation to employ the voxel-conversion approach is that image data is directly converted into a FE mesh with no user interaction, and as well as geometric information, the mesh can also include density information to represent local tissue moduli (Homminga et al. 2001). Meshes are typically generated using hexahedron elements; sometimes tetrahedron elements are used, but they typically require more elements than hexahedron meshes to represent the same geometry.

Mesh Smoothing

A disadvantage of the voxel-conversion approach is that the natural curves of bone are typically represented by hexahedron (cube-shaped) finite elements, and this can lead to stress raisers and inaccuracies in the local stresses and strains (Guldberg et al. 1998; Ulrich et al. 1998; Niebur et al. 1999). To address this problem, automated mesh smoothing approaches have been developed to reduce surface geometric discontinuities (Fig. 14.3) (Charras and Guldberg 2000; Boyd and Müller 2006); however, state-of-the-art FE solvers for solving extremely large FE models that are typical of image-based FE are limited to meshes with hexahedron elements where each element has the same shape. Smoothing a mesh requires adjusting the element shapes to generate a smooth surface, and this makes it impossible to solve these meshes using large-scale solvers (although small models can be solved using standard commercial FE packages).

Despite the limitation of the hexahedron model due to its jagged edges, these models perform extremely well in terms of representing the apparent-level mechanical properties, and this has been demonstrated through validation studies (Charras and Guldberg 2000; Su et al. 2006; MacNeil and Boyd 2008). The errors are resolution-dependent, and as a general rule it is suggested that reasonable results can be obtained when the element size represents a quarter of the mean thickness (i.e., trabeculae or cortex) (Niebur et al. 1999).

Tissue vs. Continuum Models

Clinical CT and micro-CT both provide voxel data for the generation of FE models, but the image resolution of the particular modality dictates the level of detail that the FE model can represent. For example, clinical CT does not resolve individual

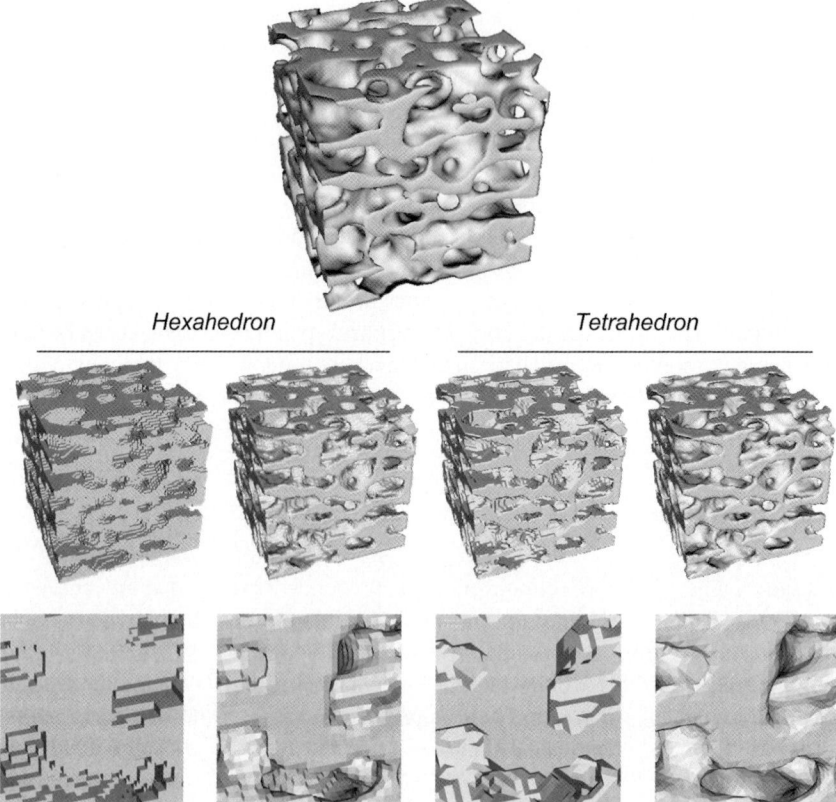

Fig. 14.3 A micro-CT scan of a cube of cancellous bone (*top*) can be meshed automatically into either hexahedron or tetrahedron elements (*middle row*). Both mesh types result in surface discontinuities, as shown in the close-ups (*bottom row*), but smoothing algorithms can be applied to the mesh to reduce these artifacts. The unsmoothed and smoothed (five smoothing iterations) are illustrated side-by-side for the hexahedron and tetrahedron meshes

trabecular elements, and so the cancellous region is represented as a continuum rather than a detailed architectural structure. Consequently, the material properties assigned to the cancellous region in the FE mesh represent the mechanical characteristics at the continuum level rather than the tissue level. Typically, the conversion from density to modulus of elasticity is based on power laws (Hodgkinson and Currey 1992). Therefore, FE models based on clinical CT are generally referred to as *continuum* models (Cody et al. 1999; Crawford et al. 2003). While the voxel-conversion approach is still used to generate the FE models, the relation between CT density information and elastic modulus is at the continuum level.

Alternatively, FE models based on micro-CT data yield meshes with highly detailed trabecular architecture, and these models are generally referred to as *large-scale* FE models (or sometimes micro-FE). "Large-scale" refers to the fact that the models consist of upwards of millions of elements, and this is in stark contrast

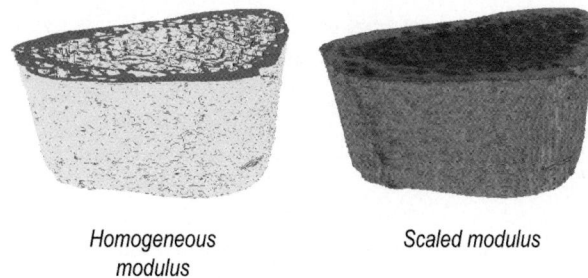

Homogeneous
modulus   Scaled modulus

Fig. 14.4 A human distal radius can be meshed using a homogeneous tissue modulus (*left*), or the density information contained in the micro-CT data can be converted into a local tissue modulus (*right*) using an exponential conversion method

to most continuum models, where mesh sizes are on the order of hundreds of thousands. In large-scale FE models, density information can be incorporated as local moduli of elasticity to refine the model representation, and this is done using exponential functions (Fig. 14.4) (Homminga et al. 2001; Bourne and van der Meulen 2004). However, although both micro-CT and clinical CT can incorporate density-to-modulus conversions, it is important to point out that they are different relations due to the difference in scale between the large-scale and continuum-level FE models. In large-scale FE models, the moduli represent the tissue properties, but in continuum models the moduli represent averaged properties (i.e., apparent properties), because each voxel contains mixtures of bone tissue and marrow. The relation between tissue modulus and density is dependent on image resolution, represented as partial volume effects and imaging artifacts such as beam hardening.

Boundary Conditions

The boundary conditions in an FE mesh define the constraints, displacements and loads applied to the structure. These are defined in the FE mesh and are applied to the nodes that make up the individual elements. In axial compression, a displacement is typically applied to the nodes on one surface of the mesh (i.e., the top surface of an excised cylinder of bone), and the nodes on the bottom surface are constrained to move within a plane (or are completely fixed). An example of a schematic representation of uniaxial compression testing boundary conditions is provided in Fig. 14.2 (middle).

There is a great deal of flexibility in the application of boundary conditions to an FE model, and this provides the opportunity to investigate the mechanical properties of a bone under many different loading conditions—something that is not possible when performing experimental mechanical testing due to the damage that would be imparted to the bone. Typical boundary conditions used in FE models range from the simplest axial loading conditions to three-point bending tests and to complex distributed loads that simulate the forces on an articular joint.

Finite Element Solvers

A finite element solver takes the defined mesh that includes the geometry, boundary conditions and material definitions and obtains a solution subsequent to the determination of equilibrium. The solution typically includes the local deformations of the mesh and reaction forces (at each node) and the stresses and strains (for each element). Large-scale finite element model solvers are specifically designed to solve very large models (Hollister and Riemer 1993; Van Rietbergen et al. 1996; Boyd et al. 2002). Typically, it is most common to solve linear models (i.e., linear deformations and elastic material properties), but it is also of interest to solve nonlinear models representing nonlinear constitutive properties (i.e., to determine the full stress–strain mechanical behavior), but these models are considerably more complex.

A linear FE model can be generally represented in terms of a large system of equations represented in the matrix form

$$\mathbf{Ax} = \mathbf{b},$$

where the element stiffnesses are represented by matrix $\mathbf{A}$, the nodal displacements represented by vector $\mathbf{x}$, and the reaction forces by vector $\mathbf{b}$. A key component of large-scale FE models is that they use an iterative solution approach and avoid assembling the full stiffness matrix $\mathbf{A}$. By using an iterative solution method (rather than direct solution method utilizing Gaussian elimination, for example) based on the method of conjugate gradients (Strang 1986), and by taking advantage of the fact that every element has the same stiffness matrix (Smith and Griffiths 1998), huge memory savings can be realized. Since every element is defined to have the identical stiffness matrix, only one copy of that element stiffness matrix needs to be stored in memory. This results in memory savings on the order of 99.9%, and thus makes it possible to solve very large systems of equations—FE models with many millions of elements can be solved! Combining this solution technique with appropriate preconditioning can accelerate the iterative solution process, and even more time savings can be realized by parallelizing the software. Thus, a typical model containing $\sim 2\,\mathrm{M}$ elements representing detailed bone architecture can be solved in only a few hours. Models of entire human femurs with nearly $100\,\mathrm{M}$ elements have been solved to investigate the loading characteristics in an osteoporotic vs. a healthy femur (Van Rietbergen et al. 2003).

Although the models typically employ an isotropic, homogeneous representation of the bone tissue, a range of tissue moduli can be represented in the FE model by scaling the individual element stiffness matrices. Typically 128 or 256 scaling levels are applied without having any significant impact on the memory requirements to solve large models. In comparison with a homogeneous tissue modulus model, where only one stiffness matrix needs to be stored in memory, a model with scaled moduli may only require up to 256 stiffness matrices to be stored—still a considerable memory saving in comparison to a model where each element stiffness matrix is stored individually.

Finite Element Applications

The application of the finite element method using image data as a basis has many practical uses, which are described in the following sections.

Direct Mechanics Approach

Large-scale FE solves are most often used for linear analyses to determine the stiffness of the bone. Although a linear FE analysis does not explicitly provide information about bone *strength*, it is well suited to providing detailed information about the elastic characteristics (*stiffness*). In the example shown in Fig. 14.2, a single axial compression test is illustrated, and this one test cannot fully describe the elastic properties (i.e., considering other loading directions or loading modes). One approach that has been developed to fully characterize the apparent properties of a cubic bone specimen is called the direct mechanics approach (Van Rietbergen et al. 1996). This involves performing six separate tests using large-scale FE on the bone, including three axial tests and three shear tests aligned with three orthogonal axes. The results from these FE tests provide the orientation and magnitude of the principal stiffness, and this information provides insight into how normal bone is organized to withstand load (i.e., in the spine, the stiffness of the bone is aligned with the main loading direction), as well as the effects of disease on bone mechanics. This characterization of the elastic properties has been shown to be highly correlated with the trabecular organization represented by the material fabric (degree of anisotropy) (Van Rietbergen et al. 1998).

Indirect Determination of Tissue Properties

One useful application of the finite element method is to indirectly determine tissue-level material properties by directly combining FE and experimental results. As input into a large-scale FE model, it is necessary to define the homogeneous, isotropic tissue properties a priori; however, this information is often difficult to obtain. This exact problem led to the creation of the so-called back-calculation procedure, which is used to estimate tissue properties (Van Rietbergen et al. 1995). The procedure involves generating an FE model from micro-CT data with an assumed modulus for the tissue (i.e., 5 GPa is a typical good starting point). The real bone specimen is then mechanically tested to provide the stress–strain curve and hence the apparent-level stiffness of the bone. The FE model can simulate the same testing conditions as used in the experiment, and it also generates an estimate for the bone stiffness based on the assumed 5 GPa tissue modulus. Although the slopes of the experimental and FE stress–strain plots will clearly differ, linear scaling of the tissue modulus can be performed until the apparent-level properties are matched. This technique is limited to studies where access to the real tissue is available (i.e., experimental animal studies or human biopsy studies). It has been used, for example, to determine that the tissue modulus remains constant in bone following a knee ligament injury despite large changes to the architecture and apparent-level strength (Boyd et al. 2002). Also, bone tissue changes have been documented as a result of age-related processes in animal models (Jacobs et at. 1997) and human patients (Day et at. 2000). It is clear that tissue modulus alters in addition to architecture, and it is an important aspect of the mechanical function of cancellous bone.

Validation

Validation of the FE method is critical to quantifying the strengths and limitations of the modeling approach. Generally, validation is performed by comparing the model results to the experimental measurements. However, when the model represents very small structures (i.e., trabecular architecture) it is difficult to obtain direct measurements. One approach has been to use the micro-CT data as input into rapid prototyping machines (e.g., fused deposition modeling) so that scaled physical models can be constructed (Fig. 14.5). This provides the opportunity to generate several identical models for experimental testing to determine the reproducibility of estimates for both apparent- and tissue-level validation (Su et al. 2006). Although the

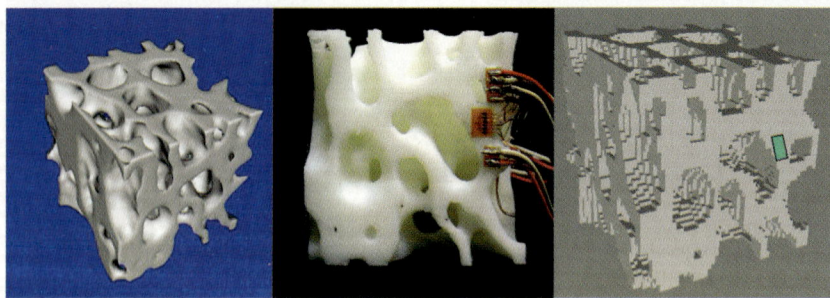

Fig. 14.5 A micro-CT scan of cancellous bone (*left*) was used to build a scaled prototype of the bone by fused deposition modeling (*middle*). Strain gauges measure local strains during axial compression, and a finite element model (*right*) can simulate the same loading conditions and validate the local strain at the corresponding position (*right; small element*)

prototypes do not have the same tissue properties as bone, this method validates the geometric representation of the bone architecture. Given appropriate tissue modulus inputs, it is possible to replicate the mechanical characteristics of a complex structure by using the image-based FE analysis approach.

Image-guided failure analysis (IGFA) is a novel method that can be used to validate finite element models by directly comparing 3D images of bone under sequential stages of loading (Nazarian and Müller 2004). A mechanical testing device that is micro-CT compatible allows compression testing to be performed inside the micro-CT, and by capturing a sequence of 3D images during loading it is possible visualize the progression of bone failure. These data provide important information about mechanisms of failure that can be used to better represent the mechanical properties of bone in FE analysis. The method of IGFA has been applied in human cadaver tissue such as the spine (Hulme et al. 2008), and most recently in the distal radius (Fig. 14.6). Furthermore, an emerging application of IGFA is to back-calculate the local strains between loading steps using deformable registration, and the resulting displacement field can be used to validate the FE model strains (Verhulp et al. 2004). This method of indirectly measuring the strains from IGFA is still under development, and promises interesting insight into bone mechanics.

A few comments in relation to accuracy and precision with regards to FE analysis are warranted. The *accuracy* represents how close a measurement (e.g., bone strength) is to the true value, and the *precision* represents the reproducibility of the result. Achieving high accuracy from FE methods is a challenging task because many of the model inputs are difficult to obtain (e.g., tissue properties). Arguably, however, precision is more critical than accuracy, as it makes it possible to monitor changes (i.e., improvements or degradation). Fortunately, achieving high precision from image-based FE analysis is possible (MacNeil and Boyd 2007b). For example,

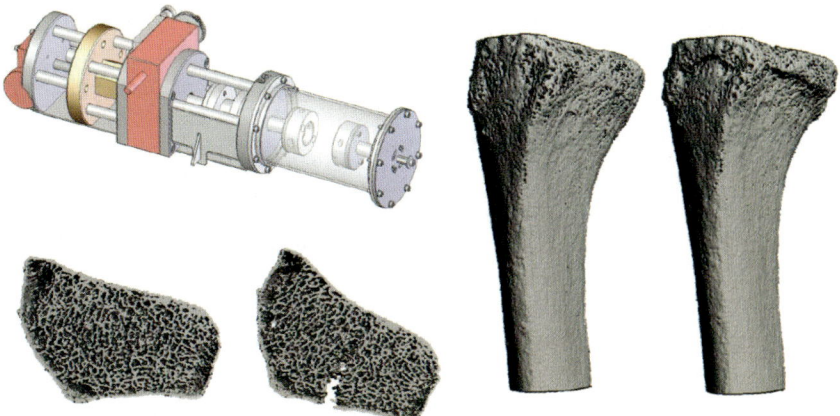

Fig. 14.6 A mechanical loading device recently developed to perform image-guided failure assessment inside a micro-CT scanner (*top left*) can apply torsion and/or compression loading to a human radius. In axial compression, the radius fails at the ultra-distal region (*right*; before and after loading); in torsion, a longitudinal crack develops (*bottom left*; before and after loading)

although a patient's baseline estimate of strength may vary from the true value (which is impossible to verify except on cadavers), follow-up measurements of that patient will incorporate the same systematic errors in the image-based FE model that caused the deviation from the true value; therefore, with high reproducibility, changes from that baseline can be interpreted with confidence.

In Vivo Determination of Bone Strength

The recent advent of in vivo micro-CT scanning has provided new-found opportunities for image-based FE analysis. Often, bone strength is an important endpoint for bone-related research, and while micro-CT has been an important contributor to those studies, it has mostly been applied in vitro. Thus, cross-sectional study designs are used for in vitro studies, and mechanical testing of the bones is performed on each bone at the end of the study. However, with in vivo micro-CT systems, it is advantageous to employ longitudinal study designs with repeated measures (Boyd et al. 2006), and therefore biomechanical testing is not possible. Therefore, image-based FE allows the assessment of mechanical properties from the series of in vivo scans so that intermediate changes in bone quality can be monitored. This has important applications in animal studies, and it can also be carried over to human studies now that there are micro-CT systems for human measurements (XtremeCT, Scanco Medical). Scans of the distal radius, for example, can be assessed to determine the relative loading capacity of the cortical bone vs. the cancellous region (Fig. 14.7), and the load sharing may be an indicator of bone quality (i.e., a thin cortex could provide insufficient load support and predispose the bone to increased fracture risk) (MacNeil and Boyd 2007a). Furthermore, it can be used to compare bones within a patient population, and to compare bone quality between family members (Fig. 14.7).

Nonlinear Finite Element Analysis

Applying the linear FE method to in vivo human data can provide the overall stiffness of the bone, but it may be advantageous to estimate bone *strength*, as it more closely is associated with fracture risk. To estimate bone strength from FE models it is necessary to employ a nonlinear solution methodology. As previously mentioned, most large-scale FE models are used to solve linear models, and thus they provide stiffness (elastic) information about the structure, not strength. A nonlinear FE analysis can provide estimations of bone strength and post-yield behavior. Developing a non-linear FE model is a major undertaking for several reasons. Typically, the computational effort required is not practical (or even possible) given the large model sizes generated from image-based meshes, so it is difficult to apply nonlinear analysis to a large cohort of subjects. Secondly, an appropriate failure criteria for the bone

Fig. 14.7 The load sharing in the human distal radius between the cortex and cancellous region can be separately assessed using in vivo micro-CT (*top*). The relation between bone size, reaction load at 1% applied strain and bone stiffness can be compared between three family members. The von Mises stress distribution at 1% applied strain is shown for each 3D image and cross-section through the distal radius (*middle*). The reaction load, average cross-sectional area, and apparent stiffness calculated here demonstrate that both male members have larger and stronger bones, and that despite the "father" having the largest bone, it is not as strong as the "son"

must be applied, and this is a nontrivial issue that is actively being investigated by several researchers.

Many different failure criteria have been proposed, and each has advantages and disadvantages (Keyak and Rossi 2000), partially depending on whether they are applied to continuum or microstructural FE models. Some examples include the von Mises (Keaveny et al. 2007), maximum principal strain (Niebur et al. 2002), and the Tsai-Wu (Fenech and Keaveny 1999) failure criteria, and recently criteria that incorporate damage accumulation (Chevalier et al. 2007). There is no consensus on which is the most appropriate criterion, but an asymmetric tissue yield strain criteria has worked well for small microstructural models (Niebur et al. 2000). Recently, the validation of a nonlinear FE approach was completed where human cadaver radii were scanned and then biomechanically tested. The image-based FE models employed the asymmetric tissue yield strain criteria to estimate the apparent yield properties of the bone (MacNeil and Boyd 2008), and correlations between experimentally and FE-determined bone strength were excellent ($R^2 > 0.9$). The application of image-based FE analysis to human in vivo micro-CT data provides an exciting new opportunity to estimate bone strength in a patient population, and this will be important for monitoring the progression of diseases and the effects of treatments.

Future Directions

Image-based FE analysis has made an important contribution to the assessment of bone quality in the past, and the advent of high-resolution in vivo scanning capabilities provides exciting new opportunities. Early results from in vivo applications have been promising, and the future development of the FE method will undoubtedly involve increasing sophistication of the representation of the constitutive properties of bone tissue, the continued development of methods to increase the efficiency of the computational methods, and the validation of the results whenever possible. The potential for image-based FE to be a pillar of bone quality assessment is high; by making it an accessible technology for health care workers, it will become an important contributor to improvements in the care and monitoring of patients.

References

Bayraktar HH, Morgan EF, Niebur GL et al. (2004) Comparison of the elastic and yield properties of human femoral trabecular and cortical bone tissue. J Biomech 37(1):27–35

Bourne BC, van der Meulen MC (2004) Finite element models predict cancellous apparent modulus when tissue modulus is scaled from specimen CT-attenuation. J Biomech 37(5):613–621

Boyd SK, Müller R (2006) Smooth surface meshing for automated finite element model generation from 3D image data. J Biomech 39(7):1287–1295

Boyd SK, Müller R, Zernicke RF (2002) Mechanical and architectural bone adaptation in early stage experimental osteoarthritis. J Bone Miner Res 17(4):687–694

Boyd SK, Davison P, Müller R et al. (2006) Monitoring individual morphological changes over time in ovariectomized rats by in vivo micro-computed tomography. Bone 39(4):854–862

Charras GT, Guldberg RE (2000) Improving the local solution accuracy of large-scale digital image-based finite element analyses. J Biomech 33(2):255–259

Chevalier Y, Charlesbois M, Varga P et al. (2007) A novel patient-specific finite element model to predict damage accumulation in vertebral bodies under axial compression. 29th American Society of Bone and Mineral Research Annual Meeting, Honolulu, HI, 16–19 Sept. 2007, 22:S484

Cody DD, Gross GJ, Hou FJ et al. (1999) Femoral strength is better predicted by finite element models than QCT and DXA. J Biomech 32(10):1013–1020

Cowin SC (2001) Mechanics of materials. In: Cowin SC (ed) Bone mechanics handbook, 2nd edn. CRC Press, Washington, DC, pp 6.1–6.24

Crawford RP, Cann CE, Keaveny TM (2003) Finite element models predict in vitro vertebral body compressive strength better than quantitative computed tomography. Bone 33(4):744–750

Eckstein F, Matsuura M, Kuhn V et al. (2007) Sex differences of human trabecular bone microstructure in aging are site-dependent. J Bone Miner Res 22(6):817–824

Fenech CM, Keaveny TM (1999) A cellular solid criterion for predicting the axial-shear failure properties of bovine trabecular bone. J Biomech Eng 121(4):414–422

Guldberg RE, Hollister SJ, Charras GT (1998) The accuracy of digital image-based finite element models. J Biomech Eng 120(2):289–295

Hodgkinson R, Currey JD (1992) Young's modulus, density and material properties in cancellous bone over a large density range. J Mater Sci: Mater Med 3:377–381

Hoffler CE, Guo XE, Zysset PK et al. (2005) An application of nanoindentation technique to measure bone tissue lamellae properties. J Biomech Eng 127(7):1046–1053

Hollister SJ, Riemer BA (1993) Digital image based finite element analysis for bone microstructure using conjugate gradient and Gaussian filter techniques. Math Meth Med Imag II, SPIE 2035:95–106

Homminga J, Huiskes R, Van Rietbergen B et al. (2001) Introduction and evaluation of a gray-value voxel conversion technique. J Biomech 34(4):513–517

Hulme PA, Ferguson SJ, Boyd SK (2008) Determination of vertebral endplate deformation under load using micro-computed tomography. J Biomech 41(1):78–85

Keaveny TM, Hayes WC (1993) A 20-year perspective on the mechanical properties of trabecular bone. J Biomech Eng 115(4B):534–542

Keaveny TM, Donley DW, Hoffmann PF et al. (2007) Effects of teriparatide and alendronate on vertebral strength as assessed by finite element modeling of QCT scans in women with osteoporosis. J Bone Miner Res 22(1):149–157

Keyak JH, Rossi SA (2000) Prediction of femoral fracture load using finite element models: an examination of stress- and strain-based failure theories. J Biomech 33(2):209–214

MacNeil JA, Boyd SK (2007a) Load distribution and the predictive power of morphological indices in the distal radius. Bone 41:129–137

MacNeil JA, Boyd SK (2007b) Improved reproducibility of high resolution peripheral quantitative computed tomography for measurement of bone quality. Med Eng Phys 29(10):1096–1105

MacNeil JA, Boyd SK (2008) Bone strength at the distal radius can be estimated from high-resolution peripheral quantitative computed tomography and the finite element method. Bone 42(6):1203–1213

Müller R, Rüegsegger P (1995) Three-dimensional finite element modelling of non-invasively assessed trabecular bone structures. Med Eng Phys 17(2):126–133

Nazarian A, Müller R (2004) Time-lapsed microstructural imaging of bone failure behavior. J Biomech 37(1):55–65

Niebur GL, Yuen JC, Hsia AC et al. (1999) Convergence behavior of high-resolution finite element models of trabecular bone. J Biomech Eng 121(6):629–635

Niebur GL, Feldstein MJ, Yuen JC et al. (2000) High-resolution finite element models with tissue strength asymmetry accurately predict failure of trabecular bone. J Biomech 33(12):1575–1583

Niebur GL, Feldstein MJ, Keaveny TM (2002) Biaxial failure behavior of bovine tibial trabecular bone. J Biomech Eng 124(6):699–705

Rho JY, Ashman RB, Turner CH (1993) Young's modulus of trabecular and cortical bone material: ultrasonic and microtensile measurements. J Biomech 26(2):111–119

Smith IM, Griffiths DV (1998) Programming the finite element method, 3rd edn. Wiley, New York

Stauber M, Huber M, van Lenthe GH et al. (2004) A finite element beam-model for efficient simulation of large-scale porous structures. Comput Methods Biomech Biomed Engin 7(1):9–16

Strang G (1986) Introduction to applied mathematics. Wellesley-Cambridge, Wellesley, MA

Su R, Campbell GM, Boyd SK (2006) Establishment of an architecture-specific experimental validation approach for finite element modeling of bone by rapid prototyping and high resolution computed tomography. Med Eng Phys 29(4):480–490

Turner CH, Burr DB (1993) Basic biomechanical measurements of bone: a tutorial. Bone 14(4):595–608

Ulrich D, van Rietbergen B, Weinans H et al. (1998) Finite element analysis of trabecular bone structure: a comparison of image-based meshing techniques. J Biomech 31(12):1187–1192

Van Rietbergen B, Weinans H, Huiskes R et al. (1995) A new method to determine trabecular bone elastic properties and loading using micromechanical finite-element models. J Biomech 28(1):69–81

Van Rietbergen B, Odgaard A, Kabel J et al. (1996) Direct mechanics assessment of elastic symmetries and properties of trabecular bone architecture. J Biomech 29(12):1653–1657

Van Rietbergen B, Odgaard A, Kabel J et al. (1998) Relationships between bone morphology and bone elastic properties can be accurately quantified using high-resolution computer reconstructions. J Orthop Res 16(1):23–28

Van Rietbergen B, Majumdar S, Newitt D et al. (2002) High-resolution MRI and micro-FE for the evaluation of changes in bone mechanical properties during longitudinal clinical trials: application to calcaneal bone in postmenopausal women after one year of idoxifene treatment. Clin Biomech (Bristol) 17(2):81–88

Van Rietbergen B, Huiskes R, Eckstein F et al. (2003) Trabecular bone tissue strains in the healthy and osteoporotic human femur. J Bone Miner Res 18(10):1781–1788

Verhulp E, van Rietbergen B, Huiskes R (2004) A three-dimensional digital image correlation technique for strain measurements in microstructures. J Biomech 37(9):1313–1320

Whiting WC, Zernicke RF (1998) Biomechanics of musculoskeletal injury. Human Kinetics, Winsor, Canada

Zysset PK, Guo XE, Hoffler CE et al. (1999) Elastic modulus and hardness of cortical and trabecular bone lamellae measured by nanoindentation in the human femur. J Biomech 32(10):1005–1012

Chapter 15
Geometric Morphometrics and the Study of Development

Benedikt Hallgrímsson(✉), Julia C. Boughner, Andrei Turinsky, Trish E. Parsons, Cairine Logan, and Christoph W. Sensen

Abstract Even though developmental biology seeks to provide developmental explanations for morphological variation, the quantification of morphological variation has been regarded as peripheral to the mechanistic study of development. In this chapter, we argue that this is now changing because the rapidly advancing knowledge of development in post-genomic biology is creating a need for more refined measurements of the morphological changes produced by genetic perturbations or treatments. This need, in turn, is driving the development of new morphometric methods that allow the rapid and meaningful integration of molecular, cellular and morphometric data. We predict that such integration will offer new ways of looking at development, which will lead to significant advances in the study of dysmorphology and also the relationship between the generation of variation through development and its transformation through evolutionary history.

Introduction

Morphometrics is the quantitative study of morphology. There are several types of morphometrics. Stereology, for instance, deals with the estimation of three-dimensional structures (such as cell counts) from two-dimensional slices. Geometric morphometrics is the quantitative assessment of form. The form of an object is the combination of its size and shape. Technically, form then refers to those geometric measurements of an object that are not changed by translation, rotation or reflection (Kendall 1977). Shape, by contrast, is form with scale removed. Formally, therefore, shape refers to those geometric measurements that are invariant to translation, rotation, reflection and scale (Kendall 1977). Form and shape of

B. Hallgrímsson
Department of Cell Biology and Anatomy and the McCaig Bone and Joint Institute, Faculty of Medicine, University of Calgary, 3330 Hospital Drive NW, Calgary, AB, Canada T2N 4N1, e-mail: bhallgri@ucalgary.ca

organisms are important aspects of the phenotypic variation that both evolutionary and developmental biology are ultimately tasked with explaining. The shape of complex skeletal structures such as the skull, the morphogenesis of the face, or the intricate morphology of the renal tubules are all developmental phenomena in which form is an important aspect of what we are seeking to explain through the study of developmental mechanisms.

The quantitative assessment of form, or morphometrics, has played only a minor role to date in developmental biology. Experimental developmental biology is concerned mainly with revealing developmental genetic mechanisms and pathways. Most studies have involved perturbation with major effects on development, and most investigators have been satisfied with a qualitative assessment at the phenotypic level. Typically, studies present a panel of images of genotypes or treatments in which the specimen shown is assumed to be typical of some larger sample examined over the course of the study. Interestingly, quantification of variation is much more common at the genetic or cellular levels; many papers that present the quantification of gene expression through RT-PCR as well as cell proliferation or apoptosis often present minimal or no quantitative assessment at the phenotypic level (Mak et al. 2008; Shuman and Gong 2007; Aioub et al. 2007). The demand for precise quantification at the molecular and cellular levels while regarding phenotypic measurements as superfluous is undoubtedly partly a cultural phenomenon, reflecting the higher value placed on molecular level data.

It has been argued that morphometrics is of value to developmental biology because morphometric tools such as those that deal with the phenotypic correlations among structures can reveal the degree to which development is modularized (Klingenberg et al. 2001a, b, 2002; Roth 2000). Morphometric analyses are also useful for QTL studies of morphological variation (Klingenberg et al. 2001a, b, 2004; Cheverud et al. 2004; Ehrich et al. 2003). However, morphometrics is likely to impact developmental biology through a much simpler and direct route. The quantification of form is likely to become more important, along with many other forms of quantification of phenotypic variation, and ironically this will happen because of the great success of the reductionist gene-focused paradigm of developmental biology. To explain genetically complex phenotypes, developmental biologists will increasingly move away from studies focused on single genes or treatments and to more complex and multifactorial studies that compare multiple variants. As we test increasingly complex hypotheses, the practice of visually comparing specimens and picking representative ones will no longer suffice. Phenotypic information will come to play a more significant role in hypothesis testing about development and this will require that such data are treated with the same rigor as molecular and cellular level data. This, for instance, is the motivation behind the Mouse Phenome Project (Bogue 2003; Grubb et al. 2004; Paigen and Eppig 2000) and the rise of "phenogenomics." The field of morphometrics, developed precisely for the rigorous quantification of morphological size and shape, will thus become increasingly important to developmental biology in the coming years.

In this chapter, we will explain the application of geometric morphometrics to the study of craniofacial development, and, at a broader level, the value of morphometric techniques to developmental biology analyses. The methods described here are relevant to organ systems other than the face. We show how geometric morphometric methods can yield useful and unique information about subtle phenotypic variation. We also argue for the integration of these methods with the quantification of cellular-level processes and show how this could be achieved.

Micro-Computed Tomography and the Morphometrics of Embryonic Morphology

The development of micro-computed tomography (micro-CT) (Feldkamp et al. 1989; Rüegsegger et al. 1996) has had a profound impact on the study of bone and skeletal disease (Boyd, this volume). Micro-CT allows for high-resolution 3D imaging of bone structure without laborious sectioning. This technology has revolutionized the study of bone architecture, and in the case of cortical bone, micro-CT based imaging has even expanded our basic understanding of the anatomy of bone (Cooper et al. 2003, 2006). Imaging of soft tissues has proven much more difficult, although this has been improved substantially with the use of better contrast agents or dual-energy scanning (Sim and Puria 2008; Vasquez et al. 2008). To date few studies have employed micro-CT for the analysis of embryonic structures, and only two have applied morphometric methods to study morphological variation in embryonic structures (Boughner et al. 2008; Parsons et al. 2008).

We have pioneered the use of computed microtomography for 3D morphometric analyses of morphology in embryos (Boughner et al. 2008; Parsons et al. 2008). A fixative that is capable of producing little distortion while leaving the tissue firm and resistant to deformation is ideal when imaging external morphology. For mouse and chicken embryos, Bouin's fixative produces minimal shrinkage artifacts in embryos compared with other fixation methods (Wang and Diewert 1992) and results in good-quality µCT data. Figure 15.1 shows micro-computed tomography scans of mouse,

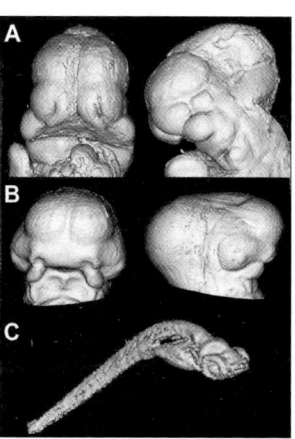

Fig. 15.1 (a–c) Micro-computed tomography scans of mouse (**a**), chick (**b**) and zebrafish (**c**) embryos

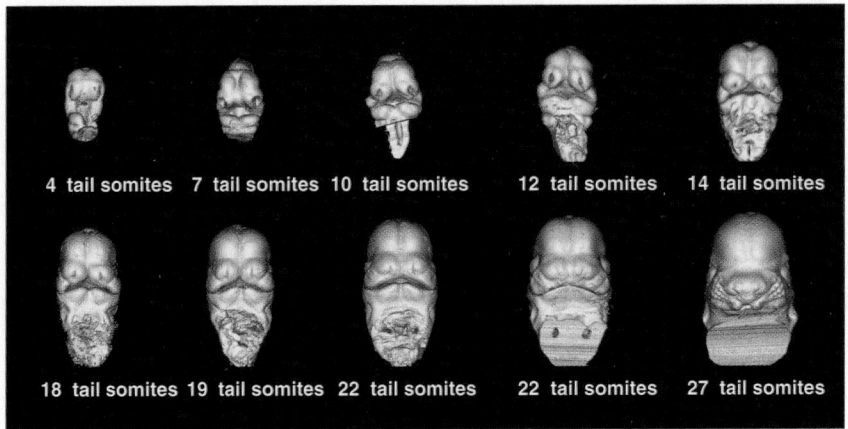

Fig. 15.2 Representative mouse embryos at different developmental stages ranging from gestational day 10 to day 13

chicken, and zebrafish embryos after Bouin's fixation; the external morphology is clearly visible and amenable to morphometric analysis. Figure 15.2 shows similar scans of representative mouse embryos at different developmental stages ranging from gestational day 10 to day 13. All of these scans were performed on a Skyscan 100 kV computed microtomograph.

Landmarks and the Quantification of Embryonic Form

Traditional morphometrics, as defined by Marcus (1990), is the univariate or multivariate analysis of linear distances, areas, volumes, and angles. Traditional morphometric methods are useful in many research contexts, but they are limited in important ways. Size and shape are usually difficult to tease apart in such analyses, and spatial relationships among measurements are typically lost (Bookstein 1991). Finally, such methods rarely yield intuitive and biologically meaningful visualizations. During the past two decades methods have been developed to overcome these shortcomings through the analysis of landmark coordinate data, rather than being based on linear measurements or angles (Bookstein 1991; Rohlf and Bookstein 1990; Dryden and Mardia 1998; Zelditch et al. 2004; Lele and Richtsmeier 1991, 2001; Lele 1993).

To be useful, landmarks must correspond biologically across individuals in a given analysis. In other words, a landmark should mark the same or homologous points in different individuals. This can be tricky in embryos due to the "soft" nature of most embryonic structures, at least as visible along the surface. Further, the substantial changes that occur during development mean that finding homologous points that persist across a developmental range of interest can be a difficult task. Another issue is that not all landmarks are the same in terms of their biological relevance. Bookstein's (1991) classification of landmarks is useful in this regard. In his classification, type I landmarks denote the location of a discrete anatomical structure. These are most useful because they contain information that is independent of other landmarks. Type II landmarks are those that demarcate extreme points of curvature on smooth features. Sets of embryonic landmarks usually contain many such features. Type III landmarks are those that are defined with respect to distant structures, such as the points at which maximum lengths occur. The points that might define the maximal widths of the face or length of limbs would be examples of such points. Such points are common in traditional morphometrics, but are problematic for use as landmarks, because their biological meaning and homology across individuals is often unclear and they contain information that is partly dependent on other landmarks. Statistically, Type III landmarks have the property that they represent a meaningful location in only one or two planes. Variation in one or two planes is, therefore, essentially arbitrary or deficient (Bookstein 1991; Gunz et al. 2005).

Confining datasets to Type I and II landmarks is statistically preferable, as this negates the substantial arbitrary variation that Type III landmarks introduce into the landmark coordinate system. However, Type I and II landmarks often do not adequately capture the morphology of interest. This is especially true of embryonic morphology, where the smooth contours of the structures of interest often contain very few discretely identifiable points. A statistical method for dealing with Type III landmarks, or semi-landmarks, has been developed for this particular reason (Bookstein 1997; Gunz et al. 2005), as explained by Mitteroecker and Gunz in this volume. This method is based on defining a few Type I or Type II landmarks and subsequently placing other equally spaced landmarks along curves or surfaces between them. The arbitrary variation of the semi-landmarks along the curves or surfaces is minimized by the specialized methods for superimposing semi-landmarks.

Figure 15.3 shows a landmark set that we have used in studies of the mouse embryonic craniofacial development. Analogous landmark sets could be used for morphometric studies related to other aspects of development and in other species. In this landmark set, no landmarks are defined with respect to other landmarks, but the difficulty involved in obtaining discrete and clearly identifiable landmarks that persist over the development of the face should be apparent from the morphology shown in Figs. 15.2 and 15.3.

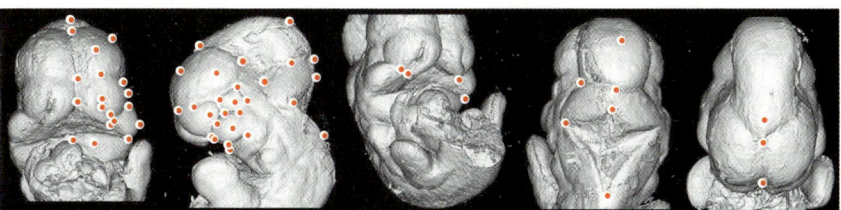

Fig. 15.3 3D landmarks for the analysis of mouse face formation shown on views of a 3D-computed microtomography reconstruction of an 11-day C57BL/6J mouse embryo

There are currently two major morphometric approaches to the quantification of form from landmark coordinate data. One of these, geometric morphometrics, is based on superimposition of the landmark configurations of the specimens to be analyzed, in order to place them in the same shape or form space. The most commonly used algorithm is called the Procrustes superimposition (Rohlf and Slice 1990). A Procrustes superimposition centers the shapes at the origin, scales them to a common size, and rotates them to minimize the differences between corresponding landmarks across individuals. This is described more fully in the chapter by Mitteroeker and Gunz in this volume. Once superimposed in this way, a sample of landmark configurations can be subjected to quantitative analysis of shape, as represented by the Procrustes coordinate data, and size, as represented by the centroid sizes of the landmark configurations.

An alternative to superimposition-based morphometrics is represented by the family of methods known as Euclidean distance matrix analysis or EDMA (Lele and Richtsmeier 1991, 2001; Richtsmeier et al. 2002). This method is based on the analyses of matrices of all the pairwise inter-landmark distances. EDMA allows comparisons of form (size + shape) and shape (once the measurements are scaled for some measure of size).

Visualization of Morphological Variation

An important advantage of geometric morphometrics is that its methods tend to lend themselves to biologically relevant visualizations. This is because, for many analyses, the results can be represented as patterns of landmark displacements, allowing one to see the morphological changes implied by the results as either deformations of a wireframe grid or deformations of an image or surface. One method that is commonly used to explore and visualize variation in Procrustes-superimposed landmark datasets is principal components analysis (PCA). This common multivariate statistical technique is used to reduce the dimensions of variation to a few more easily interpretable variables.

To visualize PCA, imagine a three-dimensional scatterplot of data. Imagine a line through the cloud of points that is as long as possible and that has the shortest distance possible between itself and the individual points in the cloud. This is the first axis or principal component (PC). The first PC is, therefore, a measure of the greatest dimension of variation. The second PC is the longest possible axis that is perpendicular to the first. Principal components are arbitrarily constrained to be orthogonal to one another and explain progressively smaller amounts of variation in the data. The components are the eigenvectors of the variance–covariance matrix of the landmark coordinates. Their eigenvalues are the variances associated with the axes along that dimension. Principal component scores are the values for individuals along these axes, which can be used to reveal patterns of variation in the data.

The eigenvectors for a specific principal component can be expressed as deviations from the mean (or some arbitrary) landmark configuration. This allows

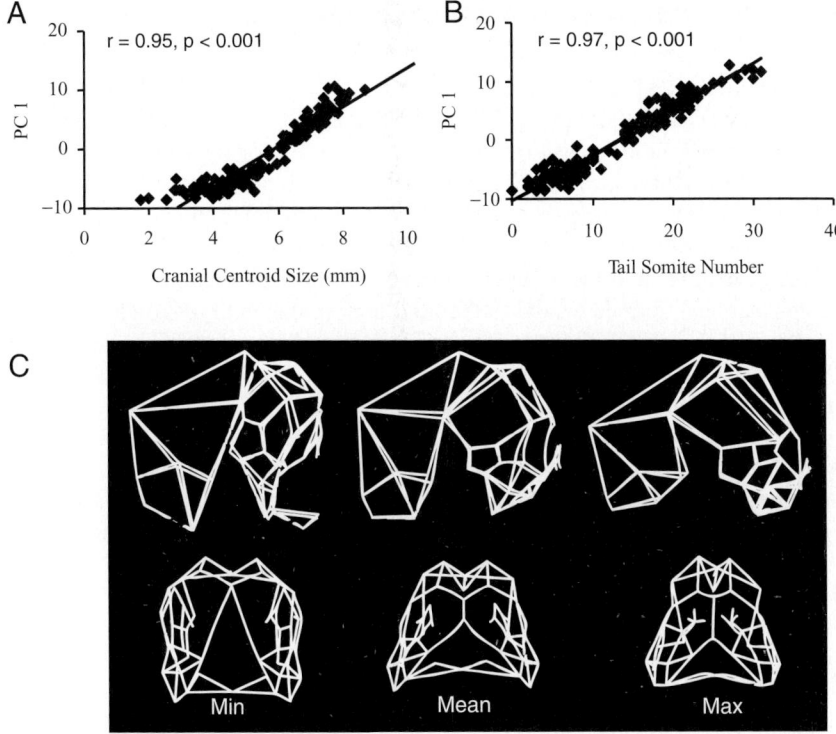

Fig. 15.4 (a–c) Visualization of ontogenetic variation based on principal components analysis of 3D landmark data from a sample of mouse embryos spanning the formation of the face ($N = 145$). (a) and (b) Scatterplots of this PC against tail-somite stage and head size. (c) A wireframe deformation depicting variation along the first principal component

intuitive visualization of the variation represented by the PCs, since these deviations of landmark positions can be depicted as deformations of wireframes, images or surfaces. Figure 15.4 shows the results from a principal components analysis of an embryo dataset. The first PC, which in this case is ontogenetic variation, is visualized as both a wireframe deformation and a deformation of a three-dimensional object map using the thin-plate spline algorithm.

Canonical variate analysis (CVA) is an exploratory technique that is closely related to PCA. CVA examines between-group differences relative to within-group variation. In CVA, the axes, or canonical variates (CVs), maximize between-group variation (relative to within-group variance) rather than the variation of the entire sample. The first CV is the axis along which groups are best discriminated. This is not necessarily the axis along which the means are the most divergent. If the within-group variation also happens to be high along the axis of greatest mean difference between the means, this axis may not discriminate between the samples as well as an axis of smaller difference which exhibits less within-group variation.

Quantifying and Accounting for Ontogeny

A crucial problem when dealing with embryonic morphology is the dramatic extent to which the size and shape of a structure change as it develops. For instance, the dramatic changes in the shape of the face during its development easily swamp differences among genotypes or treatments, which can be quite subtle by comparison to the ontogenetic variation seen in Fig. 15.2. One solution to this problem is simply to collect a sufficient number of samples such that comparisons can be made at specific developmental stages, as measured for instance by number of tail somites, somites that have formed caudal to the hind limb in the embryo. Another method, however, which is made feasible by geometric morphometric tools, is to quantify the shape change that occurs over some developmental period and then statistically standardize the sample to an arbitrary developmental stage.

In Procrustes-superimposed coordinate data, the variation associated with a particular factor can be estimated and visualized through several statistical means. In our embryonic craniofacial datasets, for instance, developmental stage can be estimated by counting tail somites. This is a better measure of developmental stage than gestational age because it is impractical to estimate precise age in most situations, and embryos are known to vary significantly in developmental stage, even within litters (Miyake et al. 1996). In a sample of embryos, the number of tail somites can then be regressed against the Procrustes-superimposed coordinate data (e.g., Fig. 15.4). Here, we used a pooled within-group regression to determine the shape variation associated with tail somite stage in embryos from two strains of mice using MorphoJ (Klingenberg 2008). Variation along the regression score can then be visualized as shown in Fig. 15.5. More importantly, datasets can then be standardized to a specific tail-somite number. The great advantage of this is that embryos don't have to be stage-matched to be compared, greatly improving the ease of sample collection and the statistical power of the data obtained from existing samples. Figure 15.6 shows two comparisons of mouse embryos from different strains. In one

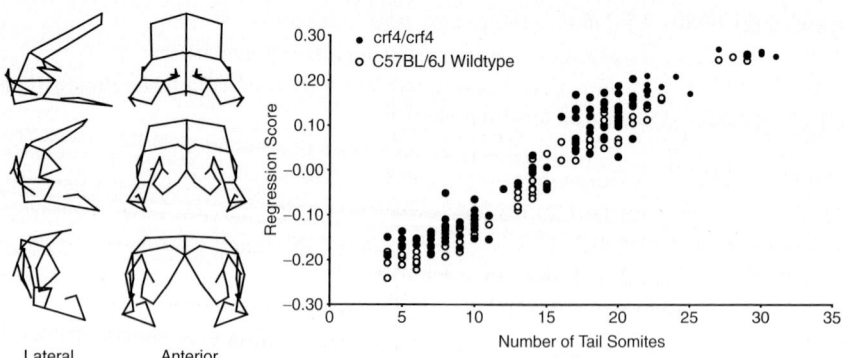

Fig. 15.5 Visualization of ontogenetic variation using pooled within-group regression of shape on tail-somite stage

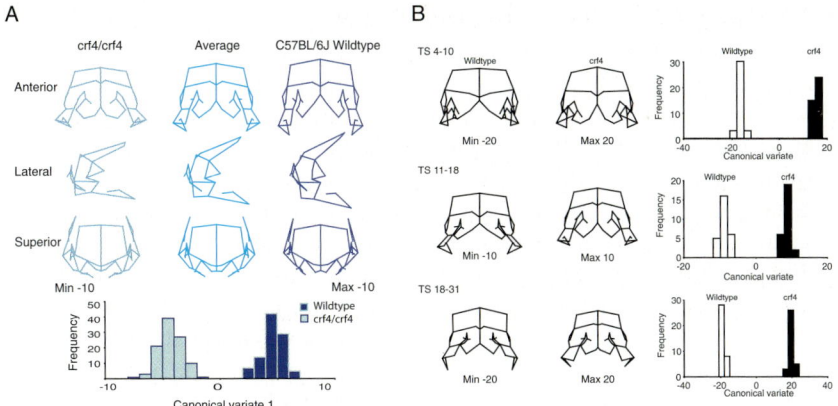

Fig. 15.6 (a) Comparison of two mouse strains by canonical variate analysis based on 3D-computed microtomography after correction for stage-related variation. **(b)** Similar analyses for the same sample after subdivision into ontogenetic subgroups and standardization for stage within subgroups

(6a), a large sample has been standardized to the median somite stage, while in the other (6b), the sample has been divided into age groups, with variation standardized to the median stage within each group.

Moreover, once quantified, ontogenetic trajectories can also be compared among strains. As is evident in Fig. 15.5, the two strains appear to differ slightly in the timing of the shape changes that occur during craniofacial development. Having quantified this variation, using either multiple regression or PCA, one can test for differences in the patterns and timing of ontogenetic shape transformations across genotypes, strains or treatments. Mutations or treatments of interest may produce changes to the overall developmental timing in addition to their primary developmental effect. The ability to separately quantify size, maturity relative to some measure of stage, and morphological shape is required in order to tease apart such complex effects.

Integrating Molecular, Cellular and Morphometric Data

One of the reasons that morphometrics has not been used extensively in developmental biology is that the quantitative assessment of form has been seen to be only remotely connected to hypotheses about developmental mechanisms. Key to bridging this divide is the ability to obtain measures that are directly linked to mechanisms from the same individuals from which morphometric data is obtained. For instance, to test a hypothesis about the role of localized patterns of cell proliferation and morphogenesis, it would be ideal to be able to record morphological and cell proliferation data from the same individual. Similarly, to relate the level of gene expression to morphology, the quantification of gene expression via RT-PCR and morphometrics would allow more direct tests of hypotheses about mechanistic cause. Ideally, this would be done longitudinally, following both kinds of data in live embryos. This is not currently possible, but methods are emerging for the integration of molecular, cellular and morphometric data at the individual level.

An objective of one of our research projects, directed at determining the basis for the development of cleft lip in a mouse model, is to relate temporospatial patterns of variation in cell proliferation within the midface to the craniofacial shape. This requires the ability to relate the two kinds of data in the same individual. Without this method, such an analysis could only be performed with genotypes or strains as units of analysis, greatly increasing the number of genotypes/strains required to test hypotheses about how histological and cellular-level variables determine craniofacial shape. Our method uses standard immunohistochemistry to determine local rates of cell proliferation in intact embryos using a commercially available antibody, anti-phospho-histone H3, a marker of cells in the M phase. Markers of the M phase may have a complex relationship to the actual mitotic rate (Aherne et al. 1977; Beresford et al. 2006). However, since we are using pHH3 to make relative comparisons among groups and not to determine the actual mitotic rate, this is not a significant issue. PHH3 is visualized via diaminobenzidine (DAB) staining, while the total number of cells present is determined via DNA-binding fluorescent stain with 4′,6-diamidino-2-phenylindole (DAPI), which stains all nuclei. The combination of the two stains thus reveals the number of cycling cells relative to the total number of cells present.

After whole-mount immunohistochemistry, whole heads can be post-fixed in Bouin's solution and then scanned and analyzed using the same methods as employed above. After scanning, the heads are sectioned to obtain the cellular level data about mitotic rate. Figure 15.7 shows representative photomicrographs (top) and microtomographs (below) from embryos stained via whole-mount immunohistochemistry for pHH3, as well as an example of a microtomograph of one such embryo, together with a representative frontal section through the same embryo showing both pHH3 and DAPI staining.

For this and similar projects, the vast amount of data generated through the whole-mount immunohistochemistry and subsequent sectioning is rapidly becoming a huge problem. Manual or even semi-automated counting of cells in individual sections is an impractical approach to this kind of dataset, given the highly labor-intensive nature of such work. For this reason, we have begun to develop software which will make obtaining cell proliferation data for the entire embryonic midface for significant samples of embryos practical. Our method is similar to that employed by Soufan et al. (2007) for the 3D quantification and visualization of cell proliferation data in the embryonic heart. Stacks of histological sections obtained as described in (D) above are superimposed onto a μCT-based 3D reconstruction of the same embryo. This is accomplished in new programs titled "Align-A-Stack" and "SliceOrientator," developed at the Sun Center of Excellence for Visual Genomics in Calgary (Fig. 15.8). Align-A-Stack refines the slice-by-slice alignment accomplished initially through an ImageJ plug-in, while SliceOrientator fits the aligned

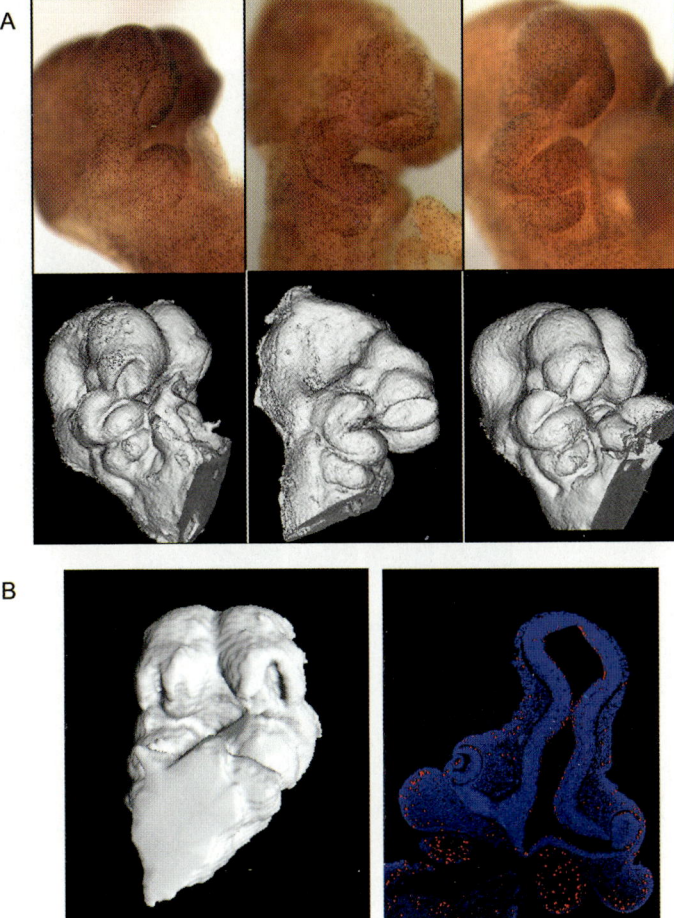

Fig. 15.7 (a–b) Sample photomicrographs of embryonic heads after whole-mount immunohis-tochemistry (**a**) and micro-CT scans of the same specimens (**b**). (**a**) HH3 and DAPI staining superimposed for a particular section. HH3-expressing nuclei are shown in *red* while all nuclei (DAPI staining) are shown in *blue*. (**b**) An external view of an object map derived from a μCT taken from this specimen into which the histological data are then superimposed

stack within the μCT-based 3D rendering of the embryo. Once the histological sections are aligned and oriented within the 3D rendering (object map) of the μCT scan, the histological data are treated as a volume by a new application titled "proliferationViz." A volumetric (3D) density gradient of pHH3-positive nuclei is then created by counting the number of such cells within a regular grid of overlapping spheres using a user-determined radius. This gradient is expressed as a 3D matrix of numbers in which the x, y, z value is the centroid of each sampling sphere. At present, the total number of cells is difficult to determine automatically from the DAPI staining at the resolution available ($1,600 \times 1,200$). However, the number of positively

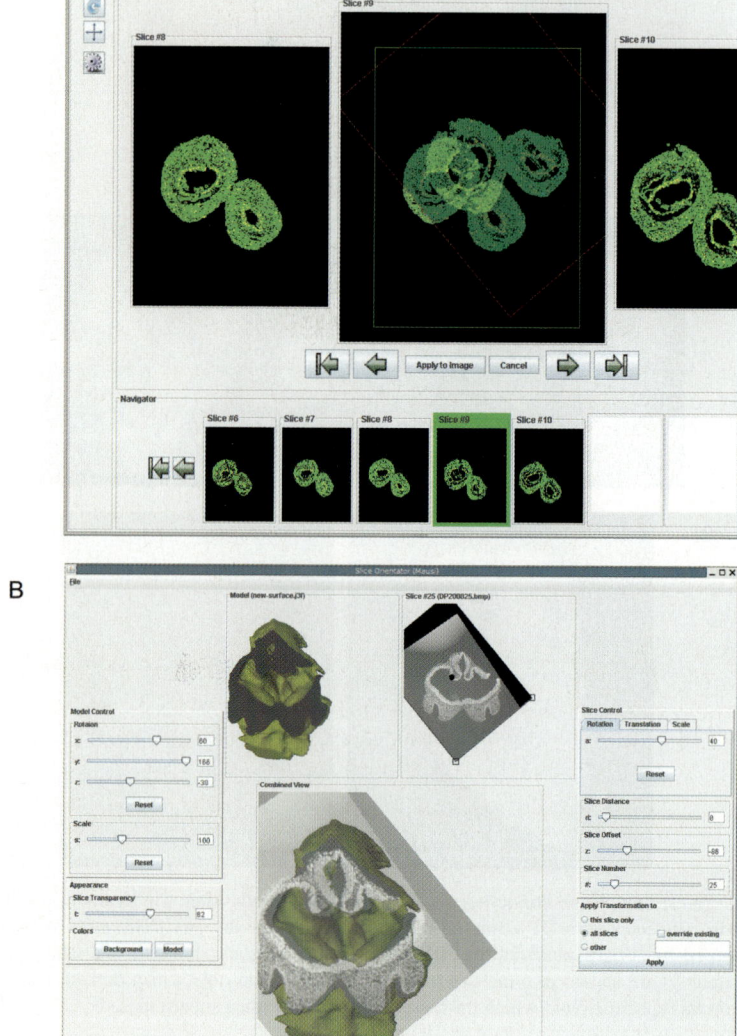

Fig. 15.8 (**a–b**) Screenshots of the AlignAStack (**a**) and SliceOrientator applications (**b**) developed by the Sensen lab

stained pixels within a sphere was used as a proxy for the total number of nuclei. We used adaptive thresholding to account for variation in lighting across the slide. The total number of cells was then expressed as a volumetric gradient (or matrix of numbers). The rate of cell proliferation was next determined by dividing the matrix

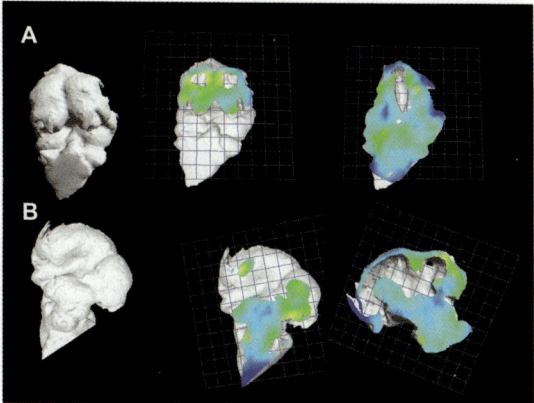

Fig. 15.9 (a–b) Screenshots of showing visualization of cell proliferation using the proliferationViz application. (**a**) An anterior view with slices cut at two coronal planes; (**b**) the same dataset in lateral view at the parasagittal and sagittal (*midline*) planes of the section

of positively stained cells by the matrix of positively stained voxels. Figure 15.9 shows an example of an embryo visualized using this technique.

To perform statistical comparisons of samples of such datasets, we need to combine data for multiple individuals. In this process, we first use a rigid superimposition (as in the high-throughput method described in preliminary data G) to superimpose the datasets for multiple individuals. Next, a set of mean landmarks is created by combining all sets of individual landmarks. Subsequently, a thin-plate spline transformation function (TPS) is calculated for each individual to morph it to the mean shape. Thereafter, this transformation function is used to transform each point (i.e., voxel) from the histological slices of an individual into the common mean space, creating average data for a group of individuals, as shown in Fig. 15.10. Once all individuals have been transformed into the mean shape for their group, the volumetric image sets or two such groups can be superimposed to show a difference map, which can be visualized as a color gradient or subjected to numerical analysis. An advantage of using this method for statistical analysis is that superimposing the data eliminates arbitrariness when defining such windows in different individuals, since the sampling volumes are defined only once for the entire dataset.

Future Directions

High-Throughput Morphometrics

A potential limitation on the application of morphometric methods is the time-intensive nature of morphometric data collection and analysis. Our experience with integrating morphometrics and developmental biology is that experiments can

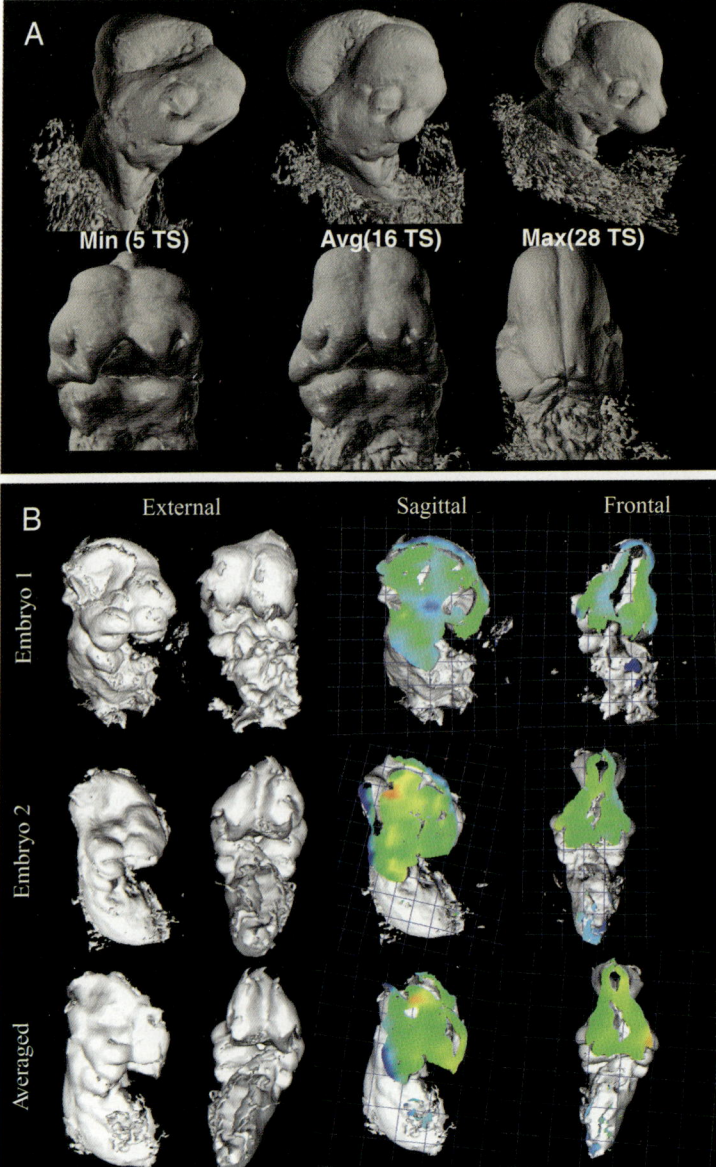

Fig. 15.10 (**a**) Morphing by 3D thin-plate spline of a single micro-CT scan of a single 16 somite stage embryo along the ontogenetic trajectory defining face formation. *a*, Lateral view; *b*, frontal view. (**b**) A mean dataset created from two C57BL/6J embryos at different stages showing data for the two original embryos and their average. Cross-sections are not in identical planes. *Blue* is low and *red* is high

generate specimens much more rapidly than we can analyze them, and this creates timing and workflow challenges for the projects. To address this issue, we have worked with Steve Boyd and his group to develop a body of visualization and morphometric methods that do not rely on labor-intensive manual landmark digitization. Our methods are semi-automated in the sense that user interaction with individual specimens is greatly reduced. Currently, visualization methods have been developed and methods for semi-automated statistical comparisons of shape are currently at the validation stage (Kristensen et al. 2008). Our method differs from traditional geometric morphometric methods in that it eliminates the need for laborious landmark selection, often the most time-consuming part of a morphometrics study. We use custom-developed software to perform intensity-based rigid image registration and image summation to create a mean shape. Image summation is accomplished by first dividing each individual voxel intensity by the number of images in the sample set and second by summing the corresponding voxels of the registered images. Regions of high shape variation are identified through the determination of edge gradients, which are defined as the changes in the voxel intensity in each of the x, y and z directions. Regions of low shape variation have a large edge gradient, and high shape variation is represented as a small edge gradient (see Boyd's chapter on computed microtomography in this volume). The differences in shape between individuals or group means are quantified by calculating surface-to-surface distances. This is analogous to the concept of volume-based thickness measurements using maximal spheres designed to calculate trabecular thickness (Hildebrand 1997). In this method, spheres are centered on each point of the surface of one object and expanded to touch the surface of the other object under analysis. The radii of these spheres are the Euclidean distances between the shapes, and they represent the mean shape differences. Figure 15.11 illustrates the basic elements of the method. This morphometrics tool has been validated using standardized 3D shapes and by comparison with standard landmark-based morphometric methods (Kristensen et al. 2008; Parsons et al. 2008).

Morphometrics of Molecular Variation

Optical projection tomography (see the chapter by James Sharpe in this volume) has opened up exciting new opportunities to integrate morphometrics and developmental biology (Sharpe et al. 2002). Using optical projection tomography (OPT) and whole-mount in situ or immunohistochemistry protocols with fluorescence labeling, it is now possible to obtain volumetric data about gene expression or the presence of specific proteins and morphometric data from the same specimen during a single scan. Examples of image sets illustrating this can be viewed at (http://www.bioptonics.com/menu_bar/Applications.htm).

The morphometric applications of this technology are interesting but are yet to be explored. The possibility of the morphometric assessment of OPT data offers a quantitative way to determine the spatiotemporal patterns of gene expression

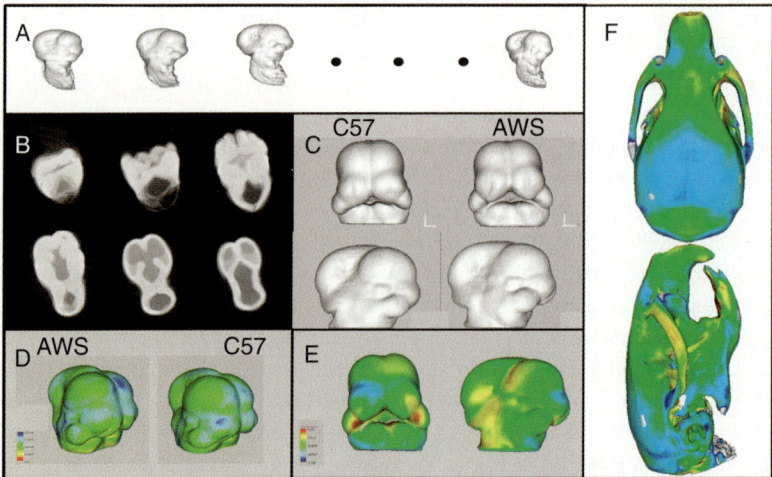

Fig. 15.11 (a–f) Schematic depiction of the high-throughput morphometric method. **(a)** Individual micro-CT scans. **(b)** Superimposed sample of many such scans. **(c)** Gradient map showing the distribution of variation within the samples. **(e)** Surface to surface gradient map showing the localization of morphological differences between samples. **(f)** A similar comparison of adult A/WySn and C57BL/6J crania. *Red* indicates regions of high difference, while *blue* indicates regions of low difference

changes produced by mutations or experimental treatments. With complex genetic designs involving comparisons of multiple genetic variants or drug treatments, morphometric quantification may be essential. Relating gene expression data to morphological variation is also possible with OPT, but this is complicated by the fact that molecular changes in development generally precede the morphological changes that they produce. Given that we are currently constrained to cross-sectional designs in most model systems and imaging modalities, research designs must consider this obvious but important fact. Morphometric statistical tools offer some solutions here, such as the use of multiple regression or principal components analysis to create shape transformation trajectories for morphological and gene expression data in the same sample. Such analyses should reveal relationships between changes in gene expression and morphological outcome, even if the effects are separated in time. Morphometric analysis of OPT data, we believe, is an important new direction that will further increase the relevance of morphometric methods and theory to developmental biology.

References

Aherne WA, Camplejohn RS, Wright NA (1977) An introduction to cell population kinetics. Edward Arnold, London

Aioub M, Lezot F, Molla M et al. (2007) Msx2 –/– transgenic mice develop compound amelogenesis imperfecta, dentinogenesis imperfecta and periodontal osteopetrosis. Bone 41:851–859

Beresford MJ, Wilson GD, Makris A (2006) Measuring proliferation in breast cancer: practicalities and applications. Breast Cancer Res 8:216

Bookstein FL (1991) Morphometric tools for landmark data. Cambridge University Press, Cambridge

Bookstein FL (1997) Landmark methods for forms without landmarks: morphometrics of group differences in outline shape. Med Image Anal 1:225–243

Bogue M (2003) Mouse Phenome Project: understanding human biology through mouse genetics and genomics. J Appl Physiol 95(4):1335–1337

Boughner JC, Wat S, Diewert VM, Young NM, Browder LW, Hallgrímsson B (2008) The Crf4 mutation and the developmental basis for variation in facial length. Anat Rec Part A. Submitted for publication

Cheverud JM, Ehrich TH, Vaughn TT, Koreishi SF, Linsey RB, Pletscher LS (2004) Pleiotropic effects on mandibular morphology II: differential epistasis and genetic variation in morphological integration. J Exp Zoolog Part B Mol Dev Evol 302:424–435

Cooper DM, Turinsky AL, Sensen CW, Hallgrimsson B (2003) Quantitative 3D analysis of the canal network in cortical bone by micro-computed tomography. Anat Rec B New Anat 274:169–179

Cooper DML, Thomas CDL, Clement JG, Hallgrimsson B (2006) Three-dimensional micro-computed tomography imaging of basic multicellular unit-related resorption spaces in human cortical bone. Anat Rec Part A 288A:806–816

Dryden IL, Mardia KV (1998) Statistical shape analysis. Wiley, Chichester

Ehrich TH, Vaughn TT, Koreishi SF, Linsey RB, Pletscher LS, Cheverud JM (2003) Pleiotropic effects on mandibular morphology I. Developmental morphological integration and differential dominance. J Exp Zoolog B Mol Dev Evol 296:58–79

Feldkamp LA, Goldstein SA, Parfitt AM, Jesion G, Kleerekoper M (1989) The direct examination of three-dimensional bone architecture in vitro by computed tomography. J Bone Miner Res 4:3–11

Grubb SC, Churchill GA, Bogue MA (2004) A collaborative database of inbred mouse strain characteristics. Bioinformatics 20(16):2857–2859

Gunz P, Mitteroecker P, Bookstein FL (2005) Semilandmarks in three dimensions. In: Slice DE (ed) Modern morphometrics in physical anthropology. Kluwer/Plenum, New York, pp 73–98

Hildebrand T, Ruegsegger P (1997) A new method for the model-independent assessment of thickness in three-dimensional images. J Microsc 185:67–75

Kendall D (1977) The diffusion of shape. Adv Appl Prob 9:428–430

Klingenberg CP (2002) Morphometrics and the role of the phenotype in studies of the evolution of developmental mechanisms. Gene 287:3–10

Klingenberg CP (2008) MorphoJ software. Faculty of Life Sciences, University of Manchester. http://www.flywings.org.uk/MorphoJ_page.htm

Klingenberg C, Leamy L, Routman E, Cheverud J (2001a) Genetic architecture of mandible shape in mice. Effects of quantitative trait loci analyzed by geometric morphometrics. Genetics 157:785–802

Klingenberg CP, Badyaev A, Sawry SM, Beckwith NJ (2001b) Inferring developmental modularity from morphological integration: analysis of individual variation and asymmetry in bumblebee wings. Am Natural 157:11–23

Klingenberg CP, Leamy LJ, Cheverud JM (2004) Integration and modularity of quantitative trait locus effects on geometric shape in the mouse mandible. Genetics 166:1909–1921

Kristensen E, Parsons TE, Gire J, Hallgrimsson B, Boyd S (2008) A novel high-throughput morphological method for phenotypic analysis. IEE Comput Graphics Appl. doi:10.1109/TBME.2008.923106

Lele S (1993) Euclidean distance matrix analysis of landmark data: estimation of mean form and mean form difference. Math Geol 25:573–602

Lele S, Richtsmeier JT (1991) Euclidean distance matrix analysis: a coordinate-free approach for comparing biological shapes using landmark data. Am J Phys Anthropol 86:415–427

Lele S, Richtsmeier JT (2001) An invariant approach to the statistical analysis of shapes. Chapman & Hall, Boca Raton, FL

Mak KK, Kronenberg HM, Chuang P-T, Mackemand S, Yang Y (2008) Indian hedgehog signals independently of PTHrP to promote chondrocyte hypertrophy. Development 135(11):1947–1956

Marcus LF (1990) Traditional Morphometrics. In: Rohlf FJ, and Bookstein FL, editors. Proceedings of the Michigan Morphometrics Workshop. Ann Arbor, Michigan: University of Michigan, Museum of Zoology

Miyake T, Cameron AM, Hall BK (1996) Detailed staging of inbred C57BL/6 mice between Theiler's [1972] stages 18 and 21 (11–13 days of gestation) based on craniofacial development. J Craniofacial Genet Develop Biol 16:1–31

Paigen K, Eppig JT (2000) A mouse phenome project. Mamm Genome 11(9):715–717

Parsons TE, Kristensen E, Hornung L et al. (2008) Phenotypic variability and craniofacial dysmorphology: increased shape variance in a mouse model for cleft lip. J Anat 212(2):135–143

Richtsmeier JT, Deleon VB, Lele S (2002) The promise of geometric morphometrics. Yearbook Phys Anthropol 45:63–91

Rohlf FJ, Bookstein FL (1990) Proceedings of the Michigan Morphometrics Workshop. U. Michigan Museum of Zoology, Ann Arbor, MI

Rohlf FJ, Slice DE (1990) Extensions of the Procrustes method for the optical superimposition of landmarks. Syst Zool 39:40–59

Roth VL (2000) Morphometrics in development and evolution. Am Zool 40:801–810

Rüegsegger P, Koller B, Müller R (1996) A microtomographic system for the nondestructive evaluation of bone architecture. Calcif Tissue Int 58:24–29

Sharpe J, Ahlgren U, Perry P et al. (2002) Optical projection tomography as a tool for 3D microscopy and gene expression studies. Science 296:541–545

Shuman JB, Gong SG (2007) RNA interference of Bmp-4 and midface development in postimplantation mouse embryos. Am J Orthod Dentofacial Orthop 131:447, e1–e11

Sim JH, Puria S (2008) Soft tissue morphometry of the malleus-incus complex from micro-CT imaging. J Assoc Res Otolaryngol 9:5–21

Soufan AT, van den Berg G, Moerland PD et al. (2007) Three-dimensional measurement and visualization of morphogenesis applied to cardiac embryology. J Microsc 225:269–274

Vasquez SX, Hansen MS, Bahadur AN et al. (2008) Optimization of volumetric computed tomography for skeletal analysis of model genetic organisms. Anat Rec (Hoboken) 291:475–487

Wang K-Y, Diewert VM (1992) A morphometric analysis of craniofacial growth in cleft lip and noncleft mice. J Craniofacial Genet Develop Biol 12:141–154

Zelditch ML, Swiderski HD, Sheets D, Fink WL (2004) Geometric morphometrics for biologists: a primer. Academic, New York

Part III
Applications

Chapter 16
Imaging in Audiology

Jos J. Eggermont

Abstract The imaging of auditory structures faces limitations posed by the location of the inner ear, requiring a combination of high-resolution computed tomography and magnetic resonance imaging (MRI). Functional imaging faces limitations posed by the adverse acoustic environment during MRI scanning, and requires a combination of passive and active noise cancellation to provide low-noise measurements. Generally speaking, up to now MRI equipment has not adequately catered for the requirements of audiological testing. Positron emission tomography (PET) scans provide a quiet environment but suffer from low spatial resolution and the need to average across participants. Functional MRI (fMRI) in high magnetic fields will allow measurements of cortical tonotopic map reorganization and provide important information about the central aspects of hearing disorders. Diffusion tensor imaging in combination with fMRI offers a currently untapped potential to study central auditory processing disorders. Simultaneous electroencephalography (EEG), evoked potentials and fMRI in audiology is feasible and allows millisecond temporal resolution and millimeter spatial resolution; this approach may provide another combination technique for probing central auditory processing disorders.

Introduction

Audiology comprises the clinical use of behavioral, electrophysiological and imaging methods to arrive at a diagnosis and to alleviate hearing problems in humans. In addition, audiologists explore brain plasticity by evaluating the long-term effects of hearing aids and cochlear implants on hearing in the widest sense. Audiology makes extensive use of the results of animal research carried out in the field of auditory neuroscience. Thus, at some point it is difficult to distinguish the basic scientific aspects

J.J. Eggermont
Department of Physiology and Biophysics and Department of Psychology, University of Calgary, Calgary, Alberta, Canada, e-mail: eggermon@ucalgary.ca

C.W. Sensen and B. Hallgrímsson (eds.), *Advanced Imaging in Biology and Medicine.*   339
© Springer-Verlag Berlin Heidelberg 2009

from the translational and clinical ones. Although there are nominal distinctions within the field, such as between clinical and experimental audiology, I will not distinguish any subfield here but instead focus on some specific challenges that the study of the auditory system poses for the suitability of imaging techniques. I will first briefly introduce the auditory system, and then I will discuss some of the challenges posed by the peripheral auditory system for structural imaging. After a brief summary of the common functional imaging techniques used in audiology, I will discuss the particularly harsh acoustic environment presented by fMRI for auditory research. The final section reviews the potential of imaging techniques for the last frontier of audiology: understanding the changes in the brain resulting from cochlear hearing loss and its implications for rehabilitation, and the further elucidation and characterization of central auditory processing disorders.

The Auditory System

The auditory system consists of two distinct parts commonly known as the peripheral and central auditory systems. The peripheral auditory system, located outside the brain cavity, comprises the external ear, the middle ear, the inner ear and the auditory nerve. The shape of the external ear allows it to extract cues about the elevation of a sound source. The external ear and middle ear are separated by the tympanic membrane or eardrum. The middle ear is an air-filled cavity which contains the middle ear bones that conduct the sound-induced vibrations of the eardrum via the stapes footplate to the fluid-filled inner ear. The tympanic membrane–middle ear–stapes system acts as an impedance transformer between air and fluid so that the efficient conversion of sound to vibrations in the inner ear can take place. Without a functioning middle ear, hearing loss of up to 40 dB occurs. The coiled inner ear (cochlea) is separated cross-sectionally into three parts by the basilar and Reissner's membranes into the scala tympani, scala media and scala vestibuli. In the triangular cross-sectioned scala media, bounded by the basilar membrane and the tectorial membrane, is the organ of Corti. This is a structure composed of outer and inner hair cells and supporting cells, where the vibrations from the cochlear fluid and the basilar membrane are transduced into voltage (membrane potential) oscillations by the hair cells. The first stage of this transduction by the hair cells resembles the mechanoelectrical action of a microphone. There are approximately 3,000 inner hair cells and approximately 10,000 outer hair cells aligned along the basilar membrane. In the inner hair cells, the membrane potential oscillations modulate the release of a neurotransmitter (glutamate) at the synapse with the auditory nerve fibers. This is the electrochemical stage of transduction and causes changes in the firing rate of the auditory nerve fibers. Each inner hair cell is innervated by $\sim$10 auditory nerve fibers, each with a different threshold and/or spontaneous firing rate. The outer hair cells have a very different function (see below).

The basilar membrane on its own performs a crude separation of sound frequency into a frequency-specific place of maximum vibration along the 35-mm-long human cochlea. This is done on the basis of its decreasing stiffness and near-constant unit mass and damping from base (high frequencies) to apex (low frequencies). The passive basilar membrane acts as a lumped system of broadly tuned mechanical frequency filters. The frequency resolution of the basilar membrane is greatly enhanced, and gains increased sensitivity to vibrations, due to a local feedback coupling of mechanical energy into the basilar membrane by the outer hair cells. As we have seen, there are roughly three times as many outer hair cells as there are inner hair cells. The outer hair cells transduce the basilar membrane motion into voltage changes and then into contractions and expansions along their length, and into periodic movements of the hairs, all of which are in phase with the basilar membrane movement. In this way the outer hair cells are able to push and pull on the basilar membrane. The active basilar membrane–outer hair cell system acts as a regenerative amplifier that produces an approximately 40 dB (100×) gain in the vibration amplitude of the basilar membrane, and this close-to-resonance action greatly sharpens its frequency resolution too. At some points along a normal cochlea, the feedback energy may be so large that spontaneous oscillations of the basilar membrane occur that set the middle ear into motion. The resulting movements of the tympanic membrane induce pressure changes in the external ear canal and can be recorded with a very sensitive microphone. These oscillations are called spontaneous otoacoustic emissions; they are ear-specific and can be used as a sensitive diagnostic for the functioning of the outer hair cells.

The ~30,000 auditory nerve fibers, which are connected only to the inner hair cells, combine to form the auditory nerve. The auditory nerve leaves the cochlea and passes through the petrous bone via the internal auditory meatus into the cerebellopontine angle to synapse with the three-part cochlear nucleus. This is where the central auditory nervous system starts (Fig. 16.1). Each auditory nerve fiber trifurcates and terminates in each subnucleus in such a way that it maintains three separate spatial mappings of frequency (tonotopic maps). Each subnucleus has a specialization; the left and right anteroventral cochlear nuclei are involved in sound localization in azimuth. The posteroventral cochlear nucleus detects and codes sound transients and pitch, and the dorsal cochlear nucleus is involved in sound localization in elevation and in integrating head and ear position and movements with sound location.

Tonotopic maps form one of the hallmarks of the central auditory system and are maintained in all brainstem and midbrain nuclei, in the medial geniculate body in the ventral thalamus, and in several auditory cortical areas. In the superior olivary complex of the brainstem, information from each of the anteroventral cochlear nuclei on the two sides (from one side via the medial nucleus of the trapezoid body, MNTB) are compared in order to localize the sound source. This is done on the basis of differences in inter-aural time of arrival, phase and sound level. In the auditory midbrain nucleus, the inferior colliculus, all information from the brainstem nuclei converges and is integrated. Binaural information is further combined with a visual map of the sound source from the superior colliculus and this results in a three-dimensional map of sound location. Note that the receptor surface of the inner ear represents only sound frequency and sound level information. The auditory system

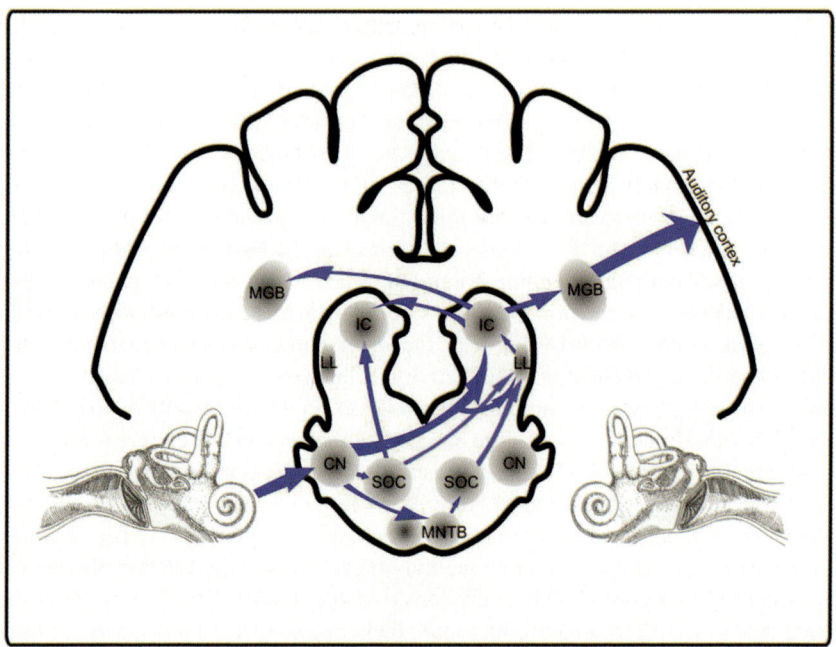

Fig. 16.1 The central auditory system pathways from one cochlea to cortex. Dominant pathways are indicated by *wide lines*. The *inset* shows the external, middle and inner ear. *CN*, cochlear nucleus; *SOC*, superior olivary complex; *MNTB*, magnocellular nucleus of the trapezoid body; *IC*, inferior colliculus; *MGB*, medial geniculate body (adapted with permission from Kral and Eggermont 2007)

has to compute the spatial map of the sound source location based on sound arrival times and spectral differences at the two ears. In contrast, the visual and somatosensory systems arrive at retinal and body maps by direct topographic mapping of their receptor spaces onto the central nuclei and cortices.

The auditory system is also distinguished from other sensory systems by its reliance on and encoding of, at least in the auditory nerve, brainstem and midbrain, the temporal aspects of sound. Sound is never static; even a constant pure tone is produced by oscillations of air molecules around the ambient barometric pressure, and the temporal aspects of hearing concern the periodic changes in that sound pressure. The encoding of sound pitch is done by those nerve fibers that fire in synchrony with the amplitude modulations (AM) and periodicity of the sound. Temporal encoding works fairly well up to 5 kHz. For higher frequencies, the release of neurotransmitter by the inner hair cells can no longer adequately be modulated by the changes in their membrane potential. This is a restriction on the transmitter release, uptake and recovery processes, because the extracellular field of the hair cells (cochlear microphonic) can follow sound frequencies of up to 100 kHz. In addition to this time-of-firing encoding, there is a also a representation of sound frequency based on the average firing rate of the auditory nerve fibers, and this firing rate is largest for those fibers that are innervating the inner hair cells at the site of maximum vibration of the basilar membrane. These frequency-dependent maxima are the basis for the tonotopic maps. For frequencies below 5 kHz there is combined time-of-firing and firing-rate encoding. The temporal accuracy of the firings in the brainstem, which is capable of making use of the minute differences in the time of arrival of the sound at the two ears (down to 10 μs in humans), and which also has the ability to follow high-frequency amplitude modulations, is gradually lost downstream and is mainly reduced to modulating carrier frequencies below a few hundred Hz in cat primary auditory cortex, and even less in other cortical areas. In the human brain there also seems to be an orderly reduction in the capacity to represent amplitude modulation frequencies. In a PET study, Giraud et al. (2000b) used amplitude-modulated (AM) noise as well as unmodulated noise and calculated the response difference between the two stimuli. While all of the auditory regions in the brain did respond to unmodulated noise (cf. Fig. 16.3), all of the regions showed increased responses to AM noise. However, they found that these increases for AM frequencies had upper limits of 256 Hz in the brainstem, 32–256 Hz in the inferior colliculus, 16 Hz for the medial geniculate body, 8 Hz for primary auditory cortex, and 4–8 Hz in the secondary auditory cortex.

The auditory midbrain is the main hub where the temporal (time-of-firing) code is converted into a rate-place code (Schreiner et al. 1988). Thus, in the midbrain, in the thalamus, and in cortex, putative spatial maps exist for attributes of sound such as pitch (Bendor et al. 2005), timbre, and amplitude modulations (Bendor et al. 2007), and definitely for sound frequency (Eggermont 2001). In the auditory brainstem, coding of sound attributes is based more on temporal aspects of sound, but the better the temporal coding, the larger the overall neural firing rate, thereby providing a basis for imaging based on metabolic activity (Eggermont 2001). Metabolic activity, as measured by most imaging techniques, does not reflect the exact timing of neural activity in individual neurons; only how much of it there is.

Structural Imaging

The peripheral auditory system poses some intricate problems for precise visualization. The combination of an air-filled middle ear and a fluid-filled inner ear embedded with soft tissue in the form of hair cells and membranes, all of which is incorporated into the dense petrous skull bone, offers distinct challenges to any imaging technique. Clinicians mostly use a combination of high-resolution computed tomography (HRCT) and MRI. Whereas HRCT offers high spatial resolution combined with high bone contrast, MRI allows a similar resolution (depending on the field strength) combined with high soft-tissue contrast. High bone contrast allows visualization of the bony wall and modiolus of the cochlea, whereas high soft-tissue contrast allows the visualization of the auditory nerve. One of the drawbacks of HRCT is the need for X-rays.

This combination of imaging techniques has been investigated by Trimble et al. (2007) to compare the usefulness of preoperative MRI and HRCT in pediatric cochlear implant candidates. In cases where children are deaf from birth, malformations of the middle- and inner ear are often found, optionally combined with the absence of a viable auditory nerve. Cochlear implantation requires both the presence of sufficient auditory nerve fibers that can be activated by the cochlear implant as well as access to them. They found that dual-modality imaging with HRCT and 1.5 T MRI of the petrous bone, auditory nerve and brainstem detected a variety of abnormalities related to deafness which would not otherwise be found using either modality alone (Fig. 16.2).

Also, for the preoperative assessment of patients being considered for cochlear implants, Lane et al. (2004) used sagittal 3D fast-recovery fast spin-echo (FRFSE) and 3D constructive interference in the steady state (CISS). Images were acquired at 3.0 T by using dual surface coils to enhance contrast to noise. Contrast-to-noise ratios for the intracanalicular auditory nerve and cerebrospinal fluid were measured in the internal auditory canal. Contrast-to-noise ratios for 3D CISS were twice those obtained with 3D FRFSE. Both techniques provided images of diagnostic quality.

These 3-T MRI techniques, including the higher field strength, are potentially superior to those used in the Trimble et al. 1.5-T study. One could thus wonder if the use of high-field scanners for clinical purposes would not make the HRCT obsolete. However, image fusion of HRCT and MRI may allow for optimized visualization of the peripheral auditory system (Seemann et al. 2005). Seemann et al. combined

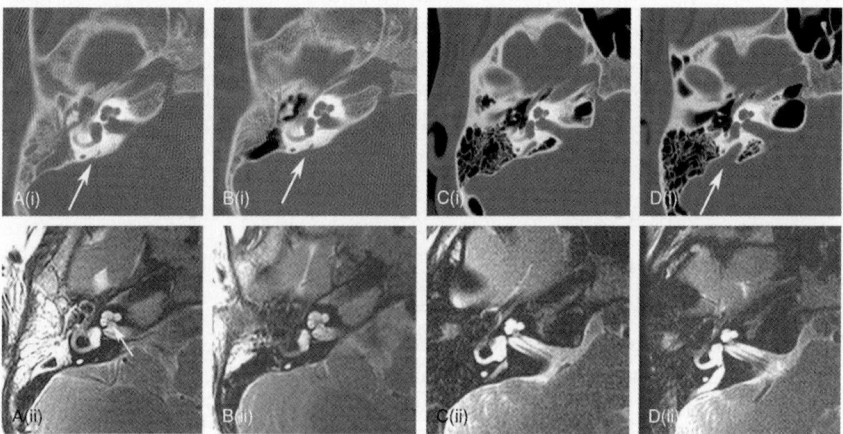

Fig. 16.2 (A–D) The *top row* shows HRCT scans, the *bottom row* T2-weighted MRI. The *four columns* show individual cases. Note that the auditory nerve and brainstem are not visible on the HRCT but are the dominant features on the T2-weighted MRI scans. **A**(i) and **A**(ii) show a normal submillimeter vestibular aqueduct (*large arrow*) and a normal modiolus (the bony structure around which the inner ear coils). Note that the modiolus on MRI is fainter than on HRCT (*thin arrow*). **B**(i) and **B**(ii) show a vestibular aqueduct that is slightly larger on HRCT than the one seen in **A**, but indistinguishable from MRI in **A**(*arrows*). **C**(i) and **C**(ii) reveal a moderately enlarged vestibular aqueduct, which is clear on HRCT, but only modestly visible on MRI. **D**(i) and **D**(ii) reveal an even larger vestibular aqueduct and endolymphatic fossa (*arrows*) (after Trimble et al. 2007, with permission)

3D CISS, T2-weighted and T1-weighted images with HRCT, and this reportedly allowed a realistic visualization of all the aspects of the peripheral auditory system.

T1-weighted images could shed some light on the myelin–gray matter content of the brain. Its evaluation is commonly based on voxel-based morphometry (VBM; e.g., Hyde et al. 2006). Sigalovsky et al. (2006b) mapped R1, i.e., the longitudinal relaxation rate for protons excited in the imaging process (R1 is the reciprocal of T1), which is predominantly affected by tissue myelin content. R1 was estimated for each voxel in the brain, and by averaging of R1 across the depth of the gray matter at finely spaced points covering the cortical surface. The resulting mappings thus do not show detailed variations in R1 across the gray matter laminae, but rather show the spatial distribution in overall R1 over the surface of the superior temporal lobe. They found that R1 was greater on the left for Heschl's gyrus, the planum temporale, the superior temporal gyrus, and the superior temporal sulcus. This suggested a greater myelination in the left auditory cortex, which may underlie the left hemisphere's specialization for the temporal processing of speech and language.

Diffusion tensor imaging (DTI), which traces the thermal motion of water flowing readily along nerve fiber tracts, is potentially an extremely useful technique for tracing the complex brainstem wiring pattern formed by the auditory nerve fibers that trifurcate at the cochlear nucleus, thereby forming the first station for parallel processing. The separate sound localization pathways and the spectral analysis pathways only combine again in the auditory midbrain and then again split into multimodal pathways (extralemniscal) and unimodal (lemniscal) tonotopic pathways. These pathways remain separate in the auditory cortex with its tonotopic and nontonotopic organized areas. So far, DTI has not been used for the purpose of tracing these subcortical pathways in the auditory system.

Functional Imaging

Functional imaging of the central auditory system has been performed using both PET and fMRI. Both methods have their advantages and disadvantages for analyzing the auditory system. First of all, the disadvantage of PET is obviously the use of radioactive tracers, which does not readily allow longitudinal studies, especially in children. The need to integrate activity over tens of minutes combined with the relatively low spatial resolution presents additional limitations. These are compensated by its major advantage for auditory research and diagnostics, namely that it is a silent technique; there is no interference from very loud scanner noise, unlike in fMRI (Ruytjens et al. 2006). A further advantage is that there is no magnetic field that interferes with the working of cochlear implants that use an implanted coil and electronics to receive radiofrequency induction of electrical signals for the up to 24 electrode contacts on the implant electrode.

The fMRI BOLD response has been used frequently in experimental audiology. A comprehensive study covering the entire auditory pathway's response to continuous noise (Sigalovsky et al. 2006a) highlights the quality that can be obtained if

346 J.J. Eggermont

the proper precautions are used (Fig. 16.3). All of the auditory nuclei shown in Fig. 16.1 produce increased activity when stimulated by continuous noise. Subcortical imaging requires cardiac gating of the response sampling to prevent blurring of the images by brainstem motion (Guimaraes et al. 1998).

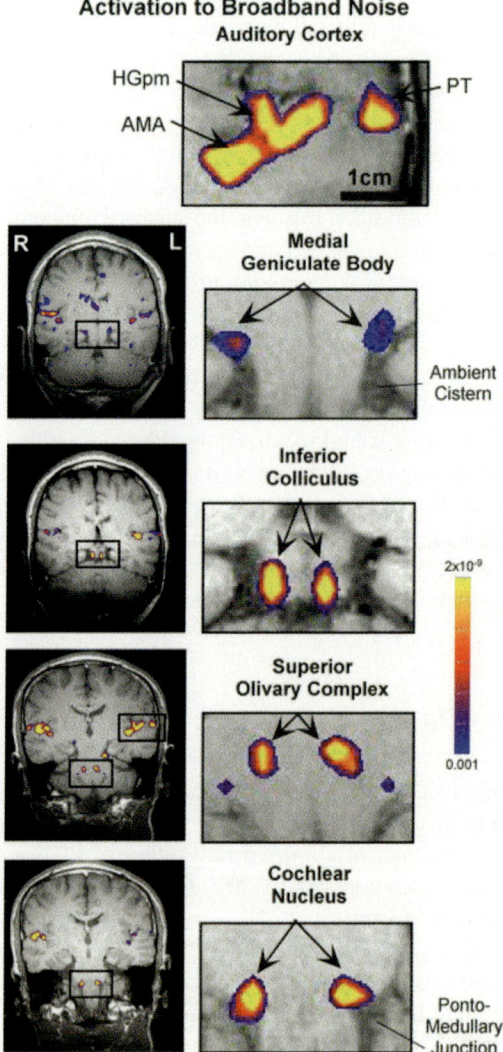

Fig. 16.3 Activation in major auditory centers throughout the human auditory pathway for binaurally presented 70 dB SL continuous broadband noise. *Panels on the right* are enlargements of the *insets on the left*. Activation (*color*) represents *p* levels below 0.001 (see *color bar*) and is superimposed on a T1-weighted anatomical (*grayscale*) image. Both the activation maps (in-plane resolution of 3.1 × 3.1 mm) and the anatomical images (1.5 × 1.5 mm) have been interpolated for these displays (from Sigalovsky and Melcher 2006, with permission)

Another silent functional imaging technique, although one that is only occasionally used, is near-infrared spectroscopy (NIRS); this is an optical imaging technique that estimates changes in cerebral blood volume and oxygen saturation induced by neuronal activity. In some ways it is the equivalent of the intrinsic optical imaging carried out in animals, and it also mirrors some of the properties of the brain that give rise to the BOLD response. In NIRS, near-infrared light is projected through the scalp and skull into the brain, and the intensity of the diffusely reflected left is recorded. The modulation of the recorded intensity by localized changes in the optical properties of the brain is used as an estimate of neural activation. Spatial resolution is relatively poor, as a result of the blurring of the light by scalp and skull, and because the sources of the vascular changes are located 2–3 cm below the skull. The technique, however, holds promise in evaluating responses of infants to sound, e.g., to speech and time-reversed speech, because head movements are tolerated since the optical fiber assembly can be fixed to the head (Pena et al. 2003; Bortfeld et al. 2007).

Logothetis et al. (2001) found that in monkeys the BOLD response directly reflects a local increase in neural activity. For the majority of the recording sites, the BOLD signal was found to be a linear function of local field potential (LFP) amplitude, and the firing rate of small neural populations. LFPs, however, were a substantially more reliable predictor than spike firing rate, and covaried with the BOLD signal even when the spike firings adapted. This suggests that the BOLD response primarily reflects the input (post-synaptic potentials, PSPs) and local processing in neural circuits rather than the output (spike) signals, which are transmitted to other regions of the brain by the principal neuron axons. However, as post-synaptic activity typically results in neural firing, the aggregate firing rate likely also reflects the BOLD signal, as elegantly demonstrated in humans, where intracranially implanted electrodes recorded spiking activity that could reliably predict the BOLD response (Mukamel et al. 2005).

In audiology, there is a long tradition of mapping the activation of the entire auditory nervous system using auditory evoked potentials (AEP) or auditory evoked magnetic fields (AEMF), not just in adults but also very successfully in infants (Shahin et al. 2004). These electrophysiological methods have several advantages over the imaging techniques reviewed so far. AEPs allow the measurement of compound and synchronized responses from the auditory nerve, brainstem, midbrain and cortex, thereby allowing a differential diagnosis for central auditory processing disorders. The structures that give rise to detectable scalp potentials are those that have a spatial alignment and sudden changes in orientation or surrounding resistivity of nerve fibers (auditory nerve and brainstem tracts; Stegeman et al. 1987) or of the dendrites (midbrain and especially cortex; Mitzdorf 1985). The thalamus does not have the proper dendrite alignment to provide for a large far-field response. The PSPs that are the origin of the long-latency AEPs are the same as those that form the basis for the BOLD response, so there should be a nice correspondence. The fact that AEPs also require precise temporal synchrony while the BOLD response does not also suggests that there could be differences in the centers of gravity for the BOLD response and AEP or AEMF activation.

The centers of gravity of the responses picked up by multielectrode scalp recordings can be estimated by reconstructing several equivalent dipoles from the scalp activity. The cortical locations and dipole moments thus obtained present an image of the center of gravity of the neural activity that gives rise to the observed scalp distribution of AEPs (Scherg et al. 2002). Using combined AEP recording and fMRI, we have shown that there is a small but systematic difference in the vertical plane between the centers of gravity for the BOLD responses and the equivalent dipoles calculated from the scalp distribution (Scarff et al. 2004b). This difference could be related to a different group of neurons that are synchronously activated (and provide the AEPs) compared to the likely larger group that combine synchronous and asynchronously firing and underlie the changes seen in the BOLD response. An alternative way of analyzing AEPs, and one that provides an image that resembles those obtained from PET scans, is called low-resolution tomography (Loreta). Loreta results from estimating the minimum distribution of large numbers of dipoles distributed over the cortical surface that also can explain the scalp recordings (Moisescu-Yiflach et al. 2005).

AEMFs are not typically sensitive to subcortical sources but, because scalp and skull impedances do not affect them, allow a more accurate reconstruction of the equivalent cortical dipoles and permit better spatial resolution than NIRS and PET. Whereas these techniques do not rival the spatial resolution provided by fMRI, and cannot resolve the small mirror-symmetric tonotopic maps in core auditory cortex (cf. Fig. 16.5) (Lütkenhöner 2003), they have the distinct advantage that they can provide millisecond-by-millisecond maps of dipole strength and location and the changes therein. As the auditory system is largely concerned with analyzing the temporal aspects of sound, AEPs have a distinct advantage, especially when exploring and diagnosing the auditory brainstem and midbrain systems. AEPs and AEMFs are in fact complementary to the functional imaging techniques based on integrated neural activity. AEPs and AEMFs are not only dependent on the number of activated neurons but, even more importantly, and as said numerous times before, on the degree of synchrony of that activation (Eggermont and Ponton 2002). This is extremely important for a system that has to resolve the fine time resolution of sound. Abnormalities in this fine time resolution are actively explored as indicators of temporal processing deficits, which are believed to underlies certain central auditory processing disorders.

Specific Problems for Imaging in Audiology

Most specific auditory-related imaging problems are associated with the fMRI recording set-up. Two aspects are highlighted here: effects of the magnetic field and effects of the scanner noise.

Magnetic Fields

Sound Delivery Systems

Magnetic fields interact with speaker and headphone systems that are based on coils and magnets. To avoid this problem, sound is typically delivered from a remote speaker or headphone via a plastic tube to the ear canal or to the inside of the ear defender. The sound quality of these devices is inherently poor, standing waves may occur in the tubes, and frequency-dependent attenuation occurs as the acoustic load of the tube to the headphone or speaker is strongly frequency-dependent. The use of electrostatic drivers mounted in ear defenders appears to provide the high-quality sound required for use in audiology research and diagnostics (Chambers et al. 2001). A clever alternative is to use electromagnetic headphones with the magnets removed and to use the scanner's magnetic field instead (Baumgart et al. 1998).

Ballistocardiogram

AEP data obtained inside the MR scanner during simultaneous fMRI and AEP recording (as introduced above) are compromised by two major artifacts. The first is the gradient artifact that reflects the switching of gradient coil currents for the magnetic fields during MRI image acquisition. This artifact signal is much larger than the AEP signal and it could saturate the EEG amplifiers and so block recording. However, regardless of the effects on the amplifiers, the signal will be superimposed on the multielectrode recordings. Since the gradient artifact is relatively invariant over time, it is possible to estimate an average artifact template and subtract it out (Scarff et al. 2004b). The second artifact is the ballistocardiogram (BCG), which is smaller but more variable, and its contribution is in the AEP frequency range (0.5–25 Hz). The BCG artifact is related to the cardiac cycle and seems to represent, among others, the pulse-related movement of the electrically conductive blood. This pulsatile current in the magnetic field causes voltage changes that are picked up by the electrodes. Similar to the gradient artifact correction procedure, a BCG correction approach is typically based on the channel-wise subtraction of a local average artifact template (Scarff et al. 2004b). More powerful techniques have been described recently (Debener et al. 2007).

Cochlear Implants

Cochlear implants are currently the device of choice to rehabilitate totally or largely deaf ears. In most devices, a flexible $\sim$20-mm-long base with up to 24 electrode contacts arranged in pairs is inserted into the $\sim$35 mm cochlea. These multichannel cochlear implants exploit the natural high to low frequency, i.e., basal to apical, organization of the cochlea and auditory nerve fibers. Whereas the hair cells are typically destroyed, in general a sufficient number of auditory nerve fibers remain to be stimulated by the implant. The electrode contacts are connected by sub-cutaneous wires to a coil-based receiver system implanted in the skull slightly above and behind the ear. A magnetic connection is then made with a transmit-ter coil placed above it on the scalp. Such trans-scalp-activated devices cannot be used properly in a MRI scanner due to the presence of the strong magnetic field, which disrupts the electronics. In a unique study using patients with an older type cochlear implant device where the electrode wires were percutaneously connected to the sound processor, it was shown that cochlear implants do activate cortex in an electrode location-specific way. The BOLD response following electrical stimulation of individual intracochlear electrodes was clearly detectable and demon-strated cochleotopic activation of the auditory cortex in a deaf subject (Lazeyras et al. 2002).

Most studies involving cochlear implants have used PET and have elucidated that different brain structures take part in speech analysis, processed via the cochlear implant, than in the acoustically based hearing of speech (Giraud et al. 2000a, 2001). Specifically, during speech, and compared to acoustically hearing controls, postlin-gually deaf cochlear implant subjects showed increased low-level phonological processing (areas BA 22/21 and BA 6) and decreased semantic processing (inferior temporal cortex). This is but one example of adaptive brain plasticity in response to a new form of activation, as typically found after several years of deafness.

Scanner Noise

Adverse Effects

The noise generated by a magnetic resonance scanner provides a major limitation of fMRI studies of the auditory system. The high-level noise is due to the flexing of the gradient coil loops in the static magnetic field as current passes through the loops during imaging. Even in the absence of this particular scanner noise, the liquid helium circulation pump and the air ventilation systems produce low-level noise that is continuously present in the magnet room (Ravicz et al. 2000). Although MR scan-ner noise contains many audible frequencies, the main power of the noise occurs in the frequency range of 1–2 kHz, although higher order resonances of lower inten-sity are also present. This scanner noise can reduce the detectable activity within the auditory system by masking it, habituation of the nervous system, saturation of acti-vation, and impairment of cognitive processing. This makes interpreting functional imaging studies of the auditory system a challenging task. Scanner-related acoustic noise induces a BOLD response in the auditory cortex. It has been shown that, simi-lar to other auditory stimuli, scanner noise induces a hemodynamic response within 2–3 s after the onset of acoustic noise that peaks after 3–8 s (hemodynamic delay) and returns to baseline in ~8 s. The hemodynamic responses to MR-related acoustic noise and stimulus-induced brain activation do not add up linearly; this implies that simple subtraction of this activity from that induced by the intended stimulus is not possible (Moelker et al. 2003).

In a study by Scarff et al. (2004a), normal hearing adults were examined using tones of five different audiometric frequencies (250, 500, 1,000, 2,000, 4,000 Hz) of equal loudness. All auditory stimuli were delivered in the presence of MR scanner noise. Using an imaging protocol with peak MR scanner noise at approximately 1.5 kHz, the perception of loudness and detectable fMRI activity in response to the 1-kHz tone was less than for the other frequencies. When the imaging protocol was changed such that peak MR scanner noise occurred at about 2 kHz, the perception of loudness and detectable fMRI activity in response to the 2-kHz tone in the same subjects was less than for the other frequencies. The reduction in the measured fMRI activity for tones near scanner frequencies may be due to an inflated scanner-induced baseline at those frequencies. In addition, fMRI activity decreased with increasing frequency, possibly due to the upward spread in the masking of low-frequency, high-level tonal stimuli or the proximity of low-frequency core and belt areas of the auditory cortex (see Fig. 16.5). Interestingly, the behavioral and fMRI data showed opposite biases (Fig. 16.4). The behavioral data were biased towards high frequencies, i.e., high frequencies were judged louder in the presence of scanner noise, and the fMRI results were biased towards low frequencies. This suggests that researchers should be cautious of using high-intensity (~85 dB SPL, needed because of the loud scanner noise) auditory stimuli when investigating correlations between perception and BOLD activation. Nonetheless, both measures detected a difference in response near peak scanner noise frequencies when the imaging protocol was changed, demonstrating a significant effect of scanner noise on auditory cortical activation and perception. These results also indicate a direct effect of scanner noise and high-intensity tonal stimuli on measurements such as auditory cortical tonotopic maps (Scarff et al. 2004a).

Ways to Avoid Adverse Effects

The most common method of avoiding some of the masking effects of the scanner noise is to use a sparse sampling method, where the desired sound stimulation starts after the scanner-noise-induced BOLD response has been reduced sufficiently (Di Salle et al. 2001; Scarff et al. 2004b). The drawback is that lengthy experiments are required with interleaved stimulus–silence or stimulus 1–stimulus 2 intervals, and this results in an effective inter-stimulus interval of ~16 s.

Ravicz and Melcher (Ravicz et al. 2001) investigated whether a device for isolating the head and ear canal from unwanted sound (a "helmet") could add to the other isolation provided by conventional hearing protection devices (i.e., earmuffs and earplugs). Both subjective attenuation (the difference in hearing threshold with vs. without isolation devices in place) and objective attenuation (the difference in ear-canal sound pressure) were measured. In the frequency range of the most intense fMRI noise (1–1.4 kHz), a helmet, earmuffs, and earplugs used together attenuated the perceived sound by 55–63 dB, whereas the attenuation provided by the

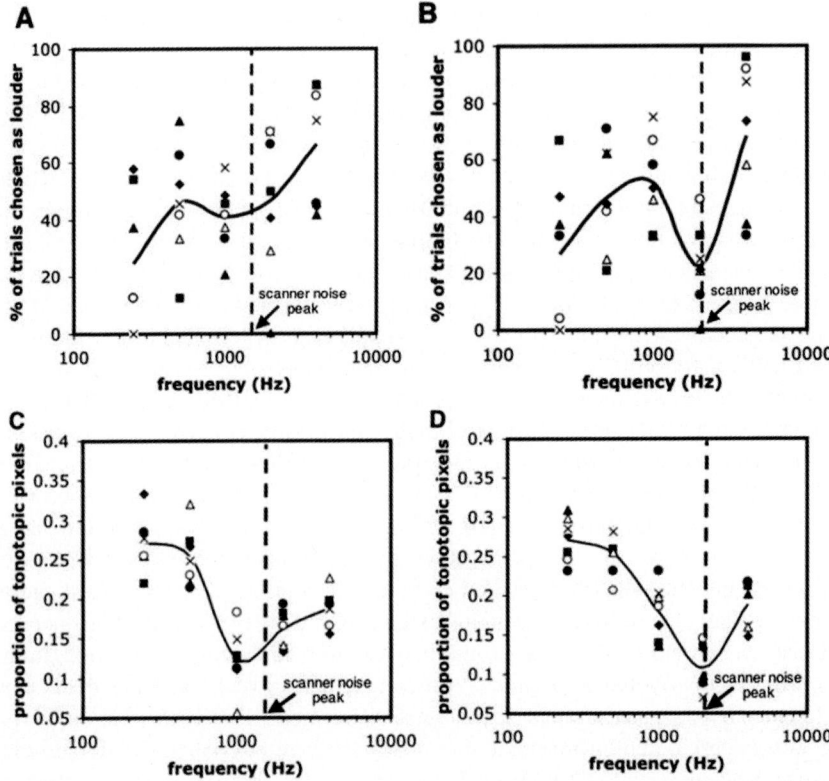

Fig. 16.4 Effects of scanner noise on behavioral performance and BOLD responses to sound of about 85 dB SPL, adjusted when necessary so that each frequency has the same loudness. Seven normal hearing subjects were tested with two imaging protocols that produced peak scanner noise at about 1.5 and 2 kHz (*dashed lines*). The *upper row* shows behavioral results, and the test frequencies near the scanner noise peak were perceived as being less loud than the others. The effect is most clear for scanner noise at 2 kHz. For the BOLD response, the number of tonotopic pixels is clearly reduced near the frequency peaks of the scanner noise. Interestingly, the overall trend in the behavioral results is that higher frequencies are judged as louder, whereas the BOLD responses tend to be lower for higher frequencies (after Scarff et al. 2004a, with permission)

conventional devices alone was substantially less: 30–37 dB for earmuffs, 25–28 dB for earplugs, and 39–41 dB for earmuffs and earplugs used together.

An alternative to these passive methods of reducing the noise would be active noise cancellation, a technique where the sound is measured locally in the earmuff and presented in phase reversal to cancel it at least partially. This technique is used commercially in quiet-listening devices for, e.g., music, and has recently become available for use in MR scanners (Chambers et al. 2001; Chapman et al. 2003). The technique requires a sound delivery system using electrostatic drivers in the ear defender, and allows an extra reduction in scanner noise of 30–40 dB (Chambers et al. 2001) or even up to 50 dB with specially designed EPI cycle pulses (Chapman et al. 2003). This suggests that by combining passive and active noise cancellation, a total attenuation of between 60 and 75 dB is not unrealistic and would allow the use of sound stimuli of approximately 50 dB SPL to probe the auditory system.

A more obvious way, at least from the point of view of the user, to reduce the noise produced by the scanner would be to change the design of it in such a way that the coils producing the sound are acoustically isolated. This would also benefit nonauditory studies such as those on attention, memory and visual perception (see Moelker et al. 2003 for a review of potential solutions).

The Final Frontier: Auditory Processing Disorders

It is still not uncommon to hear from clinical audiologists and ear surgeons that hearing disorders are a problem of the peripheral auditory system, in particular the cochlea and auditory nerve. Consequently, these disorders can be ameliorated by surgery and/or amplification, which is the expertise of surgeons and audiologists. It is of course undisputed that cochlear hearing loss causes a loss in hearing sensitivity and a corresponding reduction in frequency resolution. This is generally the result of the loss of the outer hair cells, which causes the loss of their 40 dB feedback amplification, and consequently a loss in sensitivity and a broadening of the frequency resolution by the basilar membrane. However, as a result of the frequency-dependent hearing loss and consequently frequency-specific differences in the spontaneous and stimulus-related output from the auditory nerve fibers to the central auditory system, plastic changes in the functioning of the brainstem, midbrain and especially the cortex occur over time as well. These plastic changes have been linked to phenomena such as loudness recruitment (reduced dynamic range), hyperacusis (oversensitivity to even soft sounds) and tinnitus (ringing in the ear). Sometimes all these occur in the same person at the same time. Quality of life is greatly reduced if that happens.

Adult Cortical Plasticity

Even in adulthood, the auditory cortex and thalamus remain highly plastic. This is most obvious in the tonotopic map changes that can be induced in animals by, e.g., noise-induced hearing loss (Noreña et al. 2005), long exposure to nondamaging unusual sounds (Noreña et al. 2006), perceptual learning (Fritz et al. 2007), and day-to-day experience of environmental sounds (Kilgard et al. 2007). Tonotopic map changes in animals after noise trauma typically produce a partial, high-frequency, cochlear hearing loss and show an over-representation in cortex of the so-called edge frequency of the audiogram, i.e., the highest frequency that still has near-normal thresholds (Noreña et al. 2005). The region of cortex that normally would represent the high frequencies (say 4–20 kHz) can, after noise trauma that affects hearing at those frequencies, show its most sensitive responses at 4 kHz for the entire, previously high-frequency, region. We also know that in such animals, the AEP locally recorded from the midbrain and cortex is greatly enhanced, suggesting increased neural synchrony (Salvi et al. 2000; Wang et al. 2002).

However, in humans, our subjects of interest, so far we have only been able to obtain indirect indications of changes in the frequency representation in auditory cortex as a result of, e.g., hearing loss (Weisz et al. 2005) or long-term musical training (Shahin et al. 2004). These changes in the tonotopic map are typically inferred from changes in the spatial distribution of the equivalent dipoles calculated from the scalp distribution of the AEPs or AEMFs to pure tones (Mühlnickel et al. 1998; Wienbruch et al. 2006). Whereas AEPs and AEMFs can clearly indicate changes in the dipole moment and location of the equivalent dipoles, these changes may in fact reflect local changes in neural synchrony rather than in the size or change of the activated cortical region. These two effects cannot be readily separated, as they both change the amplitude of the AEP or AEMF. Therefore, high-resolution tonotopic maps using fMRI are needed to provide independent information about potential changes in human auditory cortical maps.

High-resolution tonotopic maps can be obtained by using high-strength magnetic fields such as in the 7 T data provided by Formisano et al. (2003), of which an example is shown in Fig. 16.5. Here, a high-resolution map of mostly core auditory cortex on Heschl's gyrus is obtained. In addition to showing the gradual changes in the best frequency with position (indicated by color), it is also obvious that there are several mirror-symmetric maps. The most obvious pair is indicated by the dotted line from a to f, with the change in direction of the frequency gradient at the point marked c. This degree of spatial resolution is needed to detect changes in cortical tonotopic maps induced by, for instance, hearing loss (Sadato et al. 2004; Suzuki et al. 2002) and changes upon the use of a hearing aid (Hwang et al. 2006) or a cochlear implant (Seghier et al. 2005).

Tonotopic maps in animals are based on the responses at low sound levels, and it is not completely clear if the maps at high sound levels (as typically used in fMRI) would be similar to those near threshold. In order to obtain the most detailed maps, a relatively low level of sound stimulus is needed, as the representations of frequency at higher sound levels tend to overlap partially. This requires a quiet fMRI environment that can only be obtained by good active and passive scanner noise reduction.

Tinnitus

Tinnitus is one of the central auditory disorders, which often results from noise-induced hearing loss or oversensitivity to commonly used drugs (e.g., salicylates or quinine) but which also frequently occur following, e.g., head and neck injuries (Eggermont 2005). From the findings in animal experiments, the following neural substrates of tinnitus have been suggested: reorganization of cortical tonotopic maps, increased spontaneous firing rates, and increased neural synchrony between

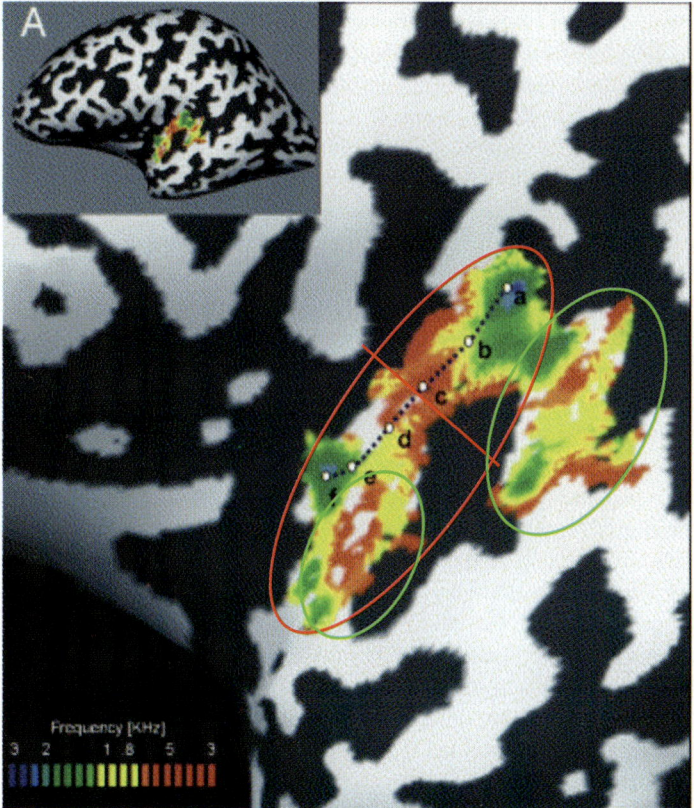

Fig. 16.5 A best-frequency map (statistical analysis based on six frequencies (0.3, *red*); 0.5, 0.8, 1, *yellow*; 2, and 3 kHz, *blue*) is displayed on an inflated representation of the subject's cortex (see insert for location of auditory cortex). The spatial arrangement of best frequencies reveals the existence in the human primary auditory cortex of two continuous representations of frequency (tonotopic maps) that present a mirror symmetry along the rostral–caudal direction. The border of the two maps is approximately at the point marked *c*. Other maps can be discerned in an approximately anteroposterior direction (from Formisano et al. 2003, with permission)

spontaneous firings of individual neurons or during sound stimulation as reflected in enhanced AEPs (Eggermont et al. 2004). Therefore, methods of imaging in humans need to provide not only accurate measurements of reliable changes in tonotopic maps, but also ways to demonstrate increased neural activity during silence (generally seen as the crucial substrate linked to tinnitus). This latter problem cannot readily be solved at the group comparison level because of high variability in resting activity levels. It has been addressed indirectly by focusing on a group of people who can modulate the strength of their tinnitus by either changing eye gaze direction (Lockwood et al. 2001) or by making orofacial movements (Cacace 2003). For these people, it has been demonstrated by using $[^{15}O]H_2O$ PET that increased tinnitus loudness corresponds to increased neural activity in several auditory areas. Group analysis of $[^{15}O]H_2O$ PET data (Plewnia et al. 2007) also showed tinnitus-related increases of regional cerebral blood flow in the left middle and inferior temporal as well as right temporoparietal cortex and posterior cingulum compared to an intravenous lidocaine-induced suppression of tinnitus. Like prior imaging studies, the group data showed no significant tinnitus-related hyperactivation of the primary auditory cortex. These findings support the notion that tinnitus-related neuroplastic changes, as also documented in the primary auditory cortex by electrophysiological methods that are more sensitive to neural synchrony, are not necessarily associated with enhanced perfusion or metabolism.

A potentially promising method of measuring resting levels of activity that would be applicable to investigations of substrates of tinnitus has been proposed by Haller et al. (2006). They used fMRI and CO_2 as a vasodilator to induce a "global" blood oxygenation level-dependent (BOLD) response. They implied, albeit indirectly, that spontaneously active areas in subjects with tinnitus will exhibit a reduced CO_2-induced change in the BOLD response due to a pre-existing tinnitus-induced BOLD response. This putative reduction in the change of the BOLD response compared to non-auditory areas might then be exploited to map the continuous neuronal activation that putatively exists in tinnitus. A comparison with the effect in non-auditory areas could provide a within-subject control.

How can we measure increased spontaneous neural synchrony? One potential way is to record the spontaneous EEG or MEG activity in the gamma frequency band (40–60 Hz). This activity in the visual cortex has been linked to visual perception, and its presence and detection requires locally increased neural synchrony. Weisz et al. (2007) have provided tantalizing evidence that in persons with tinnitus this spontaneous gamma band activity correlates with both the annoyance of tinnitus and its side of localization.

Central Auditory Processing Disorders

Typically, a central auditory processing disorder (CAPD) is diagnosed when there are abnormalities in auditory processing in the absence of a peripheral hearing loss. Examples are language-specific impairments, amusia (tone deafness), aphasias, etc. Functional neuroimaging studies with PET, fMRI, and AEMF of reading in dyslexic and non-reading-impaired adults have found less activation in a left-temporal network around the perisylvian region in dyslexics presented with a demanding phonological task (Demonet et al. 2005). Using DTI, which highlights the structural integrity of the brain wiring, Beaulieu et al. (2005) showed that regional brain connectivity in the left temporoparietal white matter correlates with a wide range of reading abilities in children as young as 8–12 years old. The maturation of the white matter may play a key role in the development of cognitive processes such as reading, as it could explain temporal processing deficits that have been suggested to play a role in dyslexia and specific language impairment.

Voxel-based morphometry (VBM) has revealed gray matter abnormalities in the caudate nucleus in individuals with severe speech and language disorders relative to normal controls. The caudate nucleus is involved in speech processes, as demonstrated by studying family members with inherited speech and language disorders (Watkins et al. 2002). Sigalovsky et al. (2006b) speculated that the use of VBM to detect changes in myelination in auditory cortex that occur over the first two decades of life (e.g., Moore et al. 2001) could replace post mortem histology and would also allow longitudinal studies. Given the crucial role of myelin in maintaining high conduction velocities and thus in the timing of neural activity, which is a crucial requirement for auditory processing, T1 patterns in individuals with auditory temporal processing deficits may be an important diagnostic. More generally, insights into the structural substrates behind human hearing and speech perception might be obtained by assessing gray matter structural variables in patients with central auditory deficits or more general communication disorders.

Individuals afflicted with tone deafness ("congenital amusia") show impaired perception and production of music, despite normal neurological history, hearing, education, intelligence and memory. A deficit in musical pitch processing appears to be a core deficit, since amusic individuals consistently demonstrate a selective impairment in the processing of musical pitch. Hyde et al. (2006) used VBM to detect brain anatomical differences in amusic individuals compared to musical controls by analyzing T1-weighted magnetic resonance images. A reduction in white matter concentration in the right inferior frontal gyrus (IFG) of amusic individuals was found. This was interpreted as a sign of an anomalous wiring of the connections of the IFG to the right auditory cortex, and was correlated with performance on pitch-based musical tasks. Genetic predisposition combined with less exposure to music might cause this reduced myelination of the connections to the frontal regions. In contrast, long-term musical training induces increases in some white matter structures (Bengtsson et al. 2005), suggesting a potential way for rehabilitation.

Combining functional imaging and DTI (Upadhyay et al. 2007) may be the technique of the future for probing central auditory processing disorders (Fig. 16.6). This elegant study combined BOLD imaging of the mirror-symmetric tonotopical areas in core cortex and DTI to probe neural pathways between areas with the same frequency-range maps (homotopic connections) and between sub-areas with different frequency ranges (heterotopic connections). They showed connectivity results that matched with the known connectivity in non-human animals (Lee et al. 2004).

Conclusions

The imaging of auditory structures faces limitations imposed by the location of the inner ear, thus requiring a combination of high-resolution computed tomography and MRI. Functional imaging faces limitations imposed by the adverse acoustic environment during MRI scanning, meaning that a combination of passive and active noise cancellation is required to achieve low-noise measurements. Generally

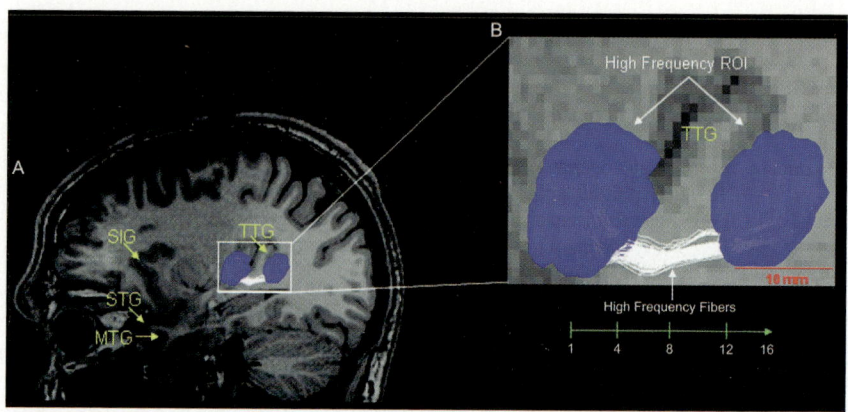

Fig. 16.6 (a–b) Isofrequency-specific fibers illustrated by DTI. **(a)** Fiber tracks connecting the caudomedial and rostrolateral high-frequency ROIs. **(b)** A zoomed view of a sagittal slice medial to high-frequency fibers and ROIs (represented in *blue*) shows the TTG in the background. Adapted from Upadhyay et al. 2007, with permission

speaking, up to now MRI equipment has not adequately catered for the requirements of audiological testing. PET scans provide a quiet environment but suffer from low spatial resolution and the need to average across participants. fMRI in high magnetic fields will allow measurements of cortical tonotopic map reorganization and provide important information about the central aspects of hearing disorders. DTI in combination with fMRI offers as-yet untapped potential for studying central auditory processing disorders. Simultaneous EEG and fMRI in audiology is feasible and allows millisecond temporal resolution and millimeter spatial resolution, and it may provide another combination technique for probing central auditory processing disorders.

Acknowledgements The author was supported by the Alberta Heritage Foundation for Medical Research, by the Natural Sciences and Engineering Research Council, by a Canadian Institutes of Health Research New Emerging Team Grant, and by the Campbell McLaurin Chair of Hearing Deficiencies.

References

Baumgart F, Kaulisch T, Tempelmann C, Gaschler-Markefski B, Tegeler C, Schindler F, Stiller D, Scheich H (1998) Electrodynamic headphones and woofers for application in magnetic resonance imaging scanners. Medical Phys 25:2068–2070
Beaulieu C, Plewes C, Paulson LA, Roy D, Snook L, Concha L, Phillips L (2005) Imaging brain connectivity in children with diverse reading ability. NeuroImage 25:1266–1271
Bendor D, Wang X (2005) The neuronal representation of pitch in primate auditory cortex. Nature 436:1161–1165

Bendor D, Wang X (2007) Differential neural coding of acoustic flutter within primate auditory cortex. Nature Neurosci 10:763–771

Bengtsson SL, Nagy Z, Skare S, Forsman L, Forssberg H, Ullen F (2005) Extensive piano practicing has regionally specific effects on white matter development. Nature Neurosci 8:1148–1150

Bortfeld H, Wruck E, Boas DA (2007) Assessing infants' cortical response to speech using near-infrared spectroscopy. NeuroImage 34:407–415

Cacace AT (2003) Expanding the biological basis of tinnitus: crossmodal origins and the role of neuroplasticity. Hearing Res 175:112–132

Chambers J, Akeroyd MA, Summerfield AQ, Palmer AR (2001) Active control of the volume acquisition noise in functional magnetic resonance imaging: method and psychoacoustical evaluation. J Acoust Soc Am 110:3041–3054

Chapman BL, Haywood B, Mansfield P (2003) Optimized gradient pulse for use with EPI employing active acoustic control. Magn Reson Med 50:931–935

Debener S, Strobel A, Sorger B, Peters J, Kranczioch C, Engel AK, Goebel R (2007) Improved quality of auditory event-related potentials recorded simultaneously with 3-T fMRI: removal of the ballistocardiogram artefact. NeuroImage 34:587–597

Demonet JF, Thierry G, Cardebat D (2005) Renewal of the neurophysiology of language: functional neuroimaging. Physiol Rev 85:49–95

Di Salle F, Formisano E, Seifritz E, Linden DE, Scheffler K, Saulino C, Tedeschi G, Zanella FE, Pepino A, Goebel R, Marciano E (2001) Functional fields in human auditory cortex revealed by time-resolved fMRI without interference of EPI noise. NeuroImage 13:328–338

Eggermont JJ (2001) Between sound and perception: reviewing the search for a neural code. Hearing Res 157:1–42

Eggermont JJ (2005) Tinnitus: neurobiological substrates. Drug Discovery Today 10:1283–1290

Eggermont JJ, Ponton CW (2002) The neurophysiology of auditory perception: from single units to evoked potentials. Audiology Neuro-otology 7:71–99

Eggermont JJ, Roberts LE (2004) The neuroscience of tinnitus. Trends Neurosci 27:676–682

Formisano E, Kim DS, Di Salle F, van de Moortele PF, Ugurbil K, Goebel R (2003) Mirror-symmetric tonotopic maps in human primary auditory cortex. Neuron 40:859–869

Fritz JB, Elhilali M, David SV, Shamma SA (2007) Does attention play a role in dynamic receptive field adaptation to changing acoustic salience in A1? Hearing Res 229:186–203

Giraud AL, Truy E, Frackowiak RS, Gregoire MC, Pujol JF, Collet L (2000a) Differential recruitment of the speech processing system in healthy subjects and rehabilitated cochlear implant patients. Brain 123(Pt 7):1391–1402

Giraud AL, Lorenzi C, Ashburner J, Wable J, Johnsrude I, Frackowiak R, Kleinschmidt A (2000b) Representation of the temporal envelope of sounds in the human brain. J Neurophysiol 84:1588–1598

Giraud AL, Price CJ, Graham JM, Frackowiak RS (2001) Functional plasticity of language-related brain areas after cochlear implantation. Brain 124:1307–1316

Guimaraes AR, Melcher JR, Talavage TM, Baker JR, Ledden P, Rosen BR, Kiang NY, Fullerton BC, Weisskoff RM (1998) Imaging subcortical auditory activity in humans. Human Brain Map 6:33–41

Haller S, Wetzel SG, Radue EW, Bilecen D (2006) Mapping continuous neuronal activation without an ON-OFF paradigm: initial results of BOLD ceiling fMRI. Eur J Neurosci 24:2672–2678

Hwang JH, Wu CW, Chen JH, Liu TC (2006) Changes in activation of the auditory cortex following long-term amplification: an fMRI study. Acta Otolaryngol 126:1275–1280

Hyde KL, Zatorre RJ, Griffiths TD, Lerch JP, Peretz I (2006) Morphometry of the amusic brain: a two-site study. Brain 129:2562–2570

Kilgard MP, Vazquez JL, Engineer ND, Pandya PK (2007) Experience dependent plasticity alters cortical synchronization. Hearing Res 229:171–179

Kral A, Eggermont JJ (2007) What's to lose and what's to learn: development under auditory deprivation, cochlear implants and limits of cortical plasticity. Brain Res Rev 56:259–269

Lane JI, Ward H, Witte RJ, Bernstein MA, Driscoll CL (2004) 3-T imaging of the cochlear nerve and labyrinth in cochlear-implant candidates: 3D fast recovery fast spin-echo versus 3D constructive interference in the steady state techniques. AJNR 25:618–622

Lazeyras F, Boex C, Sigrist A, Seghier ML, Cosendai G, Terrier F, Pelizzone M (2002) Functional MRI of auditory cortex activated by multisite electrical stimulation of the cochlea. NeuroImage 17:1010–1017

Lee CC, Schreiner CE, Imaizumi K, Winer JA (2004) Tonotopic and heterotopic projection systems in physiologically defined auditory cortex. Neuroscience 128:871–887

Lockwood AH, Wack DS, Burkard RF, Coad ML, Reyes SA, Arnold SA, Salvi RJ (2001) The functional anatomy of gaze-evoked tinnitus and sustained lateral gaze. Neurology 56:472–480

Logothetis NK, Pauls J, Augath M, Trinath T, Oeltermann A (2001) Neurophysiological investigation of the basis of the fMRI signal. Nature 412:150–157

Lütkenhöner B (2003) Single-dipole analyses of the N100 m are not suitable for characterizing the cortical representation of pitch. Audiology Neuro-otology 8:222–233

Mitzdorf U (1985) Current source-density method and application in cat cerebral cortex: investigation of evoked potentials and EEG phenomena. Physiological Rev 65:37–100

Moelker A, Pattynama PM (2003) Acoustic noise concerns in functional magnetic resonance imaging. Human Brain Map 20:123–141

Moisescu-Yiflach T, Pratt H (2005) Auditory event related potentials and source current density estimation in phonologic/auditory dyslexics. Clin Neurophysiol 116:2632–2647

Moore JK, Guan YL (2001) Cytoarchitectural and axonal maturation in human auditory cortex. J Assoc Res Otolaryngol 2:297–311

Mühlnickel W, Elbert T, Taub E, Flor H (1998) Reorganization of auditory cortex in tinnitus. Proc Natl Acad Sci USA 95:10340–10343

Mukamel R, Gelbard H, Arieli A, Hasson U, Fried I, Malach R (2005) Coupling between neuronal firing, field potentials, and FMRI in human auditory cortex. Science 309:951–954

Noreña AJ, Eggermont JJ (2005) Enriched acoustic environment after noise trauma reduces hearing loss and prevents cortical map reorganization. J Neurosci 25:699–705

Noreña AJ, Gourévitch B, Aizawa N, Eggermont JJ (2006) Spectrally enhanced acoustic environment disrupts frequency representation in cat auditory cortex. Nature Neurosci 9:932–939

Pena M, Maki A, Kovacic D, Dehaene-Lambertz G, Koizumi H, Bouquet F, Mehler J (2003) Sounds and silence: an optical topography study of language recognition at birth. Proc Natl Acad Sci USA 100:11702–11705

Plewnia C, Reimold M, Najib A, Brehm B, Reischl G, Plontke SK, Gerloff C (2007) Dose-dependent attenuation of auditory phantom perception (tinnitus) by PET-guided repetitive transcranial magnetic stimulation. Human Brain Map 28:238–246

Ravicz ME, Melcher JR (2001) Isolating the auditory system from acoustic noise during functional magnetic resonance imaging: Examination of noise conduction through the ear canal, head, and body. J Acoust Soc Am 109:216–231

Ravicz ME, Melcher JR, Kiang NY (2000) Acoustic noise during functional magnetic resonance imaging. J Acoust Soc Am 108:1683–1696

Ruytjens L, Willemsen AT, Van Dijk P, Wit HP, Albers FW (2006) Functional imaging of the central auditory system using PET. Acta Otolaryngol 126:1236–1244

Sadato N, Yamada H, Okada T, Yoshida M, Hasegawa T, Matsuki K, Yonekura Y, Itoh H (2004) Age-dependent plasticity in the superior temporal sulcus in deaf humans: a functional MRI study. BMC Neurosci 5:56

Salvi RJ, Wang J, Ding D (2000) Auditory plasticity and hyperactivity following cochlear damage. Hearing Res 147:261–274

Scarff CJ, Dort JC, Eggermont JJ, Goodyear BG (2004a) The effect of MR scanner noise on auditory cortex activity using fMRI. Human Brain Map 22:341–349

Scarff CJ, Reynolds A, Goodyear BG, Ponton CW, Dort JC, Eggermont JJ (2004b) Simultaneous 3-T fMRI and high-density recording of human auditory evoked potentials. NeuroImage 23:1129–1142

Scherg M, Ille N, Bornfleth H, Berg P (2002) Advanced tools for digital EEG review: virtual source montages, whole-head mapping, correlation, and phase analysis. J Clin Neurophysiol 19:91–112

Schreiner CE, Langner G (1988) Periodicity coding in the inferior colliculus of the cat. II. Topographical organization. J Neurophysiol 60:1823–1840

Seemann MD, Beltle J, Heuschmid M, Lowenheim H, Graf H, Claussen CD (2005) Image fusion of CT and MRI for the visualization of the auditory and vestibular system. Eur J Med Res 10:47–55

Seghier ML, Boex C, Lazeyras F, Sigrist A, Pelizzone M (2005) FMRI evidence for activation of multiple cortical regions in the primary auditory cortex of deaf subjects users of multichannel cochlear implants. Cereb Cortex 15:40–48

Shahin A, Roberts LE, Trainor LJ (2004) Enhancement of auditory cortical development by musical experience in children. Neuroreport 15:1917–1921

Sigalovsky IS, Melcher JR (2006a) Effects of sound level on fMRI activation in human brainstem, thalamic and cortical centers. Hearing Res 215:67–76

Sigalovsky IS, Fischl B, Melcher JR (2006b) Mapping an intrinsic MR property of gray matter in auditory cortex of living humans: a possible marker for primary cortex and hemispheric differences. NeuroImage 32:1524–1537

Stegeman DF, Van Oosterom A, Colon EJ (1987) Far-field evoked potential components induced by a propagating generator: computational evidence. Electroencephal Clin Neurophysiol 67:176–187

Suzuki M, Kouzaki H, Nishida Y, Shiino A, Ito R, Kitano H (2002) Cortical representation of hearing restoration in patients with sudden deafness. Neuroreport 13:1829–1832

Trimble K, Blaser S, James AL, Papsin BC (2007) Computed tomography and/or magnetic resonance imaging before pediatric cochlear implantation? Developing an investigative strategy. Otol Neurotol 28:317–324

Upadhyay J, Ducros M, Knaus TA, Lindgren KA, Silver A, Tager-Flusberg H, Kim DS (2007) Function and connectivity in human primary auditory cortex: a combined fMRI and DTI study at 3 Tesla. Cereb Cortex 17:2420–2432

Wang J, Ding D, Salvi RJ (2002) Functional reorganization in chinchilla inferior colliculus associated with chronic and acute cochlear damage. Hearing Res 168:238–249

Watkins KE, Vargha-Khadem F, Ashburner J, Passingham RE, Connelly A, Friston KJ, Frackowiak RS, Mishkin M, Gadian DG (2002) MRI analysis of an inherited speech and language disorder: structural brain abnormalities. Brain 125:465–478

Weisz N, Wienbruch C, Dohrmann K, Elbert T (2005) Neuromagnetic indicators of auditory cortical reorganization of tinnitus. Brain 128:2722–2731

Weisz N, Muller S, Schlee W, Dohrmann K, Hartmann T, Elbert T (2007) The neural code of auditory phantom perception. J Neurosci 27:1479–1484

Wienbruch C, Paul I, Weisz N, Elbert T, Roberts LE (2006) Frequency organization of the 40-Hz auditory steady-state response in normal hearing and in tinnitus. NeuroImage 33:180–194

Chapter 17
Applications of Molecular Imaging with MR

Linda B. Andersen(✉) and Richard Frayne

Abstract Molecular imaging shows great promise as a method of studying biological systems noninvasively. This chapter focuses on efforts to merge two scientific disciplines, biomedical imaging using magnetic resonance (MR) and molecular biology, for the purpose of *directly* visualizing cellular and molecular structures within their natural host environment, thereby minimizing perturbations of the system under study. A range of advanced imaging techniques and reporter probes are now becoming available that offer an alternative to the more traditional radioactive imaging agents employed. These new MR-based molecular imaging approaches capitalize on methodologies that improve specificity, a factor that will be of direct benefit to critically ill patients. Although most approaches remain in the experimental stage, primarily employing animal models, the methods and techniques discussed in the chapter are continually being refined and optimized, and are moving towards patient care.

Abbreviations

Δ_x	spatial resolution
Δ_t	temporal resolution
AD	Alzheimer's disease
Aβ	amyloid-beta
C_i	image contrast
CLIO	crosslinked iron oxide
CNS	central nervous system
CT	computed tomography

L.B. Andersen
Radiology and Clinical Neurosciences, Hotchkiss Brain Institute, University of Calgary, Calgary, Alberta T2N 1N4, Canada and Seaman Family MR Research Centre, Foothills Medical Centre, Calgary Health Region, Calgary, Alberta, T2N 2T9, Canada, e-mail: lbanders@ucalgary.ca

DTPA	diethylenetriaminepentaacetic acid
FDA	Federal Drug Administration
fMRI	functional magnetic resonance
Gd	gadolinium
Gd-DTPA	gadolinium diethylenetriaminepentaacetic acid
Gd-DTPA-DMPE	gadolinium 1,2-dimyristoyl-*sn*-glycero 3-phosphoethanolamine diethylenentriaminepentaacetic acid
HDL	high-density lipoprotein
MHz	megaHertz
MI	molecular imaging
MPIO	micron-sized iron oxide particles
MR	magnetic resonance
MRS	magnetic resonance spectroscopy
N	noise
NK	natural killer
NSF	nephrogenic systemic fibrosis
PET	positron emission tomography
PS	phosphatidylserine
$S_{background}$	signal in background tissue
S_x	signal in tissue of interest
SNR	signal-to-noise ratio
SPECT	single-photon emission computed tomography
SPIO	superparamagnetic iron oxide particle
T	Tesla
T1	characteristic MR relaxation time describing loss of energy to the lattice
T2	characteristic MR relaxation time describing affects of time-varying and stationary magnetic field
T2*	characteristic MR relaxation time describing affects of stationary magnetic field
TE	echo time
TR	repetition time
USPIO	ultrasmall iron oxide particle
VEGF	vascular endothelial growth factor

Introduction

Molecular imaging (MI) is a relatively new discipline aimed at the understanding and assessment of biochemical activity via the *direct* visualization of cellular and subcellular events. This is in contrast to other imaging techniques that focus on imaging structure and function and, at best, provide *indirect* measures of molecular activity. The ability to image macromolecules (such as proteins, carbohydrates and lipids) directly is critical, as many, if not all, healthy and diseased biological processes result from (or are accompanied by) subtle biochemical changes. In the most general sense, MI describes a broad class of techniques that link and integrate the disciplines of biomedical imaging and molecular biology for the specific purpose of visualizing normal and abnormal cellular and molecular processes in living organisms.

Success in MI is dependent upon two critical elements: (a) the development of suitable reporter probes, such as specific tracers and contrast agents, and (b) the availability of imaging instrumentation capable of a combination of high spatial resolution and sensitivity to the probe (Lecchi et al. 2007).

Conventional histology, or microscopic examination of excised tissue after staining, is readily understood, very frequently used by the biologist, and has been the predominant technique for visualizing tissue and cellular morphology. The basis of this technique relies upon the adequate fixation and suitable staining of the tissue prior to examination. Thus, this methodology requires not only an excised specimen, but represents a highly static view of the tissue. MI, on the other hand, represents a completely noninvasive method of imaging living tissue, facilitating the visualization of dynamic molecular events over time. In practice, noninvasive visualization of molecular events is the goal of nearly all MI methods currently under development.

A clear benefit of a noninvasive method of examining live tissue in situ is the potential for enhanced understanding of biological processes and for very early characterization of pathophysiological transformations within the context of the natural host environment, without perturbing the system under study. This makes for a very powerful tool for preclinical investigations, and potentially opens a pathway to the development of analogous clinical studies. A second important advantage of noninvasive MI methods is the reduced need for invasive tissue sampling to obtain tissue and cells for histological or other in vitro studies. For example, in order to characterize a tumor in the clinical situation, a biopsy is necessary. In the case of preclinical animal experiments, the subjects are often divided into groups and serially euthanized following the experiment for histological analysis. The invasiveness of these clinical and preclinical procedures is clearly undesirable and has provided the impetus for developing imaging modalities that allow the sensitive and selective visualization of cellular and subcellular structures and molecular function in living organisms in a minimally invasive fashion.

Today's MI techniques employ a range of advanced and noninvasive medical imaging methodologies. MI has traditionally utilized radioactive agents. These nuclear medicine approaches are prolific and have been exhaustively reviewed, most recently in Culver et al. (2008), Hargreaves (2008), and Weber et al. (2008). It is essential to first recognize that our collective MI knowledge draws heavily on the previous and extensive experience obtained by imaging with radioactive agents. Here, however, our presentation will focus on magnetic resonance (MR) imaging approaches, reflecting some of the exciting, emerging research efforts in this field. It should be noted that MR-based MI is a developing field, specifically in the preclinical and clinical research settings, where MI techniques have a number of putative advantages. The foremost advantage of MR-based MI is the inherent superposition of molecular information on high-quality cross-sectional (or tomographic) images of anatomy. In particular, MR can generate images with superior soft tissue contrast. It should be recognized, however, that current MR-based MI approaches, while possessing great potential for human imaging, have only demonstrated clinical application thus far.

This chapter will focus on describing and reviewing emerging MR-based MI applications. We recognize that we have not been comprehensive in our treatment; we have instead chosen to illustrate the key concepts, new ideas and recent developments in this emerging field. Here, we first provide a general introduction of MI reporter probe and MR imaging techniques, before focusing on new applications that use MR-based MI in selected diseases in the fields of, for example, oncology, cardiology and neurology.

Reporter Probe Concepts

The reconsideration and retooling of various histological concepts and methods to adapt them to either preclinical or clinical medical imaging is now underway. In particular, approaches that use nonradioactive tracers are under development and will complement and perhaps augment the better-established radioactive approaches, including single-photon emission computed tomography (SPECT) and positron emission tomography (PET). MR contrast agents seek to provide a detectable change in the MR signal that will result in acceptable image contrast. Both signal enhancement (or "positive" contrast) or signal decrease (or "negative" contrast) may be used, depending on the agent and the MR imaging technique. In general, signal-enhancing agents are preferred, as areas of signal loss can be attributable to the imaging agent, but may also be due to other factors (including air and bone) or be a result of areas of magnetic field inhomogeneity.

A large proportion of MI reporter probes have been developed for PET and SPECT, primarily due to the high sensitivity that can be obtained by these modalities. In light of the imaging advances with MR though, reporter probes have recently been created for MR molecular imaging. While PET and SPECT imaging have the advantage of high sensitivity, the disadvantages may be seen to outweigh the advantages. For example, a cyclotron is required to generate *short-lived* radioisotopes for the generation of the specific PET probes, which is not only expensive but also inconvenient if the user of the isotope is not within close traveling distance of the cyclotron. SPECT isotopes are longer lived but still require the availability of a radioactive agent from an onsite generator or an offsite reactor. Additionally, and more importantly, patients are exposed to the ionizing radiation of the isotopes during their PET or SPECT exam. As such, there is clear motivation to develop molecular probes for MR MI, and research into this area is becoming more widespread.

MI differs from more conventional imaging methods in which low-specificity contrast agents are typically used. MI agents differ in that they have high specificity and can precisely target an area or molecular mechanism of interest. For example, superparamagnetic iron oxide (SPIO) particles can be chelated to either the ligand

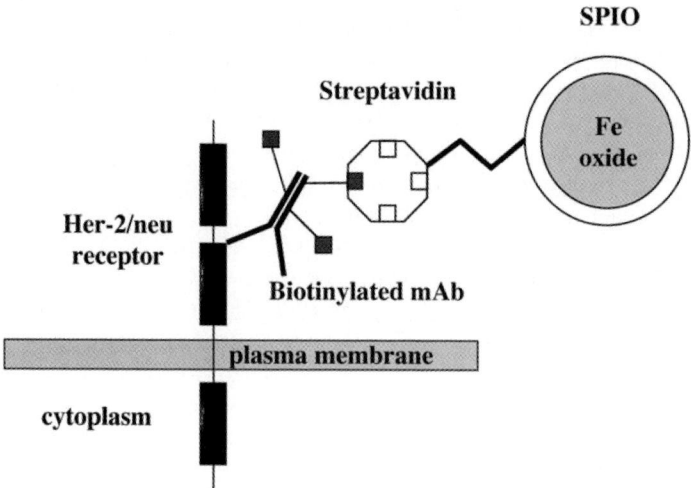

Fig. 17.1 A schematic representation of a molecular imaging agent. The imaging moiety, a superparamagnetic iron-oxide (SPIO) particle, is linked to an antibody directed against the Her-2/neu receptor. The SPIO-antibody complex represents the imaging probe that will bind to the Her-2/neu receptor on the cell surface membrane. This signal can be amplified when several biotin molecules are attached to the antibody. Streptavidin, which has a specific affinity to biotin, is chelated to the SPIO, facilitating the binding of several SPIOs to one antibody. Image from Artemov et al. (2003)

of a receptor of interest or to an antibody directed against a protein of interest, thus providing specific information about the location and/or abundance of the receptor or protein. The hypothetical probe complex (i.e., SPIO-antibody or SPIO-ligand) would be an example of a molecular imaging agent (Fig. 17.1). It is the targeted contrast agent or the imaging probe that will locate and bind to targeted molecules via a direct probe–target interaction. Individual cells may also be used as targeted contrast agents if they are loaded with paramagnetic particles, such as iron oxide, and subsequently home in to specific sites of interest, such as tumors or regions of inflammation (see the "Applications for Molecular Imaging" section). Many different types of imaging probes exist and are being developed for the full range of imaging modalities, including MR, optical, PET, SPECT and ultrasound imaging. However, this chapter limits the discussion to probes for MR imaging.

One of the important advantages of MI over conventional imaging is the ability to visualize ongoing cellular and metabolic events in longitudinal studies in live subjects. As such, it is imperative that imaging probes are designed such that they do not disturb or they only minimally perturb the system under study. In fact, the design of the probe must meet several criteria. The imaging probe must be highly specific to its target in order to eliminate or reduce background signal. The probe must not interact with the living subject outside of its intended purpose and be cleared from the system within a reasonable timeframe without residual side effects. Neurological diseases present an additional unique challenge, in that, in order to visualize the pathology in question, it is often necessary for the probe to cross the blood–brain barrier. In this case the probe-signaling compound must therefore be small enough (usually less than 500 Da) to gain unrestricted access to the brain parenchyma from the blood. Probes may also be complexed with lipid soluble components or molecules that act as specific transport systems to facilitate movement across the endothelial cell membranes that comprise the blood–brain barrier. Lastly, and most importantly, imaging probes should be designed in a manner that will facilitate eventual approval for use in humans.

Imaging modalities such as PET and SPECT are more commonly used for MI because of their excellent sensitivity and ability to generate tomographic images; however, the spatial resolution afforded by MR is superior (Dobrucki and Sinusas 2007). In addition, MR can provide exquisite anatomical detail, showing excellent soft tissue contrast that is ideal for subsequent image fusion of molecular information onto the underlying anatomy. Despite the low molecular sensitivity of MR, targeted contrast agents directed at specific molecules are being developed, thereby expanding the range of MR applications by combining noninvasiveness and high spatial resolution with the specific localization of molecular targets (Lecchi et al. 2007).

Tomographic Imaging

MR and computed tomography (CT) are two imaging modalities that form the basis of clinical cross-sectional (tomographic) imaging in humans. Over the past decade, both MR and CT have also played larger roles in preclinical medicine. The lack of risk due to ionizing radiation (from X-rays) makes MR a more preferred imaging modality compared to CT. The excellent soft-tissue image contrast obtained with MR makes it potentially more suitable for imaging organs and other non-bone structures. For these reasons, nearly all nonradioactive, cross-sectional MI efforts performed to date have focused on MR. MI is potentially possible with CT imaging, but (outside of nonspecific, typically intravascular agents) little work has been published in this area.

Signal-to-noise ratio (SNR), image contrast (C_I) and resolution are three key issues that need to be assessed when evaluating the application of a medical imaging technology. They are commonly defined as:

- Signal-to-noise ratio (SNR): the ratio of the measured signal in the tissue of interest (S_x) to the noise level (N), $SNR = S_x/N$. Typically $SNR > 5$ is required to visualize structures (the Rose criterion) (Bushberg 2002).
- Image contrast (C_I): the difference in image intensities between two tissues—normally the tissue of interest and the background tissue ($S_{background}$). Often expressed as a fractional signal enhancement relative to the background, $C_I = \left(S_x - S_{background}/S_{background}\right)$. In MI, image contrast is synonymous with the sensitivity of the targeting agent.
- Spatial and temporal resolution (Δx and Δt): the ability of the imaging technique to resolve two objects separated in space or in time, respectively. In

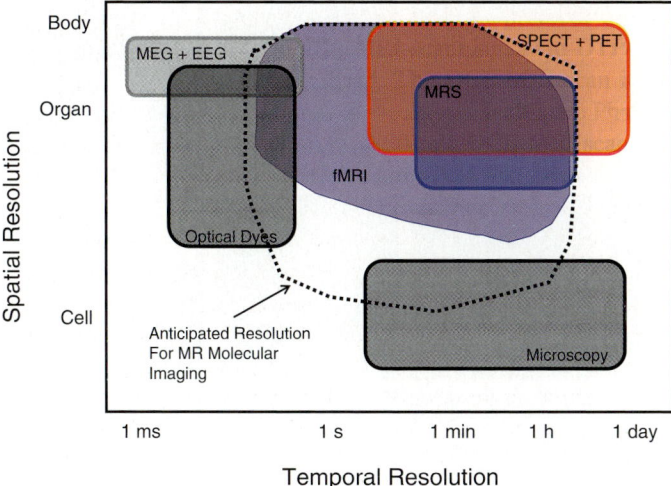

Fig. 17.2 A simplified plot of the spatial–temporal resolution occupied by various medical and nonmedical imaging modalities. The putative temporal spatial resolution for molecular MR imaging is shown as a *dashed line*

tomographic imaging, spatial and temporal resolutions are normally inversely coupled—increasing one will result in a decrease in the other for a fixed acquisition time. Figure 17.2 shows typical spatial–temporal resolution for MR and other common imaging systems.

These three key parameters are nearly always interrelated. For example, achieving a good SNR often results in the degradation of the spatial resolution. A high inherent image contrast (perhaps obtained by imaging a targeted MI agent) may offset the impact of low SNR or mitigate the affect of low spatial resolution. Contrast-detail measurements (Fig. 17.3) can help to understand the relationship between SNR, C_I and resolution (Bushberg 2002). It is important to understand that MI does not require sufficient resolution to directly visualize molecules; rather, it frequently strives to maximize the image contrast between the agent and background tissue so that molecular detail can be visualized, even at a selectively macroscopic spatial resolution (Fig. 17.3g).

The ability to visualize molecular activity in spite of relatively poor spatial resolution (i.e., with respect to molecules) is one of the key hurdles to MI. Nuclear medicine techniques, including PET, SPECT and autoradiographic approaches, employ radioactive agents to increase signal sensitivity. Radioactive MI agents generally have inherently high image contrast because (a) they detect a significant portion of the emitted signal (i.e., have a high inherent sensitivity, large S_x) and (b) minimal background signal ($S_{background} \rightarrow 0$). MR-based MI modalities in general have significantly reduced sensitivities (Table 17.1), and contrast agents typically need to be present at concentrations that are at least $10,000\times$ greater than in PET (Cassidy and Radda 2005). This observation is a key limitation and has a number of

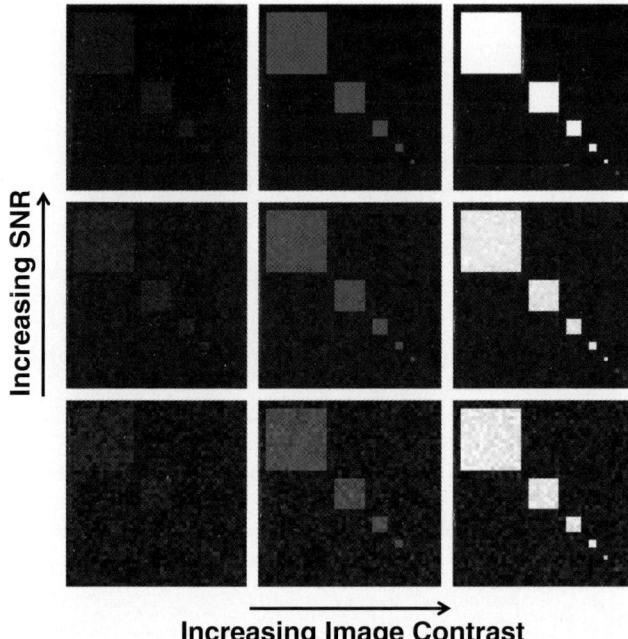

Increasing SNR →

Increasing Image Contrast →

Fig. 17.3 Contrast-detail measurements illustrating that for high image contrast it is possible to see objects smaller than the pixels in the image if the signal-to-noise (SNR) is adequate. The object in the lower right portion of each panel is 1/16 of the area of the displayed pixel gird. In the context of medical imaging, this suggests that if the effects of a MI agent are large, then it is possible to see events smaller than the image resolution

Table 17.1 Comparison of molecular agent sensitivities for MR and nuclear medicine techniques

Modality	Sensitivity
Magnetic resonance (MR)	$10^{-3}\,\mathrm{mol\,L^{-1}}$ to $10^{-5}\,\mathrm{mol\,L^{-1}}$
Single-photon emission computed tomography (SPECT)	$10^{-10}\,\mathrm{mol\,L^{-1}}$
Positron emission tomography (PET)	$10^{-11}\,\mathrm{mol\,L^{-1}}$ to $10^{-12}\,\mathrm{mol\,L^{-1}}$

MR-based MI has $>10^5$-fold reduced sensitivity. Derived from Cassidy and Radda (2005)

important implications, including (a) the potential for undesired, nonphysiologic, or otherwise confounding biological effects (i.e., receptor saturation) and (b) potential toxicological effects at a high MI agent dose.

MR Imaging

Because MR imaging technology affords good spatial resolution and very high soft-tissue discrimination, it facilitates the detection of a multitude of abnormalities including occluded vessels, tumors and stroke. Additionally, images may be acquired in any imaging plane. By changing the imaging parameters, different tissues and organs can be distinguished from one another. Injected MR contrast agents are sometimes employed to enhance this discrimination. To date, MR is widely used for anatomic and functional imaging, and is a key diagnostic imaging modality in hospitals around the world. MR has also become a preferred means for understanding animal models of disease, and for subsequent preclinical evaluation of new drugs and procedures.

Historical Review

MR is based on a phenomenon known as nuclear magnetic resonance that is exhibited by isotopes with an odd number of neutrons or an odd number of protons. Groups led by Felix Bloch and Edward Purcell independently described the phenomenon in 1946, sharing the 1952 Nobel Prize for Physics for their work. Biologically significant isotopes that exhibit nuclear magnetic resonance include 1-hydrogen (^{1}H), 13-carbon, 23-sodium, and 31-phosphorus, though ^{1}H is used in the vast majority of preclinical and clinical applications. When placed in a magnetic field, a slight imbalance in the orientations of the atoms occurs. An excess of atoms oriented in the direction parallel to the applied field results in a measurable signal. The frequency of the measured signal is directly proportional to the strength of the applied field. Prior to the 1970s, nuclear magnetic resonance was developed and used to analyze the compositions and structures of biochemical and other compounds.

In the 1970s researchers began to recognize the potential for MR to form images (Lauterbur 1973) and also to distinguish cells by their differing local chemical environments (Damadian 1971). Pioneering work by the Lauterbur and Mansfield groups led to the development of prototype MR imaging systems (Lauterbur 1973; Mansfield 1982). These systems intentionally introduced a small directional variation in the applied magnetic field in order to encode positional information into the frequency of the measured signal. During image reconstruction, the frequency information is decoded into positional information. The 2003 Nobel Prize in Medicine was awarded to Lauterbur and Mansfield for their early work in developing MR imaging.

In the early-to-mid 1980s, the first whole-body human MR systems were commercialized. These systems allowed, for the first time, tomographic imaging without the use of ionizing radiation. Initial applications focused on neuroimaging and capitalized on the superior soft tissue image contrast associated with MR imaging. Applications in body, musculoskeletal and angiographic imaging followed. In 1992 a powerful technique known as functional MRI (fMRI) was developed to facilitate the mapping of brain activity (Ogawa et al. 1990). Over the past 25 years, MR techniques have been developed that have refined the imaging process, thereby further extending the clinical usefulness and effectiveness of MR technology.

In parallel, similar developments have occurred in preclinical MR imaging, particularly over the last decade (Brockmann et al. 2007). The essential advantage of MR imaging in preclinical studies is the relatively noninvasive nature of the methodology—allowing for the same animal to be imaged repeatedly in order to monitor the progression of a disease or the response to therapy. Serial imaging of the same animal in most study designs improves the power of the study and thus can reduce the number of required experimental animals.

Key Imaging Principles

An MR imaging system consists of four key components: (a) a magnet to produce the main magnetic field, (b) a radiofrequency transmitter and receiver, (c) magnetic field gradients, and (d) controlling computer and ancillary electronics (Fig. 17.4). The bulk of the volume of the MR scanner is a cylindrical magnet, and the subject is placed into the opening (bore) in the middle of the cylinder. In most clinical and preclinical systems, this magnet produces the main magnetic field that runs parallel to the ground. Systems for human use have field strengths ranging from 0.7 to 3 T, with a few experimental systems operating at fields from 7 to 9.4 T. Preclinical systems have smaller bores, making high field strength systems (e.g., 4.7–9.4 T) more economically feasible. In general, the available MR signal increases with higher field strength and motivates the use of higher magnetic fields; though commensurate decreases in image quality and performance issues can complicate this decision.

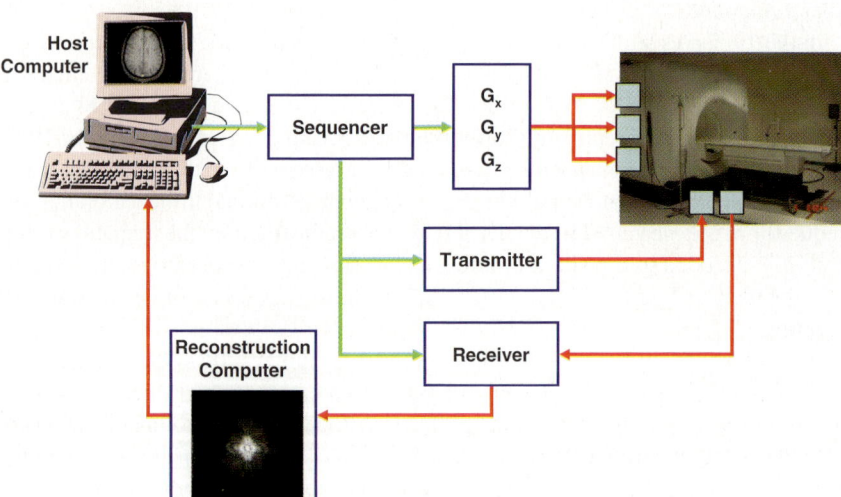

Fig. 17.4 (a–d) The MR imaging system consists of four key components: (**a**) a magnet, (**b**) a radiofrequency (RF) transmitter and receiver, (**c**) a magnetic field gradients and (**d**) a controlling computer and ancillary electronics

The main magnetic field acts to align some of the atoms of the isotope under-study; the atoms align either parallel or antiparallel to the applied field. In nearly all of clinical and most preclinical MR imaging, the ^{1}H isotope is examined. Less commonly, 13-carbon, 23-sodium and 31-phosphorus are examined, mainly in pre-clinical imaging or in higher magnetic field studies. ^{1}H imaging predominates because of its high inherent nuclear magnetic signal properties and because of the natural abundance of hydrogen in living systems.

The second component of an MR system is the radiofrequency transmitter–receiver. This component produces short pulses of energy in the radiofrequency range of the electromagnetic spectrum (i.e., transmission). The exact frequency is selected to be the same as that of the isotope under study at the applied field (the resonance frequency). For example, for ^{1}H the frequency is 64 MHz at 1.5 T and 128 MHz at 3 T. The applied energy is absorbed by some of the isotopes that are aligned parallel and antiparallel to the main magnetic field, causing these molecules to rotate (or nutate) and lie perpendicular to the main magnetic field. A weak radiofrequency signal is then emitted from the isotopes as they return (or relax) back to their rest state, which is "received" by the radiofrequency transmitter–receiver coil. One or more imaging transmitter–receiver coils are used to excite the isotopes and detect the resulting signal.

The rate of relaxation or emission of the absorbed energy by the atoms is governed by two relaxation times, known as T1 and T2 (or its relative T2*). These relaxation rates are functions of the local environments of the isotope being imaged. In general, different tissues have different relaxation times, and in conventional MR imaging sequences are the primary determinants of image contrast. T1 relaxation describes the loss of energy from the excited isotope to the surrounding tissue. T2/T2* relaxation results from a loss of coherence between adjacent isotopes due to small variations in the magnetic field. When averaged over a group of isotopes, this incoherence suggests an apparent energy loss; though at the level of an individual spin no signal is lost. T2* relaxation describes the combined effect of the stationary and nonstationary components of magnetic field inhomogeneity. This apparent signal loss is partially reversible if the MR signal is refocused into "echoes," thereby correcting for the stationary component of the magnetic field inhomogeneity. T2 relaxation describes the apparent relaxation due to the nonstationary components. T2* includes both the time-varying (or nonstationary) and time-invariant (or stationary) components of magnetic field inhomogeneity, and T2 includes only the time-varying effects, T2 $\geq$ T2*.

MR imaging is a very flexible imaging modality, and a range of image contrasts are possible simply by varying the acquisition timing (Fig. 17.5). The two primary imaging parameters are the echo time (TE), the time between excitation and when the signal echo is formed and measured, and the repetition time (TR), the time between successive excitations. The TR affects the amount of T1-weighting in the image, with shorter TRs having image contrast derived from the difference in T1 between tissues. Similarly, T2- or T2*-weighting increases with longer TE times. A potential mechanism for generating high image contrast in future MR-based MI agents could result from localized alterations in T1 or T2/T2* relaxation

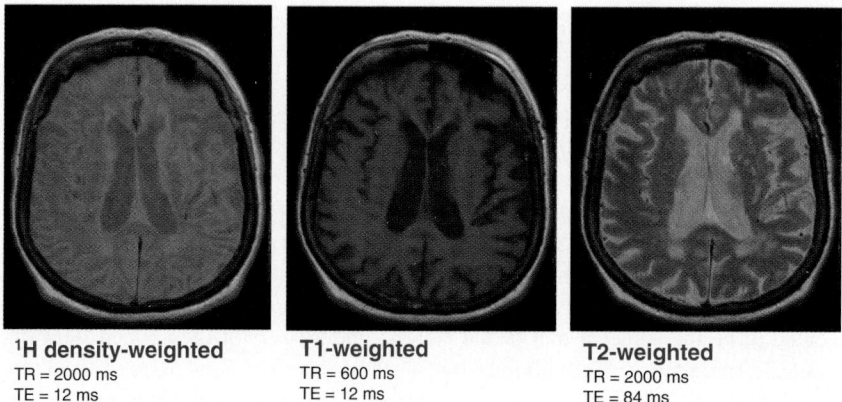

¹H density-weighted   **T1-weighted**   **T2-weighted**
TR = 2000 ms   TR = 600 ms   TR = 2000 ms
TE = 12 ms   TE = 12 ms   TE = 84 ms

Fig. 17.5 In general, different tissues have different relaxation times. These are the primary determinants of image contrast in conventional MR imaging sequences. MR imaging is a very flexible imaging modality, and a range of image contrasts can be obtained simply by varying the acquisition timing parameters (TE and TR)

times. Images with a long TR and a short TE have neither T1- nor T2/T2*-weighting and reflect the density of the atom (for ¹H imaging, this is referred to as proton density imaging).

In order to produce spatially localized maps of the relaxing ¹H atoms (i.e., images), a third component, magnetic field gradients, is needed. Spatial localization is achieved by applying additional magnetic field gradients that vary linearly across the subject. Gradients of varying strength are applied in each of the three orthogonal directions to allow for three-dimensional localization needed for tomographic imaging. During the TE interval, the gradient magnetic fields are turned on and off according to a specific and predetermined pattern. The magnetic field gradient pattern is changed between successive TR intervals to allow a full sampling of all necessary projections.

The fourth component consists of the host and reconstruction computers and ancillary electronics that coordinate the activities of the other components. Modern MR systems include a host computer and sophisticated user interface software for adjusting acquisition and reconstruction parameters such as TE and TR. Image visualization software for reformatting volumetric data and simple data analyses is also often included. The user interface and the processing software run on the controlling host computer. The reconstruction computer is optimized for generating MR images from the acquired raw data. Often reconstruction occurs on a separate (specially optimized) computer; though in some lower-cost, nonclinical systems, reconstruction can occur using the host computer hardware.

Current MR Contrast Agents

There are two main classes of contrast materials used for MI with MR: those based on gadolinium and those based on iron oxide nanoparticles or microparticles. Currently, the application of Gd-containing compounds is less common than iron oxide compounds, particularly in cellular imaging applications. This is primarily due to the comparatively low relaxivity properties of Gd and the decrease in relaxivity observed once internalized by a cell (Politi 2007). As such, high gadolinium concentrations are required (perhaps greater than 10^{-5} mol L^{-1}, see Table 17.1) in order to obtain the necessary sensitivity for the MR imaging of cells.

Gadolinium-Containing Agents

Gadolinium is a rare earth element. It has paramagnetic properties because it has seven unpaired electrons that cause local perturbations in the magnetic field and couple strongly with surrounding ^{1}H atoms, reducing their relaxation times. Changes in MR image contrast due to gadolinium are achieved by virtue of its strong effect on the T1 and T2* relaxation times. Both T1 and T2* decrease with increasing concentration of the Gd^{3+} ion, leading to desired modifications of image contrast in T1-weighted and T2*-weighted images. When injected into animals or humans, free Gd^{3+} ions are toxic (Barge et al. 2006) and are therefore first chelated to a metabolically inert complex to render them relatively stable (Wastie and Latief 2004).

One commercially available formulation chelates Gd^{3+} with diethylenetriamine pentaacetic acid (DTPA) to yield Gd-DTPA. Other formulations have also been developed, including: Magnevist® (gadopentetate dimeglumine), MultiHance® (gadobenate dimeglumine), OmniscanTM (gadodiamide), OptiMARK® (gadoversetamide), and ProHance® (gadoteridol), which have gained US Federal Drug Administration (FDA) approval for use in humans. These complexes are relatively large and cannot cross the intact blood–brain barrier. Of critical importance, however, are recent findings that have suggested that some formulations of these agents may cause an elevated risk of nephrogenic systemic fibrosis (NSF), specifically in patients with known or suspected renal disease (Broome 2008). While the exact mechanism connecting MR contrast agents and NSF is unknown, these recent findings serve to highlight the potential risk from imaging agents.

These agents, however, can all be classified as having no or relatively little specificity. To create a targeted, or more specific, contrast agent for MR imaging, gadolinium-based contrast agents are typically linked to molecules that include a moiety having a specific affinity for a protein of interest, such as a cell surface signaling receptor. For example, the protein annexin V has a high affinity for the protein phosphatidylserine (PS), which is expressed on the surface of apoptotic cells (Emoto et al. 1997). Synthesis of a PS-targeted Gd-based contrast agent was described by Hiller and colleagues (Hiller et al. 2006). Essentially, annexin V was linked via covalent binding with Gd-DTPA liposomes. The paramagnetic Gd^{3+} ions were incorporated into the membranes of the liposomes to enable interaction with surrounding protons. The Gd-DTPA-annexin V liposome complexes then effectively targeted cell-surface PS on apoptotic cells in isolated rat hearts (Hiller et al. 2006).

Agents Containing Iron Oxide

When used as a contrast medium, iron oxide particles can be divided into three classes based on their size. Ranging from large (micrometers) to small (nanometers), these classes are: micron-sized iron oxide particles (MPIOs), superparamagnetic iron oxide particles (SPIOs) and ultrasmall superparamagnetic iron oxide particles (USPIOs). SPIOs and USPIOs are in the nanometer size range. Typically, these compounds consist of an iron oxide core with various surface modifications, such as a dextran polymer coating that allows for differential cellular uptake properties, as the rate and route of cellular internalization is influenced by the coating (Fleige et al. 2002).

Iron oxide exerts its effect on the MR signal by reducing $T2^*$ relaxation. These particles have a predominant non-time-varying effect (due to locally increased magnetic susceptibility) and produce a drop in signal on $T2^*$-weighted images. Iron oxide particles are administered primarily intravenously (Trivedi et al. 2004), but can also be administered orally for gastrointestinal imaging (Johnson et al. 1996). Some commercial iron oxide agents (including Endorem® and Resovist®) have been approved by the FDA; many other agents are under development (Corot et al. 2006).

Compared to gadolinium-containing compounds, iron-oxide particles have been more extensively used for cellular labeling strategies. Iron-oxide agents are favored because a significantly lower concentration of these agents is needed compared to gadolinium agents due to the higher sensitivity of iron oxide. MR contrast agents based on iron oxide have enhanced relaxivity properties and therefore greater contrast enhancement compared to gadolinium-based agents. Primarily for this reason, iron oxide nanoparticles are now commonly used in a large number of imaging protocols (Sosnovik and Weissleder 2007).

Iron oxide compounds have another advantage in that they are biodegradable via endogenous iron metabolic pathways (Politi 2007). Once injected into the vascular system, for example, these particles are usually taken up by circulating blood-borne phagocytic cells (mostly macrophages) and transported across the vascular endothelium into adjacent organs, making iron oxide particles ideal for cell-labeling and tracking studies. This cell-labeling strategy has been used to detect rejection following organ transplantation in a rodent model of heart transplantation (Wu et al. 2006). Cells that lack substantial phagocytic capacity can also be labeled ex vivo, involving receptor-mediated endocytosis or membrane penetration, prior to injection into the circulatory system or transplantation to a specific site. This approach allows for the tracking of non-immune cells such as stem cells (Bulte et al. 2001).

Applications of MR-Based Molecular Imaging

Conventional MR imaging relies mostly on nonspecific anatomical, physiological or metabolic properties that distinguish between normal and pathophysiological tissue on a macroscopic scale. Often these changes are detected *indirectly* by imaging anatomical or functional properties. Molecular MR imaging, on the other hand, is focused on the *direct* visualization of specific populations of cells and/or molecules by virtue of their metabolic processes occurring at the submicroscopic level. Recall, as described in Fig. 17.3, that a molecular imaging agent with high sensitivity must be used to detect changes at the molecular level with an imaging technique capable of only relatively macroscopic resolution (on the order of millimeters). This statement is not inconsistent with the purpose of MI, which is not so much to define anatomical features but rather the detection of the metabolic events responsible (or thought to be responsible) for the early progression of disease. For example, appropriate MR MI reporter probes may be used to detect early changes that indicate pathological cellular transformation that leads to tumor formation, rather than simply detect the existence of a tumor that has already formed. Methods of targeting specific cells and molecules for MR imaging have been developed in a number of disease models, and key representative studies are discussed below.

Oncology

Imagine that some day cancers will no longer be characterized based on the anatomical location, such as breast or lung cancer, but rather by the underlying metabolic processes or genetic mutations within cells (Pomper 2001). This is one of the primary goals of MI in cancer research. Several cellular processes are targets in the detection of tumors, as well as in the development of anticancer therapies, such as angiogenesis (Sipkins et al. 1998; Zhao 2001; Winter 2003; Lindsey et al. 2006) cell death (apoptosis) within tumors (Zhao et al. 2001; Schellenberger et al. 2002a, b; Jung et al. 2004), and receptor targeting by growth factors (Artemov et al. 2003). The tracking of magnetically loaded cells that can home in on tumor tissue has been explored as an anticancer therapy (Kircher et al. 2003; Daldrup-Link et al. 2005; Smirnov et al. 2006; Agger et al. 2007).

MI in the field of oncology provides not only a means of detecting tumorogenic tissue (Luciani et al. 2004; Oyewumi et al. 2004; Leuschner et al. 2006), but also a means to monitor treatment and determine whether a therapy has been successful (Reichardt et al. 2005; Claes et al. 2007; Persigehl et al. 2007). However, before therapy can even begin, it is imperative that the target tissue (biomarkers) be identified for a biological treatment regimen (Schaller et al. 2007). For example, in an MR-based approach, superparamagnetic nanoparticles were used to target the tyrosine kinase Her-2/neu receptor that is present in approximately 25% of human breast cancers (Artemov et al. 2003). In this in vitro study, the nanoparticles included an iron oxide core and were conjugated to streptavidin molecules to provide specific binding to a biotinylated monoclonal antibody against the Her-2/neu receptor in vitro. The amplification of the signal was accomplished by the fact that 5–7 biotin molecules were bound to each antibody, providing multiple sites for the binding of streptavidin-conjugated iron oxide nanoparticles. Their result provides proof-of-principle evidence that MR imaging of receptor binding is possible and that the biotin–streptavidin method provides sufficient amplification for signal detection. Further, these results have the potential for screening endogenous receptor expression in cancers using noninvasive MR and could be useful for diagnosis and monitoring tumor therapy targeted against specific receptors. This study is particularly significant because the receptor-binding mechanism does not involve in vitro cellular internalization of nanoparticles prior to treatment (Choi et al. 2004), making this approach more feasible for future in vivo studies.

Angiogenesis

Angiogenesis (the formation of new blood vessels from pre-existing blood vessels) is a key process in the evolution of tumors, as cells cannot survive beyond 100–200 μm from the vascular system due to the limitations of oxygen diffusion from the vascular system (Tanabe et al. 2007). As a tumor begins to enlarge, the expanding neovasculature within the tumor therefore becomes an important target for cancer diagnosis and treatment. In this regard, an extensive amount of recent research has uncovered over 80 biomarkers involved in the regulation of angiogenesis (Daly et al. 2003; Mousa and Mousa 2004). For example, the endothelial integrin $\alpha_v\beta_3$ has been shown to correlate with tumor angiogenesis (Brooks et al. 1994, 1995; Gladson 1996). Noninvasive detection of increased microvessel density within tumors provides an attractive means to assess and monitor its growth.

Using an approach similar to that of Artemov et al. described previously, the Sipkins group (Sipkins et al. 1998) employed avidin-linked paramagnetic polymerized liposomes conjugated to a biotinylated monoclonal antibody against endothelial $\alpha_v\beta_3$. The integrin $\alpha_v\beta_3$ is expressed on the luminal surface of angiogenic vessels, which infiltrate tumorigenic tissue. This novel method enabled them to visualize the expanding tumor margin with MR, which improved the delineation of tumor from surrounding tissue (Fig. 17.6), making this approach potentially useful for distinguishing benign from aggressive lesions, and for directing therapeutic interventions.

Vascular endothelial growth factor (VEGF) is another angiogenesis marker and is well known to be important in sustaining tumor growth (McMahon 2000). As such, the inhibition of growth factor signaling involved in tumor growth is currently being explored as a treatment for some cancers (Ciardiello et al. 2003; Ekman et al. 2007; Giannelli et al. 2007). Monitoring of early treatment responses by MR can potentially determine the efficacy of anticancer drugs. Reichardt and coworkers (Reichardt et al. 2005) recently demonstrated that injecting superparamagnetic nanoparticles into the vasculature of experimental mice harboring malignant xenografts enabled an MR-based quantitative analysis of a reduction in tumor

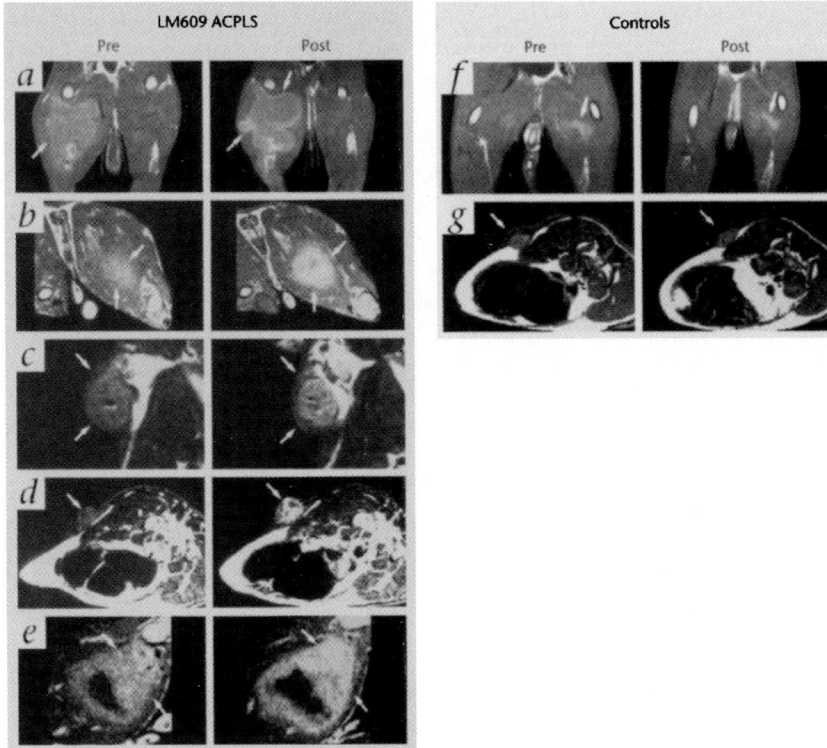

Fig. 17.6 (a–g) Antibody-conjugated paramagnetic liposomes (ACPLs) enhance the delineation of tumor margins in T1-weighted MR images compared to no contrast and control contrast administration. (**a–e**) Images of rabbit tumors pre- and post-targeted contrast administration. *Arrows* indicate the tumor periphery, the area of active angiogenesis and $\alpha_v\beta_3$ expression. Hypointense areas represent necrotic tissue (no blood supply). Angiogenic "hot spots" not seen by standard MRI were detected post-contrast in (**e**). (**f–g**) Images of rabbit tumors before and after administration of control contrast, which contained no antibody and therefore no affinity for the tumor. Only minimal enhancement of tumor is seen in controls. Image from Sipkins et al. (1998)

vascular volume in response to an experimental VEGF receptor tyrosine kinase inhibitor. This not only confirmed the antiangiogenic properties of the drug but also demonstrated the usefulness of MI in monitoring the responses to drugs for cancer therapy.

Cell Tracking

Methods of labeling and subsequent in vivo tracking of cells with MR have been described in several recent reviews (Lecchi et al. 2007; Ottobrini et al. 2005; Rogers et al. 2006; Politi 2007). Cells of the immune system such as T lymphocytes and natural killer cells (NK cells) are known to home in on and to infiltrate tumor tissue due to the host immune response to tumor antigens. The presence of lymphocytes and NK cells is associated with improved prognosis for some cancer patients (Eerola et al. 2000; Zhang et al. 2003; Fukunaga et al. 2004; Sato et al. 2005). The idea that certain cells can traffic directly to a tumor provides the basis for some innovative approaches to cancer therapy. The main goal of cancer immunotherapy studies has been to create more efficient ways of inducing the accumulation of immune effector cells into tumors (Gao et al. 2007). Magnetically labeled immune cells that home in on a tumor are therefore well suited to cancer therapy monitoring and provide useful information on disease prognosis with MR imaging (Kircher et al. 2003; Daldrup-Link et al. 2005; Smirnov et al. 2006; Agger et al. 2007).

Kircher and colleagues (Kircher et al. 2003) were the first to develop a proto-col for highly efficient in vitro cellular uptake of a novel crosslinked iron oxide nanoparticle that facilitated high-resolution imaging at the near-single-cell level (approximately two cells per voxel). They were then able to demonstrate, for the first time, the immune-specific recruitment of intraperitoneally injected labeled cells to tumors in live animals over time with high-resolution MR imaging (Fig. 17.7). This cell labeling protocol enabled them to image the recruitment of injected T cells, at approximately three cells per voxel, to intact tumors that were previously difficult to visualize due to relatively inefficient labeling methods and the dilution of injected cells.

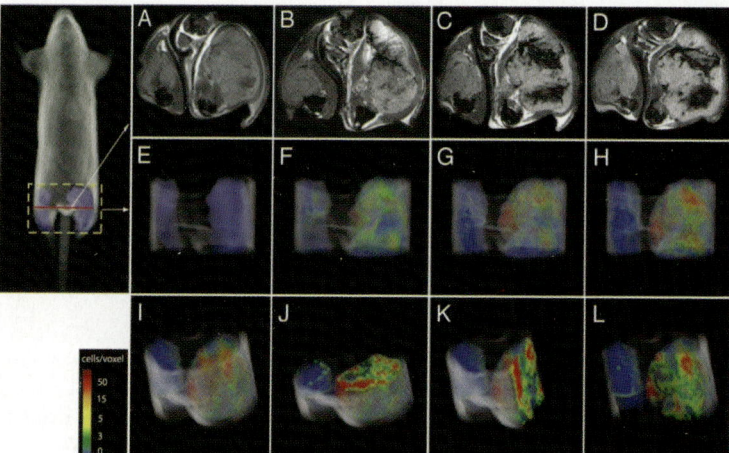

Fig. 17.7 (a–l) Time course of cytotoxic T lymphocytes, loaded with crosslinked iron oxide par-ticles, home to mouse B16-OVA tumor in vivo. Serial MR imaging was performed after systemic (i.p.) injection of CLIO-labeled cells into a mouse carrying both B16F0 (*left*) and B16-OVA (*right*) melanomas. (**a–d**) MR image slices through the thighs before injection (**a**), 12 h (**b**), 16 h (**c**) and 36 h (**d**) after injection of labeled cells. (**e–i**) The 2-D data from a to d were used to create three-dimensional color-scaled reconstructions of B16F0 (*left*) and B16-OVA (*right*) melanomas at 0 h (**e**), 12 h (**f**), 16 h (**g**) and 36 h (**h, i**) after injection. The numbers of cells/voxel are color-coded as shown in the scale. (**j–l**) Present additional axial (**j**), sagittal (**k**) and coronal (**l**) plane slices from the image in (**i**), showing the heterogeneous nature of the T-cell recruitment also seen in (**e–i**). Image from Kircher et al. (2003)

Tracking cells to tumors to visualize the reduction in tumor size is undoubtedly beneficial for monitoring purposes; however, therapeutic cell-based approaches that combine monitoring with treatment paradigms are also being developed. For example, the cytolytic effects of certain immune cells such as lymphocytes that infiltrate tumor tissue can be exploited for therapeutic purposes. It was recently shown in a model of adoptive transfer of OT-1 lymphocytes into tumor-bearing mice that the magnetically labeled lymphocytes retained their antitumoral function (after being loaded with SPIO nanoparticles and injected into mice), and subsequently induced partial tumor regression in situ (Smirnov et al. 2006). A subgroup of lymphocytes, natural killer (NK) cells, appear to have a more potent cytotoxic effect on tumor cells that otherwise escape immune surveillance (Yan et al. 1998; Tam et al. 1999; Tonn et al. 2001). These reports provide proof in principle that NK cell-based immunotherapy for several types of malignancies is feasible. In fact, the NK-92 cell line has entered phase I/II clinical trials (Tonn et al. 2001). Dalrup-Link (Daldrup-Link et al. 2005) expanded on this work and demonstrated the successful labeling of human NK-92 cells with SPIO particles followed by MR tracking of the cells to tumors in mice at a clinically relevant field strength of 1.5 T. Unfortunately, this study did not investigate any tumor-regressing activity of the injected NK cells over time. These studies however, underline the potential for using noninvasive MI strategies to evaluate both the experimental and clinical therapeutic effectiveness of methods to reduce tumor volume.

Apoptosis

Usually referred to as "programmed cell death," apoptosis is dependent upon a series of tightly regulated signaling cascades initiated by either intracellular or extracellular signals that result in the destruction of the cell. Apoptosis is necessary for tissue development and homeostasis, and thus deregulation of the intracellular chemical pathways involved can disrupt the balance between cell proliferation and death, which can result in uncontrolled tumor growth. A method of detecting apoptosis in vivo would therefore be of significant value for evaluating reduction in tumor volume and predicting early response to therapy. Previous evidence suggests that early detection of apoptosis in response to cancer therapy is a good prognostic indicator for treatment outcome (Meyn et al. 1995; Dubray et al. 1998; Chang et al. 2000). The detection of apoptosis usually requires histological examination of tumor tissue obtained from biopsy and therefore a less invasive method of detection is desirable.

The first in vivo imaging of apoptosis was performed by radiolabeling annexin V, a protein with high affinity for the outer membranes of cells that are just beginning to undergo apoptosis, with metastable technetium-99 (^{99m}Tc) in several different mouse models of apoptosis followed by imaging with a scintillation camera (Blankenberg et al. 1998). Four years later, Schellenberger and colleagues conjugated annexin V to crosslinked monocrystalline iron oxide nanoparticles (CLIOs) and a fluorescent moiety (Cy5.5), which provided a multimodal label for MR as well as optical imaging of apoptotic cells in vitro (Schellenberger et al. 2002a, b).

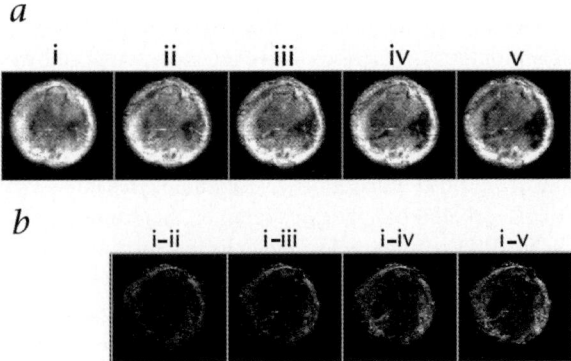

Fig. 17.8 (**a–b**) Binding of C_2-SPIO imaging probe to tumor cells in vivo that were induced to undergo apoptosis following injection into mice. (**a**) T2-weighted MR images before (*i*) C_2-SPIO injection and at 11, 47, 77 and 107 min (*ii*)–(*v*) after injection. (**b**) Subtraction images obtained by subtracting the post-contrast images from the image acquired before the injection of C_2-SPIO. In the original images (**a**), a progressive loss in signal intensity in seen in relatively well-defined areas of the tumors, which is evident as an increase in signal intensity in the subtraction images, indicating an accumulation of apoptotic cells. Image from Zhao et al. (2001)

This group later used the same nanoparticle (AnxCLIO-Cy5.5) in an elegant study to successfully image cardiomyocyte apoptosis in the reperfused infarcted beating mouse heart in vivo with MR for the first time (Sosnovik et al. 2005). Given the fact that annexin V and dextran-coated iron oxide polymers are already approved for use in humans, and that this study used a dose within the range of similar FDA-approved iron oxide particles (Combinex), AnxCLIO-Cy5.5 has a high potential for future clinical use in the imaging of apoptosis.

MR imaging of a murine model of lymphoma tumor that underwent apoptosis following drug treatment showcased the novel contrast agent C_2-SPIO (Zhao et al. 2001) (Fig. 17.8). Similar to AnxCLIO-Xy5.5, this probe detects early-stage apoptotic cells by virtue of the specific binding of the magnetically labeled C_2 domain of synaptotagmin I to externalized phospholipids on cell membranes of apoptotic cells. While both C_2-SPIO and AnxCLIO-Xy5.5 will also detect necrotic cells when their outer membranes rupture and the probe is then able to access the inner leaflet, recognition of all forms of cell death with high spatial resolution would be advantageous in practice (Zhao et al. 2001).

Cardiology

Atherosclerotic disease remains a leading cause of mortality despite numerous advances in the detection and treatment of severe arterial stenosis. Sudden rupture of vulnerable plaque (an atheromatous plaque that is particularly prone to mechanical stress) resulting in life-threatening thrombosis is a common occurrence (Falk et al. 1995), and provides the impetus for a sensitive method of direct noninvasive imaging of unstable atherosclerotic plaque formation. Although MR imaging of vulnerable plaque and arterial thrombi without contrast agents is possible in both experimental animal models (Johnstone et al. 2001; Corti et al. 2002) and humans (Moody et al. 2003), the use of conventional gadolinium chelates serves to enhance the MR signal and improves the visualization of atherosclerotic plaque development compared to imaging without contrast agents (Wasserman et al. 2002; Yuan et al. 2002). However, contrast agents targeted to specific components of the plaque or thrombus not only improve characterization but can also provide enhanced sensitivity compared to non-contrast-enhanced and non-targeted contrast MR imaging (Sirol et al. 2005).

Research has resulted in a number of molecular probes for MR imaging that are capable of detecting a variety of proteins related to vessel pathology (reviewed in: Canet-Soulas and Letourneur 2007). Due to the high levels of fibrin expressed in vessel thrombi, fibrin-targeted probes were some of the first to be developed for MR imaging. A novel fibrin-specific MR contrast agent consisting of lipid-encapsulated liquid perfluorocarbon nanoparticles capable of carrying a high gadolinium payload was developed for the purpose of the sensitive detection of microthrombi in atherosclerotic vessels (Flacke et al. 2001). Figure 17.9 shows fibrin-targeted Gd-DTPA nanoparticles bound to the surface of a surgically created thrombus, thereby enhancing the detectability of the thrombus in situ. The detectability of the thrombus was markedly enhanced by the fibrin-specific paramagnetic nanoparticles relative to control thrombus.

To confirm the clinical relevance of the probe, visualization of the nanoparticles binding directly to small fibrin deposits in a ruptured carotid plaque from human endarterectomy specimens was achieved with an MR sequence similar to that used in the canine experiments. Recently, a similar Gd-DTPA antifibrin probe (EP-2104R) has been used to demonstrate direct thrombus imaging with MR in several different experimental animal models of induced thrombosis (Botnar et al. 2004; Sirol et al. 2005; Stracke et al. 2007). To demonstrate the imaging of naturally occurring atherosclerosis, the disease was replicated in transgenic hyperlipidemic mice (apoE knockout) that develop atherosclerotic lesions, similar to those observed in humans, when fed a high-fat, high-cholesterol diet (Frias et al. 2004). Within 24 h of the intravenous injection of a high-density lipoprotein (HDL)-like nanoparticle contrast agent (GdDTPA-DMPE) that selectively targets atherosclerotic plaques, signal enhancement was detected predominantly at the plaque (Fig. 17.10). Interestingly, higher cellular content of the plaque corresponded to greater signal enhancement, affording some distinction of plaque composition.

These studies highlight the sensitive MR detection of early plaque formation without the need for invasive X-ray angiography, and demonstrate clinical applications for diagnosis and guidance of therapy following thrombosis. These contrast agents, therefore, may allow the sensitive, early detection of microthrombi that precede heart attack and stroke.

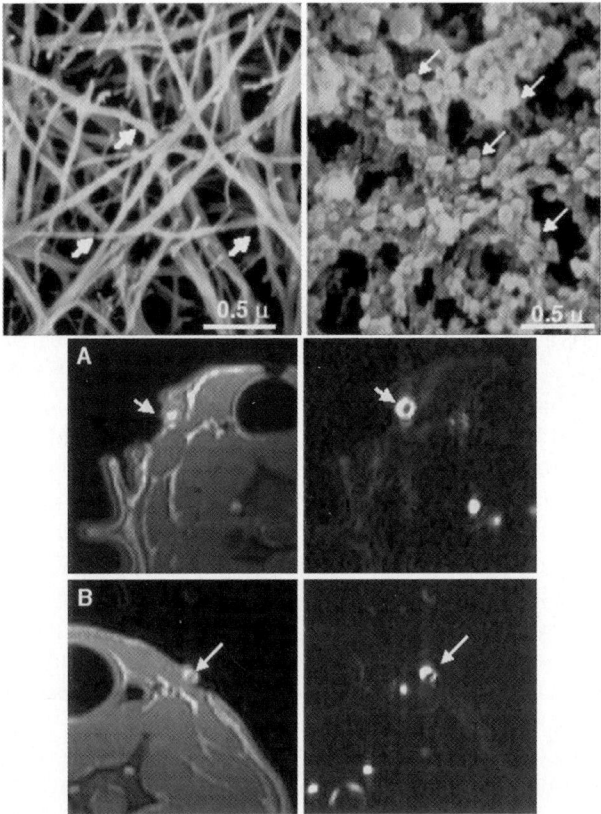

Fig. 17.9 (**a–b**) Lipid-encapsulated liquid perfluorocarbon nanoparticles coated with anti-fibrin antibodies and carrying a Gd payload that specifically targets fibrin-containing clots. The *two top images* are scanning electron micrographs of a fibrin clot (*left*) and a fibrin clot with targeted paramagnetic nanoparticles bound to the clot surface (*right*). (**a**) Thrombi in canine external jugular vein targeted with anti-fibrin Gd nanoparticles. The *left image* demonstrates T1-weighted contrast enhancement by gradient-echo MR; the *arrow* indicates flow deficit. The *right image* is a corresponding 3-D phase contrast angiogram. (**b**) Control thrombus in contralateral external jugular vein, imaged as in (**a**). Image from Flacke et al. (2001)

Neurology

Prior to the conventional brain imaging modalities currently in use, neuroimaging was an extremely invasive and unsafe procedure that required the replacement of cerebrospinal fluid with air to provide greater contrast between brain structures on an X-ray film (pneumoencephalography). However, technology has now progressed to the point where we can now carry out noninvasive imaging of structures at the cellular and molecular level. Although PET, SPECT and fMRI remain the most clinically used modalities for functional brain imaging, the development of paramagnetic molecular probes has brought MR into the molecular neuroimaging field.

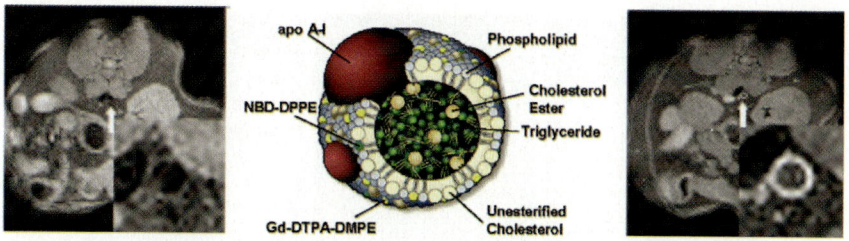

Fig. 17.10 A lipoprotein nanoparticle with a phospholipid-based contrast agent, Gd-DTPA-DMPE, incorporated into the particle selectively targets atherosclerotic plaques. A schematic showing the structure of Gd-DTPA-DMPE is flanked by MR images depicting the specific in vivo binding of the nanoparticles to atherosclerotic vessels of transgenic hyperlipidemic mice. *White arrows* point to the abdominal aorta and *insets* denote a magnification of the aorta region. *Left*: prior to injection of nanoparticles, no signal is detected in the aorta. *Right*: by 24 h after injection, contrast agent is localized predominantly at the atherosclerotic plaque. Image from Frais et al. (2004)

For example, the gold standard of accurate diagnosis of Alzheimer's disease (AD) is post mortem analysis of the brain. Therefore, finding early biomarkers of the disease and sensitive molecular probes to noninvasively detect them would greatly contribute to the in vivo diagnosis of the disease and treatment strategies.

Alzheimer's Disease

Currently, the most common biomarker utilized in the detection of AD is amyloid beta (Aβ) protein. Reduced levels of Aβ peptide in cerebrospinal fluid has been reported to distinguish between known populations of control and AD patients (Galasko et al. 1998; Andreasen et al. 1999). However, direct imaging of Aβ plaques and tau proteins in neurofibrillary tangles within the brain could prove useful in increasing the accuracy of early diagnosis, tracking the progression of the disease and monitoring therapy. Interestingly, by virtue of the high content of metal ions, including iron, in amyloid plaques (Lovell et al. 1998), visualization (in the absence of a targeted contrast agent) of individual plaques in ex vivo brain specimens from AD patients (Benveniste et al. 1999) or transgenic AD mice in vivo (Jack et al. 2005; Vanhoutte et al. 2005) with MR is possible due to the T2 and T2* relaxation effects of iron associated with the plaques. However, the use of a targeted contrast agent has the capacity to differentiate plaques from interfering structures such as blood vessels (due to the iron content in hemoglobin), and iron-enriched glial cells.

Accordingly, Poduslo and colleagues (Poduslo et al. 2002) created a gadolinium-based probe with a high affinity for Aβ-42 plaques (PUT-Gd-Aβ) that was able to cross the intact blood–brain barrier in AD transgenic mice by virtue of an attached putrescine moiety (to facilitate transport across the cell membrane) and significantly enhance MR visualization of plaques in ex vivo brain hemispheres (Fig. 17.11) at a lower spatial resolution and in less scanning time than that reported in the previous study without contrast enhancement (Benveniste et al. 1999). Several subsequent

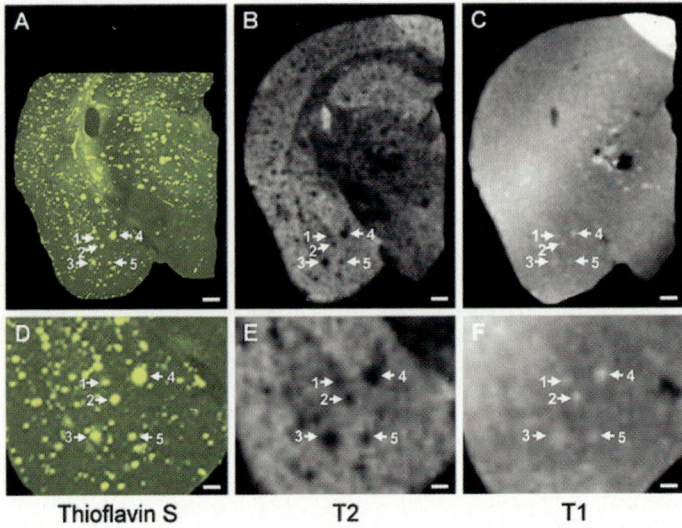

Thioflavin S   T2   T1

Fig. 17.11 (**a–e**) ß-amyloid plaques in ex vivo sections of transgenic AD mice brains are specifically targeted and visualized at high resolution (62.5 μm cubic voxels) with a Gd-containing imaging probe capable of crossing the blood–brain barrier. (**a**) Histologic section depicting thioflavin S staining of ß-amyloid deposits in AD mice following intravenous injection of the PUT-Gd-Aß probe. (**b**) Corresponding T2-weighted and (**c**) T1-weighted MR images. *White arrows* indicate examples of ß-amyloid deposits. (**d–e**) Higher magnification of (**a–c**) showing the exact colocalization of ß-amyloid deposits between histologic and MR images, indicating the selectivity of the probe. Image from Poduslo et al. (2002)

studies reported the generation of Gd-based probes with specific affinity for amyloid beta protein that were able to label plaques in AD mice in vivo (Wadghiri et al. 2003; Poduslo et al. 2004; Kandimalla et al. 2007). However, either the blood–brain barrier had to be chemically compromised to facilitate the entry of the probe into the brain (Wadghiri et al. 2003) or MR imaging was not carried out; instead, autoradiographic images of fixed brain sections confirmed the localization of the ^{125}I-Gd[N-4ab/Q-4ab]Aβ probe to the plaques in affected brain tissue of transgenic AD mice (Poduslo et al. 2004; Kandimalla et al. 2007).

Despite these advances in the development of molecular probes for biomarkers of AD, the clinical application of individual plaque visualization is still limited due to the need for a sufficiently high spatial resolution and signal-to-noise ratio, requiring a magnetic field strength of 7 T or greater (which, outside of a few research centers, is not yet available for human patients) and/or prohibitively long scan times. Until technical advances in high-field engineering are achieved, work will undoubtedly continue on the discovery of new AD biomarkers and the generation of probes that provide enhanced signal-to-noise contrast, such that their applications towards studying the etiology of AD and the monitoring of disease progression in animal models will eventually become clinically realistic.

CNS Inflammation: Stroke and Multiple Sclerosis

Other brain disorders are characterized early in the disease process by local inflammation, such as endothelial activation and macrophage infiltration following stroke (reviewed in Mehta et al. 2007) or leukocyte recruitment during the initial stages of multiple sclerosis (MS) (reviewed in Hemmer et al. 2003). With the objective of imaging brain inflammation, Laurent and coworkers developed a novel conjugate comprised of the Gd-DTPA complex and a sialyl Lewisx (sLex) mimetic (Laurent et al. 2004). sLex mimetics were developed as inhibitors against cell adhesion during the activation of vascular endothelial cells following brain injury in order to control inflammation (Huwe et al. 1999; Thoma et al. 1999) and have since been exploited to create a ligand that would target inflammation sites. Accordingly, in a mouse model of stroke, Barber and coworkers showed that early endothelial activation following transient focal cerebral ischemia could be visualized in vivo with MR using the contrast agent Gd-DTPA-sLe^{x}A (Barber et al. 2004).

Activated macrophages also play a critical role in the immune response. Under healthy physiological conditions, immune cells are restricted from entering the central nervous system. However, during pathological processes, macrophages cross the blood–brain barrier and infiltrate the brain parenchyma to the site of inflammation, where they display potent phagocytic activity. The capability of these cells to migrate towards inflammation and phagocytose particles and compounds has initiated novel MR imaging techniques such as macrophage tracking with iron oxide nanoparticles (USPIOs) in experimental animal models of both stroke (Rausch et al. 2001, 2002; Kleinschnitz et al. 2003) and MS (Rausch et al. 2003; Brochet et al. 2006; Baeten et al. 2008). The premise of these experiments is that a solution of USPIOs is injected intravenously into rodents following stroke or soon after the onset of MS symptoms. The individual iron oxide particles are phagocytosed by endogenous macrophages, which then migrate to the site of inflammation, producing signal loss on T2-weighted images. Importantly, these animal studies have recently been translated to clinical trials for stroke (Saleh et al. 2004, 2007; Nighoghossian et al. 2007) and MS (Manninger et al. 2005; Dousset et al. 2006), facilitating the characterization of inflammatory events in order to monitor patients and to potentially define the timing for therapeutic interventions (Fig. 17.12).

Summary

It is hoped that from this review the reader has gained an appreciation for the emerging concepts in the field of molecular imaging for MR. While it is recognized that PET imaging has the clear advantage of higher resolution, MR imaging reduces both the cost of development and the risk to patients from ionizing radioisotopes. In the near future, advances in MR technology will likely overcome the limitations of low resolution. At a minimum, MR MI represents a modality that is capable of augmenting other methods of molecular imaging. Relevant clinical applications of

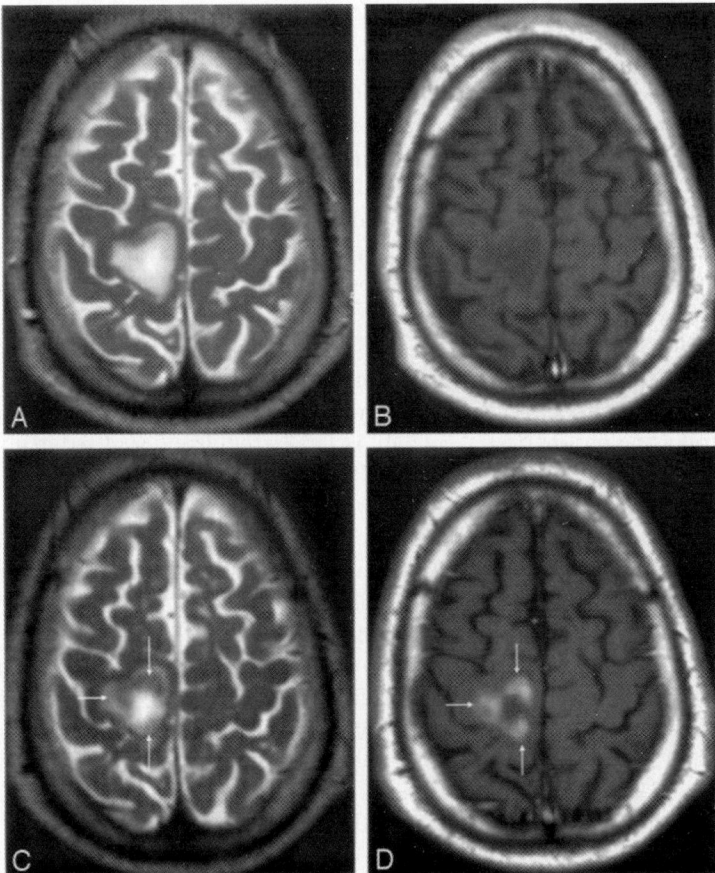

Fig. 17.12 (a–d) Injected USPIOs are taken up by macrophages in vivo, incorporated into inflammatory MS lesions in human patients, and detected by MR. In images of a patient first injected with gadolinium contrast agent, a large MS plaque is seen on a T2-weighted image (**a**) and a T2-weighted image (**b**) that was not enhanced by gadolinium. Following injection, USPIOs are taken up at the periphery of the plaque (*arrows*) and visualized as a decrease in signal intensity on the T2-weighted image (**c**) and increased signal intensity on a T1-weighted image (**d**). Image from Dousset et al. (2006)

MR-based MI probes in several disease processes are well underway. This quickly developing field will thus undoubtedly contribute significantly to effective monitoring of disease progression in oncology, cardiology, neurology and other fields, with the aim of treatment design.

References

Agger R, Petersen MS et al. (2007) T cell homing to tumors detected by 3D-coordinate positron emission tomography and magnetic resonance imaging. J Immunother 30(1):29–39

Andreasen N, Minthon L et al. (1999) Cerebrospinal fluid tau and Abeta42 as predictors of development of Alzheimer's disease in patients with mild cognitive impairment. Neurosci Lett 273(1):5–8

Artemov D, Mori N et al. (2003) MR molecular imaging of the Her-2/neu receptor in breast cancer cells using targeted iron oxide nanoparticles. Magn Reson Med 49(3):403–408

Baeten K, Hendriks JJ et al. (2008) Visualisation of the kinetics of macrophage infiltration during experimental autoimmune encephalomyelitis by magnetic resonance imaging. J Neuroimmunol 195:1–6

Barber PA, Foniok T et al. (2004) MR molecular imaging of early endothelial activation in focal ischemia. Ann Neurol 56(1):116–120

Barge A, Cravotto G et al. (2006) How to determine free Gd and free ligand in solution of Gd chelates. A technical note. Contrast Media Mol Imaging 1(5):184–188

Benveniste H, Einstein G et al. (1999) Detection of neuritic plaques in Alzheimer's disease by magnetic resonance microscopy. Proc Natl Acad Sci USA 96(24):14079–14084

Blankenberg FG, Katsikis PD et al. (1998) In vivo detection and imaging of phosphatidylserine expression during programmed cell death. Proc Natl Acad Sci USA 95(11):6349–6354

Botnar RM, Buecker A et al. (2004) In vivo magnetic resonance imaging of coronary thrombosis using a fibrin-binding molecular magnetic resonance contrast agent. Circulation 110(11):1463–1466

Brochet B, Deloire MS et al. (2006) Early macrophage MRI of inflammatory lesions predicts lesion severity and disease development in relapsing EAE. Neuroimage 32(1):266–274

Brockmann MA, Kemmling A et al. (2007) Current issues and perspectives in small rodent magnetic resonance imaging using clinical MRI scanners. Methods 43(1):79–87

Brooks PC, Clark RA et al. (1994) Requirement of vascular integrin alpha v beta 3 for angiogenesis. Science 264(5158):569–571

Brooks PC, Stromblad S et al. (1995) Antiintegrin alpha v beta 3 blocks human breast cancer growth and angiogenesis in human skin. J Clin Invest 96(4):1815–1822

Broome DR (2008) Nephrogenic systemic fibrosis associated with gadolinium based contrast agents: A summary of the medical literature reporting. Eur J Radiol 66(2):230–234

Bulte JW, Douglas T et al. (2001) Magnetodendrimers allow endosomal magnetic labeling and in vivo tracking of stem cells. Nat Biotechnol 19(12):1141–1417

Bushberg JTSJ, Leidholdt EM, Boone JM (2002) The essential physics of medical imaging. Lippincott Williams and Wilkins, Philadelphia, PA

Canet-Soulas E, Letourneur D (2007) Biomarkers of atherosclerosis and the potential of MRI for the diagnosis of vulnerable plaque. Magma 20(3):129–142

Cassidy PJ, Radda GK (2005) Molecular imaging perspectives. J R Soc Interf 2(3):133–144

Chang J, Ormerod M et al. (2000) Apoptosis and proliferation as predictors of chemotherapy response in patients with breast carcinoma. Cancer 89(11):2145–2152

Choi H, Choi SR et al. (2004) Iron oxide nanoparticles as magnetic resonance contrast agent for tumor imaging via folate receptor-targeted delivery. Acad Radiol 11(9):996–1004

Ciardiello F, Caputo R et al. (2003) Antitumor effects of ZD6474, a small molecule vascular endothelial growth factor receptor tyrosine kinase inhibitor, with additional activity against epidermal growth factor receptor tyrosine kinase. Clin Cancer Res 9(4):1546–1556

Claes A, Gambarota G et al. (2007) Magnetic resonance imaging-based detection of glial brain tumors in mice after antiangiogenic treatment. Int J Cancer 122(9):1981–1986

Corot C, Robert P et al. (2006) Recent advances in iron oxide nanocrystal technology for medical imaging. Adv Drug Deliv Rev 58(14):1471–1504

Corti R, Osende JI et al. (2002) In vivo noninvasive detection and age definition of arterial thrombus by MRI. J Am Coll Cardiol 39(8):1366–1373

Culver J, Akers W et al. (2008) Multimodality molecular imaging with combined optical and SPECT/PET modalities. J Nucl Med 49(2):169–172

Daldrup-Link HE, Meier R et al. (2005) In vivo tracking of genetically engineered, anti-HER2/neu directed natural killer cells to HER2/neu positive mammary tumors with magnetic resonance imaging. Eur Radiol 15(1):4–13

Daly ME, Makris A et al. (2003) Hemostatic regulators of tumor angiogenesis: a source of antiangiogenic agents for cancer treatment? J Natl Cancer Inst 95(22):1660–1673

Damadian R (1971) Tumor detection by nuclear magnetic resonance. Science 171(976):1151–1513

Dobrucki LW, Sinusas AJ (2007) Imaging angiogenesis. Curr Opin Biotechnol 18(1):90–96

Dousset V, Brochet B et al. (2006) MR imaging of relapsing multiple sclerosis patients using ultra-small-particle iron oxide and compared with gadolinium. AJNR Am J Neuroradiol 27(5):1000–1005

Dubray B, Breton C et al. (1998) In vitro radiation-induced apoptosis and early response to low-dose radiotherapy in non-Hodgkin's lymphomas. Radiother Oncol 46(2):185–191

Eerola AK, Soini Y et al. (2000) A high number of tumor-infiltrating lymphocytes are associated with a small tumor size, low tumor stage, and a favorable prognosis in operated small cell lung carcinoma. Clin Cancer Res 6(5):1875–1881

Ekman S, Bergqvist M et al. (2007) Activation of growth factor receptors in esophageal cancer – implications for therapy. Oncologist 12(10):1165–1177

Emoto K, Toyama-Sorimachi N et al. (1997) Exposure of phosphatidylethanolamine on the surface of apoptotic cells. Exp Cell Res 232(2):430–434

Falk E, Shah PK et al. (1995) Coronary plaque disruption. Circulation 92(3):657–671

Flacke S, Fischer S et al. (2001) Novel MRI contrast agent for molecular imaging of fibrin: implications for detecting vulnerable plaques. Circulation 104(11):1280–1285

Fleige G, Seeberger F et al. (2002) In vitro characterization of two different ultrasmall iron oxide particles for magnetic resonance cell tracking. Invest Radiol 37(9):482–488

Frias JC, Williams KJ et al. (2004) Recombinant HDL-like nanoparticles: a specific contrast agent for MRI of atherosclerotic plaques. J Am Chem Soc 126(50):16316–16317

Fukunaga A, Miyamoto M et al. (2004) CD8+ tumor-infiltrating lymphocytes together with CD4+ tumor-infiltrating lymphocytes and dendritic cells improve the prognosis of patients with pancreatic adenocarcinoma. Pancreas 28(1):e26–e31

Galasko D, Chang L et al. (1998) High cerebrospinal fluid tau and low amyloid beta42 levels in the clinical diagnosis of Alzheimer disease and relation to apolipoprotein E genotype. Arch Neurol 55(7):937–945

Gao JQ, Okada N et al. (2007) Immune cell recruitment and cell-based system for cancer therapy. Pharm Res

Giannelli G, Napoli N et al. (2007) Tyrosine kinase inhibitors: a potential approach to the treatment of hepatocellular carcinoma. Curr Pharm Des 13(32):3301–3304

Gladson CL (1996) Expression of integrin alpha v beta 3 in small blood vessels of glioblastoma tumors. J Neuropathol Exp Neurol 55(11):1143–1149

Hargreaves RJ (2008) The role of molecular imaging in drug discovery and development. Clin Pharmacol Ther 83(2):349–353

Hemmer B, Kieseier B et al. (2003) New immunopathologic insights into multiple sclerosis. Curr Neurol Neurosci Rep 3(3):246–255

Hiller KH, Waller C et al. (2006) Assessment of cardiovascular apoptosis in the isolated rat heart by magnetic resonance molecular imaging. Mol Imaging 5(2):115–121

Huwe CM, Woltering TJ et al. (1999) Design, synthesis and biological evaluation of aryl-substituted sialyl Lewis X mimetics prepared via cross-metathesis of C-fucopeptides. Bioorg Med Chem 7(5):773–788

Jack CR Jr, Wengenack TM et al. (2005) In vivo magnetic resonance microimaging of individual amyloid plaques in Alzheimer's transgenic mice. J Neurosci 25(43):10041–10048

Johnson WK, Stoupis C et al. (1996) Superparamagnetic iron oxide (SPIO) as an oral contrast agent in gastrointestinal (GI) magnetic resonance imaging (MRI): comparison with state-of-the-art computed tomography (CT). Magn Reson Imaging 14(1):43–49

Johnstone MT, Botnar RM et al. (2001) In vivo magnetic resonance imaging of experimental thrombosis in a rabbit model. Arterioscler Thromb Vasc Biol 21(9):1556–1560

Jung HI, Kettunen MI et al. (2004) Detection of apoptosis using the C2A domain of synaptotagmin I. Bioconjug Chem 15(5):983–987

Kandimalla KK, Wengenack TM et al. (2007) Pharmacokinetics and amyloid plaque targeting ability of a novel peptide-based magnetic resonance contrast agent in wild-type and Alzheimer's disease transgenic mice. J Pharmacol Exp Ther 322(2):541–549

Kircher MF, Allport JR et al. (2003) In vivo high resolution three-dimensional imaging of antigen-specific cytotoxic T-lymphocyte trafficking to tumors. Cancer Res 63(20):6838–6846

Kleinschnitz C, Bendszus M et al. (2003) In vivo monitoring of macrophage infiltration in experimental ischemic brain lesions by magnetic resonance imaging. J Cereb Blood Flow Metab 23(11):1356–1361

Laurent S, Vander Elst L et al. (2004) Synthesis and physicochemical characterization of Gd-DTPA-B(sLex)A, a new MRI contrast agent targeted to inflammation. Bioconjug Chem 15(1):99–103

Lauterbur P (1973) Image formation by induced local interactions: examples employing nuclear magnetic resonance. Nature 242:190–191

Lecchi M, Ottobrini L et al. (2007) Instrumentation and probes for molecular and cellular imaging. Q J Nucl Med Mol Imaging 51(2):111–126

Leuschner C, Kumar CS et al. (2006) LHRH-conjugated magnetic iron oxide nanoparticles for detection of breast cancer metastases. Breast Cancer Res Treat 99(2):163–176

Lindsey ML, Escobar GP et al. (2006) Matrix metalloproteinase-9 gene deletion facilitates angiogenesis after myocardial infarction. Am J Physiol Heart Circ Physiol 290(1):H232–H239

Lovell MA, Robertson JD et al. (1998) Copper, iron and zinc in Alzheimer's disease senile plaques. J Neurol Sci 158(1):47–52

Luciani A, Olivier JC et al. (2004) Glucose-receptor MR imaging of tumors: study in mice with PEGylated paramagnetic niosomes. Radiology 231(1):135–142

Manninger SP, Muldoon LL et al. (2005) An exploratory study of ferumoxtran-10 nanoparticles as a blood–brain barrier imaging agent targeting phagocytic cells in CNS inflammatory lesions. AJNR Am J Neuroradiol 26(9):2290–2300

Mansfield PMP (1982) NMR imaging in biomedicine. Academic, Orlando, FL

McMahon G (2000) VEGF receptor signaling in tumor angiogenesis. Oncologist 5(Suppl 1):3–10

Mehta SL, Manhas N et al. (2007) Molecular targets in cerebral ischemia for developing novel therapeutics. Brain Res Rev 54(1):34–66

Meyn RE, Stephens LC et al. (1995) Apoptosis in murine tumors treated with chemotherapy agents. Anticancer Drugs 6(3):443–450

Moody AR, Murphy RE et al. (2003) Characterization of complicated carotid plaque with magnetic resonance direct thrombus imaging in patients with cerebral ischemia. Circulation 107(24):3047–3052

Mousa SA, Mousa AS (2004) Angiogenesis inhibitors: current & future directions. Curr Pharm Des 10(1):1–9

Nighoghossian N, Wiart M et al. (2007) Inflammatory response after ischemic stroke: a USPIO-enhanced MRI study in patients. Stroke 38(2):303–307

Ogawa S, Lee TM et al. (1990) Brain magnetic resonance imaging with contrast dependent on blood oxygenation. Proc Natl Acad Sci USA 87(24):9868–9872

Ottobrini L, Lucignani G et al. (2005) Assessing cell trafficking by noninvasive imaging techniques: applications in experimental tumor immunology. Q J Nucl Med Mol Imaging 49(4):361–366

Oyewumi MO, Yokel RA et al. (2004) Comparison of cell uptake, biodistribution and tumor retention of folate-coated and PEG-coated gadolinium nanoparticles in tumor-bearing mice. J Control Release 95(3):613–626

Persigehl T, Bieker R et al. (2007) Antiangiogenic tumor treatment: early noninvasive monitoring with USPIO-enhanced MR imaging in mice. Radiology 244(2):449–456

Poduslo JF, Wengenack TM et al. (2002) Molecular targeting of Alzheimer's amyloid plaques for contrast-enhanced magnetic resonance imaging. Neurobiol Dis 11(2):315–329

Poduslo JF, Curran GL et al. (2004) Design and chemical synthesis of a magnetic resonance contrast agent with enhanced in vitro binding, high blood–brain barrier permeability, and in vivo targeting to Alzheimer's disease amyloid plaques. Biochemistry 43(20):6064–6075

Politi LS (2007) MR-based imaging of neural stem cells. Neuroradiology 49(6):523–534

Pomper MG (2001) Molecular imaging: an overview. Acad Radiol 8(11):1141–1153

Rausch M, Sauter A et al. (2001) Dynamic patterns of USPIO enhancement can be observed in macrophages after ischemic brain damage. Magn Reson Med 46(5):1018–1022

Rausch M, Baumann D et al. (2002) In-vivo visualization of phagocytotic cells in rat brains after transient ischemia by USPIO. NMR Biomed 15(4):278–283

Rausch M, Hiestand P et al. (2003) MRI-based monitoring of inflammation and tissue damage in acute and chronic relapsing EAE. Magn Reson Med 50(2):309–314

Reichardt W, Hu-Lowe D et al. (2005) Imaging of VEGF receptor kinase inhibitor-induced antiangiogenic effects in drug-resistant human adenocarcinoma model. Neoplasia 7(9):847–853

Rogers WJ, Meyer CH et al. (2006) Technology insight: in vivo cell tracking by use of MRI. Nat Clin Pract Cardiovasc Med 3(10):554–562

Saleh A, Wiedermann D et al. (2004) Central nervous system inflammatory response after cerebral infarction as detected by magnetic resonance imaging. NMR Biomed 17(4):163–169

Saleh A, Schroeter M et al. (2007) Iron oxide particle-enhanced MRI suggests variability of brain inflammation at early stages after ischemic stroke. Stroke 38(10):2733–2737

Sato E, Olson SH et al. (2005) Intraepithelial CD8+ tumor-infiltrating lymphocytes and a high CD8+/regulatory T cell ratio are associated with favorable prognosis in ovarian cancer. Proc Natl Acad Sci USA 102(51):18538–18543

Schaller BJ, Modo M et al. (2007) Molecular imaging of brain tumors: a bridge between clinical and molecular medicine? Mol Imaging Biol 9(2):60–71

Schellenberger EA, Bogdanov A Jr et al. (2002a) Annexin V-CLIO: a nanoparticle for detecting apoptosis by MRI. Mol Imaging 1(2):102–107

Schellenberger EA, Hogemann D et al. (2002b) Annexin V-CLIO: a nanoparticle for detecting apoptosis by MRI. Acad Radiol 9(suppl 2):S310–S311

Sipkins DA, Cheresh DA et al. (1998) Detection of tumor angiogenesis in vivo by alphaVbeta3-targeted magnetic resonance imaging. Nat Med 4(5):623–626

Sirol M, Fuster V et al. (2005) Chronic thrombus detection with in vivo magnetic resonance imaging and a fibrin-targeted contrast agent. Circulation 112(11):1594–1600

Smirnov P, Lavergne E et al. (2006) In vivo cellular imaging of lymphocyte trafficking by MRI: a tumor model approach to cell-based anticancer therapy. Magn Reson Med 56(3):498–508

Sosnovik DE, Schellenberger EA et al. (2005) Magnetic resonance imaging of cardiomyocyte apoptosis with a novel magneto-optical nanoparticle. Magn Reson Med 54(3):718–724

Sosnovik DE, Weissleder R (2007) Emerging concepts in molecular MRI. Curr Opin Biotechnol 18(1):4–10

Stracke CP, Katoh M et al. (2007) Molecular MRI of cerebral venous sinus thrombosis using a new fibrin-specific MR contrast agent. Stroke 38(5):1476–1481

Tam YK, Miyagawa B et al. (1999) Immunotherapy of malignant melanoma in a SCID mouse model using the highly cytotoxic natural killer cell line NK-92. J Hematother 8(3):281–290

Tanabe K, Zhang Z et al. (2007) Current molecular design of intelligent drugs and imaging probes targeting tumor-specific microenvironments. Org Biomol Chem 5(23):3745–3757

Thoma G, Kinzy W et al. (1999) Synthesis and biological evaluation of a potent E-selectin antagonist. J Med Chem 42(23):4909–4913

Tonn T, Becker S et al. (2001) Cellular immunotherapy of malignancies using the clonal natural killer cell line NK-92. J Hematother Stem Cell Res 10(4):535–544

Trivedi RA, U-King-Im JM et al. (2004) In vivo detection of macrophages in human carotid atheroma: temporal dependence of ultrasmall superparamagnetic particles of iron oxide-enhanced MRI. Stroke 35(7):1631–1635

Vanhoutte G, Dewachter I et al. (2005) Noninvasive in vivo MRI detection of neuritic plaques associated with iron in APP[V717I] transgenic mice, a model for Alzheimer's disease. Magn Reson Med 53(3):607–613

Wadghiri YZ, Sigurdsson EM et al. (2003) Detection of Alzheimer's amyloid in transgenic mice using magnetic resonance microimaging. Magn Reson Med 50(2):293–302

Wasserman BA, Smith WI et al. (2002) Carotid artery atherosclerosis: in vivo morphologic characterization with gadolinium-enhanced double-oblique MR imaging initial results. Radiology 223(2):566–573

Wastie ML, Latief KH (2004) Gadolinium: named after Finland's most famous chemist. Br J Radiol 77(914):146–147

Weber WA, Grosu AL et al. (2008) Technology Insight: advances in molecular imaging and an appraisal of PET/CT scanning. Nat Clin Pract Oncol 5(3):160–170

Winter PM, Caruthers SD et al. (2003) Molecular imaging of angiogenesis in nascent Vx-2 rabbit tumors using a novel alpha(nu)beta3-targeted nanoparticle and 1.5 Tesla magnetic resonance imaging. Cancer Res 63(18):5838–5843

Wu YL, Ye Q et al. (2006) In situ labeling of immune cells with iron oxide particles: an approach to detect organ rejection by cellular MRI. Proc Natl Acad Sci USA 103(6):1852–1857

Yan Y, Steinherz P et al. (1998) Antileukemia activity of a natural killer cell line against human leukemias. Clin Cancer Res 4(11):2859–2868

Yuan C, Kerwin WS et al. (2002) Contrast-enhanced high resolution MRI for atherosclerotic carotid artery tissue characterization. J Magn Reson Imaging 15(1):62–67

Zhang L, Conejo-Garcia JR et al. (2003) Intratumoral T cells, recurrence, and survival in epithelial ovarian cancer. N Engl J Med 348(3):203–213

Zhao M, Beauregard DA et al. (2001) Non-invasive detection of apoptosis using magnetic resonance imaging and a targeted contrast agent. Nat Med 7(11):1241–1244

Chapter 18
Genomic Data Visualization: The Bluejay System

Jung Soh, Paul M.K. Gordon, and Christoph W. Sensen(✉)

Abstract We have developed the Bluejay system, a visual environment for the visualization and comparison of genomes and associated biological data. The software system offers many innovative features that are essential to genome research but were not available previously. The data integration abilities allow the user to integrate gene expression data sets into a genomic context and to seamlessly invoke external Web-based services that are available for biological objects. The comparative genomics capability provides the visualization of multiple whole genomes in one context along with many crucial gene comparison features. The user can also use GPS-inspired waypoints for aligning genomes to facilitate easy comparison as well as for navigating within a genome effortlessly. Bluejay provides a unique set of customizable genome browsing and comparison features for the many biologists who want to answer complex questions using completely sequenced genomes. We expect Bluejay to be beneficial to systems biology in general and genome research in particular.

Introduction

One of the key challenges in systems biology lies in the interpretation and visualization of large volumes of data. To systematically study complex interactions in biological systems, a user-centered computational environment that also uses integrated imaging would be required. In the field of genomics, such a visually enriched environment allows biologists to explore a genomic sequence in order to extract meaningful descriptions of their data. Recent advances in DNA sequencing technologies have resulted in the proliferation of sequence and annotation data, as well as in overwhelming varieties of data types and information granularities.

C.W. Sensen
University of Calgary, Faculty of Medicine, Sun Center of Excellence for Visual Genomics, 3330
Hospital Drive NW, Calgary, Alberta Canada T2N 4N1, e-mail: csensen@ucalgary.ca

C.W. Sensen and B. Hallgrímsson (eds.), *Advanced Imaging in Biology and Medicine*.   395
© Springer-Verlag Berlin Heidelberg 2009

The image-based visual environment needs to adapt to this variation to provide the user with the best possible imagery of genomic data in a given context. As more and more Web-based tools and services become available for biologists, such a visual environment should also be able to integrate and link to those seamlessly.

The Bluejay genome browser (Soh et al. 2007; Turinsky et al. 2004) has been developed to address the challenges posed by the increasing number of data types as well as the increasing volume of data generated in genome research. In the early stages of such development, the focus was on the visualization of whole genomes consisting of annotated genes and other elements. The more recent development of Bluejay has been guided by the observation that biologists who use gene expression profiling and genome comparison gain functional insights beyond those provided by traditional per-gene analyses. As such, a key guiding principle of Bluejay development has been the visual integration of genomic and related biological information into a single unified genomic context.

Bluejay is a genome viewer that integrates genome annotation with: (a) gene expression analysis; (b) seamless invocation of numerous web services for genomic data of interest without the need for changes in source code; (c) genome comparison with an unlimited number of other genomes in the same view. Using this type of integration, the biologist can investigate a gene not just in the context of its genome, but also its regulation and its evolution. Bluejay also has rich provision for personalization by users, including various display customization features and the availability of waypoints for tagging multiple points of interest on a genome and exploiting them.

Resource Integration

Biologists have access to numerous distributed resources, such as biological databanks, software applications, and Web Services. However, when a biologist needs to investigate a particular organism or a set of related organisms, these resources will be put to good use only if they are integrated and linked in a way that is transparent to the user. Although powerful methods are available for analyzing various types of data on their own, the challenge for systems biology is to unify these analyses into a coherent model of an organism. It is a goal of Bluejay to offer a visually oriented data exploration environment that can interoperate with other tools as well as integrate heterogeneous biological information into its unified visual representations.

Gene Expression

Microarrays enable a biologist to view the expression levels of many genes in parallel. These data as a whole provide a snapshot of the transcriptional processes within a cell. With the generation of large data sets from microarray experiments, there is an increasing need for programs that can generate views of both genomic and expression data in a comprehensive manner and perform analysis on the visually represented data. To meet this need, we have integrated TIGR MultiExperiment Viewer (MeV) (Saeed et al. 2003) into Bluejay in the form of a fully embedded module. Through this integration, Bluejay can show gene expression values in a genomic context, enabling biologists to draw additional inferences from gene expression analyses (e.g., operon structures).

Bluejay internally represents the genomic data, along with gene expression values, as a Document Object Model (DOM, http://www.w3c.org/DOM). The visual display incorporating gene expression values can also be output as a high-quality Scalable Vector Graphics (SVG, http://www.w3.org/TR/SVG) image. Within Bluejay, the TIGR MeV cluster analysis results can also be displayed on the whole genome, with each cluster displayed in a unique color for easy differentiation. Figure 18.1 illustrates the interaction between the embedded TIGR MeV module and the rest of Bluejay for representing genomic and gene expression data simultaneously.

The key advantage of microarray data integration in Bluejay is the ability to show gene expression values alongside gene locations within a genome, such that they are spatially associated with the genes whose expression values they represent. This makes it much easier for a biologist to intuitively draw inferences from gene expression analysis than when the results are displayed with no relation to the genomic context, as in expression images, heat maps, or even tables of numeric

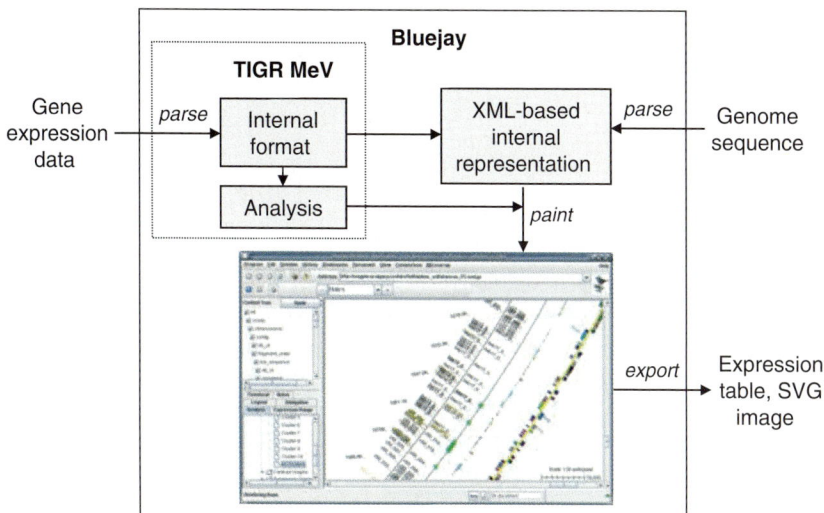

Fig. 18.1 Handling of gene expression datasets in Bluejay. Gene expression data parsing and analysis (e.g., clustering) are done by TIGR MeV. The expression and analysis data are combined with the genome data to produce a coherent visual representation. Tables of expression values and images of the genome with visualized expression values (or clustering results) can then be exported

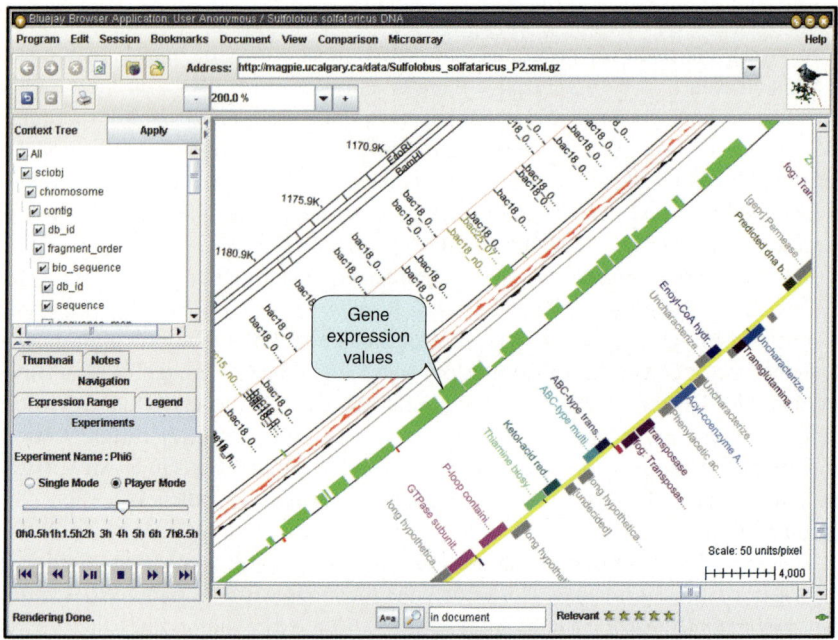

Fig. 18.2 Gene expression values shown alongside the genes. The height of a bar represents the (positive or negative) magnitude of the expression value of the corresponding gene at the outer display lane

values offered by standard gene expression viewers. An example of microarray data displayed along a genome sequence in Bluejay is shown in Fig. 18.2.

Web Services

Bluejay supports Moby (The BioMoby Consortium 2008), a protocol consisting of a common XML object ontology for biological entities and a standardized request/response mechanism for automating Web-based analysis. In Bluejay, the user can seamlessly link from the visualized data, internally represented as a DOM, to Moby-compliant Web Services by invoking the Seahawk Moby client (Gordon and Sensen 2007). Figure 18.3 shows an example of searching through and invoking Web Services from a visualized gene. The analysis results would be displayed in HTML format in a separate window.

The central services registry of Moby lists the services that work on a particular object type. Bluejay can query this registry to obtain a list of supported analysis tools for a given genomic data type. This eliminates the need to hard-code functionality based on object types into the Bluejay application. Moreover, Moby represents the data types and services based on the taxonomy of the particular

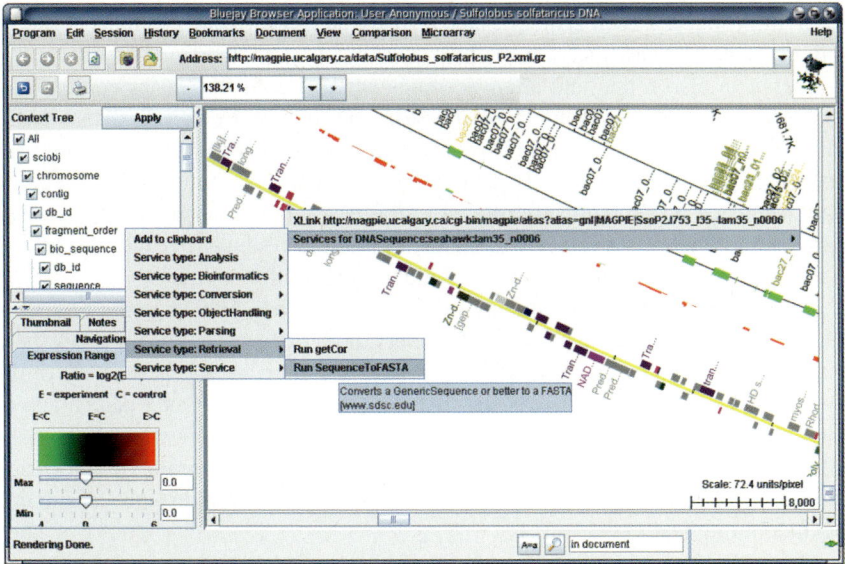

Fig. 18.3 BioMoby-compliant web services are activated by clicking on a gene: service options in cascading popup menus appear

scientific field, stored as data type hierarchies. For example, services that work on `GenericSequence` are also presented when an object of either of the more specialized data types `AminoAcidSequence` or `DNASequence` is selected, giving the user more options to choose from. Users can also link from gene expression value display in Bluejay to Moby Web Services, in order to further explore hypotheses they derive from the contextualized expression statistics.

Genome Comparison

Comparing multiple genomes is one important method of studying organism diversity, evolution, and gene functions. Bluejay is capable of visualizing multiple genomes side-by-side in a single display for direct visual comparison. The comparison mode is automatically activated if the user loads more than one sequence into Bluejay. Likewise, the comparison mode is automatically exited when only one genome remains loaded in Bluejay.

An essential feature that helps the user to compare multiple genomes is the ability to display lines to link common genes based on their Gene Ontology (GO, http://www.geneontology.org) classifications. The GO project provides a controlled vocabulary to describe gene and gene product attributes in any organism, to address the need for consistent descriptions of gene products in different databases. Using this feature alone, the user can instantly estimate how similar the compared genomes are. Within Bluejay, each gene is linked only to the nearest gene in another genome with the same GO classification, if one exists.

Users can also rotate and align genomes automatically for easier comparison. The optimal alignment is found by minimizing the sum of the angular distances for all linked pairs of genes. This amounts to minimizing linking distances to find the best global alignment of closely related genes. When this feature is enabled, the outer sequence is automatically rotated to the best-aligned position. The user can then see the functional similarity of the genes in the two sequences, with the effects of base position differences ruled out as much as possible.

An additional feature of Bluejay related to GO class linking is the ability to selectively show or hide the links. The selection is done by imposing a threshold on the linking distance. For linear genomes, a threshold is set as a percentage of the master sequence length, whereas for circular genomes it is set as a percentage of $360°$. By varying this threshold value, the user can recognize how many of the links are close in terms of gene locations. Figure 18.4 shows an example of comparing two bacterial genomes using the GO class linking, minimum distance alignment, and link hiding features.

Bluejay is also capable of comparing any number of genomes, as shown in Fig. 18.5, where four human chromosomes are compared. It is widely believed that the human chromosomes contain many instances of gene family duplications (Dehal and Boore 2005). Using Bluejay, we can visually confirm that a particular subset of human chromosomes does indeed have a number of duplicated genes. The direction of gene linking is by default from the master sequence to other sequences, but this direction can be reversed, or links in both directions can even be shown in Bluejay. For example, in the comparison snapshot of Fig. 18.5, the chromosome 17q is set as the master sequence, and the links are generated in the direction of chromosomes 17q, 12, 7, 2q (bottom to top), as well as in the opposite direction (top to bottom).

Customized Visualization

While using a genome viewer, biologists are often interested in investigating only a small portion of a genome. Navigating within a large genome to find the small region of interest usually entails scrolling around until the desired part is in view or typing in a base number to see the view around it. Either approach requires the user to go through several manipulations within a viewing environment to focus on the desired part.

With Bluejay, we now offer a highly intuitive method for exploring large genomes by using the notion of a waypoint. In global positioning system (GPS) applications, waypoints are coordinates of locations of interest on the Earth's surface. Similarly, waypoints in Bluejay are a set of flags to mark points of interest on a genome. For example, the user can set a waypoint within the sequence to highlight a specific gene or sequence feature (e.g., promoters or terminators). Figure 18.6 shows a typical use of waypoints, where the user sets two waypoints and uses one of them to quickly

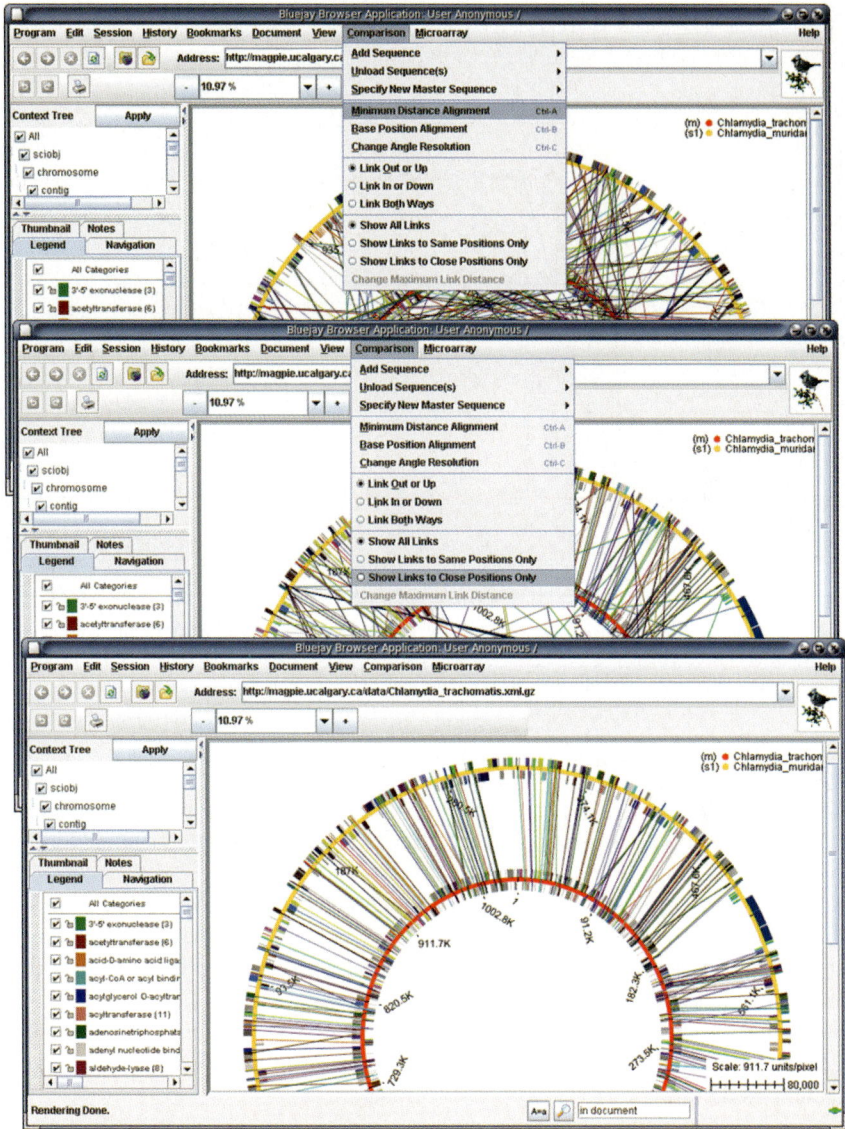

Fig. 18.4 Bacterial genome comparison. *Chlamydia trachomatis* and *Chlamydia muridarum* genomes are compared, where a pair of linked genes (one from each sequence) means that the two genes have the same Gene Ontology (GO) classification (*top*). The outer genome has been rotated to automatically align the genomes (*middle*). Only those links with an angular distance of less than 1% of 360° are shown (*bottom*). The two genomes are still fairly collinear despite gene duplications, deletions, insertions and rearrangements

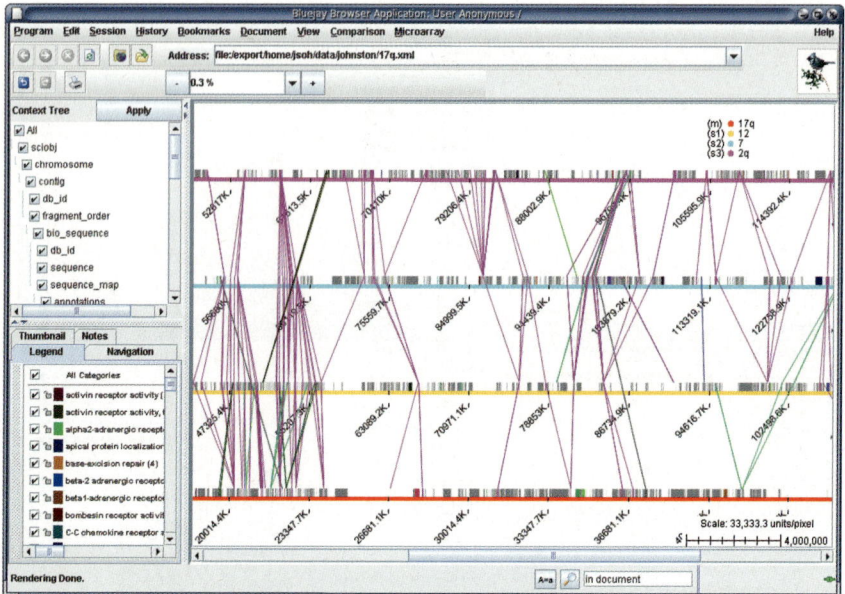

Fig. 18.5 Human chromosomes 17q, 12, 7, and 2q are compared in Bluejay to show that many gene families are duplicated in several chromosomes, as represented by the lines that link genes of the same GO categories (not all links are shown for the sake of clarity)

focus on a region of interest. Setting a waypoint enables several operations on it, such as focusing on it, cutting the genome at it, and aligning several genomes at a waypoint with the same name across multiple genomes, as demonstrated by the pop-up menu in Fig. 18.6.

Bluejay also allows the user to view a genome at the greatest level of detail, which is the nucleotide text level. In such a text view mode, a waypoint can be set not only with respect to a whole gene but also at an individual base position. This helps the user to investigate bases as well as whole genes. For example, in the "horizontal sequential" text mode, a genome is displayed as a long horizontal string of base characters. Figure 18.7 shows an example of setting multiple waypoints in that text mode.

There is often the need to align multiple genomes at specific genes to investigate how similar they are around those genes. To do that, the user would tediously try rotating the genomes by some angles until the genes of interest align exactly. Waypoints in Bluejay provide the user with the ability to flag multiple genes with the same name and align them at those flags, eliminating the need to estimate the amount of required rotation. Figure 18.8 shows an example of aligning three genomes by a gene of interest. This usage of waypoints for in combination with genome rotation for genome comparison adds greatly to the utility of waypoints.

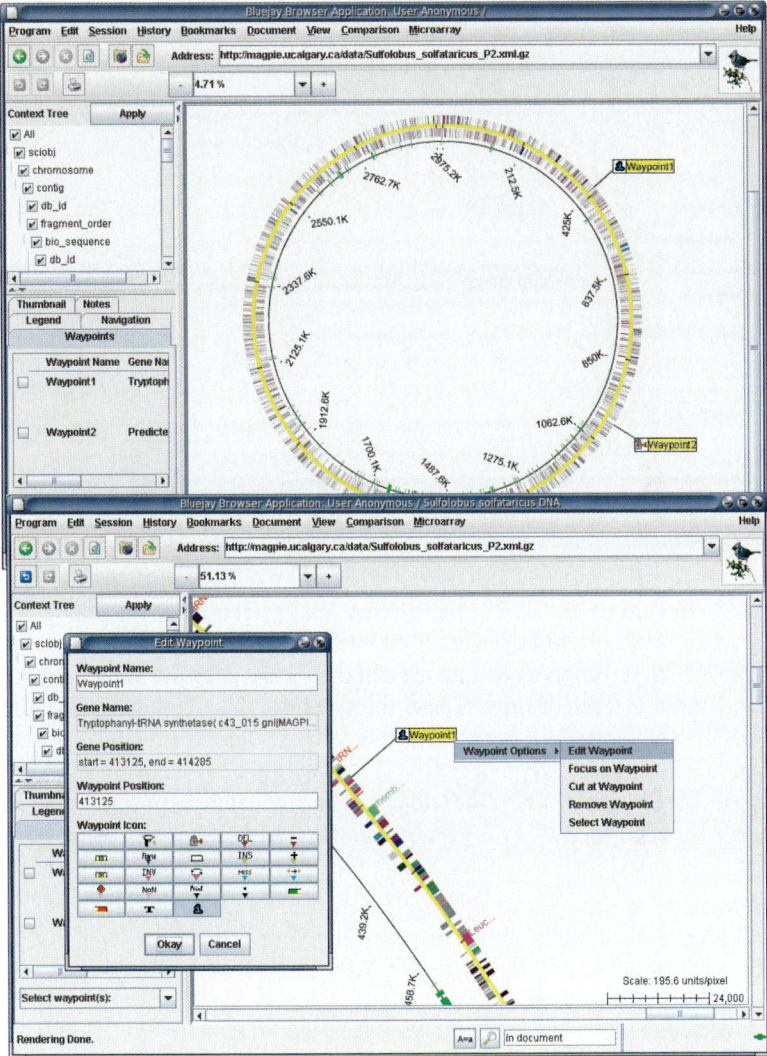

Fig. 18.6 Using waypoints for quick navigation. The user sets two waypoints (*top*) and uses them to see the sequence around *Waypoint1* in more detail (*bottom*). A right mouse click causes the waypoint operation pop-up menu to appear; the user then selects "Edit Waypoint" to change the attributes of the waypoint

Comparison with Other Genome Browsers

Recent advances in DNA sequencing technologies have generated a large amount of genomic data, which in turn requires tools for exploring biological sequences. To meet this need, several sequence browsers have been developed. The most advanced and well-known general-purpose browsers include Ensembl Genome

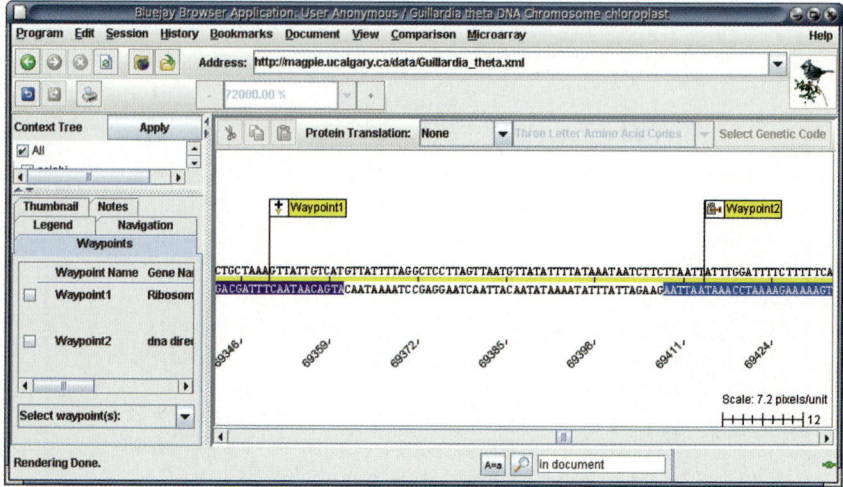

Fig. 18.7 Using waypoints in a text mode. In a text mode, the genes are shown as colored rectangles over the bases, and waypoints can be set at individual base positions, not just on genes

Browser (Hubbard et al. 2007), NCBI Map Viewer (Feolo et al. 2000), Microbial Genome Viewer (Kirkhoven et al. 2004), and UCSC Genome Browser (Kuhn et al. 2007). These browsers let the user immediately access a wealth of biological data stored at each facility. These tools deliver their content to the user as dynamically generated web forms and image maps that are rendered on a web browser.

In contrast, Bluejay has been developed to be a general-purpose genome exploration environment as well as a genome browser. Unlike most genome browsers that use pixel-based image formats, Bluejay uses Scalable Vector Graphics (SVG, http://www.w3.org/TR/SVG), a language for describing two-dimensional vector and mixed vector/raster graphics in XML. The main advantage of adopting SVG from the user's perspective is unlimited zooming of the display with no degradation in image quality. Bluejay offers provisions for processing diverse data types from many different sources and customizing the visual display to suit the user needs, rather than acting mainly as a front end to a particular data repository. To the best of our knowledge, the novel features of Bluejay described in this chapter, such as dynamic resource linking, genome comparison and waypoints capability, are not present in other genome browsers. Table 18.1 summarizes the differences between the integrative aspects of Bluejay and those of popular genome browsers.

Layout

The Ensembl Genome Browser uses a web browser-embedded program to facilitate illustration of genomes. The opening page of an organism's chromosomes shows a graphic of a whole chromosome, including details on GC percentages and repeat

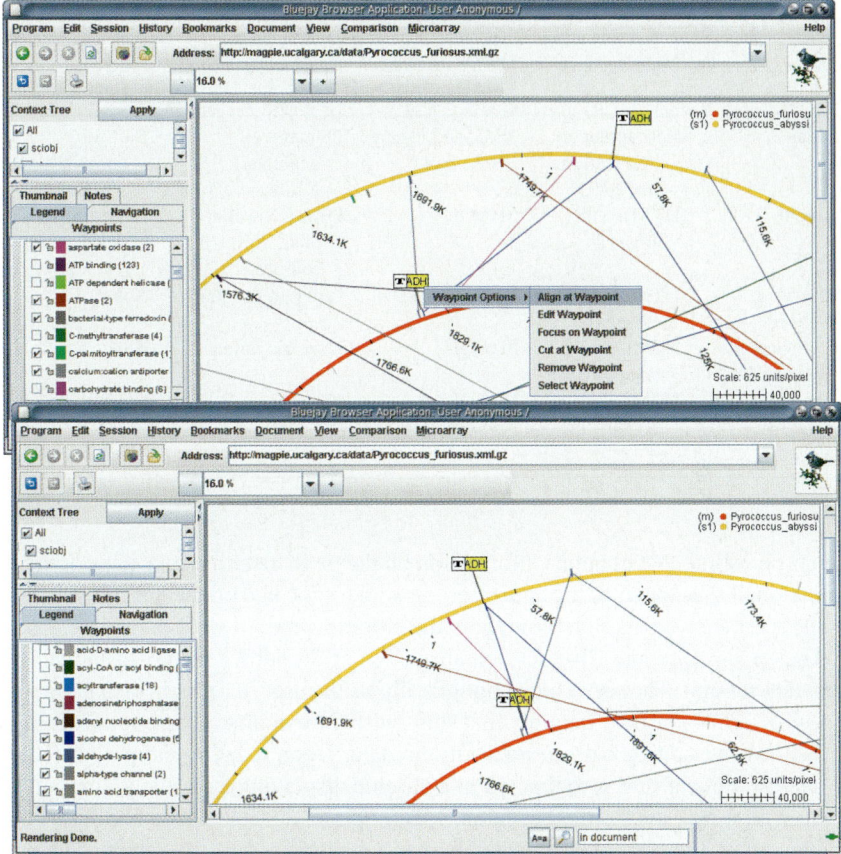

Fig. 18.8 Aligning genomes by waypoints. A waypoint named *ADH* is set at a gene of the alcohol dehydrogenase GO class in each of the two *Pyrococcus* genus genomes. The user then selects the "Align at Waypoint" action on the inner waypoint to let Bluejay align them at the *ADH* genes by rotating the outer genome by an automatically calculated angle

regions. Clicking on a portion of this chromosome image links to a second page showing four views of the region: chromosome view, overview, detailed view and base pair view. The program also allows for jumping to a specific region of the genome by inputting a base pair position. The width of the view can be adjusted with a dropdown menu to manipulate from 600 to 2,000 pixels wide. Furthermore, dropdown menus at the top allow the user to adjust the display of various data sets.

The organization of the NCBI Map Viewer website is similar to Ensembl. Organisms are located in the opening page, with links to a secondary page of chromosomes. Clicking on a chromosome will provide a general overview of the chromosome, including information on genes, descriptions of gene activity, and further links to other sources of information. In this page view, users are able to zoom into the finer features of the genome to obtain a higher-resolution image of

Table 18.1 Comparison of the major features of Bluejay with those of other genome browsers

Browser	User interface	Main data source	External data linking	External services linking
Bluejay	Java applet or application and SVG	XML and legacy text files	Dynamically generated XLinks	BioMoby-compliant web service
Ensembl	HTML and images	MySQL	DAS (Dowell et al. 2001) sources	Hyperlinks (hardcoded)
NCBI Map Viewer	HTML and images	RDBMS/ASN.1	LinkOut program	None
UCSC Genome Browser	HTML and images	MySQL	User file upload	None
Microbial Genome Viewer	HTML and SVG	MySQL	None	None

the chromosome that contains information on the gene locations and sizes. In addition, a small thumbnail of the entire chromosome is located on the left of the screen to allow for easy navigation throughout the chromosome.

The UCSC Genome Browser contains a list of dropdown menus to initiate the user data session. The set of menu options allows the user to focus in on a particular aspect of an organism's genome that they wish to examine. There is also a link that allows for adding custom tracks to the data. Upon locating the proper region of interest, the website is redirected to a graphic displaying the region horizontally along with a few tracks in that region. Menu options located at the bottom of this graphic allow the user to customize annotations displayed on the screen. The user can also click on a base to jump to the base sequence of a specific segment of the chromosome. Moving left and right is also done by clicking on the buttons adjacent to zoom and can also be done at three speeds.

Bluejay's layout closely resembles a typical Web browser (see Fig. 18.4). Dropdown menus located at the top contain many of Bluejay's functionalities and directly underneath are address fields and the back/forward buttons. The central window is allocated to displaying the genome graphic with many of the features that users can choose to display. To the left of the main display window are two additional windows; the top one is for choosing the levels of annotations to display and the bottom one is for performing other functions, including adding notes, choosing gene ontologies to be displayed, and navigating through the genome by various means.

Functionalities

Ensembl Genome Browser allows a variety of information to be selected for display, such as restriction enzyme sites. Genetic features, repeat sequences and decorations can also be added to the display. Decorative features can also be displayed, such as sequence information, codons, start and stop codons, rule and scale bars, GC percentage, a gene legend and SNP legend.

NCBI Map Viewer contains a number of interlinked services connected to genome sequences on the main view page. Journal articles pertaining to a sequence of interest is directly linked from the graphical display, providing easy access to supplemental information. This is also linked to OMIM (Online Mendelian Inheritance in Man) and GenBank for further information regarding the gene, such as diseased states, polymorphisms and nucleotide sequence information (Wang et al. 2000). Also, a "Map and Options" button opens a window for customizing the display of certain information on the main genome graphic.

UCSC Genome Browser displays many different annotations broken into several categories, including mapping and sequencing tracks, gene and gene prediction tracks, mRNA and EST tracks, expression and regulation tracks, comparative genomics, and variation and repeats. In addition to these annotations, a homology search can be performed through BLAT (Kent 2002), sequence searches using PCR primer pairs with in silico PCR, and gene table generation containing related genes based on homology, expression or genomic location with Gene Sorter (Trumbower and Jackson 2005).

Bluejay's functionalities extend beyond displaying genes within a genome. These extra functionalities include the ability to display restriction enzyme sites, AT and GC percentages, GC skew, and microarray data imposed on the main genomic image. Comparisons with multiple sequences can also be done with both genomes displayed side-by-side and lines dictating the locations of ontologically common genes relative to each other. Genomes can be drawn in circular or linear forms in order to facilitate comparisons of gene presence and the arrangement between two genomes. Bluejay also allows for a variety of data processing functions to be carried out, including exporting the current display as an image file, viewing raw sequence data, linearizing or circularizing the sequence, and customizing the genomic orientation. Moreover, there are several customization features in Bluejay that are unique when compared to the other three genome browsers.

Conclusion

Bluejay is a highly integrative visual environment for genomic and related data exploration. Among the many features of Bluejay, we described those features that are most relevant to systems biology efforts. A significant enhancement can be made to Bluejay by adding the ability to detect metabolic networks automatically and present the results to the users in a graphical fashion. Similar to the up- and downregulation of genes, the goal would be to display the change in protein concentrations along the genome and thus connect information about the gene function, the regulatory elements, the transcription and translation level within an organism. The gene expression viewing environment of Bluejay will be a basis from which to work towards such an enhanced Bluejay system. The end product of this endeavor will be beneficial to systems biology.

Bluejay is available to anyone from the project home page (http://bluejay. ucalgary.ca) as an applet, Java Web Start, or applications for several popular computing platforms. Bluejay is implemented in Java 1.5 and thus is platform-independent. It uses a number of open source libraries from the Apache Foundation (http://www. apache.org).

Acknowledgements We thank Morgan Taschuk, Anguo Dong, Andrew Ah-Seng, Andrei Turinsky, Krzysztof Borowski, and Lin Lin for their contributions to the development of Bluejay. We also thank Hong Chi Tran for his extensive comparative study of genome browsers. This work was supported by Genome Canada/Genome Alberta through Integrated and Distributed Bioinformatics Platform for Genome Canada, as well as by the Alberta Science and Research Authority, Western Economic Diversification, National Science and Engineering Research Council, Canada Foundation for Innovation, and the University of Calgary. Christoph Sensen is the iCORE/Sun Microsystems Industrial Chair for Applied Bioinformatics.

References

Dehal P, Boore JL (2005) Two rounds of whole genome duplication in the ancestral vertebrate. PLoS Biol 3(10):e314

Dowell RD, Jokerst RM, Day A, Eddy SR, Stein L (2001) The distributed annotation system. BMC Bioinformat 2:7

Feolo M, Helmberg W, Sherry S, Maglott DR (2000) NCBI genetic resources supporting immunogenetic research. Rev Immunogenet 2(4):461–467

Gordon PM, Sensen CW (2007) Seahawk: moving beyond HTML in Web-based bioinformatics analysis. BMC Bioinformat 8:208

Hubbard TJ, Aken BL, Beal K, Ballester B, Caccamo M, Chen Y, Clarke L, Coates G, Cunningham F, Cutts T, Down T, Dyer SC, Fitzgerald S, Fernandez-Banet J, Graf S, Haider S, Hammond M, Herrero J, Holland R, Howe K, Howe K, Johnson N, Kahari A,Keefe D, Kokocinski F, Kulesha E, Lawson D, Longden I, Melsopp C, Megy K, Meidl P, Ouverdin B, Parker A, Prlic A, Rice S, Rios D, Schuster M, Sealy I, Severin J, Slater G, Smedley D, Spudich G, Trevanion S, Vilella A, Vogel J, White S, Wood M, Cox T, Curwen V, Durbin R, Fernandez-Suarez XM, Flicek P, Kasprzyk A, Proctor G, Searle S, Smith J, Ureta-Vidal A, Birney E (2007) Ensembl 2007. Nucl Acids Res 35(Database issue):D610–D617

Kent WJ (2002) BLAT: the BLAST-like alignment tool. Genome Res 12(4):656–664

Kerkhoven R, Van Enckevort F, Boekhorst J, Molenaar D, Siezen RJ (2004) Visualization for genomics: the microbial genome viewer. Bioinformatics 20:1812–1814

Kuhn RM, Karolchik D, Zweig AS, Trumbower H, Thomas DJ, Thakkapallayil A, Sugnet CW, Stanke M, Smith KE, Siepel A, Rosenbloom KR, Rhead B, Raney BJ, Pohl A, Pedersen JS, Hsu F, Hinrichs AS, Harte RA, Diekhans M, Clawson H, Bejerano G, Barber GP, Baertsch R, Haussler D, Kent WJ (2007) The UCSC genome browser database: update 2007. Nucl Acids Res 35(Database issue):D668–D673

Saeed AI, Sharov V, White J, Li J, Liang W, Bhagabati N, Braisted J, Klapa M, Currier T, Thiagarajan M, Sturn A, Snuffin M, Rezantsev A, Popov D, Ryltsov A, Kostukovich E, Borisovsky I, Liu Z, Vinsavich A, Trush V, Quackenbush J (2003) TM4: a free, open-source system for microarray data management and analysis. Biotechniques 34(2):374–378

Soh J, Gordon PM, Ah-Seng AC, Turinsky AL, Taschuk M, Borowski K, Sensen CW (2007) Bluejay: a highly scalable and integrative visual environment for genome exploration. In: Proc 2007 IEEE Congr on Services, Salt Lake City, UT, 9–13 July 2007, pp 92–98

The BioMoby Consortium (2008) Interoperability with Moby 1.0: It's better than sharing your toothbrush! Brief Bioinform 9(3):220–231

Trumbower H, Jackson J (2005) Key features of the UCSC genome site. In: Proc 2005 IEEE Comput Syst Bioinform Conf Workshops 8:33–34

Turinsky AL, Ah-Seng AC, Gordon PM, Stromer JN, Taschuk ML, Xu EW, Sensen CW (2004) Bioinformatics visualization and integration with open standards: the Bluejay genomic browser. In Silico Biol 5:0018

Turinsky AL, Gordon PM, Xu EW, Stromer JN, Sensen CW (2005) Genomic data representation through images: the MAGPIE/Bluejay system. In: Sensen CW (ed) Handbook of genome research. Wiley-VCH, Weinheim, pp 187–198

Wang Y, Bryant S, Tatusov R, Tatusova T (2000) Links from genome proteins to known 3-D structures. Genome Res 10:1643–1647

Chapter 19
Anatomical Imaging and Post-Genomic Biology

Benedikt Hallgrímsson(✉) and Nicholas Jones

Abstract Anatomical imaging provides the morphological context for post-genomic biology. Volumetric imaging is central to the creation of atlases of gene expression as well as the visualization of increasingly complex molecular information in a spatiotemporal context. To become integrated into the types of hypotheses that systems biology and other post-genomic approaches generate, however, anatomical imaging must evolve further. Modes of phenotypic analyses must begin to generate databases and data repositories that will support data-mining approaches to increasingly complex and broad biological questions. For this to happen there must be increasing emphasis on high-throughput phenotypic analysis, data standardization, data sharing, and the systematic quantification of phenotypic variation. The solutions to these issues will lay the foundations for a new field of study, that of phenogenomics.

Introduction

The history of science shows us that while scientific questions often precede the technology required to answer them, the development of technology that enables novel observations also stimulates the formulation of novel questions that can open new directions of inquiry. Anton van Leeuwenhoek did not set out to ask questions about the cellular basis for life in the seventeenth century, but it is difficult to imagine the progression of biology without the invention of the microscope. The current transformation of biology, captured by the label "post-genomic," is similarly rooted in a mutually reinforcing feedback between scientific questions, practices, and new technology. The desire to map, sequence, and annotate whole genomes has driven the development of technology, which has greatly increased the rate at

B. Hallgrímsson
University of Calgary, Faculty of Medicine, Department of Cell Biology and Anatomy and the McCaig Bone and Joint Institute, 3330 Hospital Drive NW, Calgary, AB, Canada T2N 4N1, e-mail: bhallgri@ucalgary.ca

which molecular information can be obtained. Together with the standardization of databases and the development of software tools for mining them, the greatly increased pace of data acquisition has, in turn, begun to generate novel systems-level questions that would have been unthinkable a decade ago.

The impact of this post-genomic transformation, however, is not felt uniformly across biological disciplines. Even though the ultimate goal of most molecular work is to understand the genetic, developmental and physiological determinants of normal and pathological variation in organismal structure and function, our ability to measure and analyze phenotypic variation has yet to go through the kind of transformation that has driven molecular biology during the past decade. Partly this is because the problem is more difficult. Phenotypic information is more varied and more highly dimensional than genomic or proteomic data. For this reason, arriving at broadly useful standard measures of phenotypic variation is simply a larger and more difficult task than arriving at a standard methodology for sequencing a genome, difficult as that has been! The second main reason is technological. Even armed with a standard set of phenotypic variables, existing techniques for phenotypic analysis are too slow and labor-intensive to keep up with the generation of molecular data. With 25,000 genes (Lander et al. 2001; Venter et al. 2001; Waterston et al. 2002), an unknown number of *cis*-regulatory sites, around 1,000,000 proteins, an unknown number of microRNAs (Boeckmann et al. 2005), many-to-many relationships between proteins and functions, and the context dependency of gene and protein function, the potential for generating informative genetic variants for a mammalian model species such as the mouse is vast. The 3,500 mouse mutants cataloged by Jackson Labs is not even the tip of that iceberg. However, thorough phenotypic characterization of the relatively small number of mutants in the Jackson Labs database would be a monumental undertaking given current techniques.

A second and related issue is that as biological questions come to focus increasingly on the behavior of biological systems, the comparison of genotypes for a single gene will increasingly give way to larger scale and multifactorial comparisons in which the phenotypic effects of interests may be subtle. Understanding the etiology of genetically complex diseases in humans or normal phenotypic variation for almost any trait of interest will require approaches that simultaneously compare multiple genetic or environmental effects. This development will place even greater demands on our capacity to perform phenotypic characterization of model organisms or humans.

The problems described here are common to all forms of phenotypic analysis. The development of technological and methodological solutions to these issues generally falls under the label of "phenogenomics" (Bogue 2003; Grubb et al. 2004; Paigen and Eppig 2000; Henkelman et al. 2005). Anatomical imaging is only one aspect of this phenogenomic endeavor. However, it is an important component, as many diseases produce anatomical effects. For the study of morphological birth defects or of morphological variation, imaging is a primary means of phenotypic assessment.

Two of the major challenges that stand in the way of image-based phenogenomics are similar to those faced by genetics and molecular biology prior to the human genome project. These challenges are throughput and the need for data standardization, and infrastructure for data sharing. In addition, image-based phenogenomics faces two technical challenges that are somewhat different in nature. One is the need to quantify morphological variation in ways that enable large-scale systematic comparisons. This topic was discussed in the context of shape morphometrics in our other chapter in this volume. The second and related issue is the need to appreciate and adequately deal with continuous variation in phenotypic variables.

Throughput

Increases in the rate of data generation through technological innovation have been one of the major drivers of the post-genomic revolution. Although anatomical imaging technology has developed rapidly during the past decades, the rate at which phenotypic data can be generated has not increased substantially. Imaging equipment such as micromagnetic resonance machines, computed microtomography and optical projection tomography have greatly increased the amount and quality of data generated about individual specimens. Techniques for volumetric imaging have revolutionized anatomical visualization and measurement of in many areas of investigation. A problem that remains, however, is that sample processing times for many of these techniques are very long. In computed microtomography, for instance, scanning times range from half an hour to several hours per specimen, and each dataset requires substantial computer time to reconstruct into image data. Once the scans are done and reconstructed, moreover, the process of generating numerical data for analysis from the image datasets is even more labor-intensive. A single study of bone quality in a rat model, for instance, can occupy a month of personnel time just to generate the required numerical data.

When new imaging technology becomes available, the initial focus tends to be on determining the structures that can be imaged and the quality of the image data. In publications, single images for illustration are often all that is presented. As the technology matures, the focus shifts to the generation of numerical data based on image data that can support hypothesis testing. With this, there is an increasing need to image more specimens. Figure 19.1 shows an example of a dataset of a morphometric analysis of a dataset consisting of adult mouse crania. This dataset, consisting of about 400 individuals, took over 300 h of scanning and 1,000 h of personnel time to generate. Facilities with slow and labor-intensive imaging technology, such as computed microtomography, tend to be quickly swamped by the demand for high-volume scanning and analysis.

The solutions to the problem of throughput for phenotypic data analysis need to address two separate issues. The first is the technical bottlenecks in the generation of the data. The second is the generation of increasingly automated analysis techniques. Until this happens, phenotypic data cannot be meaningfully integrated into large-scale studies that address hypotheses generated within a systems biology framework.

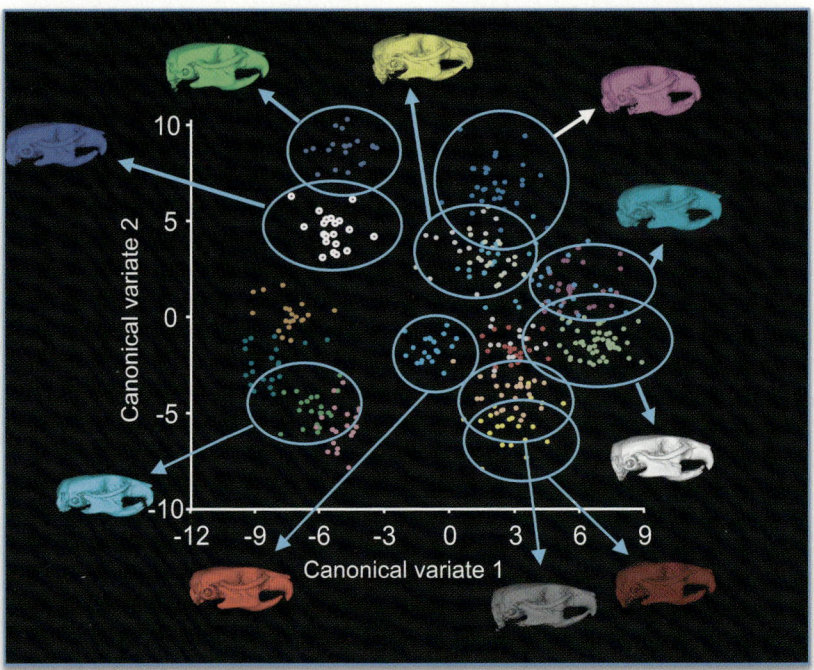

Fig. 19.1 Analysis of a large sample of adult mouse crania via canonical variates analysis of 3D landmark coordinates. The *colored skulls* represent mean shapes for some of the strains generated by scaling, volumetric superimposition and averaging of the entire samples for those strains

To date, technological development has focused more on resolution and data quality than on throughput. Over the last eight years, scanning times have not decreased substantially for either computed microtomography or micromagnetic resonance imaging, while imaging quality and resolution have increased substantially. While the hope is that this emphasis will begin to shift with the next generation of scanners, many investigators have tried to find novel ways to increase the rate of specimen processing using existing scanners. An extreme and innovative example of this is the multiple mouse holder for in vivo magnetic resonance imaging of mice at the University of Toronto Mouse Imaging Center (Fig. 19.2) (Henkelman et al. 2005). For computed microtomography scanning, some scanners allow the automated sequential scanning of multiple specimens, which allows unattended scanning of several specimens at a time.

The other major bottleneck is the quantification of morphological variation. Here, substantial progress is being made by various groups. Our group (Boyd and Hallgrimsson labs), for example, is developing a high-throughput morphological tool that can act as a screening tool to analyze large amounts of data to identify regions of shape change for further study (Kristensen et al. 2008). It differs from traditional geometric morphometric methods as it eliminates the need for laborious landmark selection, often the most time-consuming part of a morphometrics study.

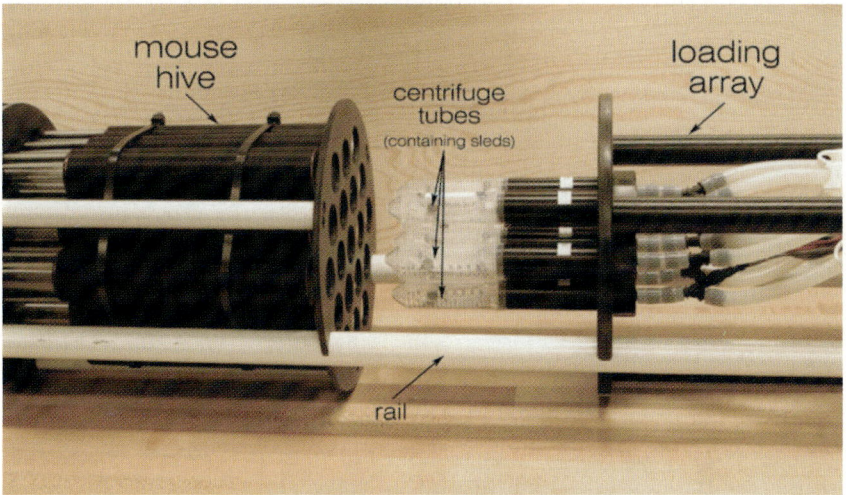

Fig. 19.2 The multiple mouse holder for in vivo micro magnetic resonance imaging at the University of Toronto Mouse Imaging Center. Image obtained with permission from the University of Toronto website

The method is demonstrated on image data from a micro-computed tomography (μCT) scanner, but the approach may be applicable to three-dimensional images obtained from any medium. This tool utilizes custom-developed software to perform intensity-based rigid image registration and image summation to create a mean shape. Mean shapes are then superimposed to create a distance map that shows the shape differences between groups after size. Size variation is removed first by scaling all individuals to centroid size. This method is described more fully in the chapters by Boyd and Hallgrímsson et al. in this volume. Figure 19.3 shows a visualization of the shape differences between the mouse strains shown in Fig. 19.1 using this method.

Related approaches are also being developed by other groups. A similar volumetric superimposition approach, for instance, is being used to visualize and measure variation in brain morphology as well as other aspects of mouse anatomy at the University of Toronto Mouse Imaging Center (Nieman et al. 2006; Dorr et al. 2008; Chen et al. 2006). An alternative approach that focuses on automated 3D landmark extraction has also been developed (Olafsdottir et al. 2007). This approach is very promising because it has the great advantage that you can then apply the existing body of geometric morphometric methods to the extracted landmark data while also preserving the 3D image data. Currently, the major impediment to the widespread application of the method developed by Olafsdottir et al. is the computational power required to process each specimen. In their study, powerful multiprocessor workstations were used to process the data, but the automated landmark extraction still took hours per specimen. That will improve, however, as either computation power increases or the method is refined.

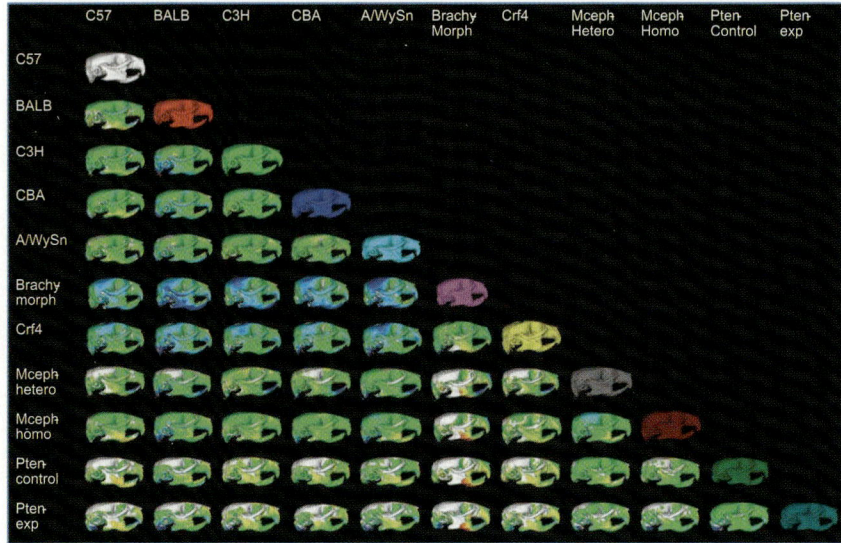

Fig. 19.3 Comparison of multiple mouse strains using volumetric superimposition and averaging. Each *skull image* shows the regions of difference between the two strains being compared (see the Boyd chapter in this volume)

The introduction of high-throughput morphometric tools like this is a first step towards bringing phenotypic analyses up to speed with fast-paced biological methods of inquiry. Although the method requires the addition of rigorous statistical analyses, it automates morphological assessments by removing much of the user interaction required by traditional geometric morphometric methods and retains the entire three-dimensional dataset.

Data Sharing

A major cultural shift that has been a major driver of the post-genomic revolution is the creation of a culture of data sharing and the creation of infrastructure to support the public dissemination of data (Peltonen and McKusick 2001). The construction of communal databases is very important because it will allow the testing of ever larger and more complex hypotheses. This is seen in the emergence of community databases such as OMIM (http://www.ncbi.nlm.nih.gov/omim). Statements and plans for the sharing of molecular data are increasingly standard requirements in grant applications from agencies such as NIH or NSF in the USA. The emerging field of bioinformatics is largely devoted to the management and mining of shared databases, and most visions for systems biology rely heavily on the availability of large amounts of accumulated data about complex systems. For imaging data to relate to genomic data, as a field we must develop standards for data collection and the infrastructure to support datasharing. A huge challenge here is that phenotypic datasets are very diverse and are usually highly specific to a particular hypothesis. A set of measurements on mouse mandibles, for instance, designed for a study on jaw biomechanics, is unlikely to be useful for answering questions about the developmental genetics of facial shape. Agreeing on standard measurements or landmarks that are broadly applicable is very difficult, although this is the approach that the Mouse Phenome Project (MPD) at Jackson Labs has taken (http://phenome.jax.org/pub-cgi/phenome/mpdcgi?rtn = docs/aboutmpd).

An alternative approach is to store phenotypic data in the rawest and most generic form that would be of general use. For volumetric image data, this would be the reconstructed image datasets. Such databases, even if originally obtained for very specific research purposes, could be mined as they grow to support large-scale phenogenomic questions. A challenge here is the large sizes of most anatomical imaging datasets, particularly those based on 3D volumetric data. A related issue is that, while searching or browsing a volumetric image dataset based on indices or metadata is relatively straightforward, being able to see the data in the form of either serial images or 3D reconstructions on the fly is more difficult. Visual interaction with the data, however, is extremely useful if one wants to be able to see what one is downloading from a database or to be able to select specimens for download based on anatomical characteristics.

We have created such a database, which we are in the process of sharing with the community. Our database, called PANDA (phenotypic analysis database) supports the storage, retrieval, maintenance and visualization of volumetric datasets. The immediate purpose of the database is to archive the 3D volumetric data generated from micro-computed tomography ($\mu - CT$) scanning in our lab and to contextualize experimental data with coherent metadata (information about the data) for the management of morphometric and other statistical outputs. Micro-CT scanning yields vast amounts of raw volumetric data; individual image file sizes can exceed several megabytes. Thus, it is necessary to reliably and efficiently store and retrieve a very large volume of 3D image data, as well as associated statistical and morphometric datasets. The key aspects of the data management software are robustness (i.e., to protect data integrity), efficiency (i.e., fast file transfer time), ease of use and maintenance. Robust data management systems demand that the metadata—the contextual information surrounding the data—must be sufficiently sophisticated in order to assist future users and collaborators with the original experimental purpose. The design of this application inherently allows for the growth of the user database to include other users who can be given usernames and passwords. We envision that data can be accessed over the web as a compelling atlas of 3D micro-CT and morphometric output. The program also integrates with the software for high-throughput analysis that is described in the chapter by Boyd in this volume. Although designed for micro-CT data, the database can support any volumetric image data type from MRI to OPT. Figures 19.4 and 19.5 show screenshots from the current version of the database.

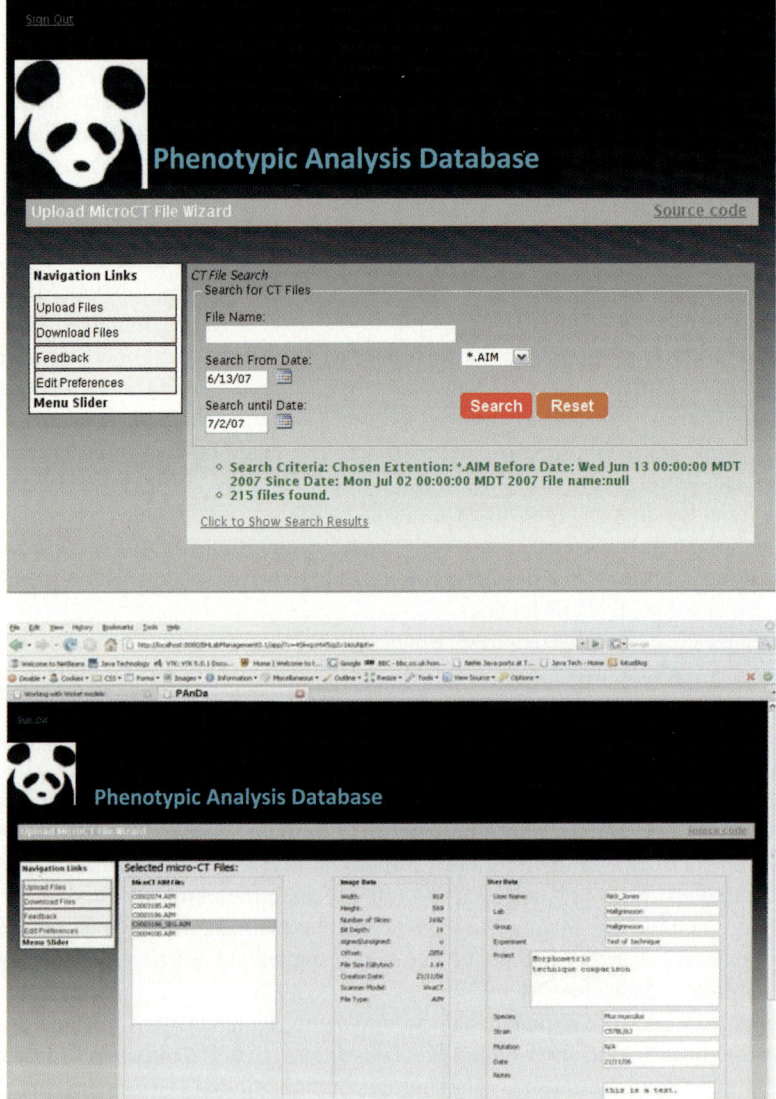

Fig. 19.4 Screenshots from the phenotypic analysis database (PAnDA). The *upper screen* shows the search dialog and the *lower screen* shows the metadata fields associated with individual scans

A
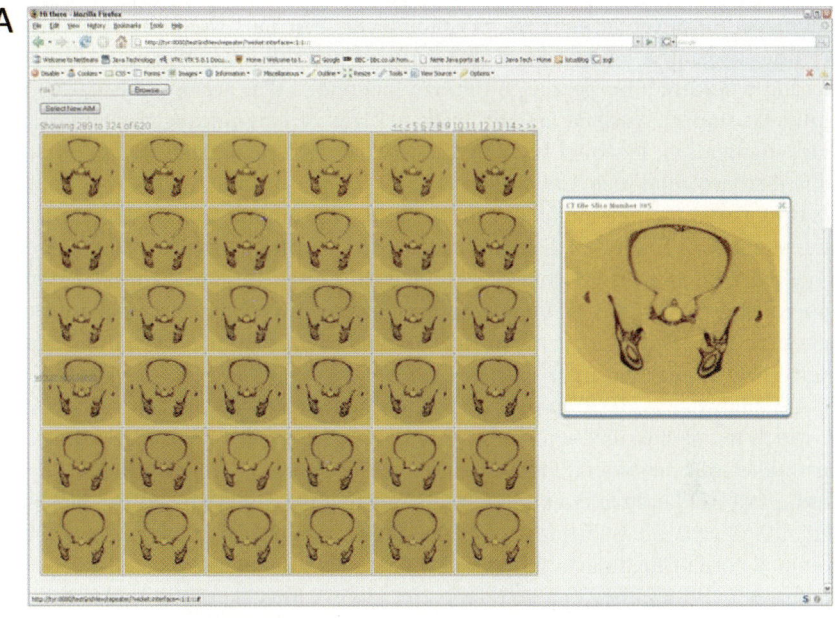

B

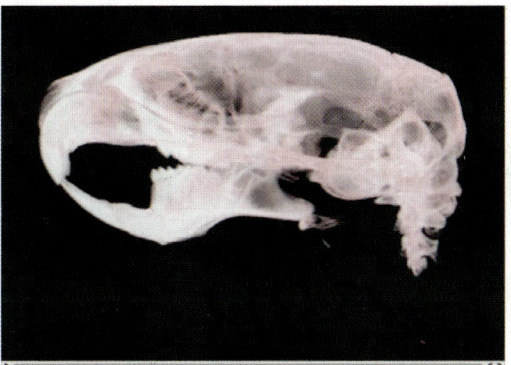

Fig. 19.5 (a–b) Screenshots showing visualizations generated on the fly for image data in PANDA. **(a)** Cross-sectional images obtained from the volumetric dataset. **(b)** A 3D reconstruction based on those data

Standardization

A challenge that is closely related to data sharing is data standardization. Like throughput and data sharing, data standardization is one of the pillars on which post-genomic biology is built (Peltonen and McKusick 2001). For shared databases to be useful, data obtained and deposited by different investigators will need to be comparable. As a community, therefore, we must agree on standard formats for data repository. One approach to this would be to agree on standard sets of 2D or 3D landmarks for different kinds of morphological data. This would be very difficult, because devising a landmark set that suits all possible research aims is impossible and would require people to digitize landmarks that are not needed for their particular studies. Further, among-observer error is a significant problem in landmark digitization. It would be very difficult to standardize landmarking protocols sufficiently to minimize among-observer error, and virtually impossible to devise ways to assess the magnitude of error in a dataset containing contributions from a large number of observers that have been added over long periods of time. A much more practical alternative would be to agree on formats for making available image and volumetric datasets. Combined with new techniques to automatically extract data from such repositories, this solution would eventually allow us to make use of shared data to address larger scale questions.

Although standardization is best done at a generic data level, there are situations in which the standardization of measurements can support long-term studies. For our studies of craniofacial variation in mice, for instance, we have defined a standard set of 3D landmarks and measurements that are always taken for each study (Fig. 19.6). Specific questions sometimes require the addition of landmarks, but a common core of landmarks remains that can be compared across studies. As such databases build up, they can be queried to address increasingly complex hypotheses. In a recent paper, for example, we pooled data from several studies to address a hypothesis about spatial packing as a determinant of craniofacial variation in mice (Lieberman et al. 2008). This hypothesis is unrelated to those that motivated the original datasets on which this study was based. This study focused on explaining

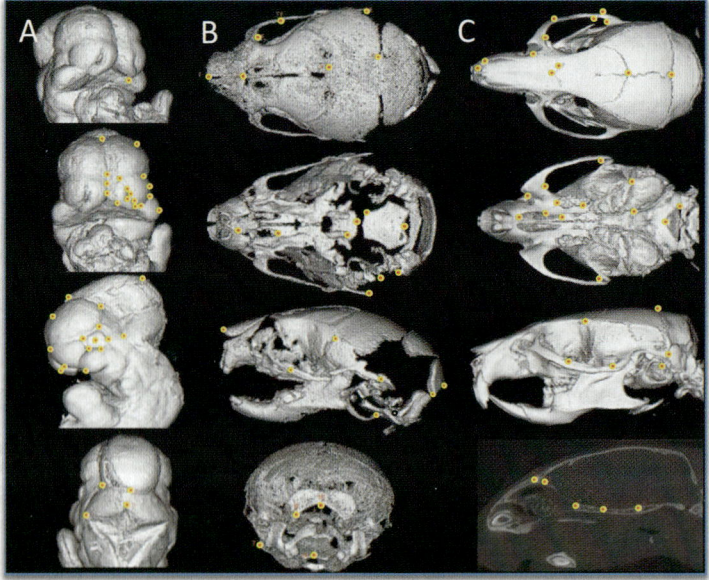

Fig. 19.6 Standard 3D landmark sets for embryonic, neonatal and adult mice

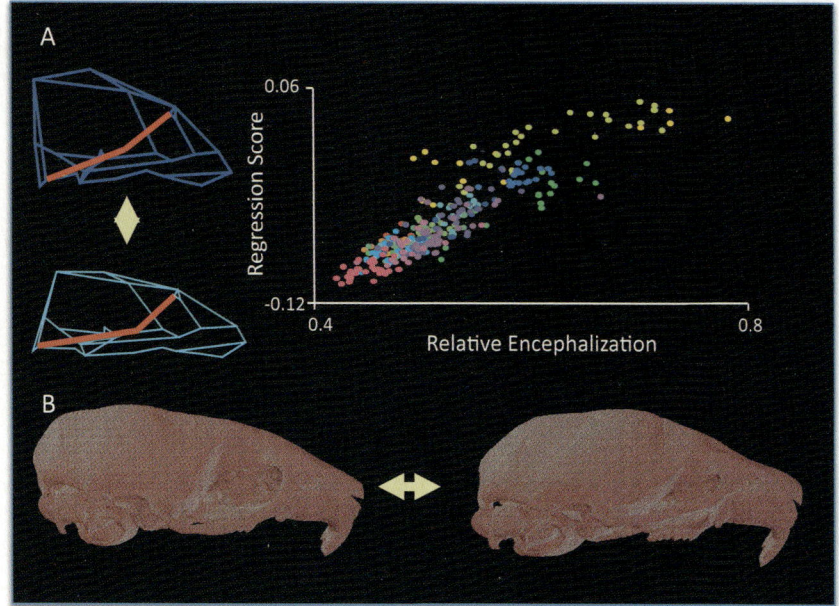

Fig. 19.7 (a–b) Multivariate regression of relative encephalization (brain size relative to basicranial length) on craniofacial shape (3D Procrustes-superimposed landmark data). **(a)** Regression plot and a wireframe deformation along the regression score. **(b)** Deformation of mouse skull along the trajectory defined by the regression

variation in the cranial base angle, which is a trait of particular importance in human evolution. Using mouse strains and mutants that vary in brain, facial and chondrocranial size, we were able to construct a model that accounts for over 90% of the variation in cranial base angle among strains. A related finding is that brain size relative to basicranial length explains over 30% of the total variation in cranial shape in mice (Fig. 19.7). These questions could not have been asked had we not decided several years back to pursue a standard format for collecting morphometric data on mouse craniofacial variation. We anticipate that as this dataset grows, it will be possible to query it in order to answer other unanticipated questions.

Quantification of Morphological Variation

When new imaging technologies arise, it is often considered sufficient to show single representative images of the structures being studied. This initial enthusiasm for the images, however, quickly gives way to a focus on the use of the quantitative data that can be obtained from the images for hypothesis testing. While the development and validation of quantification methods lags behind the development of imaging

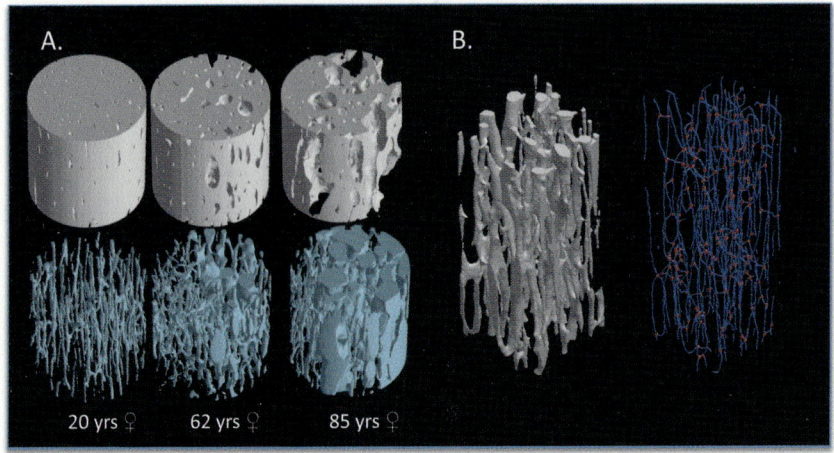

Fig. 19.8 (a–b) Imaging and quantification of cortical bone microstructure. **(a)** Scans of bone cores and visualization of the Haversian canal structure within the cores. **(b)** Skeletonization used to generate quantitative parameters from such data (work by David Cooper)

technology, the continuous development and refinement of such methods is critical to integrating imaging into the process of hypothesis testing.

In computed microtomography, the initial focus of quantitative method development was on the quantification of the trabecular bone structure (Ulrich et al. 1997; Muller et al. 1994; Hildebrand and Rüegsegger 1997a–b; Hildebrand et al. 1999). These methods have been applied extensively to the study of trabecular bone changes with conditions such as osteoporosis, and the literature in this area now tends to employ a standard suite of measures. Methods for the quantification of three-dimensional structure of cortical bone followed, which were developed by Dave Cooper and others (Cooper et al. 2003, 2004, 2006) (Fig. 19.8). Around the same time, the application of geometric morphometrics to computed tomography and microtomography data facilitated the study of shape variation (Hallgrímsson et al. 2004a, b; Wang et al. 2005).

Despite the extensive work done so far, however, the methods developed to date for the quantitative analysis of 3D volumetric datasets barely scratch the surface in terms of the potential to extract information from volumetric datasets. Some quantification of vascular morphology has been done (Bentley et al. 2002; Wan et al. 2002), but this is largely limited to luminal dimensions of individual vessels. Methods for the quantitative assessment of branching patterns, branch lengths and patterns such as those shown in (Fig. 19.9) are yet to be developed and applied in a systematic way.

Many disease processes produce complex and difficult to quantify morphological changes. Devising means to systematically quantify such changes, however, is crucial to supporting large-scale studies that compare high numbers of treatments or genetic variants. For cartilage and bone, osteoarthritis produces such a complex morphological pattern. Figure 19.10a shows computed microtomography scans

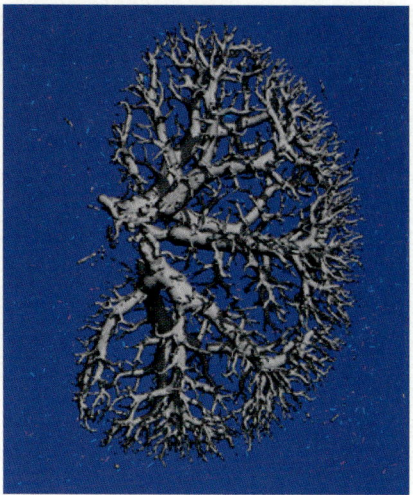

Fig. 19.9 The arterial structure of a rat kidney as imaged using computed microtomography and contrast injection

of guinea pigs at different ages with different degrees of spontaneously occurring osteoarthritis. Here, there is no individual variable that has a simple relationship to the progression of the disease. Changes in cartilage are seen much earlier than those in bone, and the nature of the morphological changes varies significantly across the progression of the disease. To tackle this problem, we recorded several bone and cartilage morphological characteristics, both qualitative and quantitative, that are likely to change with the progression of the osteoarthritis (McDougal et al. 2008). We then performed a form of data reduction—categorical principal components analysis— to obtain central tendencies in this multivariate dataset that might be related to OA. In this particular analysis, the first CATPCA component explains the majority of the variation in the multivariate dataset, correlates with both age and body mass (Fig. 19.10b), and is highly concordant with the subjective scoring of OA progression. In a dataset in which no single variable is highly correlated with OA, there is thus a central tendency that can be pulled out of the multivariate dataset which provides a good measure of OA progression as a single variable. This particular method may not apply to OA progression in humans or other animal models, but multivariate approaches like this offer one possible solution to the systematic quantification of complex morphological changes.

Conclusion

Research into the genetic and environmental basis for development and disease has been driven largely by molecular biology for the last half-century. This trend has produced a de-emphasis on phenotypic variation, and particularly on the quantitative

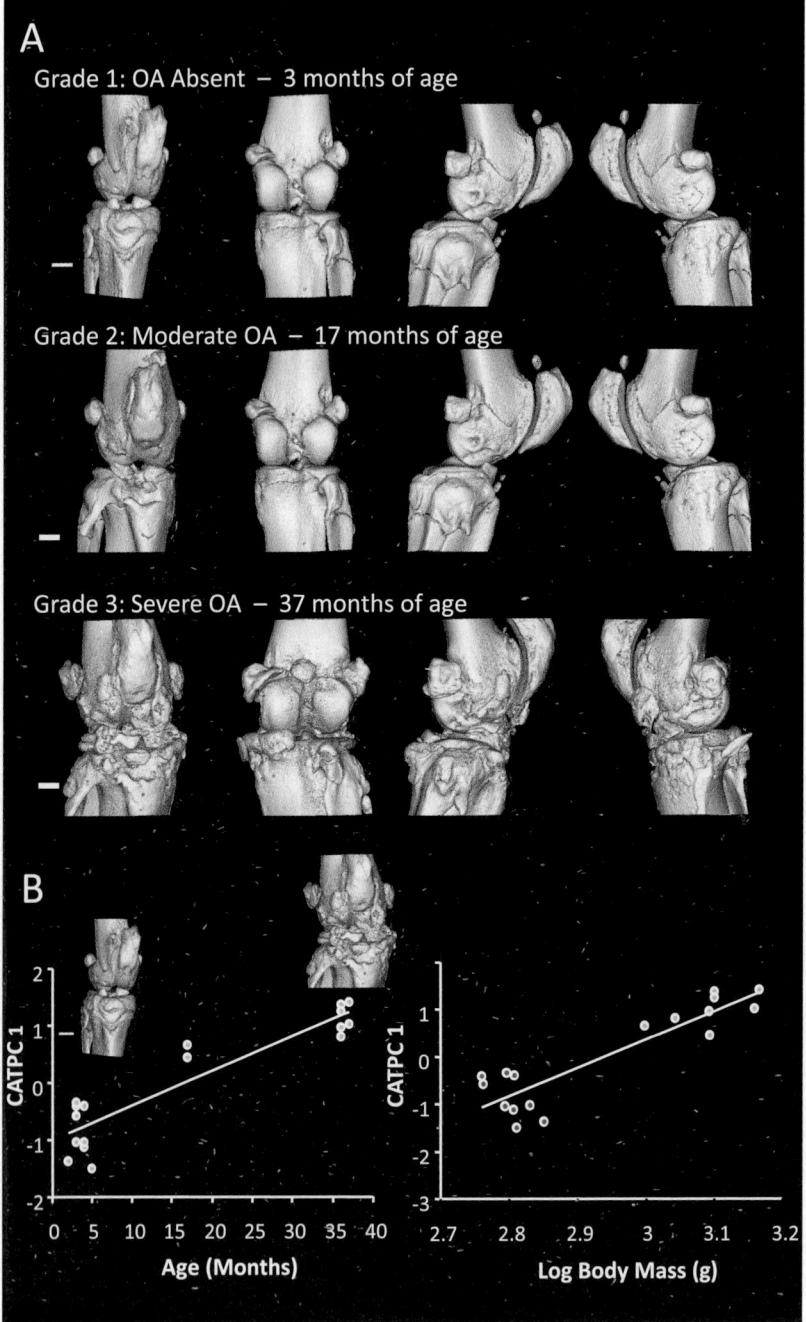

Fig. 19.10 (a) Representative computed microtomography scans of guinea pig knees with three stages of spontaneous osteoarthritis. (b) Results of a CATPCA analysis of bone and cartilage variables related to OA progression for a sample of guinea pig knees of varying ages

study of morphology. In developmental biology, for instance, research programs that focused on understanding developmental genetic pathways have paid scant attention to the measurement of morphology, often relying solely on representative individuals of genotypes or treatments. Systems approaches that are now developing in the wake of the post-genomic revolution are poised to reverse this trend. As such approaches create the need to compare multiple genotypes or treatments that vary in more subtle ways, the quantification of phenotypic information, including morphological variation, is becoming increasingly important. Imaging, particularly in three dimensions, is one of the main ways in which the study of morphology will be integrated into the study of the molecular mechanisms that generate it. To meet this need, phenotypic analysis must change in some of the same ways that molecular biology has already changed. This includes a focus on throughput, data sharing, data and analysis standardization, and the quantification of morphological variation. Only with these changes will we see the emergence of phenogenomics as the disciplinary companion of genomics, proteomics and metabolomics.

References

Bentley MD, Ortiz MC, Ritman EL, Romero JC (2002) The use of microcomputed tomography to study microvasculature in small rodents. Am J Physiol Regul Integr Comp Physiol 28 2:R1267–R1279

Boeckmann B, Blatter MC, Famiglietti L et al. (2005) Protein variety and functional diversity: Swiss-Prot annotation in its biological context. C R Biol 328:882–899

Bogue M (2003) Mouse phenome project: understanding human biology through mouse genetics and genomics. J Appl Physiol 95:1335–1337

Chen XJ, Kovacevic N, Lobaugh NJ, Sled JG, Henkelman RM, Henderson JT (2006) Neuroanatomical differences between mouse strains as shown by high-resolution 3D MRI. Neuroimage 29:99–105

Cooper DM, Turinsky AL, Sensen CW, Hallgrimsson B (2003) Quantitative 3D analysis of the canal network in cortical bone by micro-computed tomography. Anat Rec B New Anat 274:169–179

Cooper DML, Matyas JR, Katzenberg MA, Hallgrimsson B (2004) Comparison of microcomputed tomographic and microradiographic measurements of cortical bone porosity. Calcified Tissue Int 74(5):437–447

Cooper DML, Thomas CDL, Clement JG, Hallgrimsson B (2006) Three-dimensional micro-computed tomography imaging of basic multicellular unit-related resorption spaces in human cortical bone. Anatomical Record Part a-Discoveries Mol Cell Evol Biol 288A:806–816

Dorr AE, Lerch JP, Spring S, Kabani N, Henkelman RM (2008) High resolution three-dimensional brain atlas using an average magnetic resonance image of 40 adult C57Bl/6J mice. Neuroimage 42(1):60–69

Grubb SC, Churchill GA, Bogue MA (2004) A collaborative database of inbred mouse strain characteristics. Bioinformatics 20:2857–2859

Hallgrímsson B, Dorval CJ, Zelditch ML, German RZ (2004a) Craniofacial variability and morphological integration in mice susceptible to cleft lip and palate. J Anat 205:501–517

Hallgrímsson B, Willmore K, Dorval C, Cooper DM (2004b) Craniofacial variability and modularity in macaques and mice. J Exp Zoolog B Mol Dev Evol 302:207–225

Henkelman RM, Chen XJ, Sled JG (2005) Disease phenotyping: structural and functional readouts. Prog Drug Res 62:151–184

Hildebrand T, Rüegsegger P (1997a) A new method for the model-independent assessment of thickness in three-dimensional images. J Microsc 185:67–75

Hildebrand T, Rüegsegger P (1997b) Quantification of bone microarchitecture with the structure model index. Comput Methods Biomech Biomed Eng 1:15–23

Hildebrand T, Laib A, Muller R, Dequeker J, Rüegsegger P (1999) Direct three-dimensional morphometric analysis of human cancellous bone: microstructural data from spine, femur, iliac crest, and calcaneus. J Bone Miner Res 14:1167–1174

Kristensen E, Parsons TE, Gire J, Hallgrimsson B, Boyd S (2008) A novel high-throughput morphological method for phenotypic analysis. IEE Comput Graphics Appl. doi:10.1109/TBME.2008.923106

Lander ES, Linton LM, Birren B et al. (2001) Initial sequencing and analysis of the human genome. Nature 409:860–921

Lieberman D, Hallgrimsson B, Liu W, Parsons TE, Jamniczky HA (2008) Spatial packing, cranial base angulation, and craniofacial shape variation in the mammalian skull: testing a new model using mice. J Anat 23:12

McDougal JJ, Andruski B, Schuelert N, Hallgrimsson B JRM (2008) Unravelling the relationship between age, pain sensation, and joint destruction in naturally occurring osteoarthritis of Dunkin–Hartley guinea pigs

Muller R, Hildebrand T, Ruegsegger P (1994) Non-invasive bone biopsy: a new method to analyse and display the three-dimensional structure of trabecular bone. Phys Med Biol 39:145–164

Nieman BJ, Flenniken AM, Adamson SL, Henkelman RM, Sled JG (2006) Anatomical phenotyping in the brain and skull of a mutant mouse by magnetic resonance imaging and computed tomography. Physiol Genomics 24:154–162

Olafsdottir H, Darvann TA, Hermann NV et al. (2007) Computational mouse atlases and their application to automatic assessment of craniofacial dysmorphology caused by the Crouzon mutation Fgfr2(C342Y). J Anat 211:37–52

Paigen K, Eppig JT (2000) A mouse phenome project. Mamm Genome 11:715–717

Peltonen L, McKusick VA (2001) Genomics and medicine. Dissecting human disease in the postgenomic era. Science 291:1224–1229

Ulrich D, Hildebrand T, Van Rietbergen B, Muller R, Ruegsegger P (1997) The quality of trabecular bone evaluated with micro-computed tomography, FEA and mechanical testing. Stud Health Technol Inform 40:97–112

Venter JC, Adams MD, Myers EW et al. (2001) The sequence of the human genome. Science 291:1304–1351

Wan SY, Ritman EL, Higgins WE (2002) Multi-generational analysis and visualization of the vascular tree in 3D micro-CT images. Comput Biol Med 32:55–71

Wang Y, Xiao R, Yang F et al. (2005) Abnormalities in cartilage and bone development in the Apert syndrome FGFR2(+/S252W) mouse. Development 132:3537–3548

Waterston RH, Lindblad-Toh K, Birney E et al. (2002) Initial sequencing and comparative analysis of the mouse genome. Nature 420:520–562

Chapter 20
Functional Measures of Therapy Based on Radiological Imaging

David Dean(✉), Nathan Cross, Davood Varghai, Nancy L. Oleinick, and Chris A. Flask

Abstract We seek a noninvasive, functional, radiological imaging method to determine the specificity and sensitivity of an ablative treatment for deeply placed lesions that cannot be directly visualized, in our case phthalocyanine-4 photodynamic therapy (Pc 4-PDT) of brain tumors. In a preliminary study we had expected that micro-positron emission tomography (μPET) would show dramatically reduced if not negligible ^{18}F-FDG activity following Pc 4-PDT; however, our study has not found a statistically significant difference between the imaging of brain tumors in animals that underwent Pc 4-PDT and those that did not. While several magnetic resonance imaging (MRI) pulse sequences also did not discriminate tumors that had received treatment, our study of dynamic contrast enhancement magnetic resonance imaging (DCE-MRI) was able to discriminate tumors that had undergone necrosis following Pc 4-PDT. We expect that in addition to imaging therapeutic necrosis, it will also be possible to utilize other noninvasive, radiological imaging techniques to track apoptosis and/or autophagy in deeply placed lesions following treatments such as Pc 4-PDT.

Introduction

The desired outcome of many therapies, especially cancer therapies, is pathological tissue necrosis. Histological analysis of a tissue biopsy is the gold standard for the assessment of therapeutic outcomes. However, since many deep-seated tumors cannot be visualized by direct inspection, and biopsy following therapy may not be practical in some situations, it may be difficult or impossible to assess the immediate response to therapy. Noninvasive, radiological imaging may be useful for

D. Dean
Department of Neurological Surgery and Case Comprehensive Cancer Center, Case Western Reserve University, 10900 Euclid Avenue, Cleveland, OH 44106, USA, e-mail: David.Dean@Case.Edu

C.W. Sensen and B. Hallgrímsson (eds.), *Advanced Imaging in Biology and Medicine.*   427
© Springer-Verlag Berlin Heidelberg 2009

tracking the specificity and sensitivity of treatment of deeply placed or surgically inaccessible lesions during the administration of, or immediately following, therapy (Wang et al. 2008a, b).

Traditionally, there are two radiological imaging paradigms: morphological (structural) and functional imaging. The most useful morphological imaging techniques, such as conventional magnetic resonance imaging (MRI) and computed tomography (CT), provide very accurate information on the location of gene expression, various cell types, vascular structures, tissues, organs, or lesions. Functional images may also include some morphological information; however, functional imaging techniques also provide additional useful information such as cell, tissue, or organ status, function, or health, as well as levels of gene expression or metabolic function. We hypothesize that functional imaging modalities may also be useful for assessing therapeutic success by tracking physiological changes before and after treatment. As a test case, we will review our own attempts to determine a radiological method that would document the successful treatment of brain tumors following photodynamic therapy (PDT).

Functional Imaging Modalities

While optical imaging techniques are used extensively in basic science research, radiological imaging has not included optical imaging in most clinical contexts. However, that distinction has blurred in preclinical research, as there has been a revolution in the optical imaging of cellular (e.g., using confocal fluorescence technologies) and tissue and organ (e.g., optical coherence tomography, bioluminescence, and fluorescence) levels of gene expression (e.g., via molecular probes) and cellular activity (e.g., via time-lapse photography) (Czernin et al. 2006). To date, these modalities have proven more useful for learning about therapeutic mechanisms and for the screening of therapeutic potential than they have for clinical use in order to document therapeutic outcome. One reason for the lack of translation to the clinic may be that many of the preclinical models consist of tissue that is located subcutaneously (perhaps because of signal transmission depth limitations), whereas we are interested in tracking the specificity and sensitivity of treatments for deeply placed lesions (Wang et al. 2008a, b). We have concentrated on more traditional radiological methods (e.g., CT, positron emission tomography [PET], and MRI) in order to facilitate the translation of a successful technique to the clinic. Specifically, our studies have attempted to remotely track the response of photodynamic therapy (PDT), sensitized by the phthalocyanine photosensitizer Pc 4, on deeply placed lesions (i.e., malignant brain tumors).

To date, our studies have utilized an athymic nude rat model in which subcortical (brain) injections of immortalized human glioblastoma, U87 cells, produced tumors that were treated with Pc 4-PDT. We have previously established, histologically, that Pc 4-PDT brings about extensive necrosis in U87-derived tumors (George et al. 2005). Pre- and post-PDT images were taken to determine whether there was unambiguous evidence of treatment specificity and sensitivity. We first explored the use of ^{18}F-FDG (^{18}F-fluorodeoxy-glucose)-PET. We expected to see an unambiguous decrease in the amount of radionuclide uptake following PDT, owing to the loss of glucose metabolism in necrotic tissue (cf. Moore et al. 1998). It was expected that active uptake of ^{18}F-FDG would diminish dramatically, if not cease, in necrotic tissue. However, this was not what we observed; rather we saw a continued increase in ^{18}F-FDG uptake (Varghai et al. 2007).

We, therefore, have explored alternative imaging techniques, including diffusion tensor imaging, diffusion-weighted imaging, MR spectroscopy, and dynamic contrast enhancement–magnetic resonance imaging (DCE-MRI). The lattermost technique has yielded promising results (Varghai et al. 2008).

We present summaries of our research into the PET and fMR imaging of Pc 4-PDT below. Finally, we will also consider the use of radiological imaging to detect apoptosis or autophagy as indicators of therapy designed to bring about tissue necrosis in deeply placed lesions.

Nuclear Medicine

Nuclear medicine utilizes various radionuclides that are detected by PET, SPECT (single-photon emission computed tomography), and other devices that detect the location of radionuclide-labeled molecules which accumulate in lesions. ^{18}F-FDG crosses the blood–brain barrier and accumulates in cells in proportion to their metabolic activity. Cells with the highest metabolic activity (e.g., high-grade brain tumors vs. normal brain tissue) transport more ^{18}F-FDG across their membranes, where it is subsequently phosphorylated to ^{18}F-FDG-6-phosphate, a state in which it cannot leave the cell. ^{18}F-FDG is helpful for distinguishing necrotic tissue from recurrent tumor; however, it can be more difficult to distinguish between low-grade tumors and recurrent high-grade tumors with ^{18}F-FDG PET imaging than with amino acid-linked radionuclides (Chen 2007). The differential accumulation of ^{18}F-FDG is expected to result from the higher glucose metabolism of brain tumors vs. normal brain tissue. If the lesion space becomes impermeable, or at least significantly less permeable, due to necrosis, the PET or SPECT signal may be lost after therapy. However, there is concern that such a decrease may be due to reduced metabolism, not necessarily necrosis, in tumors following therapy (Krohn et al. 2005).

Advances in instrumentation have resulted in the availability of PET scanners specifically designed for imaging small animals, termed micro-PET (μPET) scanners. Prior work by Lapointe et al. (1999) suggested that ^{18}F-FDG μPET imaging *during* the administration of PDT may be useful for determining the specificity and sensitivity of the treatment. Therefore, our hypothesis was that ^{18}F-FDG μPET imaging would provide quantitative data documenting the specificity and sensitivity of Pc 4-PDT for U87-derived glial tumors in the athymic nude rat brain. Bérard et al. (2006) observed a real-time decline in ^{18}F-FDG μPET imaging activity *following* PDT. We expected that ^{18}F-FDG μPET imaging would record the difference between the high metabolic activity found prior to Pc 4-PDT in the tumor, and the loss of that activity that would result from Pc 4-PDT-induced early vascular effects (e.g., constriction or spasm) or later tumor necrosis.

^{18}F-FDG μPET Imaging of Pc 4-PDT Human Tumor Xenografts in the Athymic Nude Rat Brain

In order to create a tumor in the athymic nude rat, we inject 2.5×10^5 U87 cells into the brain. Tumors are of sufficient size ($\sim$2–5 mm in diameter) for treatment after seven days of growth. At that time, the animals are imaged functionally by ^{18}F-FDG μPET (i.e., Siemens/CTI Concorde microPET® scanner, Malvern, PA, USA) and structurally by micro-CT (i.e., Gamma Medica, X-SPECT^{TM}, Northridge, CA, USA) and/or micro-MR (μMR; i.e., 7 T Bruker Biospec^{TM}, Billerica, MA, USA). For PDT (i.e., necrosis of the tumor), the animals receive 0.5 mg kg^{-1} b.w. Pc 4 via tail-vein injection. One day later, the scalp is re-incised and the tumor is illuminated with a 30 J cm^{-2} dose of 672-nm light from a diode laser in order to bring about PDT. The next day, the animals are again imaged by ^{18}F-FDG μPET. Following the post-PDT imaging session, the animals are euthanized and their brains are explanted for hematoxylin and eosin (H&E) histology (Varghai et al. 2007).

μPET Imaging Did Not Show an Expected Post-PDT Decrease in ^{18}F-FDG μPET Activity in the U87 Tumor

The H&E histology of the tumors from Pc 4-PDT-treated animals demonstrated necrosis ranging from spotty to frank (severe), whereas the control tumors did not develop necrosis. In a preliminary analysis, the μPET images of the control and experimental groups both showed increasing brightness between the first and second scan (see Fig. 20.1). This was expected for the control group, as the tumor was expected to increase in size. However, this was not expected for the experimental group; a drop-off in ^{18}F-FDG activity was expected following tumor necrosis. Specifically, in three control animals where no tumor necrosis was observed, ^{18}F-FDG activity increased 2.28-fold following tumor photoirradiation in comparison to the pre-PDT scan. In contrast, the ^{18}F-FDG activity in five animals in the experimental group, all of which developed brain tumor necrosis after PDT, increased by 1.15-fold following PDT (Varghai et al. 2007). The level of increased ^{18}F-FDG activity seen in the two groups was not significantly (i.e., statistically significant) different.

Why Does ^{18}F-FDG Activity Increase in the U87 Tumor That Has Undergone Tissue Necrosis Due to PDT?

Our preliminary study did not reveal the expected decline in ^{18}F-FDG activity following Pc 4-PDT. Indeed, an increase in ^{18}F-FDG activity accompanying tumor necrosis was observed. That increase was, on average, lower in tumors that

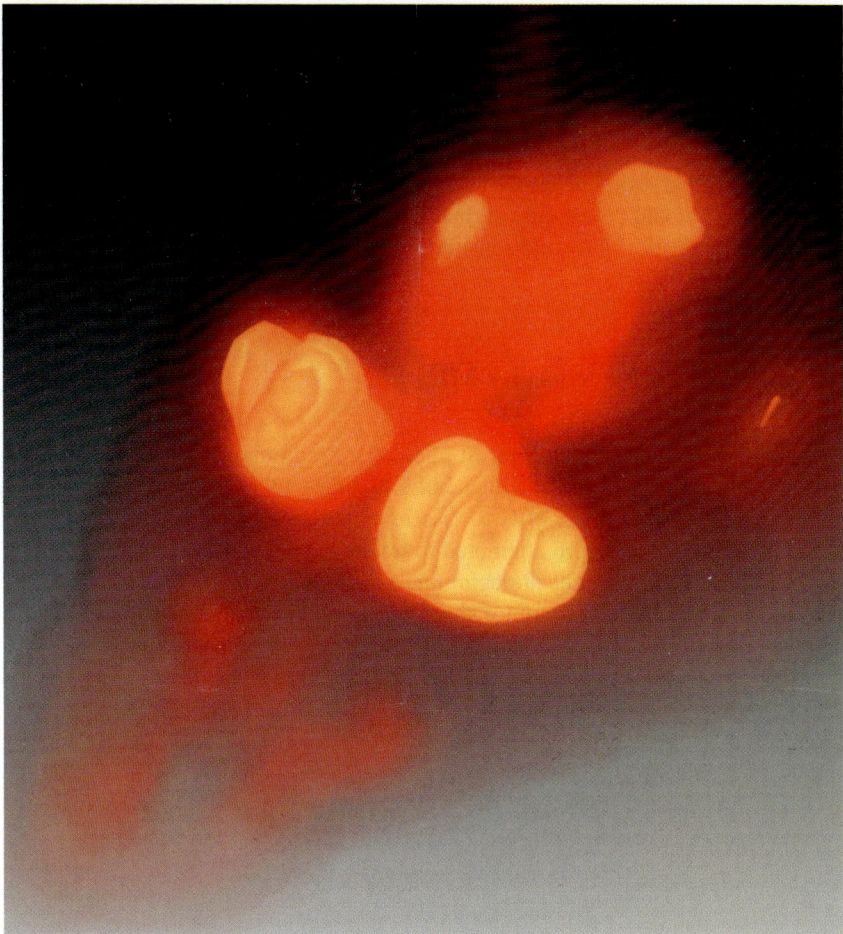

Fig. 20.1 Post-Pc 4-PDT ^{18}F-FDG μPET image of an athymic nude rat with a U87 tumor. Our hypothesis was that ^{18}F-FDG activity would decline significantly if not disappear. An increase of ^{18}F-FDG activity of 1.15 times was observed between pre- and post-PDT imaging sessions. The animal's nose is at *lower left*. The two large objects in the center of the image are the eyes. The brain is distinct, but less bright than the eyes. The tumor is the smaller of two bright spots situated within the brain. The other, more central, and further posterior, bright spot within the brain is the cerebellum

underwent PDT, but the difference between the PDT and non-PDT groups in increased ^{18}F-FDG activity was not significant. Thus, we falsified our hypothesis going into the study, that ^{18}F-FDG activity would decrease, if not disappear, following Pc 4-PDT of U87-derived tumors. Our hope was that this decline in ^{18}F-FDG activity would provide an unambiguous, noninvasive, radiological discriminator between pre- and post-Pc 4-PDT tumors (Varghai et al. 2007). We are left questioning whether the physiological changes that have occurred due to tissue ablation following PDT may include structural changes that promote vascular leakage (cf. Davis et al. 2004), and possibly pooling, of the ^{18}F-FDG radionuclide in necrotic tumor tissue (Cross et al. 2007). It would be interesting to see if these observations held true for other radionuclides such as ^{11}C-methionine or an [124]I-labeled metabolite, both of which have longer half lives than ^{18}F-FDG (Pandey et al. 2005). While these longer-lived radionuclide-bound metabolites could be given more time to clear, doing so would defeat our original purpose of developing a noninvasive measure of treatment specificity and sensitivity for use in the immediate post-operative period.

Imaging Tumor Necrosis by MRI

Because of our concern that contrast agent (i.e., radionuclide) leakage was confounding our attempt to image metabolic changes in necrotic tissue, we have investigated the use of DCE-MRI to discriminate tissue changes pre- and post-Pc 4-PDT. Specifically, we have attempted to track changes in the perfusion of a contrast agent (i.e., gadopentetate dimeglumine, Gd-DTPA – a medium-sized molecule) as it enters the peritumor and tumor space. We hypothesize that because of the dramatic cellular breakdown that occurs following PDT, the resistance to Gd-DTPA perfusion would be significantly reduced in this space. We further hypothesize that the reduced resistance would lead to increased Gd-DTPA enhancement in T1-weighted MR images, and that this could be quantitatively tracked over time by what is referred to as DCE-MRI.

DCE is based on the Tofts and Kermode (1991) proposal of a four-compartment model; those four compartments are: (1) the plasma (i.e., the vascular space), (2) the extracellular space, (3) the brain (i.e., within the blood–brain barrier), and (4) the brain lesion (i.e., brain tumor). The brain lesion space is assumed to be directly connected to the plasma compartment via a leaky blood–brain barrier. Brain lesion contrast enhancement has been seen to increase following tissue necrosis in other models of brain lesion tissue necrosis, including ischemia (Chen et al. 2007).

Our DCE-MRI protocol tracks the enhancement of specific locations in one or more of these compartments over the course of a scan after contrast agent (i.e., Gd-DTPA) is administered. We hypothesize that reduced resistance to Gd-DTPA following brain tumor necrosis would result in an increased enhancement of the tumor compartment.

U87 Cell Implantation, PDT, and μMRI Scanning

Brain tumors are generated in athymic nude rats in much the same way as in "[18]F-FDG μPET Imaging of Pc 4-PDT Human Tumor Xenografts in the Athymic Nude Rat Brain." Seven days after U87 cell implantation, the animals are DCE-MRI-scanned in a Bruker BioSpin Corporation (Billerica, MA, USA) BioSpec®

7 T microMRI (μMRI) scanner. The DCE-MRI scan protocol is described by Varghai et al. (2008). Prior to DCE-MRI μMRI-scanning, a Gd-DTPA-filled phantom is placed adjacent to the animal's head. The DCE-MRI scan itself involves the administration of Gd-DTPA (Magnevist®, Bayer Healthcare Pharmaceuticals, Leverkusen, Germany) after 4 min out of the 30-min DCE-MRI scanning protocol have elapsed. Thirty-six to forty-eight hours after Pc 4 administration, and following the initial DCE-MRI (i.e., pre-PDT) scan, each animal receives a 30 J cm^{-2} light dose from a 672-nm diode laser. The animals then undergo a post-PDT μMRI scan six days after the original μMRI scan (see Fig. 20.2). The animals are then sacrificed, and their brains are collected post mortem for H&E histological analysis (Varghai et al. 2008).

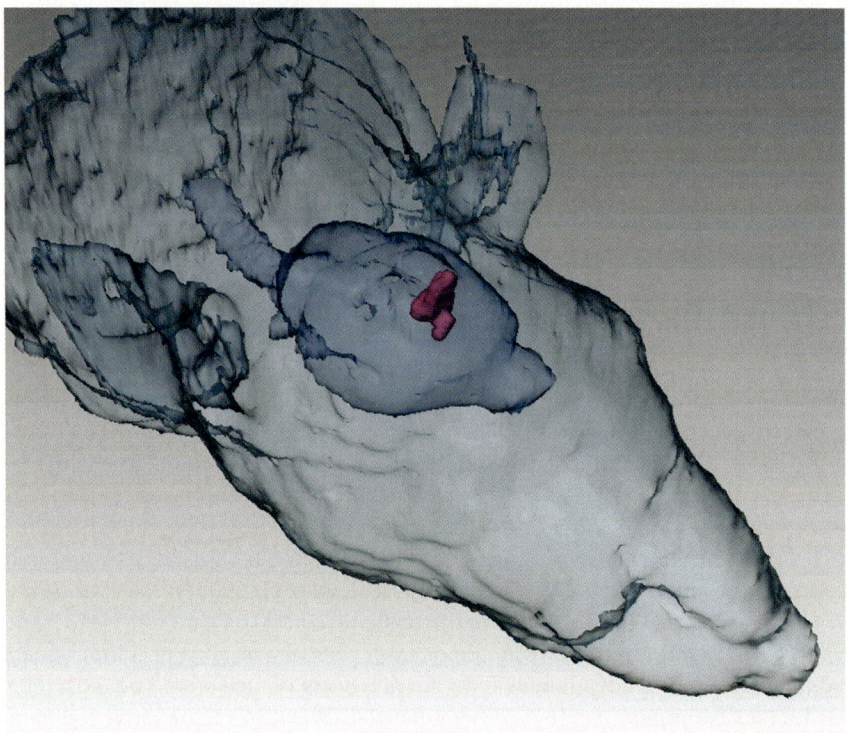

Fig. 20.2 3D MRI of Pc 4-PDT: Gd-DTPA contrast was found to significantly increase in U87-derived tumors following Pc 4-PDT. The surface of the skin is transparently shown. The cerebrum, cerebellum, and spinal cord are visible in darker, but also transparent, shades of *gray*. The tumor object is shown in *red*

DCE-MRI Demonstrates Significantly Increased Tumor Enhancement Following PDT for Unknown Duration

Our pilot study results suggest that DCE-MRI records an unambiguous increase in enhancement following PDT. Further study of this model may provide information on how soon after PDT an opening of the blood–brain barrier around the tumor can be detected, and how long it persists after brain tumor necrosis. This would be interesting to know, because Hirschberg et al. (2007) have suggested that the peri-PDT opening of the blood–brain barrier may provide a therapeutic window.

We provide two notes of caution regarding our pilot results. First, Zilberstein et al. (2001) observed a decline in enhancement following PDT of subcutaneously implanted tumors. They suggest that the lack of tracer uptake by the tumor following PDT may be due to "perfusion arrest." Perhaps there is a different response to PDT in contrast uptake between superficially and deeply placed tumors, or perhaps the two different responses are due to the use of different photosensitizers or different overall PDT doses. Second, it may be worthwhile studying how the details of contrast agent administration (e.g., bolus size, enhancing agent concentration, speed of administration, etc.) can affect the study outcome (Collins and Padhani 2004; Lopata et al. 2007).

NMR Spectroscopy of Brain Tumor Necrosis

Brain tumor ^{31}P nuclear magnetic resonance (NMR) spectroscopy has been done to detect the effects of radiation therapy (Sijens et al. 1986) and PDT (Chopp et al. 1985). More recently, multivoxel 2D proton chemical shift imaging (CSI) has been used for NMR spectroscopic tracking of choline/creatine (Cr), N-acetylaspartate (NAA)/Cr, myo-inositol/Cr, relative cerebral blood volume (rCBV), or lipid levels useful in grading tumors or, in our case, in tracking tumor proliferation and/or metabolism (Li et al. 2007; Law et al. 2003; Rock et al. 2002). Di Costanzo et al. (2008) report that high-field (i.e., 3.0 T) proton NMR spectroscopy "is useful to discriminate tumor from necrosis, oedema or normal tissue, and high from low-grade gliomas." We are not aware of, and would be most interested in, a CSI proton imaging study on the assessment of tumor status following PDT.

Radiological Imaging of Apoptosis and Autophagy in Deeply Placed Lesions

Clinically, it is not desirable to induce rapid necrosis in large intracranial lesions because this will result in unacceptable levels of edema and associated mass effect.

Therefore, it may be useful to plan treatments that bring about controlled apoptosis, or advanced autophagy, as a means of treating brain tumors.

Apoptosis is a physiological mechanism of cell death that is efficiently induced by PDT both in vitro and in vivo (Oleinick et al. 2002). The process is triggered by the immediate photodynamic damage to cells and progresses through a cascade of metabolic steps that leads to nuclear chromatin condensation, DNA cleavage, and packaging of cell constituents in membrane-limited fragments that can be removed by phagocytosis without producing extensive inflammation.

Autophagy is a process for recycling cell constituents and can be pro-survival or pro-death, depending on the level of initial cellular damage or stress. During starvation-induced autophagy, organelles are engulfed in double membrane-enclosed vacuoles, termed autophagosomes, which subsequently merge with lysosomes in the degradation of the organelle components, allowing the reuse in synthesis of essential cellular materials. PDT strongly induces autophagy which may be protective after low PDT doses but contributes to cellular demise after high PDT doses (Buytaert et al. 2006, 2008; Kessel et al. 2006; Kessel and Arroyo 2007).

Serial treatments that bring about autophagy and/or apoptosis may be desirable for the treatment of large lesions or chronic disease. By varying the degree of therapy, perhaps facilitated by radiological tracking, it may be possible to use autophagy to cause apoptosis or to directly induce apoptosis, thereby allowing tumor cells to package their contents into apoptotic bodies rather than rupturing. Assessment of serial imaging may be aided by methods for the normalization of functional images across the series.

Radiological Imaging of Apoptosis

Image-based tracking of apoptosis in tumors may be possible following the injection of a contrast agent that specifically seeks out cells undergoing this process. One method of identifying apoptotic cells in histological studies is to utilize annexin V, a protein that binds to the phosphatidylserine present on the outer leaflet of the plasma membrane of cells undergoing apoptosis. Blankenberg et al. (2006) and Tait et al. (2006) discuss the conjugation of technetium-99 M with annexin V (i.e., ^{99m}Tc-annexin V) for the SPECT imaging of cells undergoing apoptosis within a tissue. Similarly, PET imaging of apoptosis has been attempted using ^{64}Cu-labeled streptavidin following biotinylated annexin V administration (Cauchon et al. 2007; Tait 2008).

Radiological Imaging of Autophagy

No agent for identifying the surfaces of cells undergoing autophagy has been found. It may be difficult to find an agent that would do this, as autophagy is a primarily intracellular process. Probes that can be internalized and then cleaved by the set of proteases upregulated during autophagy may be useful. In many cases both

apoptosis and autophagy occur in the same cells after PDT (Xue et al. 2007), and it
is currently unclear how the two processes, autophagy and apoptosis, influence each
other following PDT. At late stages of either or both processes, the tissues will likely
appear necrotic; therefore, we expect that studies conducted early on after PDT are
more likely to distinguish the modes of cell death.

Conclusion

We review the radiological imaging, especially using MRI and PET, of therapies
that result in necrosis within deeply placed lesions. Our review centers on our experience with PDT of brain tumors. Of these techniques, DCE-MRI appears the most
promising. While it appears that these results may not hold for the treatment of some
subcutaneous lesions, this technique may allow noninvasive imaging and identification of necrosis following therapy in other deeply placed lesions. Additionally, it
would be helpful if simple-to-use radiological imaging techniques were developed
to verify apoptosis and/or autophagy in lesions. Radiological imaging of apoptosis and autophagy might allow clinicians to titrate therapies that produce delayed
necrosis, which may alleviate some if not all of the risk of edematous swelling in
patients with large brain tumors or other deeply placed lesions. Such patients would
likely benefit from serial treatment with low doses and monitoring of autophagy,
apoptosis, and/or necrosis as a measure of treatment success.

Acknowledgements This research has been partially supported by the Research Foundation of
the Department of Neurological Surgery, Case Western Reserve University (CWRU), Cleveland,
OH, and by the NCI-NIH Small Animal Imaging Resource Program (U24-CA110943). We would
like to acknowledge the collaboration, advice, and assistance provided by Ms. Kelly Covey, Ms.
Deborah Barkauskas, and Mr. John Richey, all of the Small Animal Imaging Research Center
(SAIRC), Case Center for Imaging Research (CCIR), Department of Radiology, CWRU, as well as
Drs. Raymond F. Muzic, Jr. and Mark D. Pagel and Ms. Chandra Spring-Robinson of the Depart-
ment of Radiology, CWRU, Mr. Rahul Sharma, Department of Neurological Surgery, CWRU, and
Ms. Denise K. Feyes, Department of Radiation Oncology, CWRU.

References

Bérard V, Rousseau JA, Cadorette J, Hubert L, Bentourkia M, van Lier JE, Lecomte R (2006)
    Dynamic imaging of transient metabolic processes by small-animal PET for the evaluation of
    photosensitizers in photodynamic therapy of cancer. J Nucl Med 47:1119–1126
Blankenberg FG, Vanderheyden JL, Strauss HW, Tait JF (2006) Radiolabeling of HYNIC-annexin
    V with technetium-99m for in vivo imaging of apoptosis. Nat Protoc 1:108–110
Buytaert E, Callewaert G, Hendrickx N, Scorrano L, Hartmann D, Missiaen L, Vandenheede JR,
    Heirman I, Grooten J, Agostinis P (2006) Role of endoplasmic reticulum depletion and mul-
    tidomain proapoptotic BAX and BAK proteins in shaping cell death after hypericin-mediated
    photodynamic therapy. FASEB J 20:756–758

Buytaert E, Matroule JY, Durinck S, Close P, Kocanova S, Vandenheede JR, de Witte PA, Piette J, Agostinis P (2008) Molecular effectors and modulators of hypericin-mediated cell death in bladder cancer cells. Oncogene 27:1916–1929

Cauchon N, Langlois R, Rousseau JA, Tessier G, Cadorette J, Lecomte R, Hunting DJ, Pavan RA, Zeisler SK, van Lier JE (2007) PET imaging of apoptosis with (64)Cu-labeled streptavidin following pretargeting of phosphatidylserine with biotinylated annexin-V. Eur J Nucl Med Mol Imaging 34:247–258

Chen W (2007) Clinical applications of PET in brain tumors. J Nucl Med 48:1468–1481

Chen F, Suzuki Y, Nagai N, Jin L, Yu J, Wang H, Marchal G, Ni Y (2007) Rodent stroke induced by photochemical occlusion of proximal middle cerebral artery: evolution monitored with MR imaging and histopathology. Eur J Radiol 63:68–75

Chopp M, Helpern JA, Frinak S, Hetzel FW, Ewing JR, Welch KM (1985) In vivo 31-P NMR of photoactivated hematoporphyrin derivative in cat brain. Med Phys 12:256–258

Collins DJ, Padhani AR (2004) Dynamic magnetic resonance imaging of tumor perfusion. Approaches and biomedical challenges. IEEE Eng Med Biol Mag 23:65–83

Cross N, Varghai D, Spring-Robinson C, Sharma R, Muzic RF Jr, Oleinick NL, Dean D (2007) Analysis of 18F-fluorodeoxy-glucose PET imaging data captured before and after Pc 4-mediated photodynamic therapy of U87 tumors in the athymic nude rat. In: Kollias N, Choi B, Zeng H, Malek RS, Wong BJ, Ilgner JFR, Gregory KW, Tearney GJ, Hirschberg H, Madsen SJ (eds) Proceedings of Photonics West 2007: Photonic Therapeutics and Diagnostics III (SPIE 6424). SPIE, Bellingham, WA, pp 64242F1–64242F8

Czernin J, Weber WA, Herschman HR (2006) Molecular imaging in the development of cancer therapeutics. Annu Rev Med 57:99–118

Davis TW, O'Neal JM, Pagel MD, Zweifel BS, Mehta PP, Heuvelman DM, Masferrer JL (2004) Synergy between celecoxib and radiotherapy results from inhibition of cyclooxygenase-2-derived prostaglandin E2, a survival factor for tumor and associated vasculature. Cancer Res 64(1):279–285

Di Costanzo A, Scarabino T, Trojsi F, Popolizio T, Catapano D, Giannatempo GM, Bonavita S, Portaluri M, Tosetti M, d'Angelo VA, Salvolini U, Tedeschi G (2008) Proton MR spectroscopy of cerebral gliomas at 3 T: spatial heterogeneity, and tumour grade and extent. Eur Radiol 18(8):1727–1735

George JE 3rd, Ahmad Y, Varghai D, Li X, Berlin J, Jackowe D, Jungermann M, Wolfe MS, Lilge L, Totonchi A, Morris RL, Peterson A, Lust WD, Kenney ME, Hoppel CL, Sun J, Oleinick NL, Dean D (2005) Pc 4 photodynamic therapy of U87-derived human glioma in the nude rat. Lasers Surg Med 36(5):383–389

Hirschberg H, Angell Petersen E, Spatalen S, Mathews M, Madsen SJ (2007) Increased brain edema following 5-aminolevulinic acid mediated photodynamic in normal and tumor bearing rats. In: Kollias N, Choi B, Zeng H, Malek RS, Wong BJ, Ilgner JFR, Gregory KW, Tearney GJ, Hirschberg H, Madsen SJ (eds) Proceedings of Photonics West 2007: Photonic Therapeutics and Diagnostics III (SPIE 6424). SPIE, Bellingham, WA, pp 6424B1–6424B8

Kessel D, Arroyo AS (2007) Apoptotic and autophagic responses to Bcl-2 inhibition and photodamage. Photochem Photobiol Sci 6:1290–1295

Kessel D, Vicente MG, Reiners JJ Jr (2006) Initiation of apoptosis and autophagy by photodynamic therapy. Lasers Surg Med 38:482–488

Krohn KA, Mankoff DA, Muzi M, Link JM, Spence AM (2005) True tracers: comparing FDG with glucose and FLT with thymidine. Nucl Med Biol 32(7):663–671

Lapointe D, Brasseur N, Cadorette J, La Madeleine C, Rodrigue S, van Lier JE, Lecomte R (1999) High-resolution PET imaging for in vivo monitoring of tumor response after photodynamic therapy in mice. J Nucl Med 40:876–882

Law M, Yang S, Wang H, Babb JS, Johnson G, Cha S, Knopp EA, Zagzag D (2003) Glioma grading: sensitivity, specificity, and predictive values of perfusion MR imaging and proton MR spectroscopic imaging compared with conventional MR imaging. AJNR Am J Neuroradiol 24:1989–1998

Li Y, Chen AP, Crane JC, Chang SM, Vigneron DB, Nelson SJ (2007) Three-dimensional J-resolved H-1 magnetic resonance spectroscopic imaging of volunteers and patients with brain tumors at 3 T. Magn Reson Med 58:886–892

Lopata RG, Backes WH, van den Bosch PP, van Riel NA (2007) On the identifiability of pharmacokinetic parameters in dynamic contrast-enhanced imaging. Magn Reson Med 58:425–429

Moore JV, Waller ML, Zhao S, Dodd NJ, Acton PD, Jeavons AP, Hastings D (1998) Feasibility of imaging photodynamic injury to tumours by high-resolution positron emission tomography. Eur J Nucl Med 25:1248–1254

Oleinick NL, Morris RL, Belichenko I (2002) Apoptosis in response to photodynamic therapy: what, where, why, and how. Photochem Photobiol Sci 1:1–21

Pandey SK, Gryshuk AL, Sajjad M, Zheng X, Chen Y, Abouzeid MM, Morgan J, Charamisinau I, Nabi HA, Oseroff A, Pandey RK (2005) Multimodality agents for tumor imaging (PET, fluorescence) and photodynamic therapy. A possible "see and treat" approach. J Med Chem 48:6286–6295

Rock JP, Hearshen D, Scarpace L, Croteau D, Gutierrez J, Fisher JL, Rosenblum ML, Mikkelsen T (2002) Correlations between magnetic resonance spectroscopy and image-guided histopathology, with special attention to radiation necrosis. Neurosurgery 51:912–919, discussion 919–920

Sijens PE, Bovée WM, Seijkens D, Los G, Rutgers DH (1986) In vivo 31P-nuclear magnetic resonance study of the response of a murine mammary tumor to different doses of gamma-radiation. Cancer Res 46(3):1427–1432

Tait JF (2008) Imaging of apoptosis. J Nucl Med 49(10):1573–1576

Tait JF, Smith C, Levashova Z, Patel B, Blankenberg FG, Vanderheyden JL (2006) Improved detection of cell death in vivo with annexin V radiolabeled by site-specific methods. J Nucl Med 47:1546–1553. Comment in: J Nucl Med 47:1400–1402

Tofts PS, Kermode AG (1991) Measurement of the blood–brain barrier permeability and leakage space using dynamic MR imaging. 1. Fundamental concepts. Magn Reson Med 17:357–367

Varghai D, Cross N, Spring-Robinson C, Sharma R, Feyes DK, Ahmad Y, Oleinick NL, Muzic RF Jr, Dean D (2007) Monitoring Pc 4-mediated photodynamic therapy of U87 tumors with F18-fluorodeoxy-glucose PET imaging in the athymic nude rat. In: Kollias N, Choi B, Zeng H, Malek RS, Wong BJ, Ilgner JFR, Gregory KW, Tearney GJ, Hirschberg H, Madsen SJ (eds) Proceedings of Photonics West 2007: Photonic Therapeutics and Diagnostics III (SPIE 6424). SPIE, Bellingham, WA, pp 64242G1–64242G7

Varghai D, Covey K, Sharma R, Cross N, Feyes DK, Oleinick NL, Flask CA, Dean D (2008) Monitoring Pc 4-mediated photodynamic therapy of U87 tumors with dynamic contrast enhanced-magnetic Resonance imaging (DCE-MRI) in the athymic nude rat. In: Kollias N, Choi B, Zeng H, Malek RS, Wong BJ, Ilgner JFR, Gregory KW, Tearney GJ, Hirschberg H, Madsen SJ (eds) Proceedings of Photonics West 2008: Photonic Therapeutics and Diagnostics IV (SPIE 6842). SPIE, Bellingham, WA, pp 68422P1–68422P9

Wang H, Van de Putte M, Chen F, De Keyzer F, Jin L, Yu J, Marchal G, de Witte P, Ni Y (2008a) Murine liver implantation of radiation-induced fibrosarcoma: characterization with MR imaging, microangiography and histopathology. Eur Radiol 18(7):1422–1430

Xue LY, Chiu SM, Azizuddin K, Joseph S, Oleinick NL (2007) The death of human cancer cells following photodynamic therapy: apoptosis competence is necessary for Bcl-2 protection but not for induction of autophagy. Photochem Photobiol 83:1016–1023

Zilberstein J, Schreiber S, Bloemers MC, Bendel P, Neeman M, Schechtman E, Kohen F, Scherz A, Salomon Y (2001) Antivascular treatment of solid melanoma tumors with bacteriochlorophyll-serine-based photodynamic therapy. Photochem Photobiol 73:257–266

Index

Printing: Krips bv, Meppel, The Netherlands
Binding: Stürtz, Würzburg, Germany